Analytical, Numerical and Big-Data-Based Methods in Deep Rock Mechanics

Analytical, Numerical and Big-Data-Based Methods in Deep Rock Mechanics

Editors

Shaofeng Wang
Xin Cai
Jian Zhou
Zhengyang Song
Xiaofeng Li

MDPI • Basel • Beijing • Wuhan • Barcelona • Belgrade • Manchester • Tokyo • Cluj • Tianjin

MDPI

Editors

Shaofeng Wang
Central South University
China

Xin Cai
Central South University
China

Jian Zhou
Central South University
China

Zhengyang Song
University of Science and
Technology Beijing
China

Xiaofeng Li
University of Toronto
Canada

Editorial Office
MDPI
St. Alban-Anlage 66
4052 Basel, Switzerland

This is a reprint of articles from the Special Issue published online in the open access journal *Mathematics* (ISSN 2227-7390) (available at: https://www.mdpi.com/journal/mathematics/special_issues/Analytical_Numerical__Methods_Deep_Rock_Mechanics).

For citation purposes, cite each article independently as indicated on the article page online and as indicated below:

LastName, A.A.; LastName, B.B.; LastName, C.C. Article Title. *Journal Name* **Year**, *Volume Number*, Page Range.

ISBN 978-3-0365-5761-8 (Hbk)
ISBN 978-3-0365-5762-5 (PDF)

Contents

mathematics MDPI

Editorial

Analytical, Numerical and Big-Data-Based Methods in Deep Rock Mechanics

Shaofeng Wang [1], Xin Cai [1,*], Jian Zhou [1], Zhengyang Song [2] and Xiaofeng Li [3]

1 School of Resources and Safety Enginerring, Central South University, Changsha 410083, China
2 School of Civil and Resource Engineering, University of Science and Technology Beijing, Beijing 100083, China
3 Department of Civil & Mineral Engineering, University of Toronto, Toronto, ON M5T 2S8, Canada
* Correspondence: xincai@csu.edu.cn

With the increasing requirements for energy, resources and space, numerous rock engineering projects (e.g., mining, tunnelling, underground storage, and geothermal and petroleum engineering) are more often being constructed and operated in large-scale, deep underground and complex geology environments. Meanwhile, more and more unconventional rock failures and rock instabilities (e.g., rockbursts, large-scale collapses and mine earthquakes) are occurring and severely threatening the safety of underground operations. It is well-recognized that rocks have multi-scale structures from minerals, particles, fractures, fissures, joints and stratification to faults and involve multi-scale fracture processes. In the deep earth, rocks are commonly subjected to complex high-stress and strong-dynamics disturbances simultaneously, providing a hotbed for the occurrence of unconventional rock failures. In addition, there are many multi-physics coupling processes in rock masses, such as the coupled thermo-hydromechanical interaction in fractured porous rocks. It is still difficult to understand rock mechanics and to characterize rock behaviors with complex stress conditions, multi-physics processes and multi-scale changes. Therefore, the prevention and control of unconventional instability in deep rock engineering remains a great challenge. The primary aim of this Special Issue "Analytical, Numerical and Big-Data-Based Methods in Deep Rock Mechanics" is to bring together original research discussing innovative efforts on analytical, numerical and big-data-based methods in rock mechanics. It includes 22 manuscripts that illustrate the richness and challenging nature of deep rock mechanics.

The article written by Zhang et al. [1] aims to address the difficulty in obtaining the mechanical parameters of surrounding rocks and large experimental errors, and an optimized BP neural network model is proposed in this paper. The optimized BP neural network model (MEA-BP model) takes advantage of the mind evolutionary algorithm and the neural network. It can not only avoid the local extreme value problem but also improve the accuracy and reliability of the prediction results.

The article published by Li et al. [2] aims to explore the risk of fault-induced water inrush under different mining advancing directions through numerical modeling and investigation. The findings showed that, for a water-conducting fault, the waterproof coal pillar size of mining advances from the hanging wall should be larger than that from the foot wall.

The purpose of the study proposed by Wu et al. [3] was to determine the effects of strain rate and temperature on the dynamic mechanical parameters, energy dissipation features and failure modes of granite. Their study results indicate that the dynamic compressive strength of granite increases exponentially with strain rate and decreases with increasing temperature. The dynamic elastic modulus decreased obviously with increasing temperature but did not have a clear correlation with the strain rate. Under the same incident energy, as the temperature increased, the reflected energy increased notably and the absorbed energy increased slightly, but the transmitted energy decreased. At the same

Citation: Wang, S.; Cai, X.; Zhou, J.; Song, Z.; Li, X. Analytical, Numerical and Big-Data-Based Methods in Deep Rock Mechanics. *Mathematics* **2022**, *10*, 3403. https://doi.org/10.3390/math10183403

Received: 3 August 2022
Accepted: 14 September 2022
Published: 19 September 2022

temperature, the reflected and absorbed energies increased linearly as the incident energy increased, whereas the transmitted energy increased logarithmically.

The article written by Song et al. [4] developed a reliability model to elucidate the trend of impact-related pipeline damage due to submarine slides. A probability analysis method of impact-related pipeline damage attributed to submarine slides based on the copula function was proposed. Furthermore, the copula function could reasonably characterize relevant the nonnormal distribution characteristics of risk variables and could simulate samples conforming to the distribution pattern of the risk variables.

The article published by Zhai et al. [5] investigated the mechanism of slurry diffusion in horizontal fractures of fractured aquifers; additionally, a one-dimensional seepage grouting theoretical model considering the temporal and spatial variation in slurry viscosity under a constant grouting rate was established. In this model, the grouting pressure required by the predetermined slurry diffusion radius could be obtained by knowing the grouting hole pressure and injection flow.

The article written by Zhao et al. [6] investigated the mining stress evolution law of inclined backfilled stopes in deep mining. Their study demonstrated that mining-induced stress will move to the upper stopes and the stratum below the deepest stope. The transfer range and degree of influence of mining-induced stress will increase with an increase in deep mining, resulting in the most dangerous backfilled stope occurring one or two layers above the deepest stope and the apparent stress concentration area occurring below the deepest stope.

The study by Chen et al. [7] proposed a genetic algorithm (GA) and back propagation (BP) neural network-based computational model for SWB design parameter optimization. This computational model can comprehensively reflect the relation among geological conditions, design parameters and results. Moreover, it automatically searches for the optimal blasting design parameters through the control of SWB targets.

The article published by Chen et al. [8] established a series of models to understand the fracture features of zonal disintegration and to reveal the failure mechanisms of circle tunnels excavated in deep jointed rock masses. The results demonstrate that the zonal disintegration process is induced by the stress redistribution; the dip angle of the joint set has a great influence on the stress buildup, stress shadow and stress transfer as well as on the failure mode of the surrounding rock masses; the existence of parallel and random joints lead the newly formed cracks near the tunnel surface to developing along their strikes; and the random joints make the zonal disintegration pattern much more complex and affected by the regional joint composition.

The article written by Li et al. [9] proposed a novel machine learning model, deep forest, to predict rockburst risk. The deep forest model achieved 100% training accuracy and 92.4% testing accuracy, and it has a more outstanding capability to forecast rockburst disasters compared with other widely used models.

The paper published by Liu et al. [10] studied the effect of lithology on the mechanical and damage behaviors of concrete in a concrete–rock combined specimen (CRCS). The results show that the low-strength concrete part plays a major role in the fracture behavior of CRCS. Furthermore, damage in the CRCS mainly formed in the concrete part, and the extent of the damage in the concrete part was positively correlated with the strength of the rock part.

In the paper published by Zhang et al. [11], the reactive transportation and distribution morphology of a uranium-containing solution was described, a stress-dependent reactive transport model was developed, and the simulator of FLAC3D-CFD was employed. The results show that the uranium-containing solution transport and distribution are significantly dependent on the evolution of the connected channel in a rough-walled fracture, which is significantly influenced by the confining stress and hydraulic pressure.

The article written by Wang et al. [12] proposed a method of rock drillability evaluation based on drilling process monitoring (DPM) parameters. The two-dimensional regression analysis was utilized to investigate the relationships between the drilling parameters, and the three-dimensional regression analysis was used to establish models of ROF and specific energy (SE). Finally, a prediction model for the uniaxial compressive strength (UCS) was established based on the SE and drillability index. The results show that both the regression models and the prediction models have good performances, which can serve as important guides and a source of data for field drilling and excavation processes.

The article published by Dong et al. [13] analyzed changes in the b value during a rock failure and investigated the b value characteristics of acoustic emission events. The results indicated a higher accuracy of the b value when calculated using the average amplitude setting for an interval between acoustic emission events of 200 or greater, a stress magnitude of 20 MPa or greater, and a stress proportion of 10% or greater.

The article published by Ullah et al. [14] used three approaches—t-distributed stochastic neighbor embedding (t-SNE), K-means clustering, and extreme gradient boosting (XGBoost)—to predict short-term rockburst risk. The classification accuracy of XGBoost was checked using several performance indices. The results of the proposed model serve as a great benchmark for future short-term rockburst level prediction with high accuracy.

The purpose of the article by Wang et al. [15] was to investigate the evolution of the physical and mechanical properties of coal subjected to salty solutions. The results show that the corrosion effect of a salty solution on coal samples becomes stronger with increasing immersion time. The degree of deterioration in the longitudinal wave velocity (vp) is positively correlated with the immersion time.

The article written by Sun et al. [16] investigated the criteria for determining the critical damage of rocks in a constitutive RHT, and the mechanical parameters of metamorphic sodium lava were substituted to obtain the critical damage threshold of rocks in a numerical simulation. The results show that rock clip production has an inhibitory effect on the development and propagation of blast-induced cracks. The fitting results serve as an important reference value for the design of one-time completion blasting of an upward blind shaft.

The article published by Rong et al. [17] evaluated the seismic performance of geosynthetic-reinforced soil structures (GRSSs) of high concrete face rockfill dams (CFRDs) from a stochastic perspective. The result shows that GRSSs can reduce mild damage on CFRDs during earthquakes and can restrain moderate and severe damage. The influence of vertical spacing and the length of GRSSs on the seismic performance was also obtained and provides a reference for the seismic design and risk analysis of CFRDs.

The article written by Sun et al. [18] designed a new ensemble classifier combining a random forest classifier (RF) and the beetle antennae search algorithm (BAS) that was applied to improve the accuracy of rockburst classification. The results show that BAS could tune the hyperparameters of RF efficiently, and the optimal model exhibited a high performance on an independent test set of rockburst data and new engineering projects.

In order to study the diffusion mechanism of foamed polymer slurry in rock fissures, the article published by Liang et al. [19] derived a radial diffusion model of polymer single crack grouting in consideration of factors such as grouting volume, crack width and expansion rate. The findings show that the results of the slurry diffusion radius, pressure and velocity distribution at different times under different working conditions in the model are in good agreement with the analytical solution.

In order to accurately classify the stability of the slope rock mass in an open-pit mine, the article published by Yang et al. [20] established a new stability evaluation model of the slope rock mass based on variable weight and matter–element extension theory. The results show that the classification results of the proposed model are in line with engineering practices and are more accurate than those of the hierarchical-extension model and the multi-level unascertained measure-set pair analysis model.

The article written by Yin et al. [21] focused on a numerical approach to finding the p–y curves for laterally loaded piles. The p–y curve results from this new approach were compared with the typical design equations of API (American Petroleum Institute) and Matlock. Finally, the influence of clay content on the p–y behavior was investigated using the implemented MATLAB code. When y < 0.15B, the same lateral capacity values resulted in clay contents of 27.5% and 55%, and they were higher than the ones with 0% clay content. The p–y curves showed a decreasing trend with increasing clay content after y > 0.15B.

The article published by Tan et al. [22] investigated the triangle interpolation method for the calculation of the mapping functions of plates containing an opening with arbitrary shapes with an improved method for point adjudgment during iterations. The results show that the stability and failure pattern of the rock mass is correlated with stress around the opening, which is affected by the opening shape. The existence of an opening also greatly reduces the enhancing influence of confining stress on rock specimens.

The article published by Sun et al. [23] presented two-dimensional, closed-form solutions for locating heat-concentrated sources using temperature differences for known and unknown temperature gradient systems. These proposed analytical solutions can provide a new approach to locating heat sources for more complicated conditions using temperature differences, such as the localization of geothermal sources and nuclear waste leak points.

To sum up, the guest editors hope that the selected papers will help scholars and researchers to push forward the progress in analytical, numerical and Big-Data-based methods in deep rock mechanics.

Author Contributions: Conceptualization, S.W. and X.C.; writing—original draft preparation, S.W., X.C., J.Z., Z.S. and X.L.; writing—review and editing, S.W., X.C., J.Z., Z.S. and X.L.; supervision, S.W.; project administration, S.W.; funding acquisition, S.W. All authors have read and agreed to the published version of the manuscript.

Funding: This research received no external funding.

Acknowledgments: We express our thanks to the authors of the above contributions and to the journal *Mathematics* and MDPI for their support during this work.

Conflicts of Interest: The authors declare no conflict of interest.

References

1. Zhang, J.; Li, P.; Yin, X.; Wang, S.; Zhu, Y. Back Analysis of Surrounding Rock Parameters in Pingdingshan Mine Based on BP Neural Network Integrated Mind Evolutionary Algorithm. *Mathematics* **2022**, *10*, 1746. [CrossRef]
2. Li, C.; Xu, Z. Numerical Modeling and Investigation of Fault-Induced Water Inrush Hazard under Different Mining Advancing Directions. *Mathematics* **2022**, *10*, 1561. [CrossRef]
3. Wu, Y.; Huang, L.; Li, X.; Guo, Y.; Liu, H.; Wang, J. Effects of Strain Rate and Temperature on Physical Mechanical Properties and Energy Dissipation Features of Granite. *Mathematics* **2022**, *10*, 1521. [CrossRef]
4. Song, L.; Ying, H.; Wang, W.; Fan, N.; Du, X. Reliability Modelling of Pipeline Failure under the Impact of Submarine Slides-Copula Method. *Mathematics* **2022**, *10*, 1382. [CrossRef]
5. Zhai, M.; Ma, D.; Bai, H. Diffusion Mechanism of Slurry during Grouting in a Fractured Aquifer: A Case Study in Chensilou Coal Mine, China. *Mathematics* **2022**, *10*, 1345. [CrossRef]
6. Zhao, Y.; Zhao, G.; Zhou, J.; Cai, X.; Ma, J. Mining Stress Evolution Law of Inclined Backfilled Stopes Considering the Brittle-Ductile Transition in Deep Mining. *Mathematics* **2022**, *10*, 1308. [CrossRef]
7. Chen, Y.; Chen, S.; Wu, Z.; Dai, B.; Xv, L.; Wu, G. Optimization of Genetic Algorithm through Use of Back Propagation Neural Network in Forecasting Smooth Wall Blasting Parameters. *Mathematics* **2022**, *10*, 1271. [CrossRef]
8. Chen, B.; Gong, B.; Wang, S.; Tang, C. Research on Zonal Disintegration Characteristics and Failure Mechanisms of Deep Tunnel in Jointed Rock Mass with Strength Reduction Method. *Mathematics* **2022**, *10*, 922. [CrossRef]
9. Li, D.; Liu, Z.; Armaghani, D.J.; Xiao, P.; Zhou, J. Novel Ensemble Tree Solution for Rockburst Prediction Using Deep Forest. *Mathematics* **2022**, *10*, 787. [CrossRef]
10. Liu, K.; Jin, S.; Rui, Y.; Huang, J.; Zhou, Z. Effect of Lithology on Mechanical and Damage Behaviors of Concrete in Concrete-Rock Combined Specimen. *Mathematics* **2022**, *10*, 727. [CrossRef]
11. Zhang, T.; Nie, X.; Song, S.; Hao, X.; Yang, X. Modeling Uranium Transport in Rough-Walled Fractures with Stress-Dependent Non-Darcy Fluid Flow. *Mathematics* **2022**, *10*, 702. [CrossRef]
12. Wang, S.; Tang, Y.; Cao, R.; Zhou, Z.; Cai, X. Regressive and Big-Data-Based Analyses of Rock Drillability Based on Drilling Process Monitoring (DPM) Parameters. *Mathematics* **2022**, *10*, 628. [CrossRef]
13. Dong, L.; Zhang, L.; Liu, H.; Du, K.; Liu, X. Acoustic Emission b Value Characteristics of Granite under True Triaxial Stress. *Mathematics* **2022**, *10*, 451. [CrossRef]
14. Ullah, B.; Kamran, M.; Rui, Y. Predictive Modeling of Short-Term Rockburst for the Stability of Subsurface Structures Using Machine Learning Approaches: T-SNE, K-Means Clustering and XGBoost. *Mathematics* **2022**, *10*, 449. [CrossRef]

15. Wang, M.; Guo, Q.; Tian, Y.; Dai, B. Physical and Mechanical Properties Evolution of Coal Subjected to Salty Solution and a Damage Constitutive Model under Uniaxial Compression. *Mathematics* **2021**, *9*, 3264. [CrossRef]
16. Sun, B.; Zhang, Z.; Meng, J.; Huang, Y.; Li, H.; Wang, J. Research on Deep-Hole Cutting Blasting Efficiency in Blind Shafting with High In-Situ Stress Environment Using the Method of SPH. *Mathematics* **2021**, *9*, 3242. [CrossRef]
17. Rong, Z.; Yu, X.; Xu, B.; Du, X. Reliability Analysis of High Concrete-Face Rockfill Dams and Study of Seismic Performance of Earthquake-Resistant Measures Based on Stochastic Dynamic Analysis. *Mathematics* **2021**, *9*, 3124. [CrossRef]
18. Sun, Y.; Li, G.; Yang, S. Rockburst Interpretation by a Data-Driven Approach: A Comparative Study. *Mathematics* **2021**, *9*, 2965. [CrossRef]
19. Liang, J.; Ma, S.; Du, X. Diffusion Model of Parallel Plate Crack Grouting Based on Foaming Expansion Characteristics of Polymer Slurry. *Mathematics* **2021**, *9*, 2907. [CrossRef]
20. Yang, S.; Xu, Z.; Su, K. Variable Weight Matter–Element Extension Model for the Stability Classification of Slope Rock Mass. *Mathematics* **2021**, *9*, 2807. [CrossRef]
21. Yin, K.; Li, L.; Di Filippo, E. A Numerical Investigation to Determine the p–y Curves of Laterally Loaded Piles. *Mathematics* **2021**, *9*, 2783. [CrossRef]
22. Tan, L.; Ren, T.; Dou, L.; Yang, X.; Wang, G.; Peng, H. Analytical Stress Solution and Numerical Mechanical Behavior of Rock Mass Containing an Opening under Different Confining Stress Conditions. *Mathematics* **2021**, *9*, 2462. [CrossRef]
23. Sun, D.; Wu, Y.; Dong, L.; Luo, Q. Closed-Form Solutions for Locating Heat-Concentrated Sources Using Temperature Difference. *Mathematics* **2022**, *10*, 2843. [CrossRef]

mathematics

Article

Back Analysis of Surrounding Rock Parameters in Pingdingshan Mine Based on BP Neural Network Integrated Mind Evolutionary Algorithm

Jianguo Zhang [1], Peitao Li [2], Xin Yin [2], Sheng Wang [3] and Yuanguang Zhu [3,*]

1 State Key Laboratory of Coking Coal Exploitation and Comprehensive Utilization, China Pingmei Shenma Group, Pingdingshan 467000, China; zhangjg_z@126.com
2 School of Civil Engineering, Wuhan University, Wuhan 430072, China; tm_lpt@163.com (P.L.); yinxin_engineering@163.com (X.Y.)
3 State Key Laboratory of Geomechanics and Geotechnical Engineering, Institute of Rock and Soil Mechanics, Chinese Academy of Sciences, Wuhan 430072, China; 202082010032@sdust.edu.cn
* Correspondence: ygzhu@whrsm.ac.cn

Abstract: The mechanical parameters of surrounding rock are an essential basis for roadway excavation and support design. Aiming at the difficulty in obtaining the mechanical parameters of surrounding rock and large experimental errors, the optimized BP neural network model is proposed in this paper. The mind evolutionary algorithm can adequately search the optimal initial weights and thresholds, while the neural network has the advantage of strong nonlinear prediction ability. So, the optimized BP neural network model (MEA-BP model) takes advantage of the two models. It can not only avoid the local extreme value problem but also improve the accuracy and reliability of the prediction results. Based on the orthogonal test method and finite element analysis method, training samples and test samples are established. The nonlinear relationship between rock mechanical parameters and roadway deformation is established by the BP model and MEA-BP model, respectively. The importance analysis of the three input variables shows that the ΔD is the most important input variable, while ΔBC has the smallest impact. The comparison of prediction performance between the MEA-BP model and BP model demonstrates that the optimized initial weights and thresholds can improve the accuracy of prediction value. Finally, the MEA-BP model has been well applied to predicting the mechanical parameter for the surrounding rock in the Pingdingshan mine area, which proves the accuracy and reliability of the optimized model.

Keywords: mind evolutionary algorithm; BP neural network; MEA-BP model; rock mechanical parameters; orthogonal test method

MSC: 68T07

Citation: Zhang, J.; Li, P.; Yin, X.; Wang, S.; Zhu, Y. Back Analysis of Surrounding Rock Parameters in Pingdingshan Mine Based on BP Neural Network Integrated Mind Evolutionary Algorithm. *Mathematics* **2022**, *10*, 1746. https://doi.org/10.3390/math10101746

Academic Editors: Shaofeng Wang, Jian Zhou, Xin Cai, Xiaofeng Li, Zhengyang Song and Andrey Jivkov

Received: 21 March 2022
Accepted: 18 May 2022
Published: 20 May 2022

1. Introduction

The mechanical parameters of surrounding rock are the most important indexes in underground engineering construction. The rock mass is a typical anisotropy and heterogeneity medium with fissures, fractures, joints, bedding planes, and faults [1]. Therefore, the evaluation and prediction of rock mechanical parameters is still a challenge [2]. Researchers and technicians have developed various methods for the determination of rock mechanical parameters [3]. It includes direct measurement, indirect evaluation, laboratory test, numerical analysis, etc. [2,4–8]. Due to the limitations of sample size, quantity and cost, it cannot accurately reflect the variation of rock mass mechanical parameters via laboratory tests. What is more, the surrounding rock and support structure will change slowly with stress and time. It is difficult to dynamically evaluate the mechanical properties of surrounding rock and supporting structures by the traditional method. With the application of computer technology, depth learning has become a new method to solve geotechnical problems [9–11]. It can make full use of the data easily obtained but also the dynamic prediction and evaluation. Neural networks, support vector machines, decision trees, Bayesian classifiers, and other algorithms have been used in various rock mass engineering [9,12–14].

BP neural network has the advantage of strong nonlinear prediction ability, which has been widely used [15,16]. For example, Suman et al. [17] evaluate the safety of slope with functional networks, multivariate adaptive regression splines, and multigene genetic programming. Based on the literature data, the multivariate adaptive regression splines model has the best prediction performance in comparison with other models. To evaluate the stability of a rock slope with interlayered rocks, Wu et al. [18] predict the peak shear strength by a neural network approach. It also considers the effect of joint wall strength combination, normal stress, and joint roughness. The results show a good prediction precision when compared with the experimental data. Salsani et al. [19] reveal that the most critical factor on the road headed performance is the unconfined compressive strength. The nonlinear relation between the unconfined compressive strength, Brazilian tensile strength, rock quality designation, alpha angle, and the road headed performance could be accurately predicted by an artificial neural network. Based on the data of dynamic wave velocity, point load index, slake durability index, and density, the rock strength can also be predicted by an artificial neural network [20]. The hybrid model is proposed by Dai et al. [21], which combines the improved artificial fish swarm algorithm of strong global searching ability and the back propagation algorithm of strong local search ability. The preciseness of hybrid model is better than the other models. An optimized probabilistic neural network (PNN) model is proposed by Feng et al. [22], which makes full use of the mean impact value algorithm (MIVA) and the modified firefly algorithm (MFA). Then, the proposed model shows a good prediction ability in the evaluation of rock burst for the deep tunnels.

The above studies have confirmed the application of neural networks. However, the different initial weights and thresholds of the BP neural network will lead to non-convergence or local extreme value and even a large prediction error. The main reason is that the initial weights and thresholds of BP neural network are randomly generated, which affects the performance of the model. Therefore, selecting appropriate initial weights and thresholds is very important for the accuracy of the prediction model. Mind evolutionary algorithm (MEA) draws on the idea of population and evolution in genetic algorithms. It introduces the process of convergence and alienation, which has a strong ability for global optimization. In other words, the MEA model can search for the optimal initial weights and thresholds. Therefore, some scholars use the MEA model to optimize the performance of the BP neural network. Then an MEA-BP neural network is established and applied in the prediction model, which achieves a good result [23–25]. However, the application of the MEA-BP method in the prediction of surrounding rock parameters is still less.

Based on studying the prediction method of surrounding rock mechanical parameters, the mind evolutionary algorithm and BP neural network are introduced and combined firstly. Based on the Matlab program, the optimized BP model (MEA-BP model) is proposed in this paper. Then, the important analysis and comparison for the MEA-BP method are carried out, demonstrating the validity of proposed method. Finally, based on the observed displacement of roadway in the Pingdingshan mining area, the mechanical parameters of surrounding rock are predicted and evaluated, which is expected to provide a reference and basis for the long-term stability analysis of deep surrounding rock.

2. Methodology

2.1. Prediction Method

According to the Mohr-Coulomb criterion, the mechanical parameters of surrounding rock are major factors affecting the stability of roadway surrounding rock. It includes elastic modulus (E), cohesion (c), and friction Angle (φ). On the other hand, surrounding rock deformation is highly sensitive to these three parameters, which is difficult to obtain accurately [26]. Therefore, the deformation of surrounding rock U (u_1, u_2, u_3, ...) can be defined as a function of mechanical parameters (E, c, φ). The nonlinear relationship between surrounding rock deformation and mechanical parameters can be expressed as:

$$U(u_1, u_2, u_3, \ldots) = f(E, c, \varphi) \tag{1}$$

In engineering applications, the deformation of surrounding rock can be measured easily, while it is contrary to the mechanical parameters of rock mass. Therefore, it is an innovative method to evaluate the mechanical parameters based on surrounding rock deformation. So, the relationship between surrounding rock deformation and mechanical parameters is very significant for the accuracy of prediction results. Due to the nonlinear relationship between surrounding rock mechanical parameters and deformation, it is very difficult to solve in theory.

The neural network has the advantages of strong learning ability and good plasticity, which has been widely used in nonlinear fitting and prediction. However, most neural networks are based on gradient descent algorithms. The convergence speed is slow, and the training time is too long. What is more, the selection of initial weights and thresholds seriously affects the convergence and accuracy of the neural networks, which is prone to local extreme value. Therefore, the mind evolutionary algorithm is adopted in this paper, which is aimed at the optimized initial parameters of the BP neural network. Then, the relationship between surrounding rock deformation and mechanical parameters is established by the optimized BP model. This method can avoid the local extreme value problem but also improve the accuracy and reliability of the prediction model.

2.2. The Optimized BP Model

2.2.1. BP Neural Network Model

The BP neural network model is the most intuitive and widely used among many artificial neural network models. The BP neural network model is a feed-forward multilayer perceptron neural network and error backpropagation learning algorithms. A representative BP neural network model consists of three layers: an input layer, a hidden layer, and an output layer [27], as shown in Figure 1. The learning process can be divided into two parts: model forward propagation and error back propagation. In the forward propagation, the input sample enters the input layer and then propagates to the output layer after being processed by the hidden layer. The computation process [15] between input and output can be expressed as:

$$\overline{Y} = f_{output} \sum_{j=1}^{H} w_{kj} (f_{hidden} (\sum_{i=1}^{I} w_{ji} X_i + b_j) + b_k) \tag{2}$$

Then, the error is calculated by:

$$E = \frac{1}{N} \sum_{n=1}^{N} (\overline{Y}_n - Y_n)^2 n = 1, 2, 3, \ldots, N \tag{3}$$

where, i and H represent the numbers of the input sample and hidden sample; b_j and b_k represent the basis of hidden and output layer; f_{output} and f_{hidden} represent the transfer functions for the hidden and output neurons; w_{ji} represents the weights connecting the input layer and hidden layer; w_{kj} represents the weights between the hidden layer and output layer; $\overline{Y}_n$ and Y_n represent the predicted and actual output for the training sample.

If the error exceeds the tolerance, then the output error will be propagated back to the input layer via the hidden layer. Repeated training and learning are carried out until it meets the requirement [28].

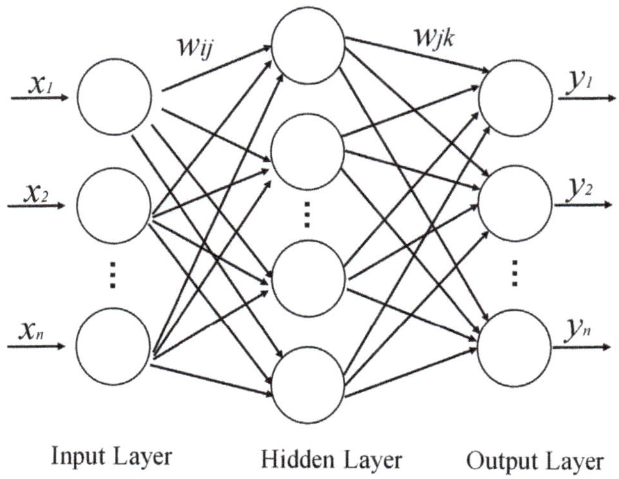

Figure 1. The structure of BP neural network.

2.2.2. Mind Evolutionary Algorithm

The mind evolutionary algorithm (MEA) inherits the concepts of "group" and "evolution" from the genetic algorithm (GA). It also has important innovations. Inspired by man's mind action attributes under certain social environments, MEA comes up with "similar-taxis" and "dissimulation" [29]. The population of MEA consists of several groups surviving around the environment. Those groups are divided into superior groups and temporary groups randomly. Each group includes some individuals according to their uniform distribution (Figure 2).

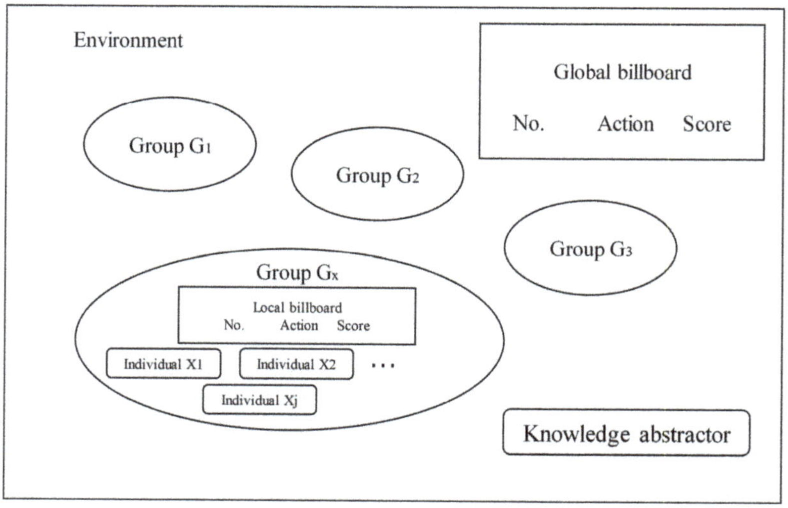

Figure 2. The structure of mind evolutionary computing.

The strategy of MEA is searching for the individual which has the highest score as a superior individual. Then, other individuals with the highest score as temporary individuals. The MEA designs two stages to realize the optimization goal [24]. Through the "similar-taxis", the local information and local optimization are obtained. While the "dissimilation" operation conducts a global search and makes "exploration" in the whole search space. Through the "similar-taxis" and "dissimilation" alternately, MEA can achieve the balance of exploration and exploitation. The evolutionary process is described as the follows:

(a) The population of MEA consists of several groups surviving around the environment. Those groups are divided into superior groups and temporary groups randomly. Each group includes several individuals according to their uniform distribution.
(b) MEA searches for the individual with the highest score as a superior individual. At the same time, other individuals which have the highest score as temporary individuals. Each superior individual and temporary individual as the center to generate the superior subgroups and temporary subgroups.
(c) The "similar-taxis" process is completed within the subgroups. The population maturity discriminant function is used to confirm whether the subgroup is mature. If it is, the mature subgroup stops this process. Then, the score of the optimal individual is regarded as the group score. The immature subgroup will produce subgroups with new centers. The above process will be repeated until they mature.
(d) The disassimilation process is completed globally. If the score of the temporary subgroup is higher than that of the optimal subgroup, this process is carried out. Then, individuals in the temporary subgroup with a high score will replace those with a low score. At the same time, the individuals in the optimal subgroup are released. A new temporary subgroup is reconstructed in the global space.
(e) Repeating the above process until the score of the optimal group is so high that it is impossible to increase. Then, the algorithm is convergent, and the winner of the superior group is just the global optimization.

Obviously, MEA takes advantage of the similar-taxis and the dissimilation to make the local search and global search alternately. Moreover, MEA designs billboards to record the evolutionary information that will guide the evolution in turn. So, MEA can make use of its directional search advantage, which is good for the search of optimal initial weights and thresholds. Currently, MEA has successfully applied to many optimization problems [25].

2.3. Combined Prediction Method

The MEA can search for the best initial weights and thresholds quickly and accurately. While the neural network has the advantages of high calculation precision and controllable error. So, the combination of two models is proposed to predict rock parameters. In other words, the optimal initial weights and thresholds value is obtained by the mind evolutionary algorithm. Then, the optimized initial value is used in the BP neural network model, which can improve the convergence speed and fitting accuracy of the BP neural network. Finally, an MEA-BP model is established based on the Matlab program.

In addition to the prediction model, the data of training and verification for the prediction model is another key point, which determined the accuracy of the predict value. The common method for the establishment of training data and verification data is field monitoring or numerical analysis [13,14,30,31]. Due to the complex engineering conditions, the filed data cannot acquired easily. Numerical analysis is an efficient method for the analysis of influence law of rock mechanical, which has been widely used in the back analysis of surrounding rock [31]. At the same time, the orthogonal experimental design method is also adopted in the establishment of training data. Orthogonal experimental design method is a representative method for the study of multi factors and levels. It can reflect the influence of all factors and levels, which is appropriate for the study of rock engineering. Thus, both orthogonal experimental design method and finite element method (FLAC3D) are adopted in the establishment of training sample and test sample.

Based on the training data and predicted model, the nonlinear relationship between mechanical parameters and deformation of surrounding rock can be established. The steps to develop a combined prediction model (MEA-BP model) are illustrated in Figures 3 and 4.

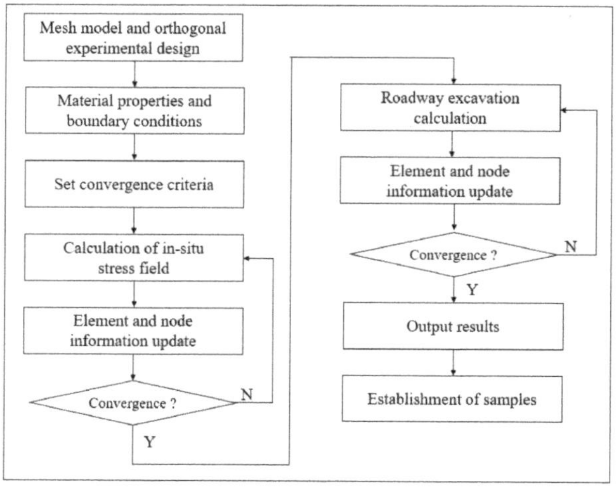

Figure 3. Flow chart of numerical analysis and sample establishment.

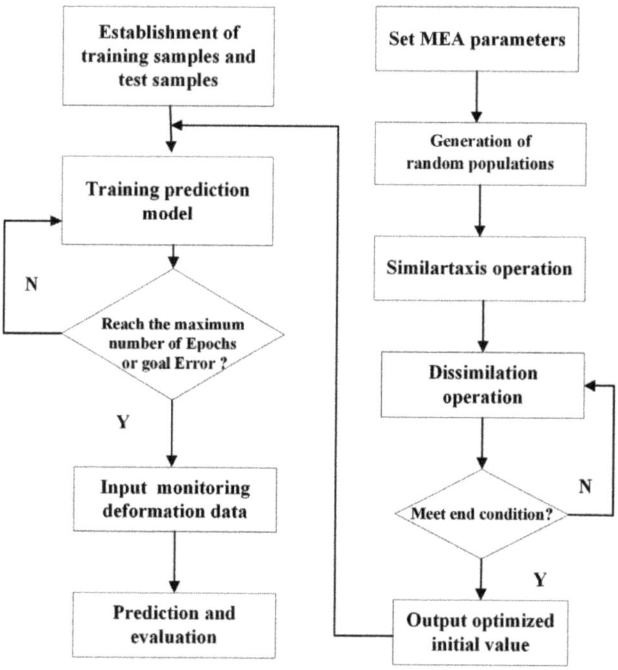

Figure 4. The flow chart of the MEA-BP prediction model.

Step 1: The establishment of a training sample. The orthogonal experimental design method is used to determine the test scheme based on three rock mechanical parameters. Then, the deformation results of surrounding rock under different mechanical parameters are simulated by the finite element method (FLAC3D). The deformation results and mechanical parameters of surrounding rock are regarded as the input and output variable values for the training sample, respectively (Figure 3).

Step 2: Determination of BP network. The parameters of the input layer, output layer, and hidden layer were determined.

Step 3: Setting MEA parameters. The number of iterations, initial population, superior subpopulation, temporary subpopulation, network weights, and threshold are determined.

Step 4: Generation of random populations. The optimized weights and thresholds are coded according to the MEA parameters. Then, the superior subgroups and temporary subgroups are generated.

Step 5: The similar taxis operation is performed in each subgroup until the subgroup is mature. The score of the optimal individual is used as the score of the subgroup.

Step 6: The dissimilation operation is performed between the superior and temporary subgroups.

Step 7: Output the superior individual when the iterations operations or the optimal global individual is found. If the above conditions are not satisfied, return to step 3.

Step 8: The optimized initial weights and thresholds are assigned to the BP network. The MEA-BP model trains the nonlinear relationship between the deformation and rock mechanical parameters.

Step 9: Prediction and evaluation. The observed displacement value is used to predict rock mechanical parameters by the MEA-BP model. Then, the predicted displacement is obtained by finite element numerical analysis methods. Finally, the comparison and evaluation between the predicted displacement and the observed value are carried out (Figure 4).

3. Application of Prediction Model

3.1. Geological and Numerical Model

The study roadway is located in the No.10 mine of the Pingdingshan mining area. The rock stratum of the roadway is a sandy mudstone layer. The overall geological structure is stable, and no obvious fissure water is found. The roadway section is a straight-wall arch, with a width of 5.6 m, a height of 4.5 m, and a radius of 1.7 m. According to the classification of surrounding rock, the surrounding rock where the roadway is located belongs to class III~V. The basic quality (BQ) rating system is most widely used as an empirical method for rock mechanical parameters [32]. Then, the range of elastic modulus (E), cohesion (c), and friction Angle (φ) is calculated by the BQ rating system. Other parameters are determined by the laboratory test results, as shown in Table 1. The measurement of in-situ stress shows that the σ_x = 30.81 MPa, σ_y = 28.78 MPa, σ_z = 27.04 MPa. To quantitatively analyze the deformation of surrounding rock, the displacement of point A and point D (ΔA and ΔD) is regarded as the vertical deformation value. At the same time, the relative displacement of point B and point C (ΔBC) is selected as the horizontal deformation value, as shown in Figure 5. The displacement of roadway surrounding rock is measured by a laser range finder (Figure 6).

Table 1. Physical and mechanical parameters of surrounding rock.

Mechanical Parameters	Density ρ/Kg·m^{-3}	Elasticity Modulus E/GPa	Poisson's Ratio/μ	Cohesion c/MPa	Friction Angle φ/(°)	Dilation Angle ψ/(°)
Value	2450	3~13.5	0.21	0.5~4	15~32.5	10

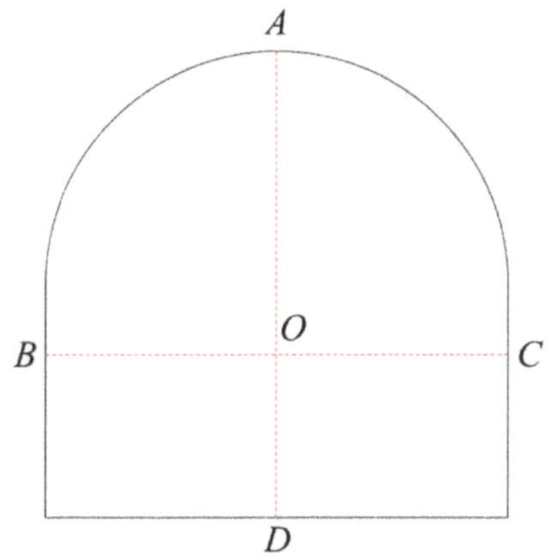

Figure 5. The monitoring points of roadway deformation.

Figure 6. Laser range finder.

The finite element analysis method (FLAC3D) is used to simulate the roadway deformation with different rock mechanical parameters to establish the training samples. The rock mechanical parameters are determined by Table 1. The Elasticity modulus, cohesion and friction angle are regarded as the variable for the establishment of training data and verification data. The other parameters are constant, which can refer to the Table 1. In addition, the Mohr-Column criterion is adopted. The model range must be sufficiently large to minimize the influence of boundary effects. Therefore, a 3D roadway numerical model with a size of 67.2 m $\times$ 60 m $\times$ 36 m is established, as shown in Figure 7. The element number of the numerical analysis model is 12,380. The upper part of the model is a free boundary, and the other boundaries are fixed. The three principal stresses are applied in the corresponding directions, and gravity is also considered. When the convergence value (ratio = 1×10^{-5}) is reached, the equilibrium state can be considered to be reached.

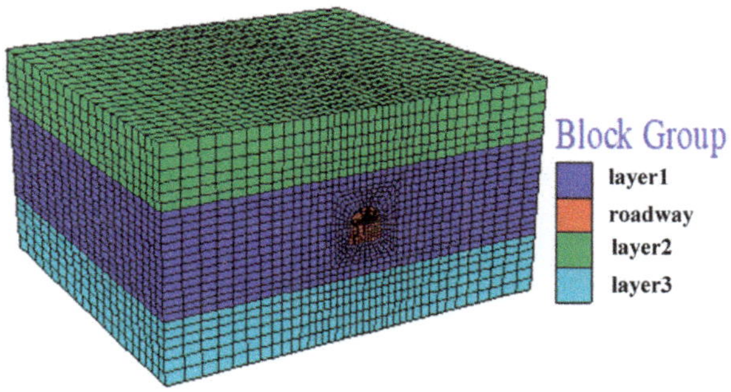

Figure 7. Three-dimensional numerical analysis model.

3.2. Establishment of Training Sample

Orthogonal experimental design is an experimental method for studying multi-factor and levels. According to the principle of orthogonal design, some representative points are selected for the test, which has the characteristics of "uniform dispersion, neat and comparable". The orthogonal test method has high efficiency, which can quickly determine the influence degree of relevant factors on deformation. So, the orthogonal experimental design method is used to construct the training samples for the prediction model. The rock mechanical parameter includes elastic modulus (E), cohesion (c), and friction angle (φ), which has eight levels, as shown in Table 2. After roadway excavation and timely support, the surrounding rock and support structure can be regarded as the overall structure. Therefore, the rock mechanical parameters described in this paper are the overall mechanical parameters of surrounding rock and support structure, which is also the basis for the analysis of surrounding rock after support.

Table 2. Rock mechanical parameters and levels for orthogonal experimental.

Level	Parameter		
	E/MPa	c/MPa	φ/°
1	3	0.5	15
2	4.5	1.0	17.5
3	6.0	1.5	20
4	7.5	2.0	22.5
5	9.0	2.5	25
6	10.5	3.0	27.5
7	12.0	3.5	30
8	13.5	4.0	32.5

Based on Table 2 and the orthogonal test design method (L64), 64 training samples are obtained by numerical analysis. Correspondingly, the deformation of roadway (ΔA, ΔD, and ΔBC) with different rock mechanical properties is regarded as the input value for the training sample (Figure 3). The mechanical parameters of orthogonal experiments are regarded as the observed value. Then, the nonlinear relationship between the deformation and rock mechanical parameters of surrounding rock is obtained by the training sample. The nonlinear relationship obtained by the MEA-BP model can be used to predict of rock mechanical parameters. Finally, the evaluation of the prediction model can be carried out (Figure 4).

3.3. Importance Analysis of Input Variables

In addition to being interested in the prediction performance of the MEA-BP model, the importance of different input variables on the prediction performance is concerned, which increases the interpretability of the model. Permutation importance (PI) is a common indicator to measure the importance of input variables [33]. It has the advantages of being easy to understand and fast to calculate. The detailed procedure of importance analysis is introduced as follows:

(1) Based on the BQ rating system, eight group mechanical parameters of test samples are generated by the random method. Then, the deformation of surrounding rock is obtained by the finite element numerical method. The data of test sample is shown in Table 3. The surrounding rock deformation results and mechanical parameters are regarded as the input value and observed value for the test samples.
(2) Train the MEA-BP model based on the original training samples (L64) Then, the prediction of the test sample (Table 3) is obtained, and the $RMSE$ is calculated by Equation (4).

$$RMSE = \sqrt{\frac{1}{N}\sum_{t=1}^{N}(observed_t - predicted_t)^2} \quad (4)$$

where, $observed_t$ represents the observed value; $predicted_t$ represents the predicted value; t represents the predicted variable, N represents the number of input variables.

(3) Shuffle the ith input variable of the training samples in reverse order, and the others remain unchanged.

(4) Train the MEA-BP model on the shuffled training samples. The prediction of the test sample and *RMSE* is calculated by Equation (4).

(5) Calculate *PI* of the ith input variable using Equation (5);

$$PI_i = |RMSE_i - RMSE*| \tag{5}$$

where, $RMSE_i$ and $RMSE^*$ are separately calculated on the shuffled training samples and original training samples; i (i = 1, 2, 3) represents the input variable (ΔA, ΔD, ΔBC).

(6) Repeat (3)~(4) until *PI* of all input variables (ΔA, ΔD, ΔBC) is obtained. To be more intuitive, *PI* is standardized by Equation (6).

$$PI_i^* = \left(PI_i / \sum_{i=1}^{n} PI_i \right) \times 100\% \tag{6}$$

Table 3. The rock mechanical parameter and deformation for test samples.

Group	Mechanical Parameters			Displacement/mm		
	E/MPa	*c*/MPa	*φ*/°	ΔA	ΔD	ΔBC
1	12.69	2.49	24.08	−19.3	16.9	−20.6
2	9.91	2.25	29.92	−18.3	21.0	15.0
3	3.01	1.86	25.28	−93.4	79.1	−126.0
4	3.84	0.78	32.10	−82.2	82.7	−115.8
5	7.25	2.24	16.40	−65.7	41.8	−110.7
6	4.01	0.63	31.15	−102.7	97.8	−143.0
7	6.76	1.82	36.40	−24.2	29.9	−14.7
8	6.34	2.54	17.38	−55.1	36.3	−87.5

Firstly, the *RMSE* of three input variables (ΔA, ΔD, ΔBC) for eight test samples is obtained (Table 4). The larger *RMSE* means a greater adverse effect. The range of *RMSE** is 0.24~1.20, with an average of 0.58. Except the group 5, the $RMSE_i$ for other groups is larger than *RMSE**. It indicates that the changes of the input value of training samples have a significant impact on the prediction results. The average of $RMSE_2$ is the largest, while the $RMSE_1$ is the second. The larger *PI* is, the more important the corresponding input variable. It can be seen that the *PI** for ΔD, ΔA, and ΔBC is 49.93%, 30.30%, and 19.77%, respectively (Figure 8). In other words, the variation of *PI** is consistent with the average of *RMSE*. Especially, the $RMSE_2$ is the biggest of six test groups. What is more, the maximum value of *RMSE* is 10.62, which also belongs to $RMSE_2$. So, it can be considered that the ΔD is the most important input variable, while the influence of ΔBC is the least.

Table 4. The RMSE for three input variables.

No.	$RMSE_1$	$RMSE_2$	$RMSE_3$	$RMSE^*$
1	6.04	7.60	3.67	0.93
2	4.66	1.49	5.13	1.20
3	1.71	8.25	1.63	0.31
4	6.41	10.62	1.35	0.49
5	0.35	1.65	0.58	0.48
6	5.16	10.07	4.60	0.24
7	2.67	2.09	2.88	0.40
8	2.01	3.45	0.86	0.57
Average	3.63	5.65	2.59	0.58

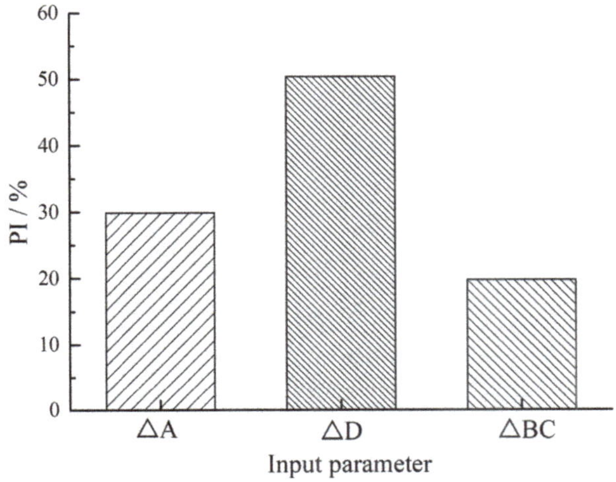

Figure 8. Importance analysis results of input variables.

3.4. Model Analysis

To verify the effectiveness of the proposed model, the mechanical parameters are predicted by the MEA-BP model and BP model, respectively. The data of test samples are shown in Table 4. The parameters of the two models are shown in Table 5. The η is the learning rate; S is population size; m is the maximal iterative number; Nt is the number of the temporary subgroup; Ns is the number of superior subgroups; g is the convergence error.

Table 5. Parameters of two prediction models.

Prediction Model	The Initial Values
BP model	η = 0.05; g = 0.001
MEA-BP model	η = 0.05; S = 200; m = 10; Nt = 5; Ns = 5; g = 0.001

The nonlinear relationship between surrounding rock deformation and mechanical parameters is established during the modeling process by 64 sets of the training samples (L64). Based on the test sample (Table 3), the *RMSE* of three predicted mechanical parameters is calculated by Equation (4). The smaller *RMSE* means the more reliable accuracy. All *RMSE* of the predicted mechanical parameters of the MEA-BP model are smaller than that of the BP model (Figure 9). It indicates that selecting initial weights and thresholds can improve the convergence and accuracy of neural networks but also avoid local extreme value problems. Compared with the BP model, the declining ratio of *RMSE* for the MEA-BP model was about 48.89~79.91%. In other words, the predicted performance of the MEA-BP model is better than that of the BP model.

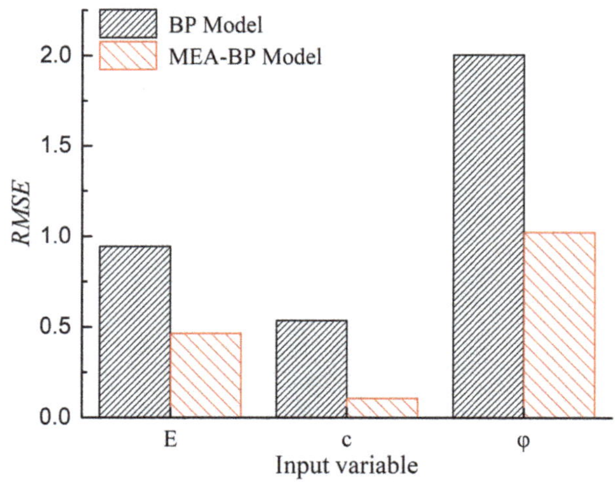

Figure 9. The variation of *RMSE* for three predicted variables.

The relative error value (*REV*) is used for the detailed comparison of MEA-BP model and BP model. The definition of relative error value is:

$$REV_x = \left| \frac{x^* - x_0}{x_0} \right| \times 100\% \quad (7)$$

where, REV_x represents the relative error value of the parameter x; x represents the parameter variable (E, c, φ); x_0 represents the initial value of a parameter; x^* represents the predicted value of a parameter.

The *REV* of three parameters for the MEA-BP model and BP model is summarized in Figure 10. Compared with the BP model, the *REV* of three mechanical parameters is smaller when the MEA-BP model is adopted. Specifically, the *REV* of elastic modulus (E) is between 0.12~10.96% when the MEA-BP model is adopted. In contrast, it is 2.37~31.32% for the BP model. The cohesion and friction angle also shows the same change trend. The *REV* of cohesion and friction angle is less than 12.26% for the MEA-BP model, which is smaller than those of BP model. What is more, the *REV* of cohesion and friction angle is smaller than that of elastic modulus (Figure 10b,c).

Although, the nonlinear relationship between mechanical property and deformation of surrounding rock is inherent. However, the importance of different variables is different. So, the REV of some point is large. What is more, the different initial input value will lead to non-convergence or local extreme value and even a large prediction error when the BP neural network is adopted. Thus, the change of REV is irregular when the BP model is adopted. It is not only the key point of the prediction model but also the advantage of MEA-BP model. The sensitivity of input variables cannot be decreased, but the accuracy of the prediction can be improved when the MEA-BP model is adopted. In other words, the application of the mind evolutionary algorithm could improve prediction accuracy. The MEA-BP model has a better prediction performance than the BP model.

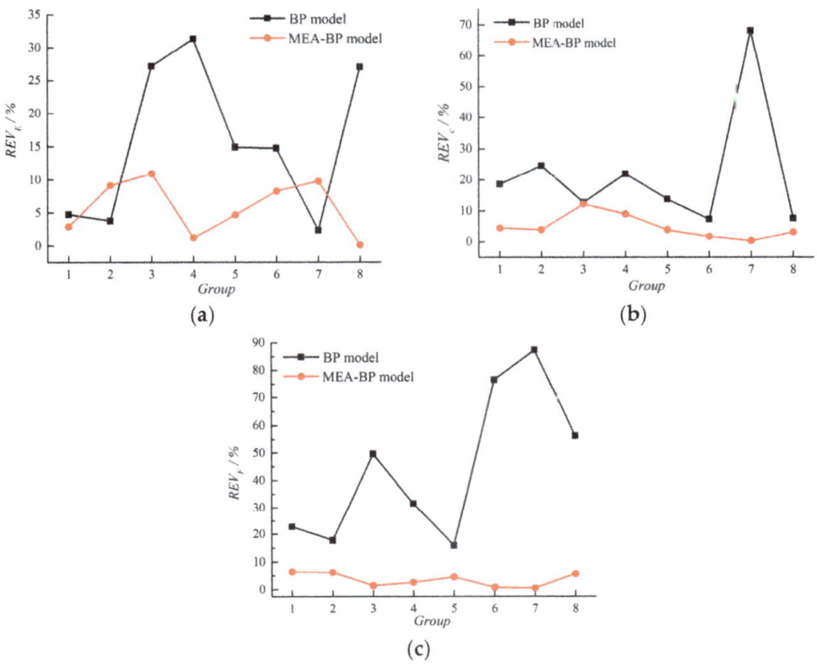

Figure 10. Comparison between the BP model and MEA-BP model. (**a**) Elastic modulus. (**b**) Cohesion. (**c**) Friction angle.

3.5. Application of MEA-BP Model

Based on Section 2, the MEA-BP model is used to predict mechanical parameters for the roadway in the Pingdingshan mine area. Firstly, two adjacent roadways are selected as monitoring roadways, and two sections are selected for each roadway. Then, the observed displacement value of roadway is shown in Table 6.

Table 6. The deformation and predicted mechanical parameters of roadway.

Measuring Point	Roadway Deformation Value			Predicted Mechanical Parameter		
	ΔA/mm	ΔD/mm	ΔBC/mm	*E*/MPa	*c*/MPa	*φ*/°
1-1	−60.9	41.4	−96.1	8.52	2.23	15.79
1-2	−64.8	41.2	−108.9	8.03	2.18	15.84
2-1	−90.7	78.5	−96.4	3.16	1.76	26.27
2-2	−83.9	75.5	−91.5	3.30	1.72	26.98

Based on the observed displacement value of roadway, the predicted rock mechanical parameter is obtained by the MEA-BP model. Then, the predicted deformation value of roadway is calculated via FLAC3D, which is used to compare the observed displacement (Figure 11). It can be seen that the *REV* of ΔA and ΔD ranges between 1.43% and 5.70%. The predicted value of ΔA and ΔD is basically consistent with the observed value. According to the important analysis of input variables (Section 3.3), the prediction result has the lowest sensitivity to ΔBC. So, it will result in a large error of ΔBC. What is more, the isotropic assumptions are adopted during the finite element analysis. It also does not consider the actual geological variation, such as fracture distribution. Therefore, the predicted deformation results of ΔBC are less than the observed values, but the *REV* of ΔBC is less than 9.1%. To sum up, the MEA-BP model can well predict the mechanical parameter.

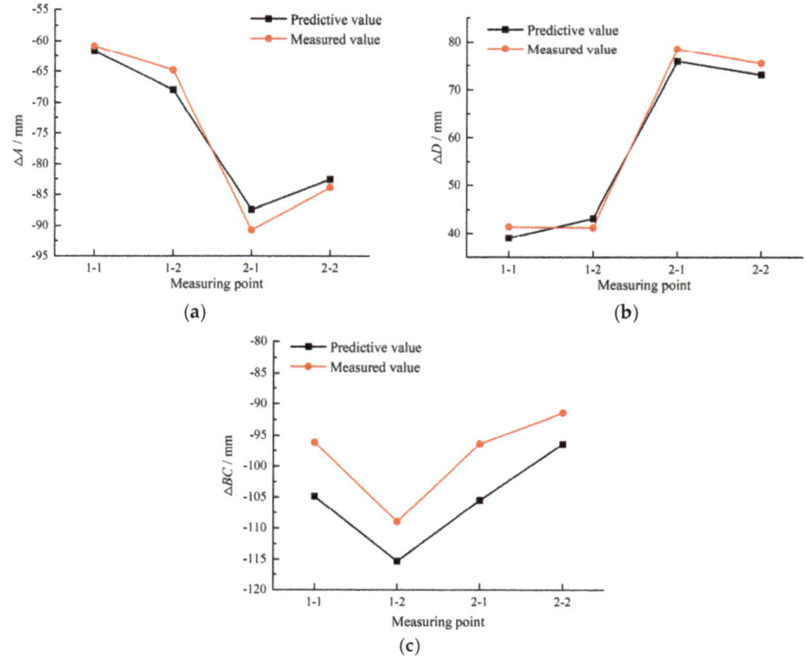

Figure 11. Comparison of surrounding rock deformation results. (**a**) the displacement of ΔA. (**b**) the displacement of ΔD. (**c**) the displacement of ΔBC.

4. Conclusions

To predict mechanical parameters for surrounding rock, an optimized BP prediction model is proposed in this paper. It makes use of the great search capability of the MEA model and the strong learning ability and plasticity of the BP model. Then, the orthogonal experimental design method and finite element method (FLAC3D) are adopted in the establishment of training sample and test sample. Thirdly, the nonlinear relationship between mechanical parameters and deformation of surrounding rock is obtained by BP model and MEA-BP model, respectively. Finally, the analysis show that the application of the mind evolutionary algorithm could improve prediction accuracy. It also has a better prediction performance in the application. The main conclusions are as follows:

(1) To avoid the local extreme value and improve the convergence speed of the BP neural network model, the initial weights and thresholds are optimized by the mind evolutionary algorithm. So, the combination of the mind evolutionary algorithm and BP neural network is proposed in this paper, which is aimed at a good prediction of rock mechanical parameters.

(2) Based on the orthogonal test method and finite element numerical method, training samples and test samples are established. The rock mechanical parameters are the observed value, while the deformation of surrounding rock (ΔA, ΔD, ΔBC) is the input value. The nonlinear relationship between rock mechanical parameters and roadway deformation is established by the BP model and MEA-BP model, respectively.

(3) The important analysis of different input variables for the MEA-BP model shows that the RMSE of ΔD is the largest. While the RMSE and PI for ΔBC are the smallest. It indicates that the ΔD is the most important input variable, while the influence of ΔBC is the least.

(4) The comparison between the MEA-BP model and BP model shows that the *RMSE* value for the MEA-BP model is reduced by about 48.89~79.91%. What is more, the

REV of the MEA-BP model is also smaller than that of the BP model. The MEA-BP model cannot decrease the sensitivity of input variables, but the accuracy of the prediction is improved. So, the prediction performance of the MEA-BP model is better than the BP model, which also demonstrates the advantages of the optimized initial weights and thresholds.

(5) The MEA-BP model is used to predict mechanical parameters for roadways in the Pingdingshan mine area. The predicted deformation results are consistent with the observed value, which demonstrated the accuracy and reliability of the optimized model.

Author Contributions: Investigation, conceptualization and resources, J.Z.; writing, validation and formal analysis, P.L. and S.W.; software and methodology, X.Y.; project administration and funding acquisition Y.Z. and J.Z. All authors have read and agreed to the published version of the manuscript.

Funding: This work was supported by the Open Research Fund of State Key Laboratory of Coking Coal Exploitation and Comprehensive Utilization, China Pingmei Shenma Group (No. 4104022017110603) and National Natural Science Foundation of China (No. 51874275).

Institutional Review Board Statement: Not applicable.

Informed Consent Statement: Not applicable.

Data Availability Statement: Not applicable.

Acknowledgments: The wish to send acknowledgments to State Key Laboratory of Coking Coal Exploitation and Comprehensive Utilization (China Pingmei Shenma Group) for their help for the data collection and measurement. The authors also are very grateful to anonymous' for their recommendations.

Conflicts of Interest: The authors declare no conflict of interest.

References

1. Zhao, H.; Chen, B.; Li, S.; Li, Z.; Zhu, C. Updating the models and uncertainty of mechanical parameters for rock tunnels using Bayesian inference. *Geosci. Front.* **2021**, *12*, 101198. [CrossRef]
2. Feng, X.-T.; Zhou, Y.-Y.; Jiang, Q. Rock mechanics contributions to recent hydroelectric developments in China. *J. Rock Mech. Geotech. Eng.* **2019**, *11*, 511–526. [CrossRef]
3. Meng, F.; Wong, L.N.Y.; Zhou, H. Rock brittleness indices and their applications to different fields of rock engineering: A review. *J. Rock Mech. Geotech. Eng.* **2021**, *13*, 221–247. [CrossRef]
4. Han, X.; Li, W.; Li, X.; Zhu, Z. Virtual reality assisted techniques in field tests and engineering application of the mechanical parameters of a horizontally layered rock mass. *Alex. Eng. J.* **2022**, *61*, 4027–4039. [CrossRef]
5. Liu, H.; Liu, Q.; Ma, H.; Fish, J. A novel GPGPU-parallelized contact detection algorithm for combined finite-discrete element method. *Int. J. Rock Mech. Min. Sci.* **2021**, *144*, 104782. [CrossRef]
6. Salmi, E.F.; Sellers, E.J. A review of the methods to incorporate the geological and geotechnical characteristics of rock masses in blastability assessments for selective blast design. *Eng. Geol.* **2021**, *281*, 105970. [CrossRef]
7. Rostamsowlat, I.; Evans, B.; Kwon, H.J. A review of the frictional contact in rock cutting with a PDC bit. *J. Pet. Sci. Eng.* **2022**, *208*, 109665. [CrossRef]
8. Schaefer, L.N.; Kereszturi, G.; Villeneuve, M.; Kennedy, B. Determining physical and mechanical volcanic rock properties via reflectance spectroscopy. *J. Volcanol. Geotherm. Res.* **2021**, *420*, 107393. [CrossRef]
9. Lawal, A.I.; Kwon, S. Application of artificial intelligence to rock mechanics: An overview. *J. Rock Mech. Geotech. Eng.* **2021**, *13*, 248–266. [CrossRef]
10. Liu, K.; Liu, B.; Fang, Y. An intelligent model based on statistical learning theory for engineering rock mass classification. *Bull. Eng. Geol. Environ.* **2018**, *78*, 4533–4548. [CrossRef]
11. Barnoud, A.; Cayol, V.; Lelièvre, P.G.; Portal, A.; Labazuy, P.; Boivin, P.; Gailler, L. Robust Bayesian Joint Inversion of Gravimetric and Muographic Data for the Density Imaging of the Puy de Dôme Volcano (France). *Front. Earth Sci.* **2021**, *8*, 575872. [CrossRef]
12. Hajihassani, M.; Jahed Armaghani, D.; Kalatehjari, R. Applications of Particle Swarm Optimization in Geotechnical Engineering: A Comprehensive Review. *Geotech. Geol. Eng.* **2017**, *36*, 705–722. [CrossRef]
13. Yin, X.; Liu, Q.; Huang, X.; Pan, Y. Perception model of surrounding rock geological conditions based on TBM operational big data and combined unsupervised-supervised learning. *Tunn. Undergr. Space Technol.* **2022**, *120*, 104285. [CrossRef]
14. Yin, X.; Liu, Q.; Huang, X.; Pan, Y. Real-time prediction of rockburst intensity using an integrated CNN-Adam-BO algorithm based on microseismic data and its engineering application. *Tunn. Undergr. Space Technol.* **2021**, *117*, 104133. [CrossRef]
15. Liu, B.; Wang, R.; Zhao, G.; Guo, X.; Wang, Y.; Li, J.; Wang, S. Prediction of rock mass parameters in the TBM tunnel based on BP neural network integrated simulated annealing algorithm. *Tunn. Undergr. Space Technol.* **2020**, *95*, 103103. [CrossRef]

16. Yin, X.; Liu, Q.; Pan, Y.; Huang, X. A novel tree-based algorithm for real-time prediction of rockburst risk using field microseismic monitoring. *Environ. Earth Sci.* **2021**, *80*, 504. [CrossRef]
17. Suman, S.; Khan, S.Z.; Das, S.K.; Chand, S.K. Slope stability analysis using artificial intelligence techniques. *Nat. Hazards* **2016**, *84*, 727–748. [CrossRef]
18. Wu, Q.; Xu, Y.; Tang, H.; Fang, K.; Jiang, Y.; Liu, C.; Wang, X. Peak shear strength prediction for discontinuities between two different rock types using a neural network approach. *Bull. Eng. Geol. Environ.* **2018**, *78*, 2315–2329. [CrossRef]
19. Salsani, A.; Daneshian, J.; Shariati, S.; Yazdani-Chamzini, A.; Taheri, M. Predicting roadheader performance by using artificial neural network. *Neural Comput. Appl.* **2013**, *24*, 1823–1831. [CrossRef]
20. Sarkar, K.; Tiwary, A.; Singh, T.N. Estimation of strength parameters of rock using artificial neural networks. *Bull. Eng. Geol. Environ.* **2010**, *69*, 599–606. [CrossRef]
21. Dai, B.; Gu, H.; Zhu, Y.; Chen, S.; Rodriguez, E.F.; Geem, Z.W. On the Use of an Improved Artificial Fish Swarm Algorithm-Backpropagation Neural Network for Predicting Dam Deformation Behavior. *Complexity* **2020**, *2020*, 5463893. [CrossRef]
22. Feng, G.; Xia, G.; Chen, B.; Xiao, Y.; Zhou, R. A Method for Rockburst Prediction in the Deep Tunnels of Hydropower Stations Based on the Monitored Microseismicity and an Optimized Probabilistic Neural Network Model. *Sustainability* **2019**, *11*, 3212. [CrossRef]
23. Kordnaeij, A.; Kalantary, F.; Kordtabar, B.; Mola-Abasi, H. Prediction of recompression index using GMDH-type neural network based on geotechnical soil properties. *Soils Found.* **2015**, *55*, 1335–1345. [CrossRef]
24. Wang, W.; Tang, R.; Li, C.; Liu, P.; Luo, L. A BP neural network model optimized by Mind Evolutionary Algorithm for predicting the ocean wave heights. *Ocean. Eng.* **2018**, *162*, 98–107. [CrossRef]
25. Gobeyn, S.; Mouton, A.M.; Cord, A.F.; Kaim, A.; Volk, M.; Goethals, P.L.M. Evolutionary algorithms for species distribution modelling: A review in the context of machine learning. *Ecol. Model.* **2019**, *392*, 179–195. [CrossRef]
26. Yu, J.; Zhang, Q.; Xu, W.; Wang, R.; Zhang, H. Study on Unloading Relaxation Characteristics of Columnar Jointed Rock Masses Based on Displacement Back Analysis. *Front. Earth Sci.* **2021**, *9*, 665275. [CrossRef]
27. Ocak, I.; Seker, S.E. Calculation of surface settlements caused by EPBM tunneling using artificial neural network, SVM, and Gaussian processes. *Environ. Earth Sci.* **2013**, *70*, 1263–1276. [CrossRef]
28. Zhang, C.; Li, J.-z.; He, Y. Application of optimized grey discrete Verhulst—BP neural network model in settlement prediction of foundation pit. *Environ. Earth Sci.* **2019**, *78*, 441. [CrossRef]
29. Jie, J.; Zeng, J.; Han, C. An extended mind evolutionary computation model for optimizations. *Appl. Math. Comput.* **2007**, *185*, 1038–1049. [CrossRef]
30. Yin, X.; Gao, F.; Wu, J.; Huang, X.; Pan, Y.; Liu, Q. Compressive strength prediction of sprayed concrete lining in tunnel engineering using hybrid machine learning techniques. *Undergr. Space* **2021**, *6*, 1–16. [CrossRef]
31. Rohola, H.; Jamal, R.; Jürgen, S.; Yilmaz, O.; Babak, S. Prediction of TBM jamming risk in squeezing grounds using Bayesian and artificial neural networks. *J. Rock Mech. Geotech. Eng.* **2021**, *12*, 21–31.
32. Cui, Z.; Sheng, Q.; Zhang, G.-m.; Liu, H. A modified rock mass classification considering seismic effects in the basic quality (BQ) system. *Bull. Eng. Geol. Environ.* **2021**, *80*, 2249–2260. [CrossRef]
33. Yin, X.; Liu, Q.; Pan, Y.; Huang, X.; Wu, J.; Wang, X. Strength of Stacking Technique of Ensemble Learning in Rockburst Prediction with Imbalanced Data: Comparison of Eight Single and Ensemble Models. *Nat. Resour. Res.* **2021**, *30*, 1795–1815. [CrossRef]

mathematics

MDPI

Article

Numerical Modeling and Investigation of Fault-Induced Water Inrush Hazard under Different Mining Advancing Directions

Chong Li [1,2,*] and Zhijun Xu [1]

1 MOE Key Laboratory of Deep Coal Resource Mining, School of Mines, China University of Mining & Technology, Xuzhou 221116, China; tb20020034b0@cumt.edu.cn
2 Jiangsu Laboratory of Mine Earthquake Monitoring and Prevention, China University of Mining & Technology, Xuzhou 221116, China
* Correspondence: lichong1923@cumt.edu.cn

Abstract: Evaluations of the risk of fault-induced water inrush hazard is an important issue for mining engineering applications. According to the characteristics of the seam floor during mining advancing, a mechanical model of fault activation is built to obtain the equations of normal stress and shear stress on the surface of fault, as well as the mechanics criterion of fault activation. Furthermore, using FLAC^{3D} numerical software, the stress variation on the surface of fault under two different mining advancing directions are numerically simulated, and the distribution characteristics of the plastic failure zone of the roof and floor near the fault are obtained. The results show that: (1) When mining advances from the hanging wall, the normal stress increases more greatly than that from the foot wall, the shear stress distribution changes drastically with a large peak, and it is more likely to cause fault activation. (2) When mining advances from the hanging wall and approaches the fault, the normal stress and shear stress within the fault first increases, and then decreases suddenly. When mining advances from the foot wall, the normal stress and shear stress increases constantly, and the fault zone stays in the compaction state where the hanging wall and foot wall are squeezed together, which is unfavorable for water inrush hazard. (3) When mining advances from the hanging wall, the deep-seated fault under the floor is damaged first, and the plastic failure zone of the floor increases obviously. When mining advances from the foot wall, the shallow fault under the floor is damaged first, and the plastic failure zone of roof increases obviously. (4) For a water-conducting fault, the waterproof coal pillar size of the mining advancing from the hanging wall should be larger than that from the foot wall. (5) The in-situ monitoring results are in agreement with the simulation results, which proves the effectiveness of the simulation.

Keywords: fault; water inrush; mechanical behavior; mining advancing direction

MSC: 86A60

Citation: Li, C.; Xu, Z. Numerical Modeling and Investigation of Fault-Induced Water Inrush Hazard under Different Mining Advancing Directions. *Mathematics* **2022**, *10*, 1561. https://doi.org/10.3390/math10091561

Academic Editors: Elena Benvenuti, Shaofeng Wang, Jian Zhou, Xin Cai, Xiaofeng Li and Zhengyang Song

Received: 20 February 2022
Accepted: 14 April 2022
Published: 5 May 2022

1. Introduction

Mining-induced water inrush is one of the main kinds of mine water hazards in China, especially in northern China where water inrush hazards from the floor frequently occur [1,2]. According to the statistics, more than 55% of the mine water hazards are caused from the floor, of which about 80% of the water inrush is related with a fault [3–8].

It is of importance to know how water inrush could develop during mining advances [1,5,9–13]. As shown in Figure 1a, intact rock mass has good mechanical properties with smaller permeability, but small fractures in the floor are caused by mining [14–17]. However, fault not only changes the mechanical properties of rock mass [18–21], but also reduces the intensity and modulus of deformation and seriously affects the permeability properties of rock mass, as shown in Figure 1b. The water in the fault produces the physical, chemical reaction with the rock mass, and the change of hydraulic stress will cause the change of stress distribution in rock mass [22]; meanwhile, the stress change causes a change in the rock pores [23–27], which in turn affects the groundwater flow and water pressure. In this way, the water in the fault and rock mass are mutually influenced. If there is a fault in the practical project, we need to consider the influence that mining pressure brings to fault activation, especially the inrush problems of fault, or it will cause property loss and casualties [21,28,29]. Fault is an important water inrush channel due to the crushed rocks in it. At present, the study on water inrush mechanism is mainly concentrated on either damage of floor strata or reactivation of fault [30–36]. They did not consider the mining advance direction on failure analysis of the surrounding rocks near faults. In particular, the stress variation and failure characteristics with the formation of water inrush channel adjacent to faults in floor strata are not completely understood.

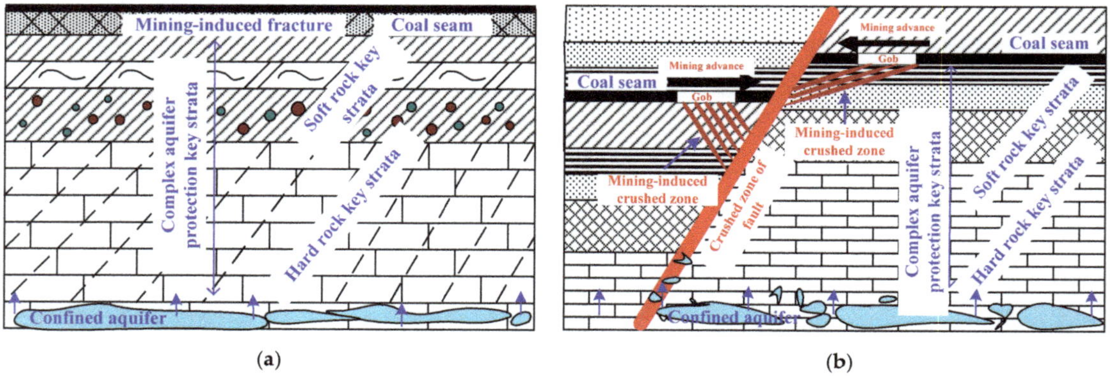

Figure 1. Sketch of mining-induced crushed zone development. (**a**) Intact rock strata. (**b**) Floor within fault.

This paper aims to analyze the stress variation and failure characteristics of the rock surrounding the fault by advancing in different directions. It focuses on the variation of normal stress and shear stress on the fault plane, as well as failure characteristics of the roof and floor by mechanical model and numerical simulation in the process of mining advancing. The influence that mining advance direction brings to fault activation is obtained, and a certain theoretical basis is provided to the design of a waterproof coal pillar. Finally, in-situ monitoring is conducted to verify the effectiveness of simulation results.

2. Mechanical Behavior of Mining Advancing Direction on Fault Activation

2.1. Mechanical Analysis of Fault Activation

As shown in Figure 2, before coal mining, the stress of the rock mass stays in the original equilibrium, while the rock stress will redistribute after mining. According to the theory of mining pressure control [37,38], in the mining advancing direction, the peak abutment pressure of coal floor $n\gamma H$ appeared in the working face within a certain distance to the coal wall. Because the floor of the gob is compacted by roof caving rocks, the abutment pressure gradually recovers to the original stress γH, where γ is the bulk density of the rock, H is the depth of the buried coal seam, and n is the stress concentration factor.

In order to study the influence that the mining advancing direction brings to fault activation, according to the above-mentioned abutment pressure distribution law of mining advance direction, we take the surrounding rock mass along the mining advancing direction of the working face in the central field as the research object in longwall mining. Here it can be treated as a plane strain problem. Assume that the floor rock mass is elastic; abutment pressure applied on the floor is simplified as a linear distribution load; the original rock stress as a uniform distribution load; the stress in stress-concentrated area as linear increase; the stress in stress-relaxed area as linear decrease; and the stress concentration factor ahead of the working face is n. Therefore, the mechanical model is established as Figure 3.

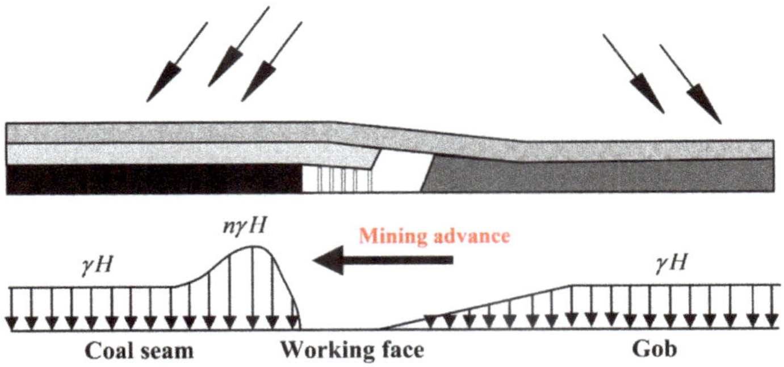

Figure 2. Distribution of abutment pressure in floor.

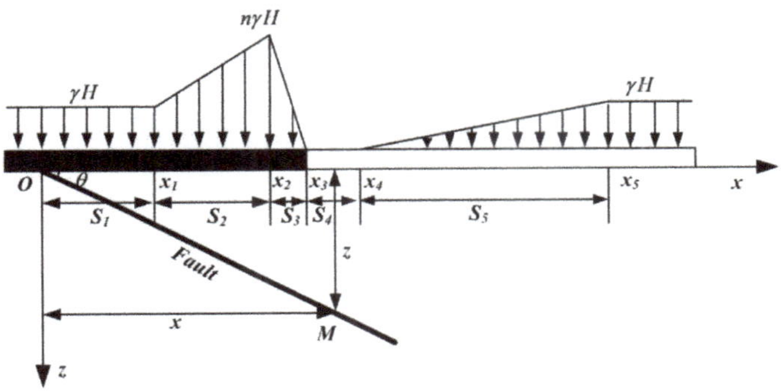

Figure 3. Mechanical model of fault activation.

According to stress analysis of the boundary half plane applied by normal stress in elastic mechanics theory [33], the stress equations of σ_x, σ_z and τ_{xz} of any point on the fault are obtained as Equation (1).

$$\left\{\begin{array}{l}\sigma_x=\frac{2\gamma H}{\pi}\left\{\int_{-\infty}^{x_1}\frac{z(x-\xi)^2d\xi}{\left[z^2+(x-\xi)^2\right]^2}+\int_{x_1}^{x_2}\frac{[(n-1)(\xi-x_1)+S_2]z(x-\xi)^2d\xi}{S_2\left[z^2+(x-\xi)^2\right]^2}+\right.\\ \left.\int_{x_2}^{x_3}\frac{n(x_3-\xi)z(x-\xi)^2d\xi}{S_3\left[z^2+(x-\xi)^2\right]^2}+\int_{x_4}^{x_5}\frac{(\xi-x_4)z(x-\xi)^2d\xi}{S_5\left[z^2+(x-\xi)^2\right]^2}+\int_{x_5}^{x+\infty}\frac{z(x-\xi)^2d\xi}{\left[z^2+(x-\xi)^2\right]^2}\right\}\\ \sigma_z=\frac{2\gamma H}{\pi}\left\{\int_{-\infty}^{x_1}\frac{z^3d\xi}{\left[z^2+(x-\xi)^2\right]^2}+\int_{x_1}^{x_2}\frac{[(n-1)(\xi-x_1)+S_2]z^3d\xi}{S_2\left[z^2+(x-\xi)^2\right]^2}+\right.\\ \left.\int_{x_2}^{x_3}\frac{n(x_3-\xi)z^3d\xi}{S_3\left[z^2+(x-\xi)^2\right]^2}+\int_{x_4}^{x_5}\frac{(\xi-x_4)z^3d\xi}{S_5\left[z^2+(x-\xi)^2\right]^2}+\int_{x_5}^{x+\infty}\frac{z^3d\xi}{\left[z^2+(x-\xi)^2\right]^2}\right\}\\ \tau_{xz}=\frac{2\gamma H}{\pi}\left\{\int_{-\infty}^{x_1}\frac{z^2(x-\xi)d\xi}{\left[z^2+(x-\xi)^2\right]^2}+\int_{x_1}^{x_2}\frac{[(n-1)(\xi-x_1)+S_2]z^2(x-\xi)d\xi}{S_2\left[z^2+(x-\xi)^2\right]^2}+\right.\\ \left.\int_{x_2}^{x_3}\frac{n(x_3-\xi)z^2(x-\xi)d\xi}{S_3\left[z^2+(x-\xi)^2\right]^2}+\int_{x_4}^{x_5}\frac{(\xi-x_4)z^2(x-\xi)d\xi}{S_5\left[z^2+(x-\xi)^2\right]^2}+\int_{x_5}^{x+\infty}\frac{z^2(x-\xi)d\xi}{\left[z^2+(x-\xi)^2\right]^2}\right\}\end{array}\right. \quad (1)$$

According to the stress equation on the oblique section in elastic mechanics, the normal stress and shear stress equations on the fault planes under the abutment pressure are obtained as Equation (2).

$$\left\{\begin{array}{l}\sigma_N=\sigma_x\sin^2\theta+\sigma_z\cos^2\theta+2\tau_{xz}\sin\theta\cos\theta\\ \tau_N=\sin\theta\cos\theta(\sigma_z-\sigma_x)+\left(\sin^2\theta-\cos^2\theta\right)\tau_{xz}\end{array}\right. \quad (2)$$

Considering the Mohr–Coulomb criterion, the shear strength of fault plane is available:

$$\tau_f = c + \sigma_N \tan\varphi \quad (3)$$

and the condition of fault activation is:

$$\tau_N \geq \tau_f \quad (4)$$

Combined with Equation (3), the mechanics criterion of fault activation after mining is obtained as:

$$\tau_N \geq c + \sigma_N \tan\varphi \quad (5)$$

where c is the cohesive force of the rock, and φ is the internal friction angle of the rock.

2.2. *Effect of Mining Advancing Direction to Fault Activation*

Due to the difficult simplification of Equation (1) substituted into Equation (2), numerical analysis therefore is adopted to these equations. Considering the actual situation, S_1 = 20 m, S_2 = 25 m, S_3 = 5 m, S_4 = 10 m, S_5 = 100 m, n = 2.5, H = 400 m, $\gamma = 2.5 \times 10^4$ N/m^3, the distance between the working face to the fault is 50 m. When the mining advances from the hanging wall, then θ = 60°; when the working face advances from the foot wall, then θ = 120°. The normal stress and shear stress distribution can be obtained by Equation (2) when the distance between the working face to the fault is 50 m, as seen in Figures 4 and 5.

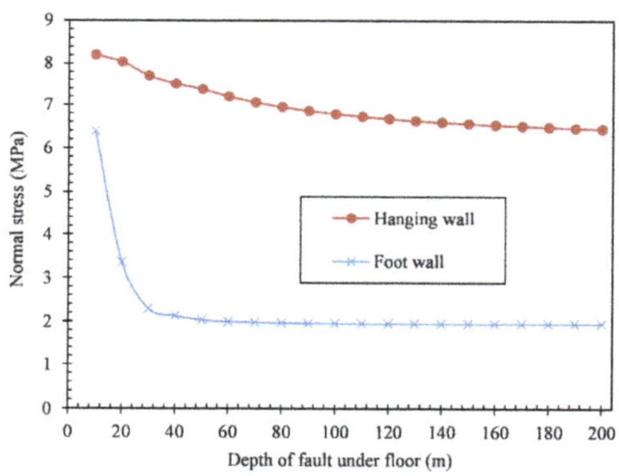

Figure 4. Distribution of normal stress on fault.

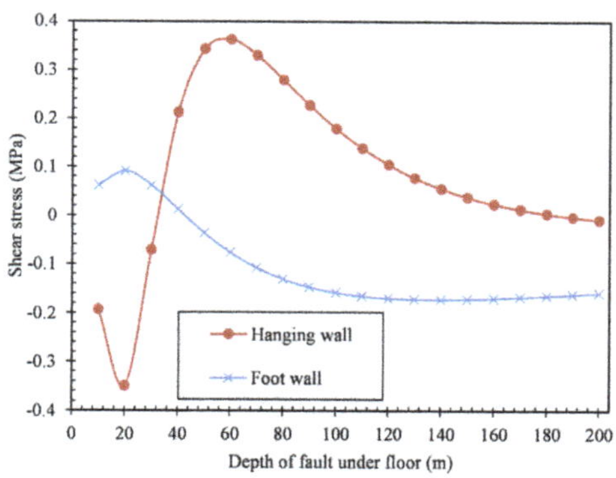

Figure 5. Distribution of shear stress on fault.

It can be seen from Figures 4 and 5 that when the working face is 50 m apart from the fault, the normal stress on a fault below the floor range of 0~40 m has the largest change, which brings the largest disturbance to fault; when the depth is more than 40 m, the normal stress on a fault keeps a certain value, the hanging wall advance stays around 6.5 MPa, and the foot wall advance stays around 2.0 MPa. The normal stress on the fault advancing from the hanging wall, which causes larger disturbance, is always greater than that from the foot wall. The variation range of shear stress distribution on a fault advancing from the hanging wall is larger than that from the foot wall, i.e., −0.35~0.36 MPa from the hanging wall, and −0.17~0.09 MPa from the foot wall, respectively. With the increase of depth, the shear stress appears to be stable from foot wall advance. Before the mining advances to a fault, hanging wall advance is easier to cause the fault activation, because there are dramatic changes of the shear stress distribution, large peak values, and wide ranges of influence on the fault.

3. Numerical Simulations

To understand the stress variation on faults and the distribution characteristics of floor and roof plastic failure near the fault, the simulation software of FLAC^{3D} is applied in this study for further numerical work. FLAC^{3D} is a finite difference numerical simulation software. There are twelve elastic and plastic constitutive models and five calculation modes built in the software, which can realize the coupling between different modes. In addition, the built-in FISH program language can obtain the coordinates, displacement, stress, strain and other parameters of nodes and units in the calculation process. It can well simulate geological materials' mechanical behaviors, such as plastic flow or damage, when they reach the yield limit or strength limit. Besides, it can also analyze the gradual damage and instability, and track the gradual failure of materials. Therefore, this software has been widely used in the field of geotechnical engineering.

3.1. Engineering Background

Buliangou coal mine is located in the northeast of Zhungeer coalfield in Inner Mongolia, China. The development of a fault in the minefield brings a great challenge to mining safety. According to the preliminary exploration, a normal fault goes through the middle of the F6210 working face. The F6210 is buried at a depth of 398–405 m, and the thickness of the coal seam is about 4 m. The fault has a strike of N46°E, a dip of N51°W, a dip angle of 55–65°, a zone width of 2.2–3.5 m, and an average drop of 4.0 m. To reduce the risk of water inrush hazard, as well as improve the mining efficiency, the mining advanced from the hanging wall and foot wall of the fault, respectively, and advanced from far and near to the fault fracture zone.

3.2. Numerical Calculation Model

Through the proper and effective simplification, the geometric model is built with the size of 300 m × 200 m × 200 m (length × width × height), with 98,400 divided units and 105,493 nodes. The coal seam is buried at a depth of 400 m and with a thickness of 4 m. A fault with a drop of 4 m, an inclination angle of 60°, and a fault zone width of 3 m is selected in this study. There is horizontal displacement constraint on the sides of the model, vertical displacement constraints on the bottom, and the upper surface is treated as free surface. Overlying rock applies about 10 MPa of uniform loading on the upper surface. Mining advances from far and near, and gradually advances to the fault with an advancing step of 10 m. At the advancing interval, artificial filling is made to simulate the mining caving for the last advancing step. A numerical calculation model is established as Figure 6, the meshing model as Figure 7, and the rock physical and mechanical parameters are shown in Table 1.

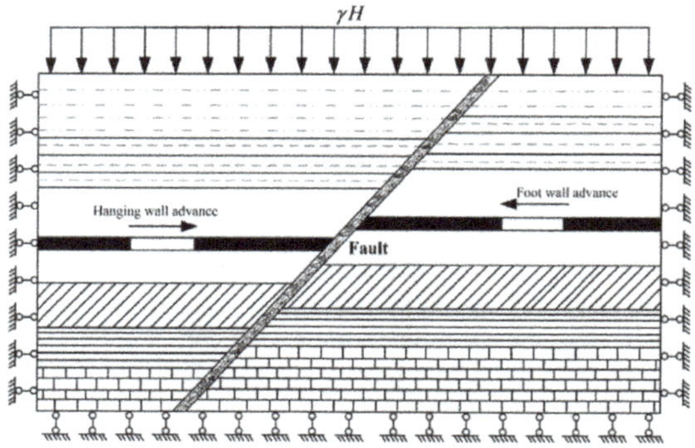

Figure 6. Mechanical model of the numerical simulation.

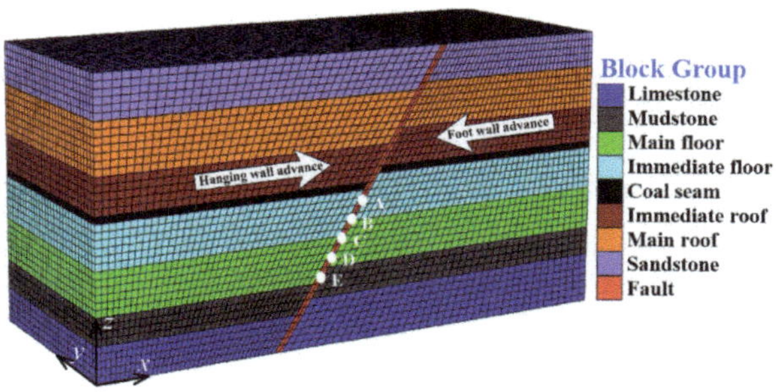

Figure 7. Sketch of the FLAC^{3D} mesh of the fault.

Table 1. Physical and mechanical parameters of rock strata.

Rock Property	Thickness h (m)	Bulk Modulus K (GPa)	Shear Modulus G (GPa)	Internal Friction Angle φ (°)	Cohesion c (MPa)	Tensile Strength σ_c (MPa)	Density ρ (kg·m^{-3})
Sandstone	20	15.3	9.2	36	3.5	3.5	2550
Main roof	22	16.7	10.0	35	3.6	2.0	2500
Immediate roof	15	15.9	8.2	29	2.5	2.5	2450
Coal seam	3	9.2	3.7	25	1.5	1.1	1500
Immediate floor	14	17.4	9.0	32	2.9	2.4	2500
Main floor	16	12.3	8.1	35	3.3	2.3	2580
Mudstone	12	12.6	9.1	33	2.6	1.8	2500
Limestone	18	19.0	9.8	38	3.9	3.5	2550
Fault	~	8.3	3.0	18	0.5	0.5	1900

3.3. Results and Analysis of Numerical Simulation

3.3.1. Stress Distribution Characteristics on the Fault

The monitoring points are put in the fault zone that lies about 5 m below the coal seam, 10 m, 15 m, 20 m and 25 m, respectively, in this simulation displayed as the white dots in Figure 7, which are listed as A, B, C, D and E, respectively. As the mining working face approaches to the fault, the normal stress and shear stress change curves of monitoring points are shown as Figures 8 and 9, respectively.

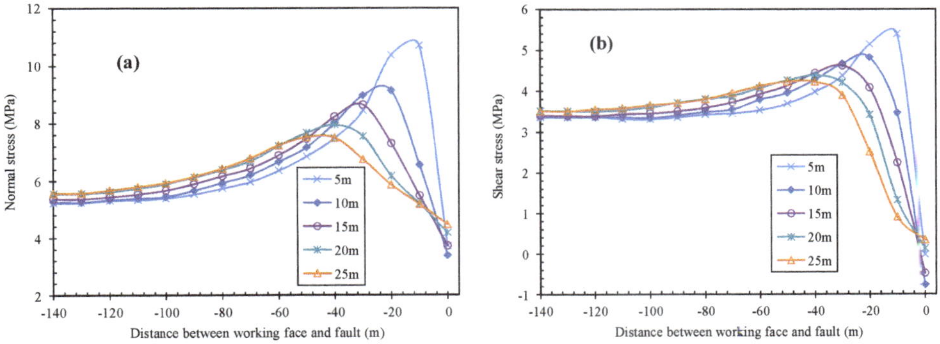

Figure 8. Stress change curves of the fault when mining advances along the hanging wall: (**a**) normal stress; (**b**) shear stress.

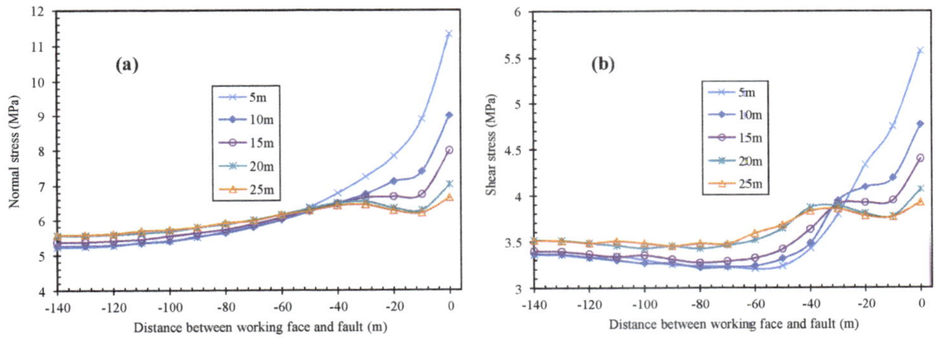

Figure 9. Stress change curves of the fault when mining advances along the foot wall: (**a**) normal stress; (**b**) shear stress.

As is shown in Figures 8 and 9:

When mining advances from the hanging wall, the normal stress and shear stress within the fault zone under the floor first increases and then decreases. When the working face is more than 10 m apart to fault, the normal stress within the fault zone that is 5 m under the floor always increases, from 5.2 MPa to 10.7 MPa, an increase of 5.5 MPa; as the working face advances to the fault continuously, the normal stress starts to decrease, and eventually reduces to 3.7 MPa. The change law of shear stress in the fault zone is similar to that of normal stress. The shear stress in the fault zone that is 5 m below the floor increases from 3.4 MPa to 5.4 MPa, an increase of 2.0 MPa, after which the shear stress decreases. The increased range of the shear stress is less than that of normal stress, so the working face advance brings a larger disturbance to normal stress in the fault zone. The normal stress and shear stress on the fault that is 25 m below the floor starts to decrease when it is 50 m away from the working face, which indicates that the fault under the floor and far away from the mining layer is disturbed first.

When mining advances from the foot wall, the normal stress and shear stress within the fault zone under the floor increases gradually. The normal stress within the fault zone that is 5 m under the floor increases from 5.2 MPa to 11.3 MPa, while the shear stress increases from 3.4 MPa to 5.6 MPa. When the mining working face is 0~20 m apart from the fault, there is the largest increase and the largest disturbance to the fault. Comparing the stress change of 5 m and 25 m below the floor within the fault zone, it can be seen that the fault under the floor and closest to the mining layer is disturbed first when the working face advances. Similarly, with the mining advance from the hanging wall, the change range of shear stress is less than that of normal stress within the fault zone.

Comparing the working face advance from the hanging wall and the foot wall, we can find that there is an obvious difference in stress change within the fault zone. When mining advances from the hanging wall, the stress first increases, after which it decreases; the instantaneous release of stress tends to cause the fault activation. When the working face advances from the foot wall, the fault zone stays in the compaction state where the hanging wall and foot wall are squeezed together, which is unfavorable for water inrush.

3.3.2. Plastic Failure Characteristics of Roof and Floor

Affected by mining, plastic failure can occur in the surrounding rock of the working face. When the plastic failure zone is linked with the plastic zone near the fault, there is a hidden water inrush in the water-conductive fault. Figures 10 and 11, respectively, show the plastic failure distribution zone that occurs when the working face advances from the hanging wall and the foot wall.

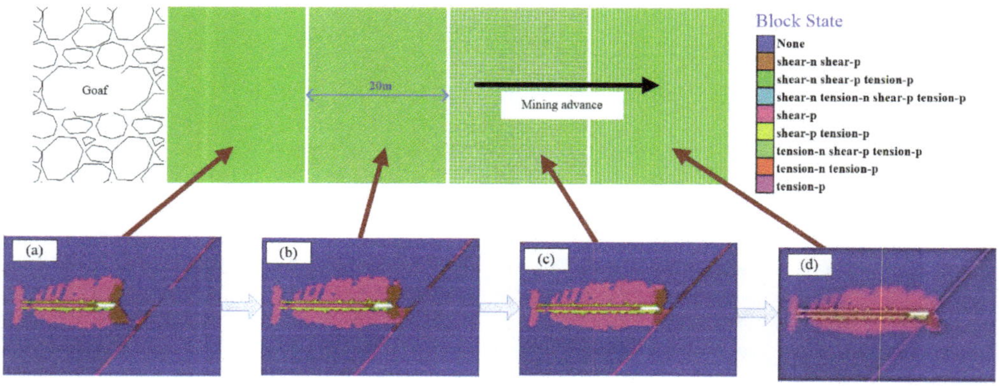

Figure 10. Distribution of plastic zone with mining advance along the hanging wall. Distance between working face and the fault: (**a**) 60 m; (**b**) 40 m; (**c**) 20 m; (**d**) 0 m.

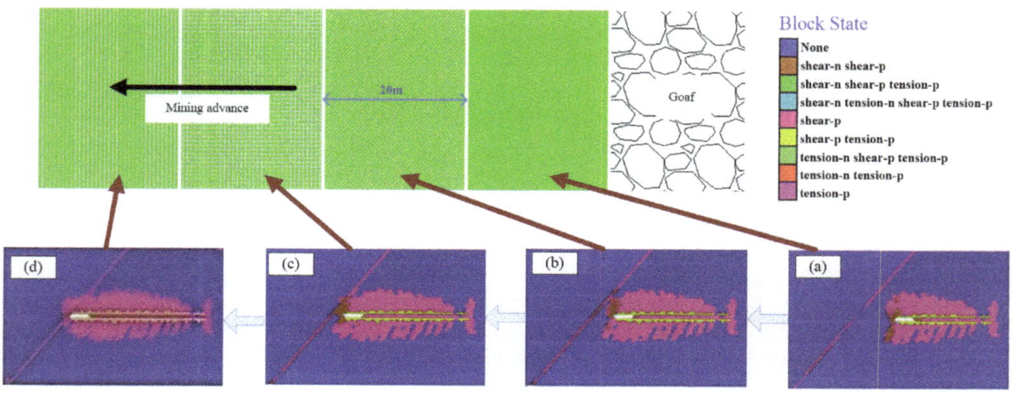

Figure 11. Distribution of plastic zone with mining advance along the foot wall. Distance between working face and the fault: (**a**) 60 m; (**b**) 40 m; (**c**) 20 m; (**d**) 0 m.

As is shown in Figures 10 and 11:

When mining advances in the initial stage, the plastic zone that is caused by mining is relatively small and far from the fault; with the advance of the working face, the distance between the plastic failure zone and the fault becomes small. As for the hanging wall, when the working face is 60 m apart from the fault, the plastic failure zone caused by mining will be linked with the plastic zone near the fault; as for the foot wall, when the working face is 40 m away from the fault, the plastic failure zone caused by mining will be linked to the plastic zone near the fault. If the fault is conducted with an aquifer, a water-conductive canal is easily formed, which will lead to water inrush. Therefore, a waterproof coal pillar should be reserved. In this simulation study, the size of the waterproof coal pillar for the hanging wall is no less than 40 m, and for the foot wall it is no less than 20 m.

When mining advances from the hanging wall, the plastic failure zone caused by the floor is linked with the plastic zone near the fault first; on the other hand, when mining advances from the foot wall, the plastic failure zone caused by the roof is linked first. Therefore, when the working face approaches to the fault, plastic failure in the floor is largely affected by the hanging wall advance, and plastic failure in the roof is largely affected by the foot wall advance.

3.4. Verification by In-Situ Monitoring

To verify the validity of simulation results, the water injection leak detection method [39] is designed to observe the development of the plastic zone during the mining advancement. In this method, a higher flow rate of leakage means more severe plastic damage to the borehole, and the rapid increase in flow rate represents the connection of fractures near the borehole. As shown in Figure 12, Points A, B, C and D are set to monitor the plastic zone development in both roof and floor of the hanging wall and foot wall 60 m apart from the fault. In each monitoring point, four boreholes for different minoring depths were arranged, numbered 1#, 2#, 3# and 4#, respectively. The horizontal distance between the boreholes was 2 m, and the vertical distance was 5 m.

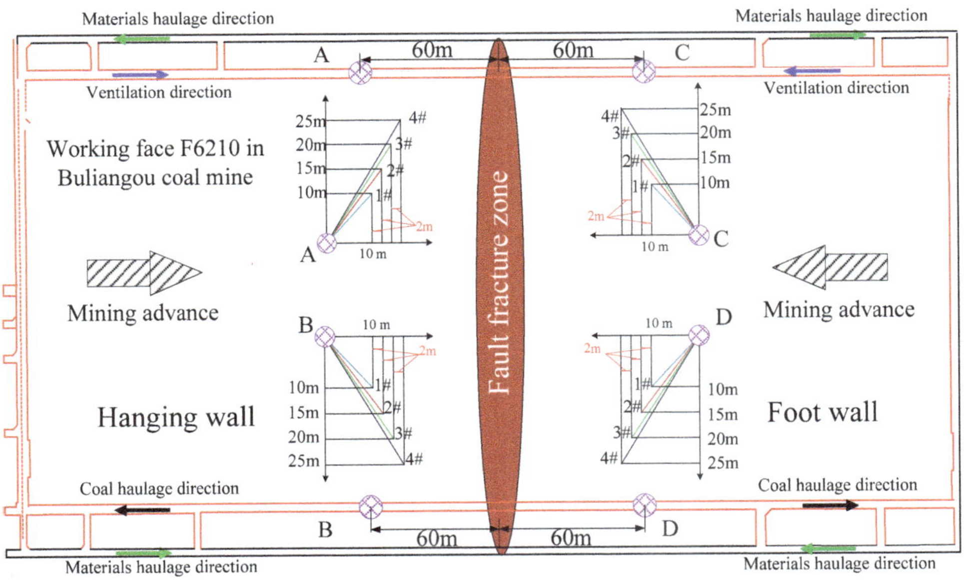

Figure 12. Layout of plastic zone monitoring of working face F6210 in Buliangou coal mine.

As is shown in Figure 13, with the advancement of the working face in the hanging wall, the size of the plastic zone increases gradually. As shown in Figure 13a, when the working face is 120 m apart from the fault, the flow rate of the A-1# borehole increases to 2.38 m^3/h, while the data in A-2# stays at a low value, which means the height of the plastic zone is 10–15 m. When the working face is 70 m apart from the fault, the height of the plastic zone is more than 25 m. Similarly, as shown in Figure 13b, the flow rate of the B-1# borehole increases to 2.49 m^3/h when the working face is 130 m apart from the fault, which proves that the plastic zone depth is 10–15 m. When the working face is 100 m apart from the fault, the plastic zone depth is more than 25 m. By comparing Figure 13a,b, the plasticas zone height in the floor is larger than the plastic zone depth in the roof at the same advancing distance; that is, the advance in the hanging wall has a more significant influence on the floor. This monitoring result is consistent with that of simulation.

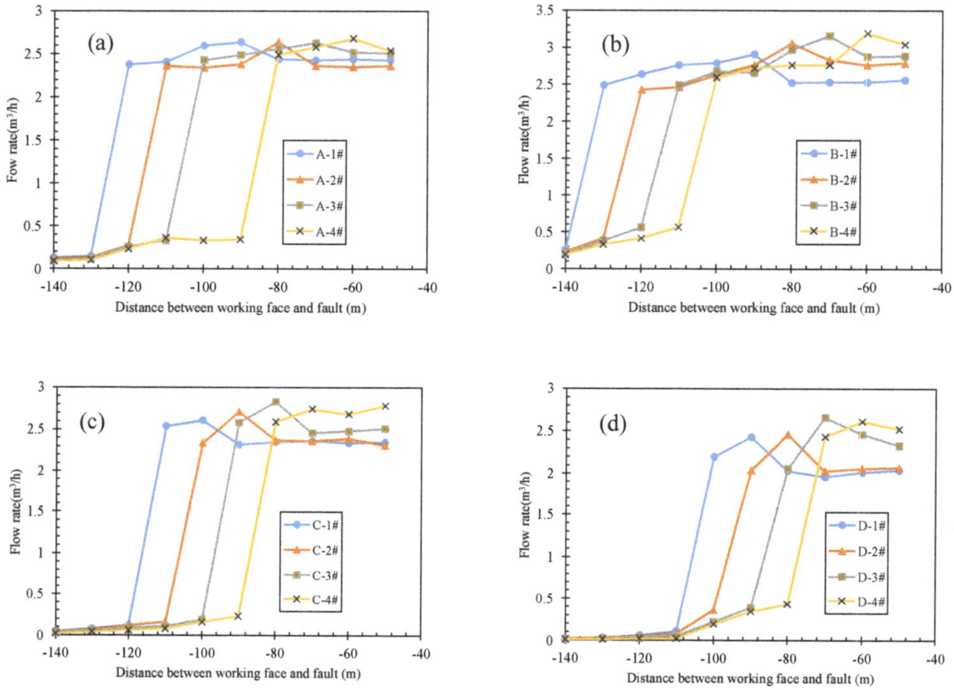

Figure 13. In-situ monitoring results of mining-induced plastic zone in different monitoring points: (**a**) Point A; (**b**) Point B; (**c**) Point C; (**d**) Point D.

As the working face advances in the foot wall, the size of the plastic zone in the roof and floor also increases. As shown in Figure 13c, when the working face is 110 m apart from the fault, the flow rate of the C-1# borehole increases to 1.18 m^3/h, which indicates that the height of the plastic zone in the roof is 10–15 m. When the working face is 80 m apart from the fault, the height of the plastic zone is more than 25 m. Similarly, as shown in Figure 13d, the plastic zone depth is 10–15 m when the working face advances 100 m apart from the fault. When the working face is 70 m apart from the fault, the depth of the plastic zone in the roof reaches 25 m. In the comparison of Figure 13c,d, the advance from the foot wall had a greater impact on the roof, which corresponds with the simulation results.

Comparing Figure 13a,c, when the distance between the working face and fault is −100 m, the height of the plastic zone in the hanging wall is 20–25 m, while that in the foot wall is 15–20 m. In the comparison of Figure 13b,d, when the working face advances to 100 m away from the fault, the depth of the plastic zone in the hanging wall (more than 25 m) is much larger than that in the foot wall (5–10 m). It is concluded that, at the same advancing distance, the height and depth of the plastic zone in the hanging wall is larger than that in the foot wall, which is also in line with the simulation results.

To sum up, the in-situ monitoring results are in good agreement with the simulation results, which verifies the validity of the numerical simulation.

4. Conclusions

To analyze the effect of mining advancing direction of the working face on fault activation, a mechanical model of fault activation is built, and FLAC^{3D} numerical software is used to simulate the stress variation on the surface of fault under two different mining advancing directions, and the distribution characteristics of a plastic failure zone of the roof and floor near the fault are obtained. According to the mechanical behavior analysis, when the working face advances from the hanging wall, the normal stress has a large increase in the fault zone, and the shear stress distribution changes drastically with a large peak. Advancing from the foot wall more easily causes fault activation. Based on the numerical results, as the working face gets close to the fault from the hanging wall, the normal stress and shear stress increase first and then decrease, the instantaneous release of stress easily increases the risk of fault activation; from the foot wall, the normal stress and shear stress always increase, then the hanging wall and foot wall are squeezed together, which is unfavorable for water inrush in the fault zone. When the working face advances from the hanging wall, the floor is largely affected, while the roof is largely affected by the foot wall advance. In this simulation study, the size of the waterproof coal pillar for the hanging wall is no less than 40 m, and for the foot wall no less than 20 m. The size of the waterproof coal pillar for the hanging wall should be larger than that of the foot wall. The in-situ monitoring results in Buliangou coal mine show good consistency with the simulation results.

Author Contributions: C.L. conceived and designed the numerical model; Z.X. performed the numerical simulation; C.L. and Z.X. analyzed the data and wrote the paper. All authors have read and agreed to the published version of the manuscript.

Funding: This work was supported by the National Natural Science Foundation of China (51874277).

Institutional Review Board Statement: Not applicable.

Informed Consent Statement: Not applicable.

Data Availability Statement: Not applicable.

Acknowledgments: The authors would like to thank the numerical assistance from China University of Mining and Technology.

Conflicts of Interest: The authors declare no conflict of interest.

References

1. Wang, J.A.; Park, H.D. Coal mining above a confined aquifer. *Int. J. Rock Mech. Min. Sci.* **2003**, *40*, 537–551. [CrossRef]
2. Ma, D.; Duan, H.; Zhang, J. Solid grain migration on hydraulic properties of fault rocks in underground mining tunnel: Radial seepage experiments and verification of permeability prediction. *Tunn. Undergr. Sp. Tech.* **2022**, *125*, 104525. [CrossRef]
3. Zhou, Q.; Herrera-Herbert, J.; Hidalgo, A. Predicting the risk of fault-induced water inrush using the adaptive neuro-fuzzy inference system. *Minerals* **2017**, *7*, 55. [CrossRef]
4. Zhang, R.; Jiang, Z.; Zhou, H.; Yang, C.; Xiao, S. Groundwater outbursts from faults above a confined aquifer in the coal mining. *Nat. Hazards* **2013**, *71*, 1861–1872. [CrossRef]
5. Wu, Q.; Wang, M.; Wu, X. Investigations of groundwater bursting into coal mine seam floors from fault zones. *Int. J. Rock Mech. Min. Sci.* **2004**, *41*, 557–571. [CrossRef]
6. Li, L.; Yang, T.; Liang, Z.; Zhu, W.; Tang, C. Numerical investigation of groundwater outbursts near faults in underground coal mines. *Int. J. Coal Geol.* **2011**, *85*, 276–288. [CrossRef]
7. Huang, Z. Analytical and experimental study of water seepage propagation behavior in the fault. *Acta Geodyn. Geomater.* **2014**, *11*, 361–370. [CrossRef]
8. Ma, D.; Duan, H.; Zhang, J.; Feng, X.; Huang, Y. Experimental investigation of creep-erosion coupling mechanical properties of water inrush hazards in fault fracture rock masses. *Chin. J. Rock Mech. Eng.* **2021**, *40*, 1751–1763.
9. Zhang, J.C.; Shen, B.H. Coal mining under aquifers in China: A case study. *Int. J. Rock Mech. Min. Sci.* **2004**, *41*, 629–639. [CrossRef]
10. Yang, T.H.; Tham, L.G.; Tang, C.A.; Liang, Z.Z.; Tsui, Y. Influence of heterogeneity of mechanical properties on hydraulic fracturing in permeable rocks. *Rock Mech. Rock Eng.* **2004**, *37*, 251–275. [CrossRef]
11. Yang, T.H.; Liu, J.; Zhu, W.C.; Elsworth, D.; Tham, L.G.; Tang, C.A. A coupled flow-stress-damage model for groundwater outbursts from an underlying aquifer into mining excavations. *Int. J. Rock Mech. Min. Sci.* **2007**, *44*, 87–97. [CrossRef]

12. Donnelly, L.J. A review of coal mining induced fault reactivation in Great Britain. *Q. J. Eng. Geol. Hydrogeol.* **2006**, *39*, 5–50. [CrossRef]
13. Ma, D.; Duan, H.; Liu, W.; Ma, X.; Tao, M. Water–Sediment Two-Phase Flow Inrush Hazard in Rock Fractures of Overburden Strata During Coal Mining. *Mine Water Environ.* **2020**, *39*, 308–319. [CrossRef]
14. Liang, D.-X.; Jiang, Z.-Q.; Zhu, S.-Y.; Sun, Q.; Qian, Z.-W. Experimental research on water inrush in tunnel construction. *Nat. Hazards* **2015**, *81*, 467–480. [CrossRef]
15. Dou, L.; Yang, K.; Chi, X. Fracture behavior and acoustic emission characteristics of sandstone samples with inclined precracks. *Int. J. Coal Sci. Technol.* **2021**, *8*, 77–87. [CrossRef]
16. Huang, Z.; Jiang, Z.; Tang, X.; Wu, X.; Guo, D.; Yue, Z. In situ Measurement of Hydraulic Properties of the Fractured Zone of Coal Mines. *Rock Mech. Rock Eng.* **2015**, *49*, 603–609. [CrossRef]
17. Huang, Z.; Jiang, Z.; Fu, J.; Cao, D. Experimental measurement on the hydraulic conductivity of deep low-permeability rock. *Arab. J. Geosci.* **2014**, *8*, 5389–5396. [CrossRef]
18. Ma, D.; Zhang, J.; Duan, H.; Huang, Y.; Li, M.; Sun, Q.; Zhou, N. Reutilization of gangue wastes in underground backfilling mining: Overburden aquifer protection. *Chemosphere* **2021**, *264*, 128400. [CrossRef]
19. Liang, D.-X.; Jiang, Z.-Q.; Guan, Y.-Z. Field Research: Measuring Water Pressure Resistance in a Fault-Induced Fracture Zone. *Mine Water Environ.* **2014**, *34*, 320–328. [CrossRef]
20. Zhang, S.; Lu, L.; Wang, Z.; Wang, S. A physical model study of surrounding rock failure near a fault under the influence of footwall coal mining. *Int. J. Coal Sci. Technol.* **2021**, *8*, 626–640. [CrossRef]
21. Islam, M.R.; Shinjo, R. Mining-induced fault reactivation associated with the main conveyor belt roadway and safety of the Barapukuria Coal Mine in Bangladesh: Constraints from BEM simulations. *Int. J. Coal Geol.* **2009**, *79*, 115–130. [CrossRef]
22. Ma, D.; Kong, S.; Li, Z.; Zhang, Q.; Wang, Z.; Zhou, Z. Effect of wetting-drying cycle on hydraulic and mechanical properties of cemented paste backfill of the recycled solid wastes. *Chemosphere* **2021**, *282*, 131163. [CrossRef] [PubMed]
23. Si, G.; Cai, W.; Wang, S.; Li, X. Prediction of Relatively High-Energy Seismic Events Using Spatial–Temporal Parametrisation of Mining-Induced Seismicity. *Rock Mech. Rock Eng.* **2020**, *53*, 5111–5132. [CrossRef]
24. Ma, D.; Wang, J.; Cai, X.; Ma, X.; Zhang, J.; Zhou, Z.; Tao, M. Effects of height/diameter ratio on failure and damage properties of granite under coupled bending and splitting deformation. *Eng. Fract. Mech.* **2019**, *220*, 106640. [CrossRef]
25. Xue, D.; Zhou, J.; Liu, Y.; Gao, L. On the excavation-induced stress drop in damaged coal considering a coupled yield and failure criterion. *Int. J. Coal Sci. Technol.* **2020**, *7*, 58–67. [CrossRef]
26. Wang, S.; Li, X.; Yao, J.; Gong, F.; Li, X.; Du, K.; Tao, M.; Huang, L.; Du, S. Experimental investigation of rock breakage by a conical pick and its application to non-explosive mechanized mining in deep hard rock. *Int. J. Rock Mech. Min. Sci.* **2019**, *122*, 104063. [CrossRef]
27. Wang, J.; Ma, D.; Li, Z.; Huang, Y.; Du, F. Experimental investigation of damage evolution and failure criterion on hollow cylindrical rock samples with different bore diameters. *Eng. Fract. Mech.* **2022**, *260*, 108182. [CrossRef]
28. Cao, Y.X.; He, D.D.; Glick, D.C. Coal and gas outbursts in footwalls of reverse faults. *Int. J. Coal Geol.* **2001**, *48*, 47–63. [CrossRef]
29. Ma, D.; Duan, H.; Zhang, J.; Liu, X.; Li, Z. Numerical simulation of water-silt inrush hazard of fault rock: A three-phase flow model. *Rock Mech. Rock Eng.* **2022**, *in press*.
30. McLellan, J.G.; Oliver, N.H.S.; Schaubs, P.M. Fluid flow in extensional environments; numerical modelling with an application to Hamersley iron ores. *J. Struct. Geol.* **2004**, *26*, 1157–1171. [CrossRef]
31. Gudmundsson, A.; Berg, S.S.; Lyslo, K.B.; Skurtveit, E. Fracture networks and fluid transport in active fault zones. *J. Struct. Geol.* **2001**, *23*, 343–353. [CrossRef]
32. Evans, J.P.; Forster, C.B.; Goddard, J.V. Permeability of fault-related rocks, and implications for hydraulic structure of fault zones. *J. Struct. Geol.* **1997**, *19*, 1393–1404. [CrossRef]
33. Barton, C.A.; Zoback, M.D.; Moos, D. Fluid-flow along potentially active faults in crystalline rock. *Geology* **1995**, *23*, 683–686. [CrossRef]
34. Babiker, M.; Gudmundsson, A. The effects of dykes and faults on groundwater flow in an arid land: The Red Sea Hills, Sudan. *J. Hydrol.* **2004**, *297*, 256–273. [CrossRef]
35. Wong, T.F.; David, C.; Zhu, W.L. The transition from brittle faulting to cataclastic flow in porous sandstones: Mechanical deformation. *J. Geophys. Res.-Solid Earth* **1997**, *102*, 3009–3025. [CrossRef]
36. Wang, S.; Sun, L.; Li, X.; Wang, S.; Du, K.; Li, X.; Feng, F. Experimental Investigation of Cuttability Improvement for Hard Rock Fragmentation Using Conical Cutter. *Int. J. Geomech.* **2021**, *21*, 06020039. [CrossRef]
37. Ma, D.; Duan, H.; Zhang, Q.; Zhang, J.; Li, W.; Zhou, Z.; Liu, W. A Numerical Gas Fracturing Model of Coupled Thermal, Flowing and Mechanical Effects. *Comput. Mater. Contin.* **2020**, *65*, 2123–2141. [CrossRef]
38. Si, G.; Belle, B. Performance analysis of vertical goaf gas drainage holes using gas indicators in Australian coal mines. *Int. J. Coal Geol.* **2019**, *216*, 103301. [CrossRef]
39. Yin, S.; Zhang, J.; Liu, D. A study of mine water inrushes by measurements of in situ stress and rock failures. *Nat. Hazards* **2015**, *79*, 1961–1979. [CrossRef]

mathematics

MDPI

Article

Effects of Strain Rate and Temperature on Physical Mechanical Properties and Energy Dissipation Features of Granite

Yangchun Wu, Linqi Huang *, Xibing Li, Yide Guo, Huilin Liu and Jiajun Wang

School of Resources and Safety Engineering, Central South University, Changsha 410083, China; wuyangchun1995@csu.edu.cn (Y.W.); xbli@csu.edu.cn (X.L.); guoyide@csu.edu.cn (Y.G.); lhlblack@csu.edu.cn (H.L.); jjwang@csu.edu.cn (J.W.)

* Correspondence: huanglinqi@csu.edu.cn

Abstract: Dynamic compression tests of granite after thermal shock were performed using the split Hopkinson pressure bar system, to determine the effects of strain rate and temperature on the dynamic mechanical parameters, energy dissipation features and failure modes of granite. The results indicate that the dynamic compressive strength increased exponentially with strain rate and decreased with increasing temperature. Temperature and incident energy can equivalently transform for the same dynamic compressive strength. Dynamic elastic modulus of granite decreased obviously with increasing temperature but did not have a clear correlation with strain rate. As the impact gas pressure increased, the stress-strain curves changed from Class II to Class I behavior, and the failure modes of specimens transformed from slightly split to completely pulverized. The critical temperature at which the stress-strain curves changed from Class II to Class I was determined to be 300 °C, when the impact gas pressure is 0.6 MPa. As the applied temperature increased, density, wave velocity and wave impedance all decreased, meanwhile, the degree of granite specimen crushing was aggravated. Under the same incident energy, as the temperature increased, the reflected energy increased notably and the absorbed energy increased slightly, but the transmitted energy decreased. For the same temperature, the reflected and absorbed energies increased linearly as the incident energy increased, whereas the transmitted energy increased logarithmically. The SEM images of the thermal crack distribution on the granite specimen surface at different temperatures can well explain the essence of mechanical parameters deterioration of granite after thermal shock. This work can provide guidance for impact crushing design of high temperature rocks during excavations.

Keywords: strain rate; temperature effect; mechanical properties; energy dissipation features; failure modes

MSC: 74R10; 74A15

Citation: Wu, Y.; Huang, L.; Li, X.; Guo, Y.; Liu, H.; Wang, J. Effects of Strain Rate and Temperature on Physical Mechanical Properties and Energy Dissipation Features of Granite. *Mathematics* **2022**, *10*, 1521. https://doi.org/10.3390/math10091521

Academic Editors: Fernando Simoes and Andrey Jivkov

Received: 29 March 2022
Accepted: 28 April 2022
Published: 2 May 2022

1. Introduction

The crushing mechanism and stability control of a rock mass after thermal shock have a widespread application in the engineering of geothermal drilling [1], nuclear waste disposal [2], and tunnel repairs after a fire [3]. Therefore, the study of the dynamic mechanical behavior and energy dissipation features of rocks after thermal shock is critical to the solution of the engineering problems of high temperature rock mechanics.

Facilitating the exploitation of mineral resources deep within the earth, underground engineering construction in high temperature and pressure conditions face many challenges. For example, efficient drilling and wellbore stability control under high temperature and pressure are the main challenges in geothermal drilling engineering. Given these challenges, the effects of temperature and external loads on rock have been extensively studied. Results have shown that the mechanical properties of high temperature rocks are different from those of normal temperature rocks [4–7]. However, the mechanical responses of high temperature rock are not only related to temperature [8] but also heating method [9,10], cooling method [11,12], heating rate [13,14] and loading method [15]. The influence of heating rate on the mechanical properties of granite was studied by conducting uniaxial compression tests [13] and the results indicated that 5 °C/min was the critical heating rate for producing thermal cracking in Beishan granite. Therefore, the heat treatment applied in this study was carried out at a heating rate of 2 °C/min. The physical mechanical parameters (C_s, E_s and UCS) of sandstone decreased when temperature increased, but the mechanical parameters of sandstone cooling in liquid nitrogen were lower than that of cooling in water [16].

Previous literature has classified three types of high-temperature rock mechanic behaviors [10]: real-time high temperature, high temperature thermal shock cooling (rapid cooling in a liquid such as water or liquid nitrogen), and high temperature slow cooling (air cooling). Under real-time high temperature and pressure, granite permeability was observed to increase rapidly above the critical temperature (350 °C), but the critical temperature of permeability decreased with increasing confining pressure [17]. And thermal cracks have been found to appear in inter-granular and grow with the increasing of temperature [18], especially above 500 °C, thermal cracks appeared in intra-granular [17,19]. Under slow cooling, the UCS and C_s of marble [20] and limestone [21] decreased with increasing temperature. The mechanical properties deterioration of granite after rapid cooling were observed to accelerate as the applied temperature increased [22], and UCS decreases significantly with the cooling cycles increased [23]. The order of UCS and E_s of granite at the same temperature from high to low was real-time high temperature, slow cooling, and rapid cooling [10,24,25].

As mechanical rock breaking devices typically use water to cool the frictional heat generated by rock breaking, it is critical to study the mechanical behaviors of rock after thermal shock. Indeed, as the dynamic mechanical properties of rock must be considered during any engineering rock excavation, researchers have widely studied the dynamic mechanical properties of rock using the SHPB system. Post-peak behaviors of stress strain curves determine the macroscopic failure modes of the rock, so post-peak behavior is of great significance for estimating engineering instability [26]. According to the value of the post-peak elastic modulus, the dynamic stress-strain curves can be divided into Class I (with a post-peak elastic modulus less than zero) and Class II (with a post-peak elastic modulus greater than zero) [26,27]. Under the former behavior, stability failure occurs under rigid conditions, whereas under the latter behavior, unavoidable spontaneous failure occurs under rigid conditions. The dynamic mechanical behaviors and energy dissipation features have been widely studied in terms of temperature [9], moisture content [28], pre-existing cracks [29,30] and cyclic impact [31]. The dynamic compressive strength of shale with different bedding dips after heat treatment was studied, the result found that dynamic compressive strength of shale at 0°and 30° bedding dips increased with the temperature increasing, but the strength at 45°, 60° and 90° bedding dips are opposite [32]. Both the dynamic strength and dissipated energy density of rock specimens have been found to increase significantly with increasing strain rate [29], but the relationship between the elastic modulus and strain rate has not been determined [28,33,34]. Furthermore, though the selection of a reasonable impact frequency can effectively reduce energy dissipation and improve rock drilling efficiency [35], the corresponding strain rate response of rock after thermal shock has been rarely studied. The mechanical properties and damage constitutive model of granite under coupling of temperature and dynamic loading were studied [33], and the damage model was found to accurately reflect the effects of the strain rate and temperature on the dynamic strength. However, the relationships between reflected, transmitted, absorbed and incident energies have been neglected to date. Furthermore, all previous results were obtained under real-time high temperature conditions, though some engineering excavations have instead been constructed in rock subjected to thermal shock. Therefore, it is critical to study the strain rate effect and energy dissipation features of rock after thermal shock.

Dynamic compression tests of granite specimens after thermal shock were performed using the SHPB system in this study. Firstly, the effects of strain rate and temperature on strength, elastic modulus, energy dissipation, and failure modes of specimens were analyzed. Secondly, effects of temperature and incident energy on the reflected, transmitted, and absorbed energies were explained using the theory of elastic wave propagation at the interface between different media. Finally, the essential reasons for the observed deterioration of the physical and mechanical properties of granite were revealed by SEM images of the surface cracks on the specimens after thermal shock. This work can provide guidance for the design of high-temperature impact rock drilling procedures.

2. Experimental Methods

2.1. Specimen Preparation

The rock employed in this study was a gray–white, medium–coarse–grained biotite granite block obtained from a quarry in Changsha, China. This block was processed into two types of specimens (Figure 1)—a series of Φ 50 × 100 mm cylinders for the uniaxial compression tests and a series of Φ 50 × 25 mm discs for the dynamic compression tests—that met the ISRM requirements for processing accuracy [36]. Each of these tests was conducted three times under the same conditions. The ρ_s of the specimen is about 2.64 to 2.65 g/cm^3. The XRD spectrum (Figure 2) shows that the granite was mainly composed of quartz (28.6%), feldspar (55.1%), mica (11.5%), and chlorite (4.8%) minerals.

Figure 1. Specimens photograph.

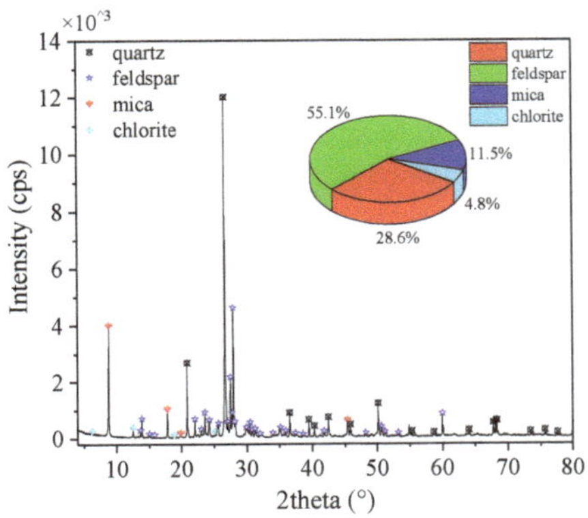

Figure 2. XRD spectrum of granite (20 °C).

2.2. Experiment Device

The dynamic compression test was performed using the SHPB system (Figure 3) comprising a gas gun, a spindle-shaped striker, elastic bars (an incident, transmission, and absorption bar) and a damper [37]. The elastic bars and spindle-shaped striker were both made of 40 Cr steel with a ρ_e and C_e of 7.810 g/cm^3 and 5410 m/s, respectively. For other technical parameters, please refer to the previous research from our group [38,39]. The spindle-shaped striker produced a stable half-sine waveform to eliminate Pochhammer-Chree (P-C) oscillations and reduce the wave dispersion effect [35,40].

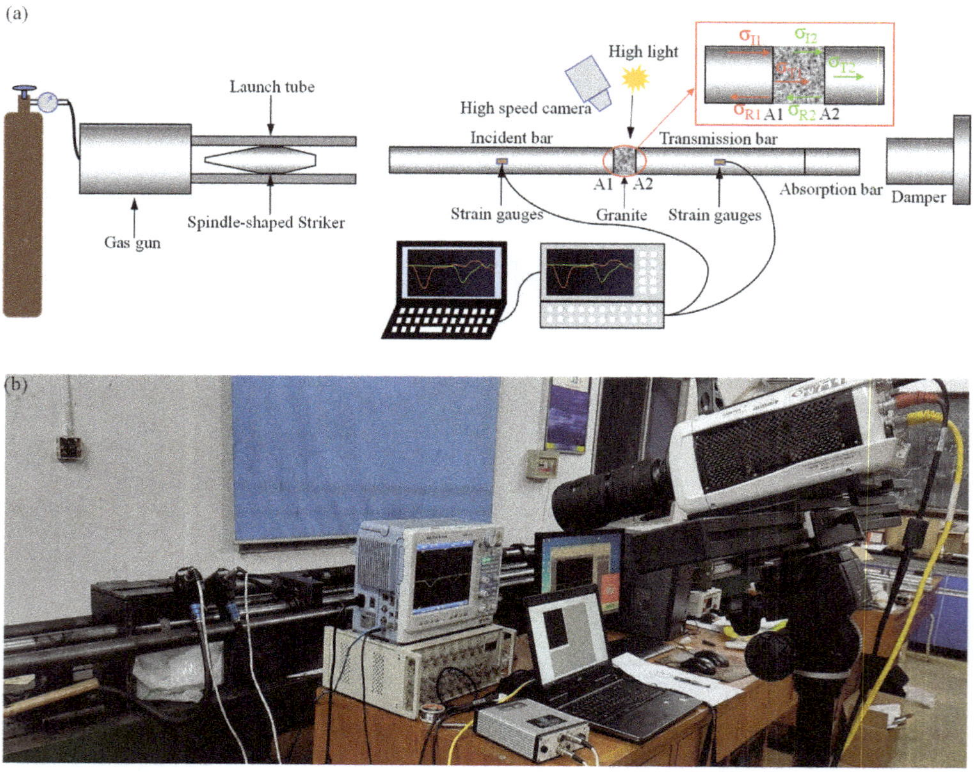

Figure 3. SHPB device. (**a**) Schematic diagram; (**b**) Scene photo.

2.3. Experiment Principle

When the target system satisfies the three basic assumptions (one-dimensional stress wave propagation, stress balance, and ignoring friction), the stress, strain rate and strain of the specimen can be calculated using the incident, reflected and transmitted wave signals as follows [41]:

$$\begin{cases} \sigma(t) = \frac{A_e E_e}{2A_s}[\varepsilon_I(t) + \varepsilon_R(t) + \varepsilon_T(t)] \\ \varepsilon(t) = \frac{C_e}{L_s}\int_0^{\tau}[\varepsilon_I(t) - \varepsilon_R(t) - \varepsilon_T(t)]\mathrm{d}t \\ \dot{\varepsilon}(t) = \frac{C_e}{L_s}[\varepsilon_I(t) - \varepsilon_R(t) - \varepsilon_T(t)] \end{cases} \quad (1)$$

where σ, $\dot{\varepsilon}$, ε are respectively the stress, strain rate and strain. A_e and A_s are respectively the cross-section areas of the elastic bars and specimens. C_e is the wave velocity of the elastic bars, L_s is the length of the specimen, and τ is the stress wave duration.

2.4. Testing Procedure

The mass, volume, and C_s of the granite specimens were measured before and after thermal treatment, and mass divided by volume equals density. The specimen mass was weighed by an electronic balance with an accuracy of 0.01 g, and the specimen volume was measured using a vernier caliper. It is important to note that a suitable quantity of petroleum jelly was applied to the contact surfaces between the transducer and specimens to reduce error when using the HS-YS4A rock ultrasonic testing analyzer measure the C_s. The specimens were divided equally into six groups, each of which included fifteen discs specimens and three cylinders specimens. Five specimen groups were heated to target temperatures of 200 °C, 300 °C, 400 °C, 500 °C and 600 °C at a heating rate of 2 °C/min using a muffle furnace. The target temperature was then held steady for 3 h to ensure an even distribution of specimen temperature. Next, the specimen was quickly removed from the muffle furnace and cooled in a water tank with a sufficient amount of water at 20 °C to ensure that the water temperature did not change significantly during the 20 min cooling process, so that the specimen completely cooled. The post-thermal shock specimens were kept in a cool and dry place until the mass of the sample tend to be stable. Static and dynamic compression tests of granite specimens after thermal shock were then conducted on a mechanical testing machine and SHPB device, respectively. In order to maintain a constant incident energy under the same impact gas pressure, it was necessary to ensure that the distance between the incident bar and the spindle-shaped striker remained equal in each test. The impact gas pressure was also changed from 0.5 MPa tc 0.9 MPa in 0.1 MPa increasements to study the effect of strain rate on the behaviors of granite after thermal shock.

3. Static Physical and Mechanical Test Results

3.1. Physical Properties of Granite

Thermal shocking can induce many defects inside a rock specimen, and change its physical and mechanical properties. The initial mass and volume of each specimen are different due to the processing accuracy and heterogeneity, so the change rate of physical parameters (mass, volume and density) at different temperatures can better quantify the effect of heat treatment on physical parameters of the specimen. The change rate of physical parameters of the specimen can be calculated as follow:

$$\delta = \frac{P_2 - P_1}{P_1} \times 100 \tag{2}$$

where P_1 and P_2 are respectively the physical parameters (such as mass, volume and density) of the specimen before and after heat treatment. δ is the change rate.

It can be observed in Figure 4a that an increase in temperature resulted in negative increases in the mass and density change rates, whereas the volume change rate exhibited the opposite trend. This indicated that the specimen mass loss gradually increased, the volume expanded, and the density decreased with increasing applied temperature. These phenomena are consistent with the other people's findings [42,43]. According to previous literature [44], the reduction in specimen mass can be divided into two components: the escape and evaporation of the attached water and strongly bound water below 400 °C and the escape of internal weakly bound water and structural water above 400 °C. The volume change rate was observed to increase significantly with increasing applied temperature. Particularly at 573 °C, the angle of any two Si-O tetrahedron changed from 150° (α phase) to 180° (β phase) in the quartz crystal [45]. Because the mass decreased and volume increased with increasing temperature, the granite density gradually decreased. Furthermore, Figure 4b shows that C_s gradually decreased from 4558.4 m/s at 20 °C to 950.9 m/s at 600 °C, a loss of 79.1%. The full results are reported in Table 1. A strong correlation was noted between ρ_s and C_s indicating that both can reflect the degree of thermal damage to the specimen to some extent [46].

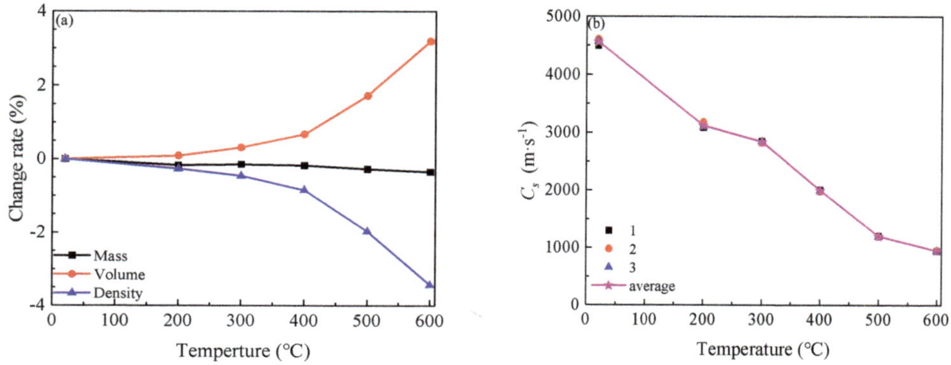

Figure 4. Relationship between physical properties and temperature. (**a**) Change rate of mass, volume and density; (**b**) Wave velocity of specimen.

Table 1. Static physical and mechanical parameters of specimen after thermal shock.

Temperature (°C)	Change Rate of Physical Parameters (%)			C_s (m/s)	UCS (MPa)	Peak Strain (%)	E_s (GPa)
	Mass	Volume	Density				
20	0	0	0	4558.4	103.5	1.73	9.30
200	−0.170	0.092	−0.261	3124.8	96.6	1.72	8.99
300	−0.144	0.312	−0.454	2840.9	85.2	1.75	8.13
400	−0.174	0.677	−0.844	1995.3	74.8	1.82	7.17
500	−0.272	1.726	−1.964	1202.7	66.5	2.07	5.95
600	−0.346	3.190	−3.427	950.9	55.1	2.49	4.30

3.2. Static Mechanical Properties of Granite

Figure 5 shows that the stress-strain curves of specimen exhibited obvious stages of compaction, elastic and yield failure. The UCS of specimens gradually changed from 103.5 MPa at 20 °C to 55.1 MPa at 600 °C, a 46.8% decrease. Meanwhile, the peak strain increased from 1.73% at 20 °C to 2.49% at 600 °C, a 43.9% increase. Notably, as the peak strain increased sharply above 300 °C, the critical temperature at which granite transforms from brittle to ductile behavior was determined to be 300 °C. Furthermore, the E_s of granite gradually decreased with increasing applied temperature. These phenomena indicate that mechanical properties of granite will obviously deteriorate after thermal shock. The full uniaxial compression test results are shown in Table 1.

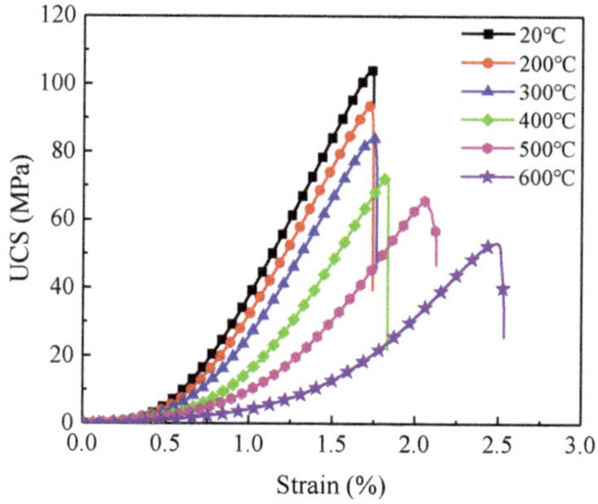

Figure 5. Stress-strain curves of specimens after thermal shock.

4. Dynamic Compression Test Results and Analysis

4.1. Investigation of Dynamic Stress Balance

The dynamic compression test cannot be initiated until the incident waveform is completely transmitted (no reflected waveform) in the pre-impact test when the incident bar and transmission bar are fully concentrically aligned. During the test, specimens were clamped between incident and transmission bars, and a suitable quantity of petroleum jelly was symmetrically applied to interfaces between the specimen and elastic bars to ensure effective contact. Figure 6 shows that the sum of reflection stress and incident stress at the incident interface was equal to the transmission stress at the transmitted interface, so the stress was balanced at both ends of the specimen, and the test results were considered effective. The reflected wave primarily reflects the strain rate history of the specimen, and has a very long plateau that represents the constant strain rate loading, and the transmitted wave primarily reflects the stress variation in the specimen. Note that all data that did not meet the assumed stress balance were eliminated to ensure reliable results.

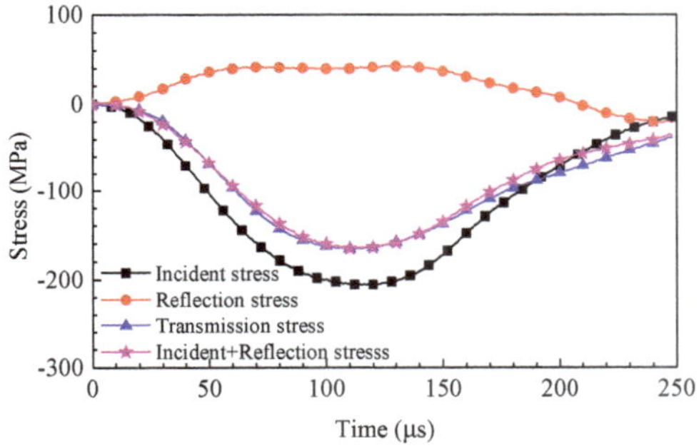

Figure 6. Dynamic stress balance checkout of specimen (20 °C, 0.8 MPa).

4.2. Dynamic Stress-Strain Curves

According to the theory of stress wave propagation, the dynamic mechanical parameters (σ_d, E_d and $\dot{\varepsilon}$) of the specimens were calculated with the results listed in Table 2. Because the speed with which the internal micro-cracks closed was much lower than the loading rate, the dynamic stress-strain curves exhibited no obvious compaction stage compared with the static stress-strain curves (Figure 7). On the one hand, the dynamic strain exhibited a rebound phenomenon at the lower impact gas pressure, when the stress-strain curves are defined as Class II (indicating that unloading occurred in the post-peak stage), and the specimen exhibited a spontaneous failure. On the other hand, the strain increased monotonically at higher impact gas pressure, and the stress-strain curves were defined as Class I (indicating a strain softening process in the post-peak stage). Thus, at the same temperature, the stress strain curves gradually transformed from Class II to Class I with increasing impact gas pressure. When the impact gas pressure was 0.6 MPa, the post-peak modulus of the stress-strain curve was greater than zero below 300 °C (Class II), but the post-peak modulus was less than zero above 300 °C (Class I) [26]. Therefore, 300 °C can be considered as the critical temperature at which stress-strain curves of granite transition from Class II to Class I behavior at a 0.6 MPa impact gas pressure (Figure 7). Furthermore, for the same applied temperature, the peak stress and peak strain of specimens have a tendency to increase with increasing impact gas pressure.

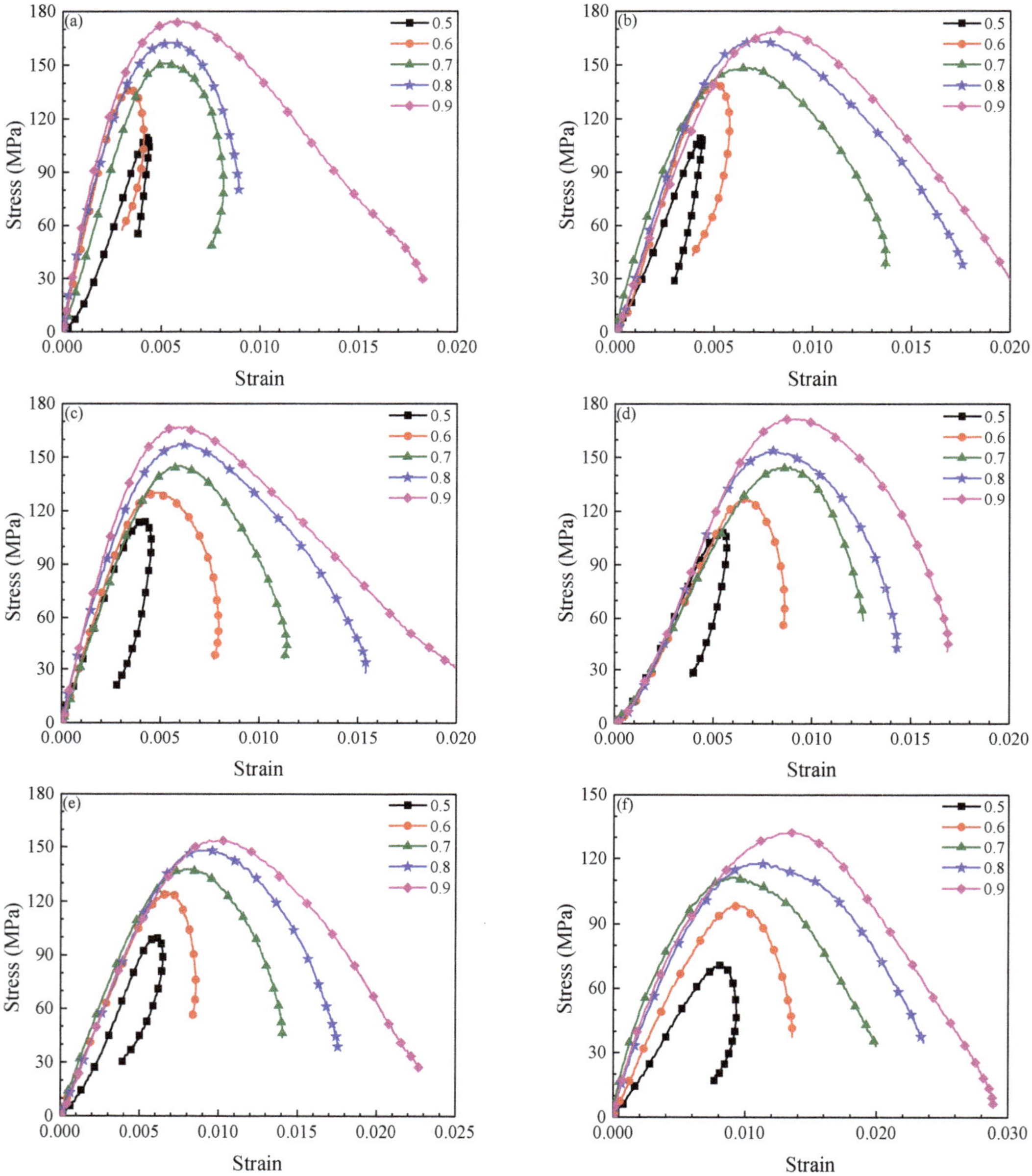

Figure 7. Dynamic stress-strain curves of granite at different temperature and impact gas pressure. (**a**) 20 °C, (**b**) 200 °C, (**c**) 300 °C, (**d**) 400 °C, (**e**) 50 °C, (**f**) 600 °C.

Table 2. Dynamic mechanical parameters of specimens at different temperature and gas pressure.

T (°C)	No.	Impact Gas Pressure (MPa) 0.5			0.6			0.7			0.8			0.9		
		$\dot{\varepsilon}_d$	E_d	σ_d	$\dot{\varepsilon}_d$	E_d	σ_d	$\dot{\varepsilon}_d$	E_d	σ_d	$\dot{\varepsilon}_d$	E_d	σ_d	$\dot{\varepsilon}_d$	E_d	σ_d
20	1	25.0	32.8	110.1	31.8	47.4	136.5	45.1	39.8	150.8	52.5	43.5	162.9	70.5	46.7	174.5
	2	26.5	33.4	118.4	30.5	46.1	138.5	46.2	41.3	155.5	50.6	42.8	158.1	72.4	45.6	178.6
	3	23.9	32.5	102.6	31.6	47.5	136.6	45.0	39.0	146.6	53.7	43.1	167.2	67.2	46.9	170.7
	Ave	25.1	32.7	110.4	31.3	47.0	137.2	45.3	40.0	151.0	52.3	43.1	162.7	70.0	46.4	174.6
200	1	27.1	30.0	109.5	45.3	36.1	138.5	58.3	32.7	148.1	71.4	33.9	163.8	80.7	28.6	169.2
	2	28.5	33.7	111.2	46.3	35.5	140.1	60.1	33.2	150.3	73.8	34.8	167.5	78.3	29.1	167.3
	3	26.6	32.1	106.3	44.1	35.0	136.0	57.6	33.5	146.4	69.4	34.5	157.6	81.1	30.9	171.1
	Ave	27.4	31.9	109.0	45.2	35.5	138.2	58.7	33.1	148.3	71.5	34.4	163.0	80.0	29.5	169.2
300	1	32.2	34.5	114.3	46.3	36.9	130.6	56.7	34.4	146.1	74.5	34.9	157.4	87.6	39.7	166.8
	2	33.4	33.1	116.7	47.9	36.1	132.9	55.3	35.1	145.6	76.2	35.9	158.9	89.4	37.2	169.1
	3	31.7	34.7	113.1	45.9	35.9	129.2	54.6	32.9	144.4	73.8	34.1	156.7	87.3	38.4	165.1
	Ave	32.4	34.1	114.7	46.7	36.3	130.9	55.5	34.1	145.4	74.8	35.0	157.7	88.1	38.4	167.0
400	1	47.1	25.4	108.5	52.1	26.1	126.7	70.1	24.4	144.9	81.3	29.5	154.0	94.8	27.4	169.1
	2	48.9	25.9	110.8	51.4	27.3	125.3	72.1	25.6	147.1	80.6	32.1	150.4	97.4	29.1	174.2
	3	46.1	24.4	107.4	54.6	26.8	129.1	67.8	25.3	142.0	83.0	28.7	156.4	95.5	27.9	171.9
	Ave	47.4	24.8	108.9	52.7	26.7	127.0	70.0	25.1	144.7	81.6	30.1	153.6	95.9	28.1	171.7
500	1	52.1	22.0	99.6	60.0	21.0	125.1	82.0	22.2	138.1	99.3	21.7	147.9	104.3	21.5	153.4
	2	53.7	22.0	103.4	62.9	22.3	128.6	83.9	24.0	141.6	96.9	23.4	146.5	103.1	23.4	150.2
	3	50.3	20.1	97.3	58.8	20.4	122.9	81.0	21.3	135.8	102.5	21.0	150.3	107.6	20.8	157.5
	Ave	52.0	21.7	100.1	60.6	21.2	125.5	82.3	22.5	138.5	99.0	22.0	148.2	105.0	21.9	153.7
600	1	71.3	10.3	70.5	84.2	13.4	98.5	110.5	15.4	110.6	130.9	13.9	117.1	142.1	13.2	133.2
	2	72.9	12.1	73.5	86.4	13.1	100.4	114.3	16.0	114.1	133.1	14.2	120.4	144.5	14.4	135.5
	3	68.1	9.7	68.3	83.2	11.9	96.1	109.2	14.1	109.3	127.1	13.6	115.3	140.8	12.9	128.9
	Ave	70.8	10.7	70.8	84.6	12.8	98.3	111.3	15.2	111.3	130.4	13.6	117.6	142.5	13.5	132.5

Note: T (Temperature), NO. (Number), $\dot{\varepsilon}_d$ (s^{-1}), E_d (MPa), σ_d (GPa) and Ave (average value).

4.3. Strain Rate Effect on Granite after Thermal Shock

At the same temperature, the σ_d of the specimens exhibited an obvious strain rate response as the impact gas pressure increased (Figure 8). Temperature was also observed to play an important role in the σ_d under similar strain rates. For example, when the strain rate was around 70 s^{-1}, the σ_d decreased from 174.6 MPa at 20 °C to 70.8 MPa at 600 °C, a 103.8 MPa reduction. Therefore, the relationship between σ_d and strain rate must consider the effect of temperature. The σ_d and strain rate can be great fitted into an exponential form as follows:

$$\begin{cases} \sigma_{T_1} = 33.031\dot{\varepsilon}_{T_1}{}^{0.3948} & R^2 = 0.9316 \\ \sigma_{T_2} = 28.689\dot{\varepsilon}_{T_2}{}^{0.4062} & R^2 = 0.9929 \\ \sigma_{T_3} = 30.833\dot{\varepsilon}_{T_3}{}^{0.3797} & R^2 = 0.9859 \\ \sigma_{T_4} = 11.688\dot{\varepsilon}_{T_4}{}^{0.5894} & R^2 = 0.9754 \\ \sigma_{T_5} = 12.988\dot{\varepsilon}_{T_5}{}^{0.5344} & R^2 = 0.9191 \\ \sigma_{T_6} = 2.8098\dot{\varepsilon}_{T_6}{}^{0.7766} & R^2 = 0.9201 \end{cases} \quad (3)$$

where T_1, T_2, T_3, T_4, T_5 and T_6 are 20 °C, 200 °C, 300 °C, 400 °C, 500 °C and 600 °C, respectively.

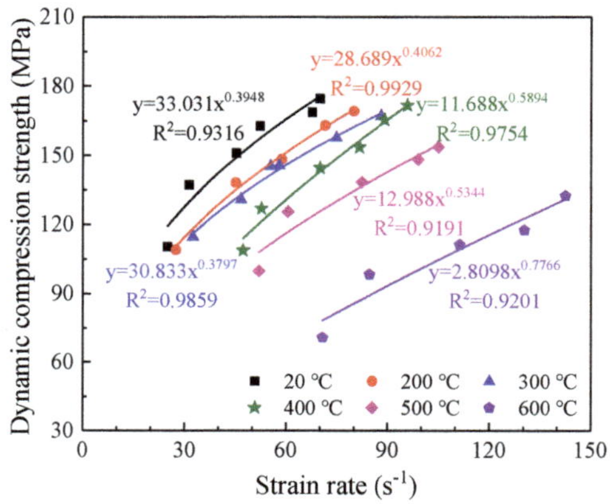

Figure 8. Relationship between strain rate and dynamic compressive strength of granite after thermal shocking.

The *DIF* is generally used as an index for the evaluation of the relationship between σ_d and strain rate, and is defined as follows [33]:

$$DIF = \sigma_d / \sigma_s \tag{4}$$

where σ_s is static compressive strength and is equal to UCS, and σ_d is dynamic compressive strength.

Figure 9 shows that the *DIF* exhibited a clear linear relation with the $\lg(\dot{\varepsilon})$ at the same temperature. Similarly, this result was found by Wang [33]. However, the curves slopes clearly differ according to temperatures. On the one hand, thermally induced cracks were not visible below 300 °C [19], so the slopes of the corresponding curves can be observed to be approximately equal. On the other hand, the slopes of the curves are significantly different above 300 °C. Because the thermally induced cracks on the surface of the specimen grow rapidly above 300 °C [17], the deformation resistance capacity of the specimen decreased. Therefore, $\lg(\dot{\varepsilon})$ increased significantly with increasing temperature under the same incident energy. The distribution of thermally induced surface cracks observed in the granite specimens will be discussed in Section 5.

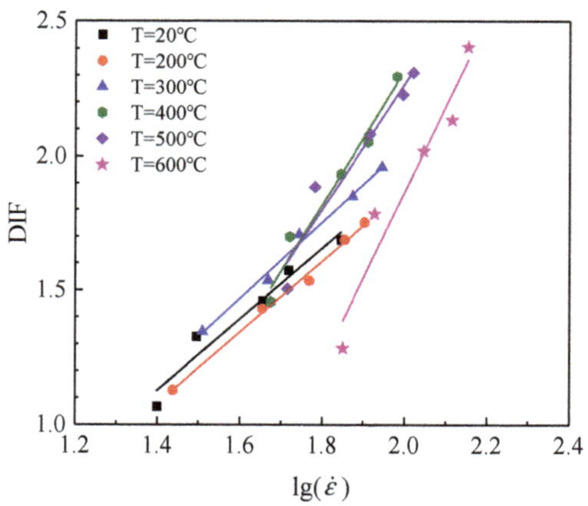

Figure 9. Relationship between *DIF* and $\lg(\dot{\varepsilon})$ at different temperatures.

To characterize the relationship between the *DIF* and $\lg(\dot{\varepsilon})$, most scholars have used piecewise functions to fit the strain rate [47,48], and only a few consider the temperature effect [33]. The relationships between the *DIF* and the $\lg(\dot{\varepsilon})$ at different temperatures were given:

$$\begin{cases} DIF_{T=20^{\circ}\mathrm{C}} = 1.3213\lg(\dot{\varepsilon}) - 0.7224 & R^2 = 0.9554 \\ DIF_{T=200^{\circ}\mathrm{C}} = 1.3229\lg(\dot{\varepsilon}) - 0.7738 & R^2 = 0.9947 \\ DIF_{T=300^{\circ}\mathrm{C}} = 1.4257\lg(\dot{\varepsilon}) - 0.8129 & R^2 = 0.9913 \\ DIF_{T=400^{\circ}\mathrm{C}} = 2.4966\lg(\dot{\varepsilon}) - 2.6745 & R^2 = 0.9742 \\ DIF_{T=500^{\circ}\mathrm{C}} = 2.3362\lg(\dot{\varepsilon}) - 2.4034 & R^2 = 0.9344 \\ DIF_{T=600^{\circ}\mathrm{C}} = 3.2006\lg(\dot{\varepsilon}) - 4.5351 & R^2 = 0.9366 \end{cases} \tag{5}$$

4.4. Temperature Sensitivity of Granite Behaviors

Figure 10 shows that temperature had an obvious effect on dynamic stress-strain curves of the granite specimens. Taking an impact gas pressure of 0.5 MPa as an example (Figure 10a), as the temperature increased, the starting point of strain recovery gradually moved away from the peak strain, causing the curves to exhibit Class II behavior. The strain rates of specimens increased from 25.1 s^{-1} at 20 °C to 70.8 s^{-1} at 600 °C, rising 182.1%, but σ_d decreased gradually from 109.0 MPa at 20 °C to 70.8 MPa at 600 °C, falling 35.0%. This result indicates that under the same impact gas pressure, the strain rate increased with increasing temperature, whereas the σ_d decreased.

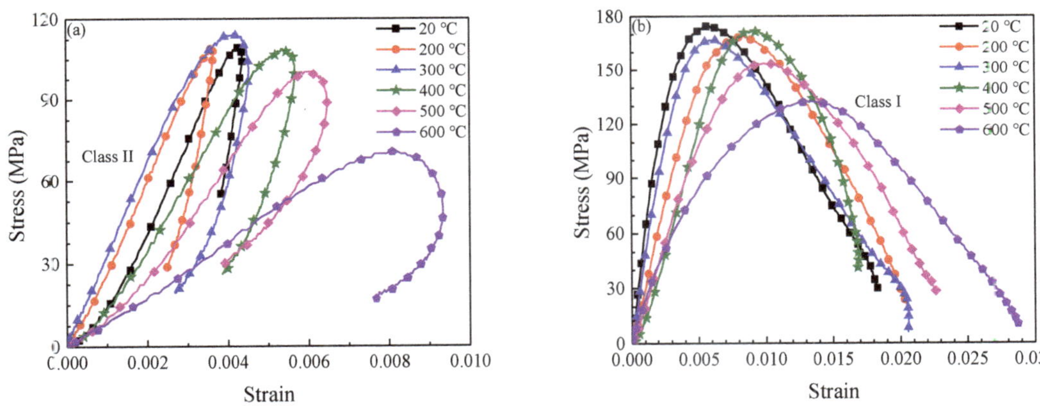

Figure 10. Dynamic stress-strain curves of specimens at different temperature. (**a**) 0.5 MPa, (**b**) 0.9 MPa.

Figure 10b shows that the dynamic stress-strain curves all exhibited Class I behavior for an impact gas pressure of 0.9 MPa. The strain rate and σ_d exhibited similar laws to those for an impact gas pressure of 0.5 MPa, but the plastic failure of rock was more obvious. Meanwhile, the slopes of the elastic segment of the curves can be observed to decrease as the temperature increases, this result is consistent with that of the static compression test.

4.5. Dynamic Elastic Modulus of Granite under Thermal Shocking

The dynamic elastic modulus equation is given as follows [49].

$$E_d = \frac{\sigma_b - \sigma_a}{\varepsilon_b - \varepsilon_a} \tag{6}$$

where σ_b and σ_a are 60% and 40% of peak stress, respectively. ε_b and ε_a are strains corresponding to σ_b and σ_a, respectively.

Figure 11 and Table 2 show that under the same applied temperature, the E_d changed slightly with increasing impact gas pressure. Wang [33] et al. also found that strain rate has no significant effect on elastic modulus. Under the same impact gas pressure, the E_d exhibited an obviously decreasing trend with increasing temperature. Indeed, the E_d decreased from 32.7 GPa at 20 °C to 10.7 GPa at 600 °C when impact gas pressure was 0.5 MPa, a 67.3% reduction. Meanwhile, when impact gas pressure was 0.9 MPa, the E_d decreased from 46.4 GPa at 20 °C to 13.5 GPa at 600 °C, a 70.9% reduction. These results indicate that the attenuation of E_d value as the applied temperature increased from 20 °C to 600 °C increased slightly with increasing impact gas pressure (Figure 11b).

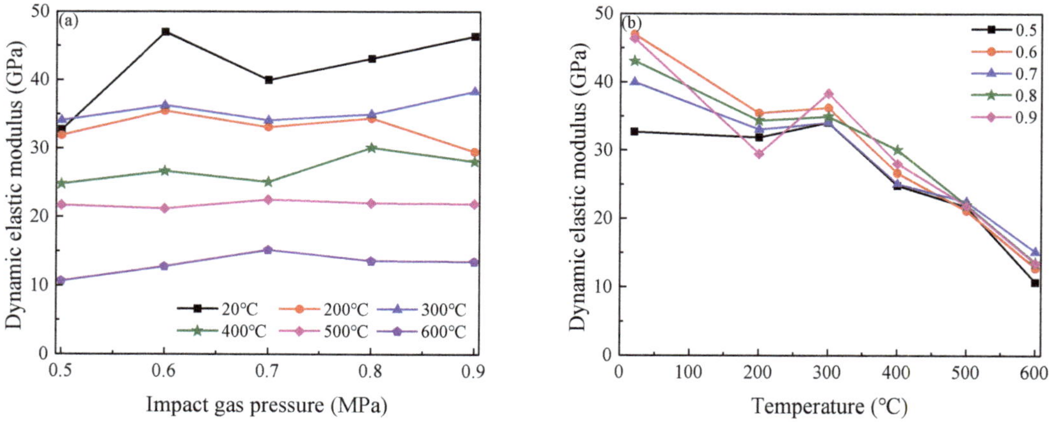

Figure 11. Dynamic elastic modulus of specimens. (**a**) Impact gas pressure, (**b**) Temperature.

4.6. Energy Dissipation Features of Specimens under Thermal Shock

According to the elastic stress wave transmission theory, the expressions of the incident energy E_I, reflected energy E_R and transmitted energy E_T are as follows [31]:

$$\begin{cases} E_I = \frac{A_e}{\rho_e C_e}\int_0^\tau \sigma_I^2(t)\mathrm{d}t \\ E_R = \frac{A_e}{\rho_e C_e}\int_0^\tau \sigma_R^2(t)\mathrm{d}t \\ E_T = \frac{A_e}{\rho_e C_e}\int_0^\tau \sigma_T^2(t)\mathrm{d}t \end{cases} \quad (7)$$

where ρ_e is the density of the elastic bars.

According to the law of energy conservation, the absorbed energy E_A can be calculated as follows:

$$E_\mathrm{A} = E_I - E_R - E_T \quad (8)$$

At the same temperature, the E_R will increase linearly with E_I (Figure 12); under the same E_I, the higher the temperature, the greater the E_R. These phenomena can be explained by the laws of reflection and transmission of elastic waves at the interface between different media (Figure 3a). The corresponding reflection (R) coefficient and transmission (T) coefficient are presented as follows [50]:

$$\begin{cases} \sigma_R = R\sigma_I \\ \sigma_T = T\sigma_I \\ R = \frac{1-n}{1+n} \\ T = \frac{2}{1+n} \end{cases} \quad (9)$$

where n is the ratio of the wave impedance of two materials, $n = \rho_1 C_1/\rho_2 C_2$.

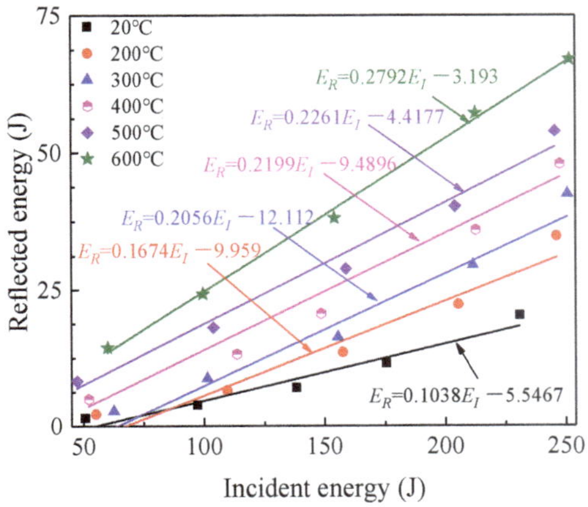

Figure 12. Relationship between reflected and incident energy.

The following conclusions can be drawn according to the ratios of wave impedances between the specimen and elastic bars at interfaces A1 and A2 shown in Figure 3a:

At interface A1, $n = \rho_e C_e / \rho_s C_s > 1$, $R < 0$, $T < 1$; At interface A2, $n = \rho_s C_s / \rho_e C_e < 1$, $R > 0$, $T > 1$.

At the same temperature, the C_s and ρ_s of the samples can be considered to be constant according to Table 1, so the wave impedance of granite was nearly the same regardless of applied impact gas pressure. The R at the incident end was invariable, combining Equations (7) and (9), it can well explain that the E_R increased linearly with increasing E_I. Meanwhile, the C_s and ρ_s gradually decreased with increasing temperature, leading to wave impedance of the specimen decrease. Therefore, n increased at the interface A1 ($n > 1$), and the corresponding $|R|$ increased, essentially explaining that E_R increased with increasing temperature under the same E_I.

A part of the E_T at the interface A1 is consumed by the energy dissipation of the rock fracture, and the other part of E_T is transmitted and reflected as E_I at the interface A2 (Figure 3a). Figure 13 shows that the E_T logarithmically increased with increasing E_I, indicating that the growth rate of E_T slowed. Meanwhile, the E_T decreased with increasing temperature under the same E_I. In a word, the magnitude of ET was related to transmission coefficient of interface and energy absorption capacity of the specimen.

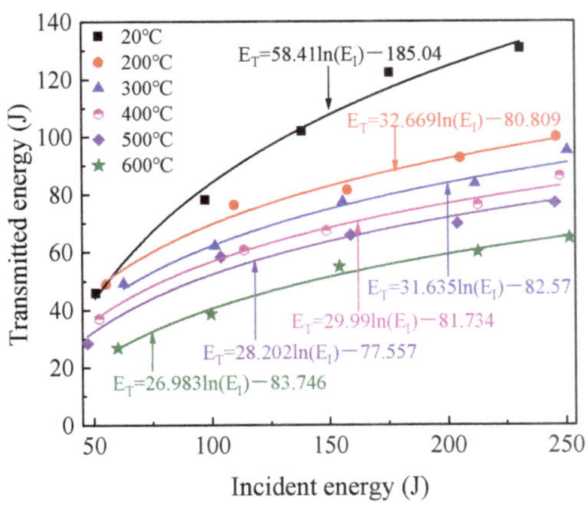

Figure 13. Relationship between transmitted and incident energy.

Figure 14 shows that the E_A exhibited a linear increase with the E_I for the same applied temperature. The greater the value of the E_I, the more broken the specimen owing to the quantity of the energy absorbed. The curves slopes were all determined to be around 0.5, indicating that nearly half of the impact incident energy was reflected and harmlessly transmitted in the form of elastic waves.

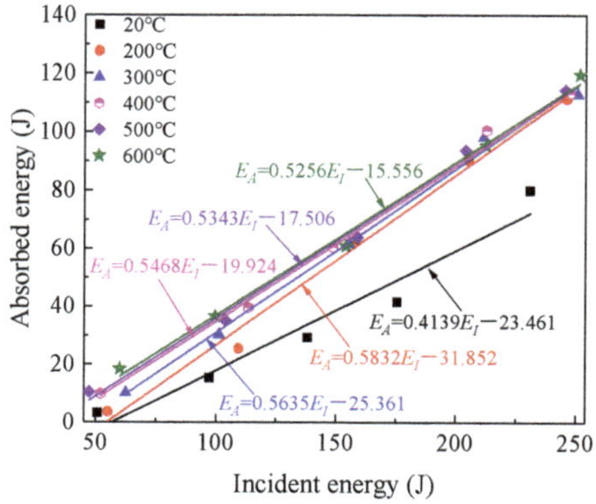

Figure 14. Relationship between absorbed and incident energy.

5. Discussion

5.1. Dynamic Failure Modes of Specimens after Thermal Shock

Macroscopic spontaneous destruction occurs in the post-peak failure stage, so any stability analysis should pay careful attention to the post-peak failure properties of the rock. As discussed in Section 4.2, the stress–strain curves obtained in this study were divided into Class I and Class II behaviors according to previous literature [26,27]. Figure 15a shows that for an impact gas pressure of 0.9 MPa, the strain rate reached a stable state before achieving the peak stress, and there was an obvious plateau reflecting the constant strain rate loading. The strain rate still increased for some time after the peak stress was achieved, and the residual strain continued to increase until becoming stable. At this time, the specimen broke rapidly and pulverized, and the stress-strain curve exhibited Class I behavior. Figure 15b shows that the strain initially increased to a peak value before gradually decreasing, and the strain recovery led to the post-peak unloading, indicating stored energy was not high enough to fracture the sample in such a case. When the loading peak strain rate was 40.0 s^{-1}, the stress was about one-third of the peak stress. The unloading process enabled the elastic recovery of the specimen, the unloading peak strain rate was -25.8 s^{-1}, rapidly releasing the stored elastic energy and leading to a small split in the specimen. This stress-strain curve exhibited Class II behavior as it showed an obvious rebound.

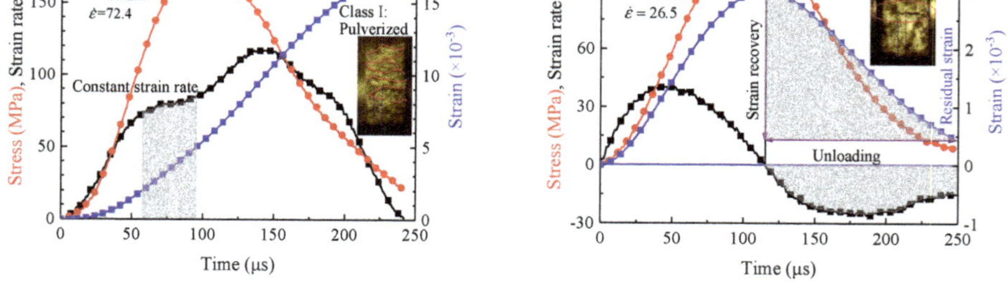

Figure 15. Variation curves of stress, strain rate and strain of specimens at 20°C with time. (**a**) 0.9 MPa; (**b**) 0.5 MPa.

Figure 16 shows that the specimens exhibited increasingly severe destruction as the impact gas pressure increased; the distribution ratios of small pieces and powder increased while the average size of the fragment decreased. The specimens were slightly broken with some debris for an impact gas pressure less than 0.7 MPa, but clearly absorbed more energy for an impact gas pressure greater than 0.7 MPa, when the specimen fragments were mostly debris and powder as the unit energy consumption increased. When the strain rate was about 45.1 s^{-1} (for an impact gas pressure of 0.7 MPa), the average size of the specimen fragment exhibited an obvious variation corresponding to the transition of the stress-strain curve from Class II to Class I in Figure 7a. The spontaneous failure at a low strain rate belongs to tensile spalling, and the fragments are comparatively intact. The specimen was crushed by excessive incident energy at a high strain rate, causing the smaller fragments to increase.

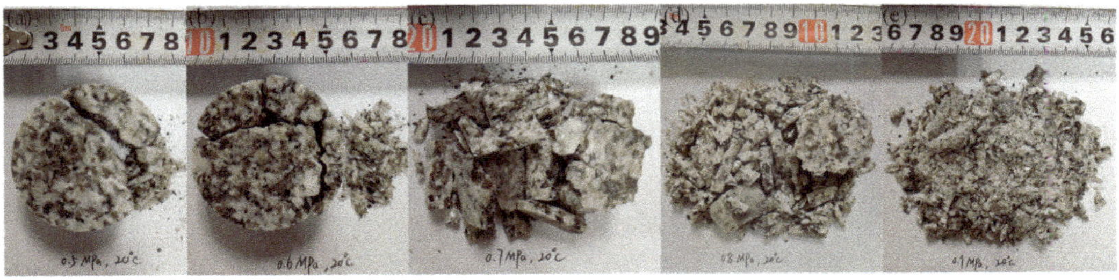

Figure 16. Relationship between fragment degree of specimens and impact gas pressure at 20 °C. (**a**) 0.5 MPa; (**b**) 0.6 MPa; (**c**) 0.7 MPa; (**d**) 0.8 MPa; (**e**) 0.9 MPa.

Under the same impact gas pressure, the failure modes of the specimen (Figure 17) changed significantly with increasing temperature, which transformed from a slight split at 20 °C to pulverization at 600 °C. The failure modes of the specimens were in complete agreement with the dynamic stress-strain curve transition from Class II to Class I behavior, as shown in Figure 7. This indicates that the increase of temperature reduced the critical impact gas pressure describing the stress-strain curves transition. Indeed, the rock failure modes are very sensitive to temperature, so the temperature effect must be considered in the design of structures to be constructed in rock mass. Furthermore, in addition to controlling the stability of the surrounding rock, heat treatment can be considered to reduce rock strength and thereby more effectively crush the rock. For example, the strain rate effect of high temperature rocks can be considered to improve the geothermal drilling effect.

Figure 17. Relationship between fragment degree of specimens and temperature at 0.6 MPa. (**a**) 20 °C, (**b**) 200 °C, (**c**) 300 °C, (**d**) 400 °C, (**e**) 500 °C, (**f**) 600 °C.

The degree of specimen crushing was found to be aggravated by increases in incident energy and temperature, but the σ_d decreased as temperature increased or E_I decreased (Figure 18). The relationship among the σ_d, temperature and E_I are fitted as follows:

$$\sigma(T, E_I) = 84.47 + 0.039T + 0.617E_I - 0.000155T^2 - 0.00098E_I^2 - 0.000083TE_I R^2 = 0.937 \quad (10)$$

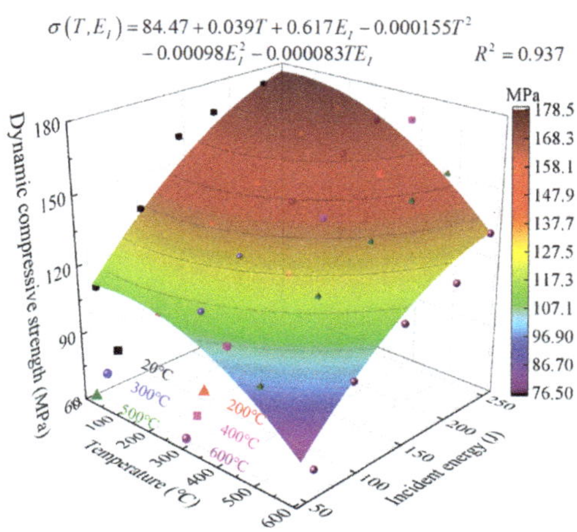

Figure 18. Relationship between temperature, incident energy and dynamic compressive strength of granite.

Figure 18 and Equation (10) show that the same σ_d corresponded to multiple temperature and E_I composite states. In a sense, the temperature and E_I could equivalently transform for the same σ_d, but the degree of granite fragmentation in these states varied considerably. Therefore, the E_I can be appropriately adjusted to account for the large differences among the ground temperatures of the different strata encountered in the process of rock drilling for deep excavations, thereby reducing the energy consumption required for rock crushing and improving rock drilling efficiency.

5.2. Thermal Damage Assessment based on SEM

The heat-treated granite specimens were sprayed with gold using an ion sputterer, then scanned by SEM to assess the thermal damage degrees of the specimen surface. Each SEM image was magnified 60 times to ensure comparability. Figure 19 shows the SEM images of the specimen surfaces after thermal shock. Firstly, it can be observed that the surfaces of the specimen were complete without obvious cracks for applied temperatures of less than 300 °C. Two intersecting intergranular cracks occurred on specimen surface at 300 °C, which can be considered as the critical temperature for thermal shock crack initiation. Finally, the surface cracks on the specimens developed quickly for applied temperatures above 300 °C, and the extent of the crack network increased. Especially for applied temperatures of 500 °C and 600 °C, obvious loose areas were present in addition to a lot of intergranular and transgranular cracks. These phenomena essentially reflect the decrease in ρ_s and C_s with increasing temperature. This is also why the higher the temperature, the more easily broken the specimen under the same E_I. Therefore, for the same E_A, the specimen subjected to higher temperatures before thermal shock were more uniformly broken and the powder was finer, primarily because the internal cracks in the granite were found to be more intensive after exposure to a high temperature. In other words, the energy required for crack propagation is relatively smaller. Taken together, these results indicate that thermal shock plays an important role in rock mass crushing and stability control.

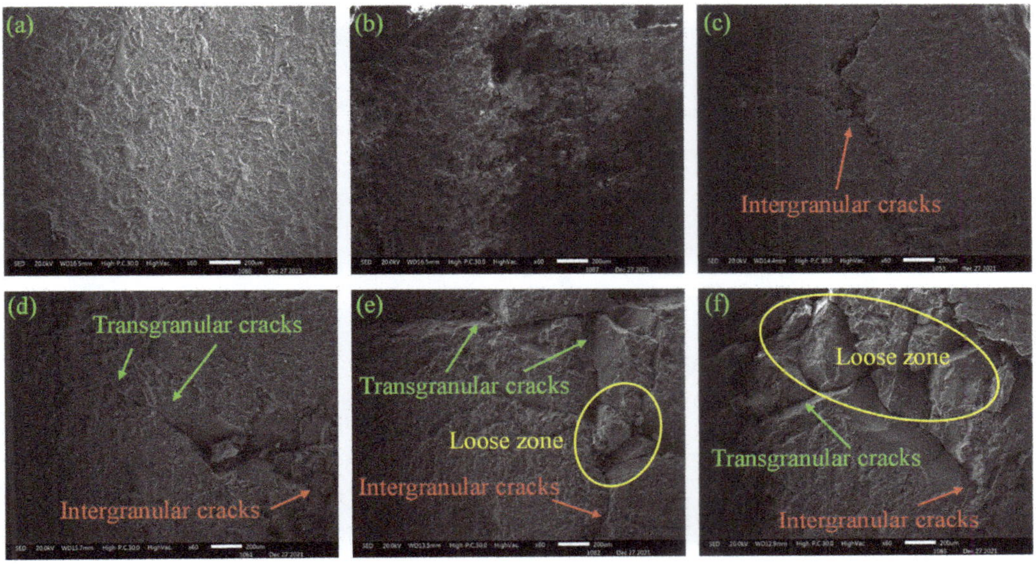

Figure 19. SEM images of specimens at different temperatures. (**a**) 20 °C, (**b**) 200 °C, (**c**) 300 °C, (**d**) 400 °C, (**e**) 500 °C, (**f**) 600 °C.

6. Conclusions

In this study, dynamic compression testing of granite specimens under six temperatures and five impact gas pressures were performed using the SHPB system. Then, the effects of temperature and strain rate on the physical mechanical parameters, energy dissipation features and failure modes of granite were analyzed. The specific conclusions are summarized below:

(1) The physical and mechanical parameters (mass, ρ_s, C_s, UCS and E_s) of granite specimen decreased with increasing temperature, except for volume increased.

(2) At the same temperature, the σ_d of granite specimens increased exponentially with increasing strain rate, and the stress-strain curves gradually transformed from Class II to Class I behavior. For the same strain rate or incident energy, the dynamic compressive strength decreased with increasing temperature. Therefore, the temperature and E_I can equivalently transform for the same σ_d, but the degree of the granite fragmentation in these states varied considerably.

(3) The *DIF* exhibited an obvious linear correlation with the $\lg(\dot{\varepsilon})$ at the same temperature. The E_d of granite was determined to be more sensitive to temperature than impact gas pressure and decreased obviously as the temperature increased. Meanwhile, the higher the temperature, the greater the strain rate and the more severe the specimen failure, and the failure modes of the granite changed from brittle to ductile.

(4) When the impact gas pressure was 0.6 MPa, the critical temperature for the stress-strain curve transition from Class II to Class I behavior was determined to be 300 °C.

(5) The critical temperature of the crack initiation of granite was determined to be 300 °C. As the temperature increased, the initiation cracks in the granite specimen surface developed from intergranular cracks to transgranular cracks. Furthermore, a number of loose areas were observed on the granite specimens subjected to higher temperatures, especially those greater than 500 °C.

(6) The E_R and E_T were found to be only related to the ratio of the wave impedances of the elastic bars and the specimen. For the same temperature, as the E_I increased, the E_R and the E_A increased linearly, whereas E_T increased logarithmically. For a similar E_I, the E_R increased but the E_T decreased as the temperature increased.

Author Contributions: Conceptualization, Y.W. and L.H.; methodology, Y.W.; software, Y.W.; validation, Y.W., X.L. and L.H.; formal analysis, Y.W. and Y.G.; investigation, Y.W. and H.L.; resources, L.H.; data curation, L.H.; writing—original draft preparation, Y.W.; writing—review and editing, Y.W.; visualization, Y.W. and J.W.; supervision, J.W.; project administration, L.H.; funding acquisition, X.L. All authors have read and agreed to the published version of the manuscript.

Funding: This work was supported by the National Natural Science Foundation of China (51904335, 52174098, 51927808), the Postgraduate Scientific Research Innovation Project of Hunan Province (No. CX20210297), and the Fundamental Research Funds for Central Universities of the Central South University (No. 2021zzts0276).

Institutional Review Board Statement: Not applicable.

Informed Consent Statement: Not applicable.

Data Availability Statement: The data presented in this study are available on request from the corresponding author.

Conflicts of Interest: The authors declare no conflict of interest.

Nomenclature

σ_d, σ_s	dynamic compressive strength, static compressive strength, MPa;
E_I, E_R, E_T, E_A	incident, reflected, transmitted, absorbed energy, J;
ρ_s, ρ_e	specimen density, elastic bar density, g/cm^3;
C_s, C_e	longitudinal wave velocity of the specimen, longitudinal wave velocity of the elastic bar, m/s;
E_s, E_d	static elastic modulus of the specimen, dynamic elastic modulus of the specimen, GPa;
UCS	uniaxial compressive strength, MPa;
DIF	dynamic increase factor;
SHPB	split Hopkinson pressure bar;
SEM	scanning electron microscope;
XRD	X-ray diffraction;
$\lg(\dot{\varepsilon})$	logarithm of strain rate, s^{-1};
R, T	the reflection coefficient, the transmission coefficient;
n	the ratio of wave impedance of two materials;
$\dot{\varepsilon}$	strain rate, s^{-1}.

References

1. Delong, Z.; Jun, J.; Yuwen, H.; Wei, W.; Wenjian, Z. Study on well control technology of high temperature geothermal drilling. *Procedia Eng.* **2014**, *73*, 337–344. [CrossRef]
2. Xu, H.; Rutqvist, J.; Birkholzer, J. A study of thermal pressurization and potential for hydro-fracturing associated with nuclear waste disposal in argillaceous claystone. *Int. J. Rock Mech. Min. Sci.* **2020**, *136*, 104536. [CrossRef]
3. Hua, N.; Elhami Khorasani, N.; Tessari, A.; Ranade, R. Experimental study of fire damage to reinforced concrete tunnel slabs. *Fire Saf. J.* **2022**, *127*, 103504. [CrossRef]
4. Xu, X.L.; Karakus, M. A coupled thermo-mechanical damage model for granite. *Int. J. Rock Mech. Min. Sci.* **2018**, *103*, 195–204. [CrossRef]
5. Ge, Z.; Sun, Q. Acoustic emission (AE) characteristics of granite after heating and cooling cycles. *Eng. Fract. Mech.* **2018**, *200*, 418–429. [CrossRef]
6. Fan, L.F.; Wu, Z.J.; Wan, Z.; Gao, J.W. Experimental investigation of thermal effects on dynamic behavior of granite. *Appl. Therm. Eng.* **2017**, *125*, 94–103. [CrossRef]
7. Huang, L.; Guo, Y.; Li, X. Failure characteristics of shale after being subjected to high temperatures under uniaxial compression. *Bull. Eng. Geol. Environ.* **2021**, *81*, 33. [CrossRef]
8. Kumari, W.G.P.; Ranjith, P.G.; Perera, M.S.A.; Shao, S.; Chen, B.K.; Lashin, A.; Arifi, N.A.; Rathnaweera, T.D. Mechanical behaviour of australian strathbogie granite under in-situ stress and temperature conditions: An application to geothermal energy extraction. *Geothermics* **2017**, *65*, 44–59. [CrossRef]
9. Yin, T.; Wu, B.; Wang, C.; Wu, Y. Determination of dynamic tensile strength of microwave-induced basalt using Brazilian test. *Rock Mech. Rock Eng.* **2022**, *55*, 1429–1443. [CrossRef]
10. Kumari, W.G.P.; Ranjith, P.G.; Perera, M.S.A.; Chen, B.K.; Abdulagatov, I.M. Temperature-dependent mechanical behaviour of Australian Strathbogie granite with different cooling treatments. *Eng. Geol.* **2017**, *229*, 31–44. [CrossRef]

11. Guo, Y.; Huang, L.; Li, X.; Chen, J.; Sun, J. Experimental investigation on the effects of thermal treatment on the physical and mechanical properties of shale. *J. Nat. Gas Sci. Eng.* **2020**, *82*, 103496. [CrossRef]
12. Shao, S.; Wasantha, P.L.P.; Ranjith, P.G.; Chen, B.K. Effect of cooling rate on the mechanical behavior of heated Strathbogie granite with different grain sizes. *Int. J. Rock Mech. Min. Sci.* **2014**, *70*, 381–387. [CrossRef]
13. Chen, S.; Yang, C.; Wang, G. Evolution of thermal damage and permeability of Beishan granite. *Appl. Therm. Eng.* **2017**, *110*, 1533–1542. [CrossRef]
14. Shu, R.; Yin, T.; Li, X. Effect of heating rate on the dynamic compressive properties of granite. *Geofluids* **2019**, *2019*, 8292065. [CrossRef]
15. Guo, L.; Zhang, Y.; Zhang, Y.; Yu, Z.; Zhang, J. Experimental investigation of granite properties under different temperatures and pressures and numerical analysis of damage effect in enhanced geothermal system. *Renew. Energy* **2018**, *126*, 107–125. [CrossRef]
16. Li, Q.; Yin, T.; Li, X.; Zhang, S. Effects of rapid cooling treatment on heated sandstone: A comparison between water and liquid nitrogen cooling. *Bull. Eng. Geol. Environ.* **2020**, *79*, 313–327. [CrossRef]
17. Zhao, Y.; Feng, Z.; Zhao, Y.; Wan, Z. Experimental investigation on thermal cracking, permeability under HTHP and application for geothermal mining of HDR. *Energy* **2017**, *132*, 305–314. [CrossRef]
18. Zhang, W.; Wang, Z.; Du, Y.; Zhang, S.; Shi, Z.; Li, F. Effect of high temperature on pore characteristics, yield stress, and deformation property of sandstone. *Bull. Eng. Geol. Environ.* **2021**, *81*, 43. [CrossRef]
19. Zhao, Y.S.; Wan, Z.J.; Feng, Z.J.; Xu, Z.H.; Liang, W.G. Evolution of mechanical properties of granite at high temperature and high pressure. *Geomech. Geophys. Geo-Energy Geo-Resour.* **2017**, *3*, 199–210. [CrossRef]
20. Peng, J.; Rong, G.; Cai, M.; Yao, M.-D.; Zhou, C.-B. Physical and mechanical behaviors of a thermal-damaged coarse marble under uniaxial compression. *Eng. Geol.* **2016**, *200*, 88–93. [CrossRef]
21. Zhang, Y.; Sun, Q.; Cao, L.; Geng, J. Pore, mechanics and acoustic emission characteristics of limestone under the influence of temperature. *Appl. Therm. Eng.* **2017**, *123*, 1237–1244. [CrossRef]
22. Han, H.; Peng, J.; Guo, Y.; He, Q.; Zhou, J.; Gou, X. Experimental research on the mechanical properties of heated granite after rapid cooling. *J. Eng.* **2021**, *2021*, 8838520. [CrossRef]
23. Zhang, B.; Tian, H.; Dou, B.; Zheng, J.; Chen, J.; Zhu, Z.; Liu, H. Macroscopic and microscopic experimental research on granite properties after high-temperature and water-cooling cycles. *Geothermics* **2021**, *93*, 102079. [CrossRef]
24. Xiao, P.; Zheng, J.; Dou, B.; Tian, H.; Cui, G.; Kashif, M. Mechanical behaviors of granite after thermal shock with different cooling rates. *Energies* **2021**, *14*, 3721. [CrossRef]
25. Wu, X.; Huang, Z.; Song, H.; Zhang, S.; Cheng, Z.; Li, R.; Wen, H.; Huang, P.; Dai, X. Variations of physical and mechanical properties of heated granite after rapid cooling with liquid nitrogen. *Rock Mech. Rock Eng.* **2019**, *52*, 2123–2139. [CrossRef]
26. Tarasov, B.G.; Stacey, T.R. Features of the energy balance and fragmentation mechanisms at spontaneous failure of class I and class II rocks. *Rock Mech. Rock Eng.* **2017**, *50*, 2563–2584. [CrossRef]
27. Wawersik, W.R.; Fairhurst, C. A study of brittle rock fracture in laboratory compression experiments. *Int. J. Rock Mech. Min. Sci. Geomech. Abstr.* **1970**, *7*, 561–575. [CrossRef]
28. Ma, L.; Wu, J.; Wang, M.; Dong, L.; Wei, H. Dynamic compressive properties of dry and saturated coral rocks at high strain rates. *Eng. Geol.* **2020**, *272*, 105615. [CrossRef]
29. Feng, P.; Xu, Y.; Dai, F. Effects of dynamic strain rate on the energy dissipation and fragment characteristics of cross-fissured rocks. *Int. J. Rock Mech. Min. Sci.* **2021**, *138*, 104600. [CrossRef]
30. Yan, Z.; Dai, F.; Liu, Y.; Du, H. Experimental investigations of the dynamic mechanical properties and fracturing behavior of cracked rocks under dynamic loading. *Bull. Eng. Geol. Environ.* **2020**, *79*, 5535–5552. [CrossRef]
31. Dai, B.; Shan, Q.; Chen, Y.; Luo, X. Mechanical and energy dissipation characteristics of granite under cyclic impact loading. *J. Cent. South Univ.* **2022**, *29*, 116–128. [CrossRef]
32. Huang, L.; Guo, Y.; Li, X. Mechanical response to dynamic compressive load applied to shale after thermal treatment. *J. Nat. Gas Sci. Eng.* **2022**, *102*, 104565. [CrossRef]
33. Wang, Z.L.; Shi, H.; Wang, J.G. Mechanical behavior and damage constitutive model of granite under coupling of temperature and dynamic loading. *Rock Mech. Rock Eng.* **2018**, *51*, 3045–3059. [CrossRef]
34. Yin, T.; Chen, Y.; Li, X.; Li, Q. Effect of high temperature and strain rate on the elastic modulus of rocks: A review. *Int. J. Earth Sci.* **2021**, *110*, 2639–2660. [CrossRef]
35. Li, X.B.; Lok, T.S.; Zhao, J. Dynamic characteristics of granite subjected to intermediate loading rate. *Rock Mech. Rock Eng.* **2005**, *38*, 21–39. [CrossRef]
36. Zhou, Y.X.; Xia, K.; Li, X.B.; Li, H.B.; Ma, G.W.; Zhao, J.; Zhou, Z.L.; Dai, F. Suggested methods for determining the dynamic strength parameters and mode-I fracture toughness of rock materials. *Int. J. Rock Mech. Min. Sci.* **2012**, *49*, 105–112. [CrossRef]
37. Yin, T.; Shu, R.; Li, X.; Wang, P.; Dong, L. Combined effects of temperature and axial pressure on dynamic mechanical properties of granite. *Trans. Nonferrous Met. Soc. China* **2016**, *26*, 2209–2219. [CrossRef]
38. Guo, Y.; Li, X.; Huang, L.; Liu, H.; Wu, Y. Effect of water-based working fluid imbibition on static and dynamic compressive properties of anisotropic shale. *J. Nat. Gas Sci. Eng.* **2021**, *95*, 104194. [CrossRef]
39. Huang, L.; Wang, J.; Momeni, A.; Wang, S. Spalling fracture mechanism of granite subjected to dynamic tensile loading. *Trans. Nonferrous Met. Soc. China* **2021**, *31*, 2116–2127. [CrossRef]

40. Li, X.B.; Lok, T.S.; Zhao, J.; Zhao, P.J. Oscillation elimination in the Hopkinson bar apparatus and resultant complete dynamic stress–strain curves for rocks. *Int. J. Rock Mech. Min. Sci.* **2000**, *37*, 1055–1060. [CrossRef]
41. Liu, X.; Liu, Z.; Li, X.; Gong, F.; Du, K. Experimental study on the effect of strain rate on rock acoustic emission characteristics. *Int. J. Rock Mech. Min. Sci.* **2020**, *133*, 104420. [CrossRef]
42. Jin, P.; Hu, Y.; Shao, J.; Zhao, G.; Zhu, X.; Li, C. Influence of different thermal cycling treatments on the physical, mechanical and transport properties of granite. *Geothermics* **2019**, *78*, 118–128. [CrossRef]
43. Inserra, C.; Biwa, S.; Chen, Y. Influence of thermal damage on linear and nonlinear acoustic properties of granite. *Int. J. Rock Mech. Min. Sci.* **2013**, *62*, 96–104. [CrossRef]
44. Sun, Q.; Zhang, W.; Xue, L.; Zhang, Z.; Su, T. Thermal damage pattern and thresholds of granite. *Environ. Earth Sci.* **2015**, *74*, 2341–2349. [CrossRef]
45. Ohno, I. Temperature variation of elastic properties of α-quartz up to the α-β transition. *J. Phys. Earth* **1995**, *43*, 157–169. [CrossRef]
46. Li, Q.; Li, X.; Yin, T. Factors affecting pore structure of granite under cyclic heating and cooling: A nuclear magnetic resonance investigation. *Geothermics* **2021**, *96*, 102198. [CrossRef]
47. Tedesco, J.W.; Ross, C.A. Strain-rate-dependent constitutive equations for concrete. *J. Press. Vessel Technol.* **1998**, *120*, 398–405. [CrossRef]
48. Li, Q.M.; Meng, H. About the dynamic strength enhancement of concrete-like materials in a split Hopkinson pressure bar test. *Int. J. Solids Struct.* **2003**, *40*, 343–360. [CrossRef]
49. Yin, T.; Shu, R.; Li, X.; Wang, P.; Liu, X. Comparison of mechanical properties in high temperature and thermal treatment granite. *Trans. Nonferrous Met. Soc. China* **2016**, *26*, 1926–1937. [CrossRef]
50. Li, X.B. *Rock Dynamics Fundamentals and Applications*; Science Press: Beijing, China, 2014.

mathematics

Article

Reliability Modelling of Pipeline Failure under the Impact of Submarine Slides-Copula Method

Laifu Song [1,2], Hao Ying [2], Wei Wang [2], Ning Fan [2,*] and Xueming Du [3]

1 State Key Laboratory of Water Resources and Hydropower Engineering Science, Wuhan University, Wuhan 430072, China; songlaifu_jia@wzu.edu.cn
2 College of Civil Engineering and Architecture, Wenzhou University, Wenzhou 325035, China; wzuyinghao@163.com (H.Y.); wzuwangwei@yeah.net (W.W.)
3 School of Water Conservancy Engineering, Zhengzhou University, Zhengzhou 450001, China; 2007-dxm@163.com
* Correspondence: fanning@wzu.edu.cn

Abstract: The instability of seabed slope sediments is the main factor influencing the safety of marine resource development. Therefore, to ensure the safe operation of submarine pipelines under complex and uncertain seabed rock and soil conditions, a reliability model was developed to elucidate the trend of impact-related pipeline damage due to submarine slides. Then, a risk assessment of the damage process of submarine slides impacting pipelines was conducted, which is of great significance for the in-depth safety assessment of pipelines impacted by submarine slides. Based on the copula function, a joint probability distribution model considering the correlation among risk variables was established for rational correlation characterization. A probability analysis method of impact-related pipeline damage attributed to submarine slides based on the copula function was proposed. The Monte Carlo simulation (MCS) method was employed to simulate the random uncertainty in limited observation values and accurately determine the reliability of safe pipeline operation under the action of submarine slides. The conclusions were as follows: (1) Based on the copula function, a joint probability distribution model of risk variables with any marginal distribution function and related structure could be developed. (2) The copula function could reasonably characterize relevant nonnormal distribution characteristics of risk variables and could simulate samples conforming to the distribution pattern of the risk variables. (3) The failure probability calculated with the traditional independent normal distribution model was very low, which could result in a notable overestimation of the reliability of submarine pipelines.

Keywords: submarine slides; submarine pipelines; copula function; reliability; slide–pipeline interaction

MSC: 76-10

Citation: Song, L.; Ying, H.; Wang, W.; Fan, N.; Du, X. Reliability Modelling of Pipeline Failure under the Impact of Submarine Slides-Copula Method. *Mathematics* **2022**, *10*, 1382. https://doi.org/10.3390/math10091382

Academic Editors: Shaofeng Wang, Jian Zhou, Xin Cai, Xiaofeng Li and Zhengyang Song

Received: 26 February 2022
Accepted: 18 April 2022
Published: 20 April 2022

1. Introduction

With the continuous progress in development and exploration technology, the exploitation of oil and natural gas has been increasingly promoted from land to sea. In recent years, the number of submarine pipelines has significantly increased, and the exploitation of offshore oil and gas resources has become a new field of oil and gas exploration worldwide. However, offshore oil and gas development faces more risks and challenges than those associated with onshore oil and gas development. Among these issues, the most important problem affecting the safety of marine resource exploitation is the instability of submarine slope sediments. Under the action of earthquakes and faults, gas hydrate disassociation, waves and currents, rock and soil masses, and sediments are susceptible to sliding, thereby forming submarine slides that could impact submarine pipelines [1,2]. Hance [3] noted that a submarine slide can be characterized by a large volume, large distance, and high speed. For example, the maximum value could reach approximately 20,331 km^3, the maximum sliding distance could reach 850 km, and the maximum speed could reach 10,100 km/h. Submarine slides could easily impact submarine pipelines, destroy oil and gas transportation pipelines and exploitation facilities, and threaten the safety of offshore oil and gas development [4,5]. Therefore, to ensure the safe operation of submarine pipelines located on the ocean floor under highly complex and uncertain rock–soil mass conditions, a reliability model was constructed to elucidate the pattern of pipeline impact damage due to submarine slides, after which a risk assessment of the damage process of submarine slides impacting pipelines was conducted, which can provide an important theoretical basis for an in-depth safety evaluation of the effect of submarine slides on pipelines.

Since the 1970s, many scholars in China and abroad have studied the impact force of submarine slides exerted on pipelines, such as Demars [6], Randolph and Houlsby [7], and Zakeri and Hawlader [8], and various bearing capacity calculation equations have been proposed based on geotechnical mechanics theory. The impact force of a landslide was effectively combined with the rock and soil mass strength. Pazwash and Robertson [9], Chehata et al. [10], and Liu et al. [11] regarded the landslide mass as a fluid and analysed the impact force of submarine slides based on classical cylindrical fluid flow theory. Randolph and White [12], Dong [13], Dutta et al. [14], and Fan et al. [15] systematically analysed the impact force of submarine slides on pipelines by combining the theory of rock and soil mechanics with fluid mechanics theory. Although achievements have been made in recent years, the attained progress has promoted the development of submarine landslide–pipeline interaction research to a certain extent. However, the marine environment of submarine slides is complex and changeable. Compared to landsides, many uncertain factors exist, and the formation location is difficult to determine. In addition, field monitoring and sampling are extremely difficult and costly operations, and the obtained test data are limited, which further increases the uncertainty in seabed rock and soil mass parameters. Under the condition of limited data and a large number of uncertain factors, it remains difficult to apply the traditional analysis method to analyse the damage impact of submarine slides exerted on pipelines. Reliability theory can quantitatively consider multiple uncertain factors in a scientific and reasonable manner and can effectively overcome the limitation of a single index for structural safety evaluation. This approach has received increasing attention in the field of civil engineering and has been widely employed in structural safety design and analysis in bridge, structural, and other engineering fields [16–19]. However, in the marine engineering field, reliability research involving the impact of submarine slides on pipelines remains lacking. Therefore, the combination of reliability analysis theory and a method to evaluate the safety of pipelines under the damage impact of submarine slides could provide an important theoretical and scientific basis for disaster prevention and a mitigation design of marine energy exploitation systems.

The vertical force of submarine slides exerted on pipelines constitutes one of the important indexes used to evaluate the safety of submarine oil and gas pipelines. The vertical force fluctuates, which could pose a potential resonance risk to a given pipeline. When there exists a narrow gap between the pipeline and seabed, the load fluctuation magnitude is large, which matches that of the horizontal load. Moreover, the vertical force component could affect the pipeline vertical position, causing horizontal force fluctuations [20]. Fan et al. [21,22] found, through physical model tests and a large number of numerical calculations, that the impact force of submarine slides exerted on pipelines can be divided into two mechanical stages, including the instantaneous stage (the peak impact force is considered to represent the destructive effect of submarine slides) and the stable stage (the steady impact force is considered to represent the continuous effect of submarine slides on pipelines), and there exists a certain correlation between these two types of destructive forces. Through numerical analysis, a prediction model of the impact-related damage of pipelines due to submarine slides could be established, which could provide a reference for the safety and disaster prevention design of submarine pipelines.

However, the correlation between two or more risk factors has seldom been considered in numerical analysis methods, and risk variables are not simple and isolated quantities in complex geological environments. Moreover, if the correlation among risk variables is ignored and variables are assumed to follow independent normal distributions to simplify the research problem, the obtained reliability analysis results cannot truly reflect the impact of the correlation among the considered risk variables. Therefore, it is necessary to establish a reasonable joint probability distribution model of the risk variables for reliability analysis improvement. However, due to the high uncertainty in the submarine geological environment, there exists no unified opinion on the expression of the relationship between the peak impact force of submarine slides on pipelines and the steady impact force. At present, the primary task involves the urgent introduction of an accurate and reliable method to characterize the relationship between risk variables and construct the optimal joint probability distribution function in a scientific, reasonable, and comprehensive manner. In recent years, with the deepening of mathematical theory, the development of copula theory has provided a new method to establish joint probability distribution models of relevant nonnormal variables. The core idea of copula theory entails the separate construction of the marginal distribution function and copula function. There exists an unlimited variety of marginal distribution functions and corresponding structural types, and varied joint distribution models under arbitrary combinations can be established within the framework of copula theory. Copula theory has been widely applied in finance [23,24], hydrology [25,26], ecological sciences [27,28], reliability analysis [29,30], geotechnical engineering, and other fields [31–36] due to its incomparable flexibility and applicability in the establishment of a joint distribution model of variables. Currently, no study has fully considered the correlation among risk variables or has proposed a framework to analyse the reliability of safe pipeline operation under the action of submarine slides.

In summary, modelling was performed while considering the correlation among risk variables under the condition of incomplete probability information, the reliability of observations was evaluated through visualization methods, and the accuracy of the obtained reliability evaluation results was ensured. This paper fully considered two risk variables (peak and stable vertical force values) in evaluating the importance of submarine slides to pipeline safety. Based on the copula function, a joint probability distribution model considering the correlation among risk variables was established for reasonable correlation characterization. A probability analysis method for pipeline impact-related damage due to submarine slides based on the copula function was proposed. The Monte Carlo simulation (MCS) method was used to simulate the random uncertainty in limited observation values, and combined with big data analysis and visualization technology, the reliability of safe pipeline operation under the action of submarine slides was accurately analysed.

2. Joint Distribution Model of the Risk Variables Based on the Copula Function

In 1959, Sklar [37] first proposed the copula function and suggested that any multidimensional joint distribution function could be divided into a copula function and a corresponding number of marginal distribution functions. The copula function represents the correlation among variables (including the correlation coefficient and correlation structure). Its essence is the bridge function connecting the marginal distribution function of the variables with the joint distribution function and is often referred to as the combination function, bond function, or connection function. For a detailed theoretical introduction to copulas, please refer to Joe [38], Durante and Sempi [39], and Salvadori and De Michele [40].

The basic concept of the copula function is as follows: under N-dimensional conditions, the copula function can be defined as an N-dimensional joint distribution function with the marginal distribution function in $[0,1]^N$ space uniformly distributed in [0,1]. The Sklar theorem [37] can be described as follows: let $F(x_1, x_2, \ldots, x_N)$ be an N-dimensional joint distribution function with N marginal distribution functions $F_1(x_1), F_2(x_2), \ldots, F_N(x_N)$. Then, there exists a copula function $C(u_1, u_2, \ldots, u_N)$ connecting the marginal distribution function $F_1(x_1), F_2(x_2), \ldots, F_N(x_N)$ and the joint probability distribution function $F(x_1, x_2, \ldots, x_N)$.

Considering the Sklar theorem and a bivariate distribution [37], the joint distribution function comprises two parts: the distribution function of the variables and the copula function characterizing these variables. If $H(x,y)$ is the joint distribution function with marginal distribution functions $G(x)$ and $Q(y)$, a copula function must exist. For any variables x and y, the copula function satisfies the following:

$$H(x,y) = C(G(x), Q(y); \theta) \tag{1}$$

where $G(x)$ and $Q(y)$ are the distribution functions of the risk variables, and θ denotes the parameters of the copula function.

If the marginal distributions $G(x)$ and $Q(y)$ are continuous, the copula function C is unique, and the joint probability density function can then be given as:

$$h_{X,Y}(x,y) = D(G(x), Q(y); \theta) g_X(x) q_Y(y) \tag{2}$$

where $g_X(x)$ and $q_Y(y)$ are the density functions of $G(x)$ and $Q(y)$, respectively, and $D(G(x), Q(y); \theta)$ is the density function of $C(G(x), Q(y); \theta)$.

The process of constructing a multivariate joint probability distribution function based on the copula function can be divided into two steps: (1) the marginal distribution function of the variables is determined based on the original data; and (2) the type of copula function properly representing the correlation among the variables is selected. These two steps are independent. Thus, the advantage of establishing a joint probability distribution model based on the copula function is that a normal or nonnormal distribution separates the marginal distribution from related structures, which can overcome the limitations of the traditional model. Moreover, a joint probability distribution model can be constructed with an arbitrary marginal distribution function and the related structure type of the joint probability distribution function, which can reveal the internal regularity of the original data.

3. Reliability Modelling Method of Pipeline Impact-Related Damage Due to Submarine Slides

To solve the problem whereby the traditional numerical analysis approach based on a single safety evaluation index cannot consider the limitations of various uncertain factors, and to reasonably characterize the correlation among the risk variables of pipeline impact-related damage due to submarine slides, a method based on the copula function was proposed to determine the correlation between submarine slides and pipeline damage risk variables and to evaluate the reliability of pipeline safety. Figure 1 shows a flowchart of the proposed method, which comprises three main steps, as described below.

Step 1: Determination of the optimal marginal distribution function of the risk variables.

The marginal distribution function can accurately describe the probability distribution of the variables. The primary task of the establishment of a joint probability distribution model based on the copula function entails the determination of the optimal marginal distribution function types of the variables. Since the destructive impact force of submarine slides exerted on pipelines is positive, this paper selected five marginal distribution functions commonly considered in engineering, namely, the normal distribution, log-normal distribution, truncated extremum type I distribution, Weibull distribution, and gamma distribution. Table 1 lists the various probability density functions and cumulative distribution functions of the different distribution types, where μ is the mean value and σ is the standard deviation.

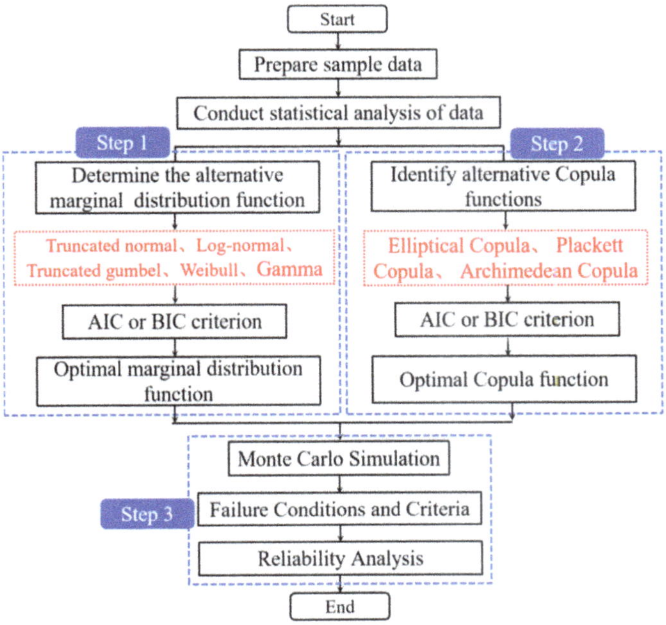

Figure 1. Flowchart of the developed reliability analysis approach

Table 1. Five alternative marginal distribution functions.

Distribution Type	Probability Distribution Function	Probability Density Function	Note
Truncated normal	$\left[\Phi\left(\frac{x-p}{q}\right) - \Phi\left(\frac{0-p}{q}\right)\right] / \left[1 - \Phi\left(\frac{0-p}{q}\right)\right]$	$\varphi\left(\frac{x-p}{q}\right) / \left[1 - \Phi\left(\frac{0-p}{q}\right)\right]$	$p = \mu$ $q = \sigma$
Log-normal	$\Phi\left(\frac{\ln x - p}{q}\right)$	$\frac{1}{\sqrt{2\pi}qx} \exp\left[-\frac{1}{2}\left(\frac{\ln x - p}{q}\right)^2\right]$	$p = \ln \frac{\mu}{\sqrt{1+\sigma^2/\mu^2}}$ $q = \sqrt{\ln\left(1 + \frac{\sigma^2}{\mu^2}\right)}$
Truncated Gumbel	$\frac{\exp\{-\exp[-q(x-p)]\} - \exp[-\exp(pq)]}{1 - \exp[-\exp(pq)]}$	$\frac{q \exp\{-q(x-p) - \exp[-q(x-p)]\}}{1 - \exp[-\exp(pq)]}$	$\mu = p + \frac{0.5772}{q},\ \sigma^2 = \frac{\pi^2}{6q^2}$
Weibull	$1 - \exp\left[-\left(\frac{x}{p}\right)^q\right]$	$\frac{q}{p}\left(\frac{x}{p}\right)^{q-1} \exp\left[-\left(\frac{x}{p}\right)^q\right]$	$\mu = p\Gamma\left(1 + \frac{1}{q}\right)$ $\sigma^2 = p^2\left[\Gamma\left(1 + \frac{2}{q}\right) - \Gamma^2\left(1 + \frac{1}{q}\right)\right]$
Gamma	$\frac{1}{\Gamma(q)} \int_0^{px} t^{q-1} e^{-t} dt$	$\frac{p^q x^{q-1}}{\Gamma(q)} e^{-px}$	$\mu = \frac{q}{p} - \sigma^2 = \frac{c}{p^2}$

Notes: $\Phi()$ denotes the standard cumulative distribution function, $\phi()$ is the probability density function of the normal distribution, and $\Gamma()$ denotes the factorial.

In engineering, the Akaike information criterion (*AIC*) [41] and Bayesian information criterion (*BIC*) [42] are commonly adopted to determine the optimal marginal distribution function. These criteria require that the marginal distribution function with the lowest calculated *AIC* or *BIC* value is the optimal marginal distribution function of the fitting variable. The above two criteria are simple in principle, provide a suitable stability, can be easily implemented in calculations, are widely applied in engineering, and can facilitate accurate and reliable data fitting. Therefore, the optimal marginal distribution function of the risk variables was identified and determined with the above method, and the specific expressions are as follows:

$$AIC = -2\sum_{i=1}^{n} \ln f(x; p, q) + 2k_1 \tag{3}$$

$$BIC = -2\sum_{i=1}^{n} \ln f(x; p, q) + 2\ln n \tag{4}$$

where p and q are distribution parameters associated with μ and σ, respectively, n is the total number of data, and k_1 denotes the number of parameters of the alternative distribution types. Moreover, $f(x_i; p, q)$ is the probability density function of the alternative distribution types. Table 1 indicates that all five distribution types contain two distribution parameters. Therefore, $k_1 = 2$, which is the minimum value considered in the calculation results, determines the optimal distribution type.

Based on the above principles, the steps to determine the optimal marginal distribution function of the risk variables are as follows: (1) the mean and variance in the original data are calculated; (2) the parameters of the different types of marginal distribution functions are determined; (3) according to Equations (3) and (4), the optimal marginal distribution function of each risk variable is obtained based on the minimum calculated *AIC* or *BIC* value. When the alternative distribution types consider the same parameter samples, the identification results based on the *AIC* and *BIC* are the same. In this section, the *AIC* is used to identify the optimal edge distribution types.

Step 2: Determination of the optimal copula function fitting the correlation among the risk variables.

The correlation between parameters can be captured with the correlation coefficient and correlation structure type. In terms of the correlation coefficient, the Pearson linear correlation coefficient and Kendall rank correlation coefficient are mainly adopted. The Pearson linear correlation coefficient is an index used to measure the degree of linear correlation between the considered parameters. The Kendall rank correlation coefficient is based on the rank of the original parameter data and can describe the correlation between the parameters. The related structure types can be described according to the different copula functions. The θ parameter is the key to copula function determination and can be obtained based on the Pearson correlation coefficient and Kendall rank correlation coefficient θ [43]. According to the definition of the correlation coefficient, the relationship between parameter θ of the copula function and the Pearson correlation coefficient ρ is:

$$\rho = \int_{-\infty}^{\infty}\int_{-\infty}^{\infty}\left(\frac{x_1-\mu_1}{\sigma_1}\right)\left(\frac{x_2-\mu_2}{\sigma_2}\right) f_1(x_1)f_2(x_2)D(F_1(x_1), F_2(x_2); \theta)\mathrm{d}x_1\mathrm{d}x_2 \tag{5}$$

where μ_1 and μ_2 are the average values of the two variables x_1 and x_2, respectively, and σ_1 and σ_2, respectively, are the standard deviations of these two variables.

With the above equation, parameter θ of the copula function can be determined. However, except for the Gaussian copula function, most copula functions are difficult to solve via integration. According to a previously reported method in the literature [44,45], parameter θ of the Gaussian copula function can be computed through the Pearson correlation coefficient, as follows:

$$\rho = \int_{-\infty}^{\infty}\int_{-\infty}^{\infty}\left(\frac{u_1-\mu_1}{\sigma_1}\right)\left(\frac{u_2-\mu_2}{\sigma_2}\right)\frac{f_1(u_1)f_2(u_2)}{\sqrt{1-\theta^2}}\exp\left\{-\frac{\zeta_1{}^2\theta^2 - 2\theta\zeta_1\zeta_2 + \zeta_2{}^2\theta^2}{2(1-\theta^2)}\right\}\mathrm{d}u_1\mathrm{d}u_2 \tag{6}$$

where $\zeta_1 = \Phi^{-1}(u_1)$ and $\zeta_2 = \Phi^{-1}(u_2)$ are variables of the standard normal distribution. $\Phi()$ is the standard normal distribution function, and $\Phi^{-1}()$ is the inverse of the standard normal distribution function.

After parameter θ of the Gaussian copula function has been obtained, the Kendall rank correlation coefficient τ can be calculated with the following equation:

$$\tau = \frac{2\arcsin(\theta)}{\pi} \tag{7}$$

Finally, the following equation can be employed to determine the parameter θ values of the different copula functions:

$$\tau = 4\int_0^1 \int_0^1 C(u_1, u_2; \theta) dC(u_1, u_2; \theta) - 1 \tag{8}$$

In copula theory, many copula functions are available [43] that can be employed to describe variable correlation structure types, but most types can only facilitate simulations within a limited range of correlation coefficient values [46], e.g., elliptic copula functions such as the Gaussian copula function and t copula function, the Plackett copula function, and Archimedean copula functions, such as the Frank, Clayton, CClayton, Gumble, No. 16 and No. 17 copula functions.

The copula function couples the marginal distribution function and joint distribution function. After determination of the optimal marginal distribution function, the optimal copula function can be obtained based on the *AIC* and *BIC*. The corresponding calculation methods are expressed as Equations (9) and (10), respectively. Considering the above principles, the process of optimal copula function fitting is as follows: (1) the Pearson correlation coefficient and Kendall rank correlation coefficient are calculated; (2) parameter θ of the copula function is determined according to Equation (8); (3) *AIC* and *BIC* values can be computed with Equations (9) and (10), respectively, to determine the optimal copula function.

$$AIC = -2\sum_{i=1}^{n} \ln D(u_i, v_i; \theta) + 2k_2 \tag{9}$$

$$BIC = -2\sum_{i=1}^{n} \ln D(u_i, v_i; \theta) + k_2 \ln n \tag{10}$$

where $(u_i, v_i), i = 1, 2, \ldots, n$ denotes the test data of the parameters, n is the total number of data, $D(u_i, v_i; \theta)$ is the probability density function of the alternative copula function, k_2 is the number of parameters of the alternative copula function, and the minimum value in the calculation results determines the optimal copula function. The evaluation criterion is the same as that in step 2, and the *AIC* can be used to determine the optimal copula function.

Step 3: Reliability analysis based on simulations.

In practical engineering, due to the limitations of engineering technology and economic conditions, the available test and measurement data are very limited, and the joint probability distribution function of variables, which requires complete probability information, cannot be obtained. Only the marginal distribution function and correlation coefficient of the considered variables can be determined under limited data conditions, i.e., incomplete probability information. Especially in ocean engineering, it is very difficult to collect a large amount of high-quality test data. The MCS method can randomly generate sufficient samples based on the characteristics of the original data and has become a robust statistical tool [47]. This method is widely applied in probability analysis and provides an important technical means for the reliability analysis of submarine slide-impacted pipelines. Reliability analysis based on the MCS method directly calculates the failure probability by combining random simulation and statistical tests, and the calculation equation is as follows:

$$P_f = \frac{n_f}{n} \tag{11}$$

where n_f is the number of samples in the failure domain and n is the total number of samples. The method exhibits a clear concept, simple application, and few limitations. With an increasing sample number, the calculation results become increasingly accurate and reliable. Under the condition of extremely limited data, the joint probability model of the considered risk variables was established based on the copula function, and relevant variables were effectively simulated with the MCS method. The corresponding probability was determined according to a large amount of simulation data. For example, Wang and Kulhawy [48] analysed the reliability of the normal service state and limit state of a given structure based on the MCS method. Aladejare and Wang [49] adopted the MCS method to generate a large amount of rock strength data and combined this method with correlation analysis to evaluate the reliability of rock slope stability. Pan et al. [50] examined the reliability of the tunnel driving face with the MCS method. Therefore, this study aimed to establish a joint probability model of the relevant risk variables based on the copula function and employed the MCS method to simulate a large amount of simulation data to approximate the actual variable distribution characteristics and process the limited available observation data.

4. Failure Probability Estimation under Submarine Slide-Induced Pipeline Impact Damage

In the structural system of civil engineering, Equations (12) and (13) can be applied to calculate the failure probability P_f. It can be considered that $Y(Q)$ denotes the minimum limit value, $Y - Y(Q) < 0$ is the failure condition, and the failure probability can be calculated with Equation (12). Alternatively, $Y(Q)$ denotes the maximum limit value, and $Y - Y(Q) > 0$ is the failure condition. The failure probability can be calculated according to Equation (13).

$$P_f = P[Y - Y(Q) < 0] \tag{12}$$

where $Y(Q)$ denotes the minimum limit value and Y is the actual observed value.

$$P_f = P[Y - Y(Q) > 0] \tag{13}$$

where $Y(Q)$ is the maximum limit value and Y denotes the actual observed value.

Compared to one-way flow through a cylinder, the process of submarine slide impacting pipelines is more complicated, and there are many risk factors. Determination of the risk level posed by submarine slides to pipelines enables the safety control of submarine pipelines. In this study, the computational fluid dynamics (CFD) method was adopted to simulate a series of multiphase flows, and peak and stable values of the vertical sliding pipe forces were determined under different landslide velocity conditions. Considering the peak and stable values of vertical forces, the failure probability of submarine slides impacting pipelines can be expressed as:

$$P_f = P[G(X_1) < 0 \cup G(X_2) < 0] \tag{14}$$

where $G(X_1) = X_1 - \overline{X}_1$, $G(X_2) = X_2 - \overline{X}_2$, X_1, and X_2 are the actual peak and stable values, respectively, and $\overline{X}_1$ and $\overline{X}_2$ are the maximum limits of the peak and stable values, respectively. For $G(X_1) > 0$ or $G(X_2) > 0$, the submarine pipeline is considered to be invalid. By establishing the joint probability distribution model of the peak and stable values considering different related structure types, the varied failure probabilities of pipelines under the impact of submarine slides can be determined.

5. Case Study

5.1. Numerical Model

Induced by earthquakes, sedimentation, hydrate decomposition, and waves, a given submarine slope first becomes unstable, and the landslide mass begins to slide upon detachment from the unstable area. Under the influence of the water environment, the landslide mass gradually changes into a flow and continues to slide across a certain distance before stopping. In the whole landslide flow process, the flow velocity of the landslide mass gradually increases, reaching a maximum of up to 30 m/s [51]. Compared to the trigger start-up stage, the strength of the landslide mass at the flow slide stage is lower and the function rate is higher, which can significantly impact submarine pipelines. Therefore, based on the flow slip stage of submarine slides, this study employed the commercial CFD platform ANSYS-CFX to conduct numerical simulations of the impact of submarine slides on pipelines. The pre- and postprocessing tools required for CFD simulations include a

3D model, meshing scheme (ICEM-CFD), and a CFX solver based on the finite volume (FV) method. The CFD numerical method is very useful in fluid–structure interaction analysis. This study employed CFD software ANSYS 14.5 (CFX 2010a: CFX solver models; CFX program (version 13.0) physical modelling documentation, Canonsburg, PA, USA; ANSYS Inc. 2010b: CFX solver theory, CFX program (version 13.0) theory documentation, Canonsburg, PA, USA), which is a general-purpose CFD program including a solver based on the FV method for unstructured grids. A Euler–Euler multiphase flow model with nonuniform two-phase separation was applied to simulate the interaction between the submarine slide mass and seawater.

The established submarine slide–pipeline interaction numerical model is shown in Figure 2. The numerical calculation domain exhibits dimensions of 15.5 m × 8 m × 0.5 m (height × width × thickness), the pipeline diameter D_{pipe} is 0.5 m, and the distance from the horizontal entrance is 2.5 m ($5D_{\mathrm{pipe}}$). Moreover, the gap between the pipeline and seabed is H_{ps}, and the landslide mass enters from a height of 10.5 m. The pipeline was considered fixed (or pipeline position variation could be ignored). The centre of the pipe is 2.75 m away from the inlet. The grid adopts tetrahedral elements, and the maximum mesh size depends on the pipe diameter. The maximum mesh size is $0.5D_{\mathrm{pipe}}$, and the number of elements in numerical analysis exceeds 270,000. The grid in the area within a radius of 0.5–0.75 m ($1.5D_{\mathrm{pipe}}$) was refined, and five layers of refined grids (with a total thickness of 0.05 m) were set up near the pipeline. The inlet was set as a velocity boundary, and the outlet was defined as an open boundary. The top and upper boundaries of the entrance were set as free-slip surfaces, and the surfaces of the pipeline and seabed were rough, each with an equivalent roughness k_s of 0.0015 mm. The submarine slide flow was assumed to involve continuous free surface flow considering buoyancy and was simulated as an incompressible two-phase flow. All high-speed water and slide flow motions were determined based on the extended standard k-ε turbulence model. The landslide entrance was defined as the velocity boundary, the exit was set as an open boundary, the top of the computational domain was established as a free-slip boundary, the bottom and pipe surface were defined as rough no-slip boundaries, and the surface equivalent roughness k_s values were 0.5 and 0.0015 mm, respectively. When the sliding distance reached 48 m ($96D_{\mathrm{pipe}}$), the simulation calculation was terminated, and the calculation process adopted the second-order, high-precision upwind difference format. By varying the flow velocity and Reynolds number, the peak and stable vertical force values of 67 groups of submarine slides impacting pipelines were obtained.

Due to the complex marine environment and numerous factors influencing the stability of submarine slopes, submarine slides have become high-frequency geological disasters with a wide impact and potential threats. The highly notable impact produced seriously threatens the stability and safety of submarine pipelines. Therefore, it is of great significance to effectively simulate the impact of submarine slides on pipelines, reasonably determine the impact force of submarine slides exerted on pipelines, especially the vertical impact force must be improved, and study the correlation between the peak and stable vertical force values of submarine pipelines to accomplish more convincing safety and reliability evaluations of submarine pipeline projects.

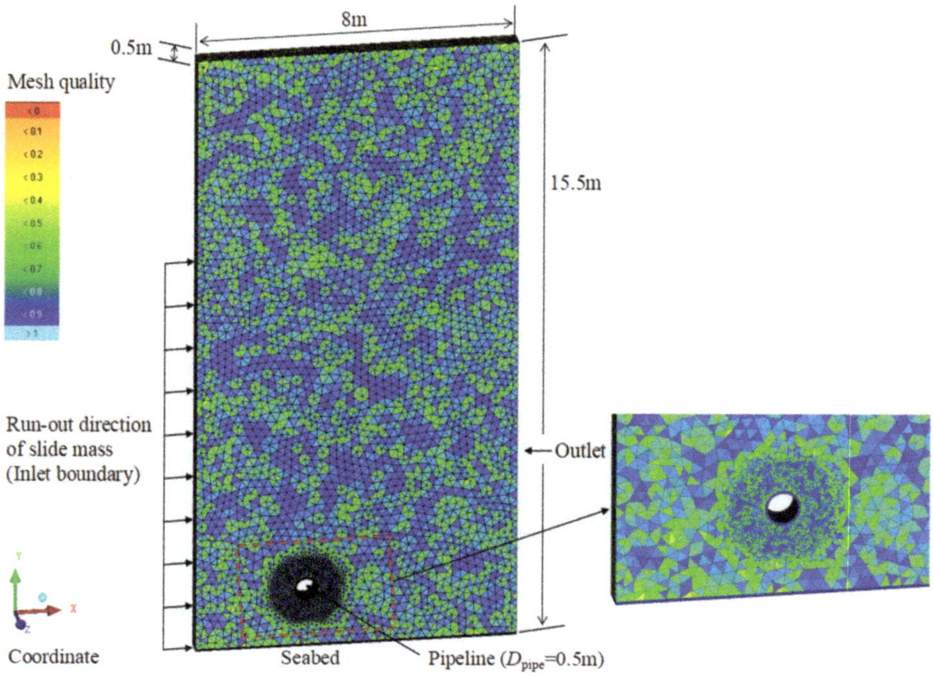

Figure 2. Numerical computational model.

5.2. Joint Distribution Model of the Risk Variables

Through calculation, 67 sets of data samples of peak and stable vertical force values were obtained (linear correlation coefficient R = 0.8004). Calculated peak and stable vertical force values over time, considering a flow rate of V = 1.0 m/s and Reynolds number of $Re_{\text{non-Newtonian}}$ = 8.7, are shown in Figure 3. Based on the original data scatter plot depicted in Figure 4a, the peak and stable vertical force values exhibited a lower tail correlation and were linearly positively correlated, but the linear relationship was not sufficiently obvious. The original sample data could be converted into uniformly distributed data with the semiparametric method based on maximum likelihood estimation, and the calculation process is expressed in Equation (14).

$$\begin{cases} u_i = \frac{Rank(x_i)}{N+1} \\ v_i = \frac{Rank(y_i)}{N+1} \end{cases} i = 1, 2, \ldots, N \quad (15)$$

where (x_i, y_i) is the original data sample value, *Rank* is a sorting function, which can be used to arrange the original sample data in ascending order, and (u_i, v_i) is a standard uniformly distributed random variable after transformation.

Figure 4b shows the standard uniformly distributed random variable after transformation. Compared to Figure 4a, Figure 4b reveals a more obvious linear positive correlation. Therefore, the candidate copula function selected in this study should be symmetric and must provide a good ability to describe the positive correlation structure of the random variables.

Many types of two-dimensional copula functions exist. The common two-dimensional copula functions can be divided into three types: (1) Gaussian copula functions; (2) two-dimensional Plackett copula function; (3) two-dimensional Archimedean copula functions (for example, the Frank, Clayton, CClayton, No. 16. and No. 17 copula functions). To select the optimal copula function capturing the correlation among the risk variables, a copula function with a similar correlation structure to that of the measured data is usually selected in advance as the alternative copula function. Therefore, the copula function type selected in this paper can not only capture all copula function types, but can also capture top- and bottom-tail correlations to comprehensively analyse the original data and determine all possible correlation distribution types of the original data. The five considered alternative copula functions, i.e., the Gaussian, Plackett, Frank, CClayton, and No. 16 copula functions, are classic copula function families, and these functions can suitably describe both positive and negative correlations among variables. The absolute values of the correlation coefficients of these five copula functions all approached 1, which can meet the requirements depicted in Figure 4. Details of these five copula functions are summarized in Table 2.

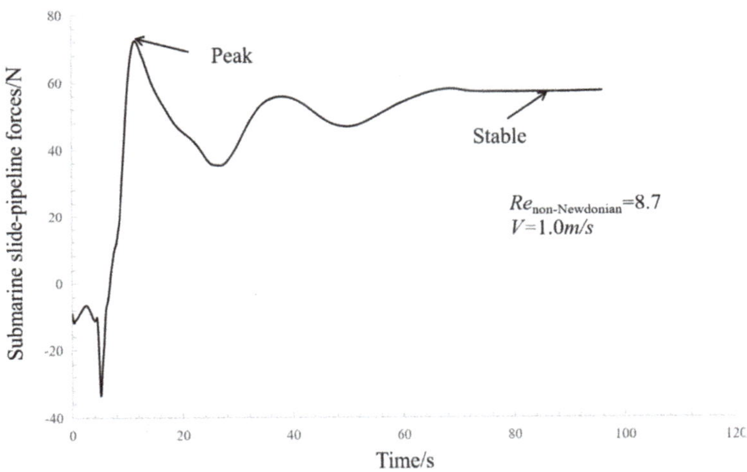

Figure 3. Variation in the vertical slide–pipeline forces over time.

Table 2. Five types of 2D copula functions.

Copula Function Type	Copula Distribution Function $C(u_1,u_2;\theta)$	Copula Density Function $D(u_1,u_2;\theta)$	$\varphi_\theta(t,\theta)$
Gaussian	$\Phi_\theta\left(\Phi^{-1}(u_1),\Phi^{-1}(u_2);\theta\right)$	$\frac{\varphi_2\left(\Phi^{-1}(u_1),\Phi^{-1}(u_2);\theta\right)}{\varphi\left(\Phi^{-1}(u_1)\right)\varphi\left(\Phi^{-1}(u_2)\right)}$	/
Plackett	$\frac{S-\sqrt{S^2-4u_1u_2\theta(\theta-1)}}{2(\theta-1)}$; $S=1+(\theta-1)(u_1+u_2)$	$\frac{\theta[1+(\theta-1)(u_1+u_2-2u_1u_2)]}{\left\{[1+(\theta-1)(u_1+u_2)]^2-4u_1u_2\theta(\theta-1)\right\}^{3/2}}$	/
Frank	$-\frac{1}{\theta}\ln\left[1+\frac{\left(e^{-\theta u_1}-1\right)\left(e^{-\theta u_2}-1\right)}{e^{-\theta}-1}\right]$	$\frac{-\theta\left(e^{-\theta}-1\right)e^{-\theta(u_1+u_2)}}{\left[\left(e^{-\theta}-1\right)+\left(e^{-\theta u_1}-1\right)\left(e^{-\theta u_2}-1\right)\right]^2}$	$-\ln\left[\frac{e^{-\theta t}-1}{e^{-\theta}-1}\right]$
CClayton	$u_1+u_2-1+\left(W_1^{-\theta}+W_2^{-\theta}-1\right)^{-1/\theta}$; $W_i^{-\theta}=1-u_i$	$(1+\theta)(W_1W_2)^{-\theta-1}\left(W_1^{-\theta}+W_2^{-\theta}-1\right)^{-2-1/\theta}$; $W_i=1-u_i$	$\frac{1}{\theta}\left(t^{-\theta}-1\right)$
No. 16	$\frac{1}{2}\left(S+\sqrt{S^2+4\theta}\right)$, $S=u_1+u_2-1-\theta\left(\frac{1}{u_1}+\frac{1}{u_2}-1\right)$	$\frac{1}{2}\left(1+\frac{\theta}{u_1{}^2}\right)\left(1+\frac{\theta}{u_2{}^2}\right)S^{-0.5}\left\{-S^{-1}\left[u_1+u_2-1-\theta\left(\frac{1}{u_1}+\frac{1}{u_2}-1\right)\right]^2+1\right\}$ $S=\left[u_1+u_2-1-\theta\left(\frac{1}{u_1}+\frac{1}{u_2}-1\right)\right]^2+4\theta$	$\left(\frac{\theta}{t}+1\right)(1-t)$

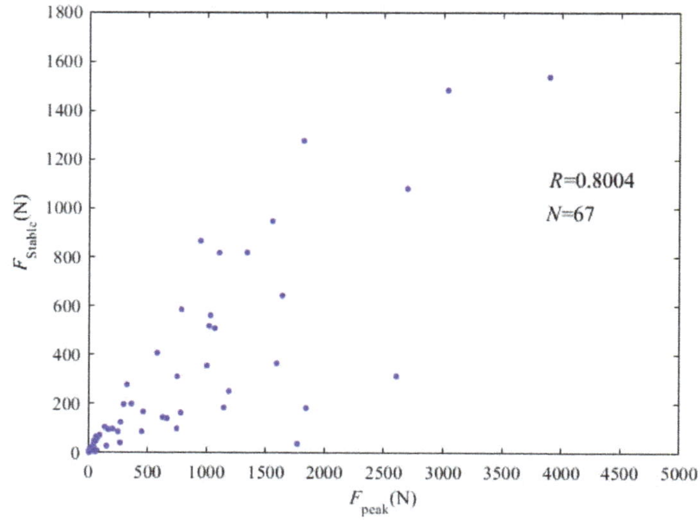

(**a**) Raw data scatter plot

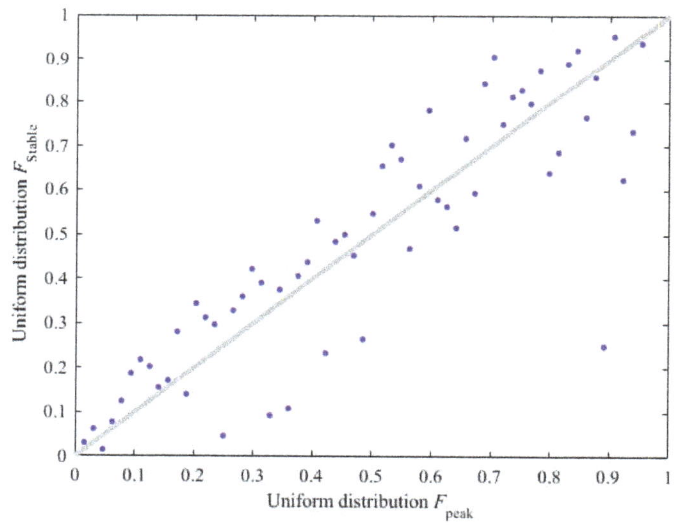

(**b**) Uniformly distributed data scatter plot

Figure 4. Raw data scatter plot (**a**) and scatter plot of the uniformly distributed data (**b**).

In the third section, a method was introduced to determine the optimal marginal distribution function and copula function. Data statistics constituted the basis for the determination of the optimal marginal distribution function, and an important theoretical basis was provided to analyse data distribution characteristics. Table 3 lists statistical information on the peak and stable vertical force values of the 67 groups of submarine slides impacting pipelines. As indicated in Table 3, the mean value of the peak forces was 2.65 times that of the stable forces, and the standard deviation was 2.25 times that of the stable forces. The fluctuation range was larger than that of the stable forces, and the variation coefficient value of the peak forces was much lower than that of the stable forces.

Therefore, the change in the stable forces was significantly greater than that in the peak forces.

Table 3. Risk variable statistics (units: N).

Parameter	Mean	Standard Deviation	Maximum	Minimum	Coefficient of Variability
Vertical forces (peak)	668.21	822.38	3894	6	1.23
Vertical forces (stable)	251.88	366.29	1539	2.7	1.45

Notes: Coefficient of variability = standard deviation/mean.

5.2.1. Optimal Marginal Distribution Function

The optimal marginal distribution function was determined by comparing five marginal distribution functions (truncated normal, log-normal, truncated Gumbel, Weibull, and gamma distribution functions). Equations (3) and (4) were applied to determine the optimal marginal distribution function, and the results are provided in Table 4. Table 4 reveals that the optimal marginal distribution function for both peak and stable variable fitting was the Weibull distribution. To understand the fit between the original data sample and the marginal distribution function more intuitively, Figure 5 shows a histogram of the original data and the five considered marginal distribution functions. It is evident that the shape of the optimal marginal distribution function was better than that of the other marginal distribution functions, which indicates a better agreement with the distribution characteristics of the risk variables. The results in Figure 6 are consistent with those provided in Table 4, confirming the effectiveness of the *AIC* in determining the optimal marginal distribution function via fitting.

Table 4. Calculation results for the optimal marginal distribution function of the risk variables.

Parameter		Truncated Normal	Log-Normal	Truncated Gumbel	Weibull	Gamma
Vertical forces (peak)	*AIC*	1231.44	1141.46	1180.91	**1100.72**	1117.02
Vertical forces (stable)	*AIC*	1073.21	1016.71	1029.85	**953.69**	1117.15

To further verify the fitting effect of the marginal distribution function, the Kolmogorov–Smirnov (K-S) method was implemented to assess the fitting degree of the alternative marginal distribution function to the sample data. The K-S test is a probability distribution type test method suitable for small sample data sizes. By measuring the distance D between the known hypothesis probability distribution and the empirical distribution of the measured data, this method evaluates whether the distance occurs within the confidence interval [52]. The specific process of the K-S test method can be summarized as follows: let $A_1(x)$ denote the theoretical distribution function assumed in advance, while $A_2(x)$ denotes the actual cumulative distribution function of sample group A. Moreover, D is the maximum value of the gap between $A_1(x)$ and $A_2(x)$, i.e., $D = \max|A_1(x) - A_2(x)|$. For $D \geq D_{n,\alpha}$ ($D_{n,\alpha}$ is the rejection threshold), the original hypothesis can be rejected, and, conversely, the original hypothesis can be accepted.

Table 5 summarizes the K-S test results for the risk variables (the peak and stable vertical force values). In regard to the risk variable of the peak vertical forces, the D value of the Weibull distribution is 0.0168, and, compared to the other marginal distributions, the D value is the smallest, i.e., the K-S distance is the smallest. According to the basic principle of the K-S test method, when the D value is smaller, it indicates that the two distributions are very similar and that the fitting degree is high, which further verifies the rationality of the Weibull distribution for peak variable fitting. Similarly, the Weibull distribution can be effectively used for stable variable fitting.

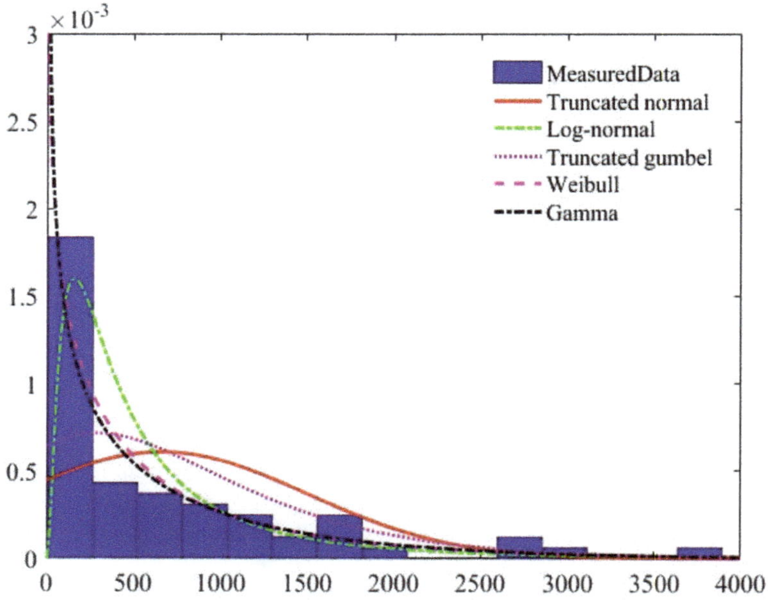

(**a**)

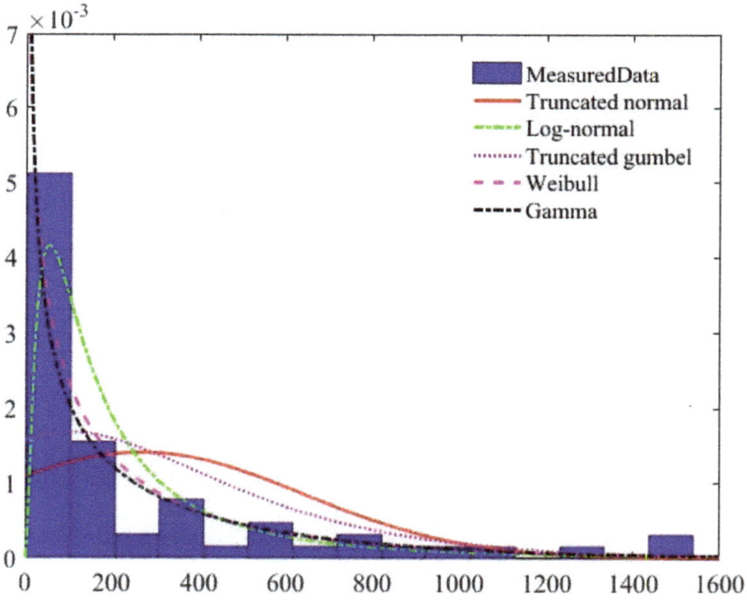

(**b**)

Figure 5. Frequency histogram of the raw data and probability density function of the marginal distribution: (**a**) Vertical forces (peak); (**b**)Vertical forces (stable).

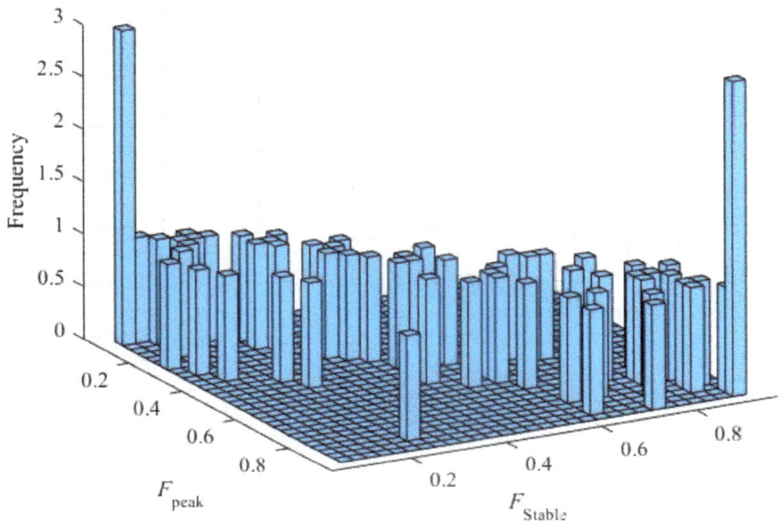

Figure 6. Frequency histogram of the raw data.

Table 5. K-S test results for the five marginal distributions.

Parameter	Marginal Distribution	*D* Value
Vertical forces (peak)	Truncated normal	0.0389
	Log-normal	0.0263
	Truncated Gumbel	0.0317
	Weibull	**0.0168**
	Gamma	0.0217
Vertical forces (stable)	Truncated normal	0.0156
	Log-normal	0.0090
	Truncated Gumbel	0.0135
	Weibull	**0.0078**
	Gamma	0.0291

5.2.2. Optimal Copula Function

Figure 4b shows that the risk variables exhibited a significant positive correlation. If their correlation were ignored, a simple independent distribution could not represent the real distribution characteristics of the original data. Therefore, it is necessary to characterize the correlation among the risk variables based on the copula function and establish a joint probability distribution model. First, Kendall rank correlation coefficient values were calculated with Equation (8) to obtain the parameters of the copula functions with the different structures.

Then, the *AIC* or *BIC* was considered to determine the optimal copula function. The results are listed in Table 6. The table demonstrates that, among the five copula functions, the Plackett copula function yielded the lowest *AIC* values. Hence, the Plackett copula function could be effectively employed to fit the risk variable correlation structure.

Table 6. Identification of the optimal copula function.

Copula Function	Gaussian	Plackett	Frank	Clayton	No. 16
AIC	−88.1813	**−92.9037**	−92.7052	−56.7529	−48.0144

To further verify that the Plackett copula function is the optimal copula function, Figure 6 shows a frequency histogram of the original data sample, while Figure 7 shows a probability density plot and corresponding contour plot of the Plackett copula function. Figures 6 and 7a clearly exhibit the same shape overall. These two graphs indicate a trend of high values at both ends and low values in the middle, and both graphs exhibit symmetrical tails, suggesting that these graphs are sensitive to changes in the tail correlation between the random variables and reveal a high correlation between the variables. Hence, the symmetric tail correlation between the random variables can be better captured, thus confirming that the Plackett copula function can reasonably represent the structure of the correlation among the risk variables (peak and stable values). Moreover, Figure 7b shows that, in the contour map of the Plackett copula function, the risk variables exhibited a significant symmetry and a positive phase along the diagonal direction. Therefore, both Figures 6 and 7 verify that the Plackett copula function could reasonably represent the correlation characteristics of the original data and could be employed to establish a joint probability distribution model of the risk variables, thereby laying a foundation for subsequent reliability analysis.

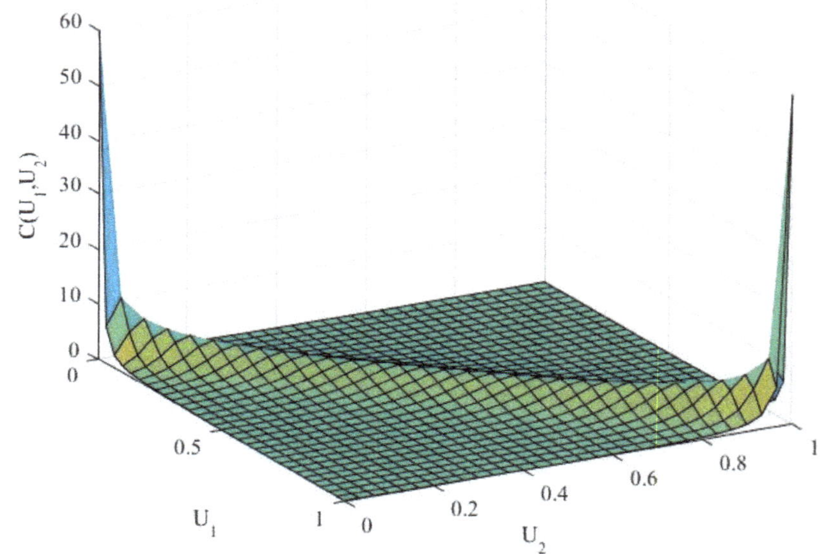

(**a**) Probability density map

Figure 7. *Cont.*

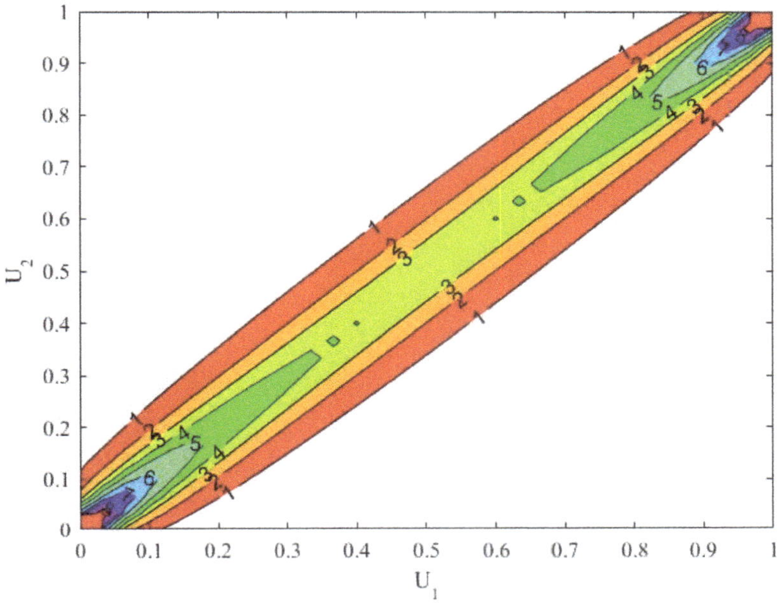

(**b**) Contour map

Figure 7. Probability density map (**a**) and contour map of the Plackett copula function (**b**).

5.3. Reliability Analysis

After determination of the optimal marginal distribution function and copula function, the joint probability density function of the risk variables can be obtained with Equations (1) and (2). To verify the superiority and importance of establishing a joint probability distribution model of the risk variables based on the copula function, the MCS-copula simulation method was applied to evaluate the reliability of pipelines under impact damage due to submarine slides. This method is helpful to accurately evaluate the safety status of submarine pipelines under the influence of landslide impact processes and provides an important theoretical basis for marine pipeline engineering design.

In this paper, a joint probability distribution model of the risk variables was established based on the copula function, and 10^5 samples were randomly generated with the MCS method, as shown in Figure 8. The figure shows that, without considering the correlation among the variables, the risk variables obey a normal distribution, whereas the simulated data exhibit a discrete uniform distribution, which is significantly different from the distribution characteristics of the original data. Moreover, due to the high variance in the original data sample and wide dispersion range, many negative values occur, which is inconsistent with the actual simulated data. The samples generated based on the above five alternative copula functions can describe the correlation among the variables. The simulated data were roughly distributed along the 45° diagonal line, which is similar to the distribution characteristics of the original data. The simulated data were matched to the original data. Compared to the other four copula functions, the distribution characteristics of the simulated data obtained with the Plackett copula function were the closest to those of the original data, and this copula function could be adopted to accurately fit the distribution characteristics of the original data. This confirms that the Plackett copula function is the optimal copula function to fit the risk variables.

In this study, two risk variables (peak and stable vertical force values) were adopted as evaluation objects, and a series system [53,54] was employed as a criterion to evaluate the structural failure risk. In other words, when these two conditions were simultaneously satisfied, the structural system was considered to occur in the failure state. Figure 8 shows that, under a safety standard of 2500 N for the vertical peak forces and a stable force safety standard of 1000 N, the area simultaneously meeting these two risk standards is the safe area, indicated as the shaded green area, and the failure point occurs outside the shaded area. The failure probability is the ratio of the number of points in the failure area to the total number of points, as expressed in Equation (14).

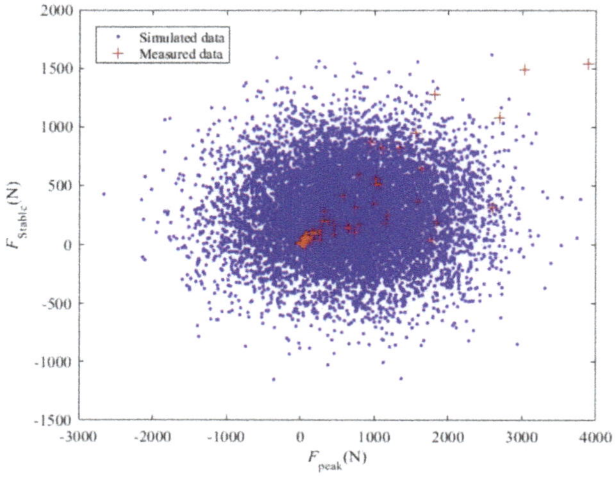

(a)

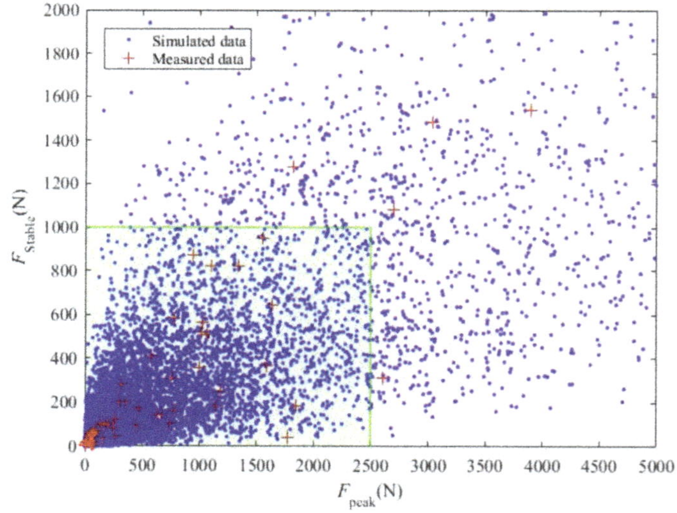

(b)

Figure 8. *Cont.*

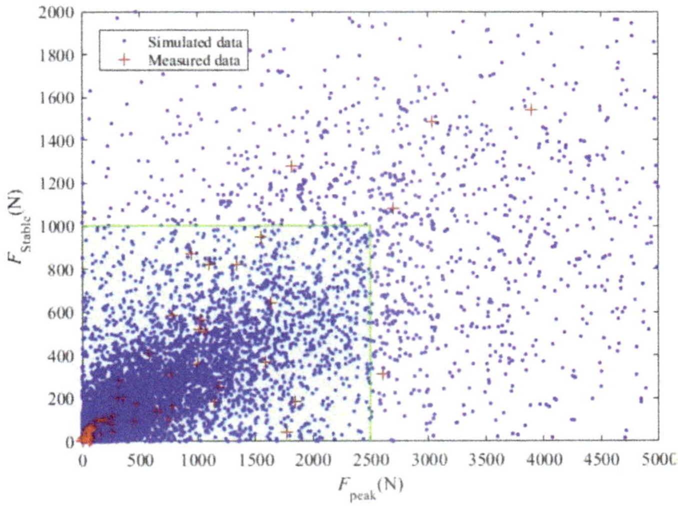

(**c**)

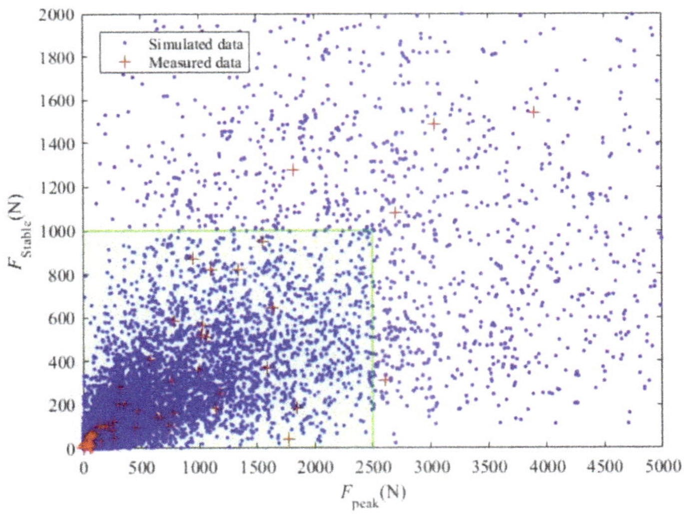

(**d**)

Figure 8. *Cont.*

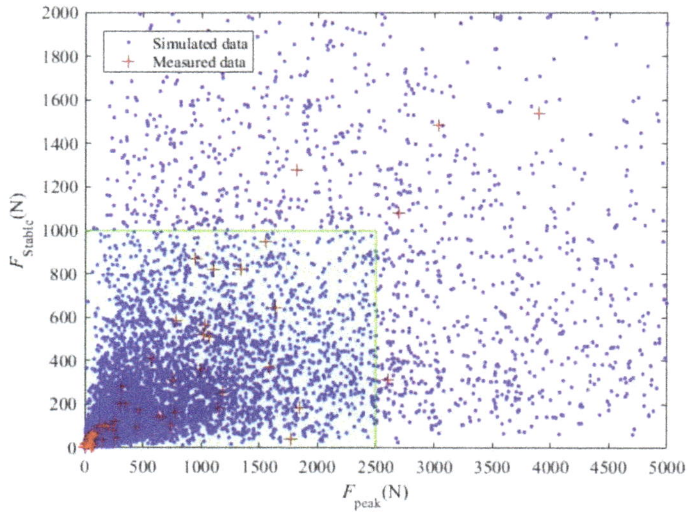

(e)

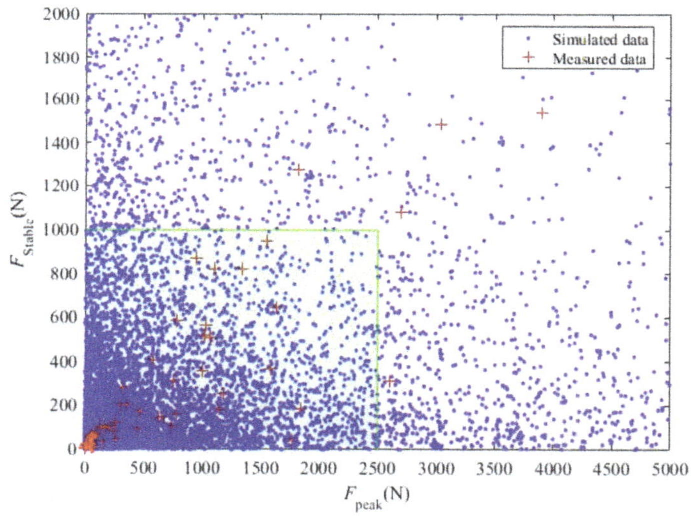

(f)

Figure 8. Scatter diagram of the different copula functions and security areas: (**a**) Independent normal distribution; (**b**) Gaussian copula; (**c**) Plackett copula; (**d**) Frank copula; (**e**) Clayton copula; (**f**) No. 16 copula.

The selection of the copula function directly determines the joint probability distribution model of the risk variables and thus greatly influences the structural reliability analysis results. Therefore, based on Section 5.2.2, and with the use of the Plackett copula function as an analysis standard, Table 7 lists the relative errors of the failure probability calculated with the different distribution models. Moreover, to analyse the change pattern of the risk variables and failure probability, this study simplified the problem and only considered the change in one variable. Figure 9 shows the change curve of the submarine pipeline failure probability under the influence of the risk variables. The results revealed that (1) it is unreasonable to employ the traditional independent normal distribution model, which ignores the correlation among the variables, and that the established joint probability distribution model could not provide accurate and reliable simulation data for reliability analysis. This could lead to inaccuracy of the calculated failure probability, and the error could even reach 82.54%. (2) The failure probability calculated based on the traditional independent normal distribution model was very low, which could result in a serious overestimation of the structural reliability and could bias the resultant structure design towards danger. The model established based on the copula function could accurately describe the correlation in the original data and could reasonably characterize the distribution characteristics of the original data, which could provide an accurate analysis model for structural reliability analysis and improve the calculation accuracy of the structural failure probability. (3) Due to the different correlation structures of the various copula functions, the calculated failure probability values notably differed. The failure probability error calculated with the No. 16 copula function was the largest, and the failure probability error calculated with the Gaussian copula function was the smallest. As shown in Figure 7, the different types of copula functions notably affected the distribution of the simulated data, which in turn influenced the reliability analysis results. (4) With increasing risk variable value (peak and stable vertical force values), the failure probability of submarine pipelines significantly increased. The five copula functions basically exhibited the same variation trend, but there occurred significant differences in the numerical values, among which the Gaussian copula function yielded the smallest difference.

Table 7. Relative error of the failure probability calculated with the different copula functions.

Model	Gaussian	Plackett	Frank	Clayton	No. 16	Independent
Failure probability	0.07419	0.07561	0.07124	0.06257	0.06038	0.0132
Relative error	1.88%	0.00	5.78%	17.25%	20.14%	82.54%

Note: The control conditions for the vertical force peak and stable values are 2000 and 1000 N, respectively.

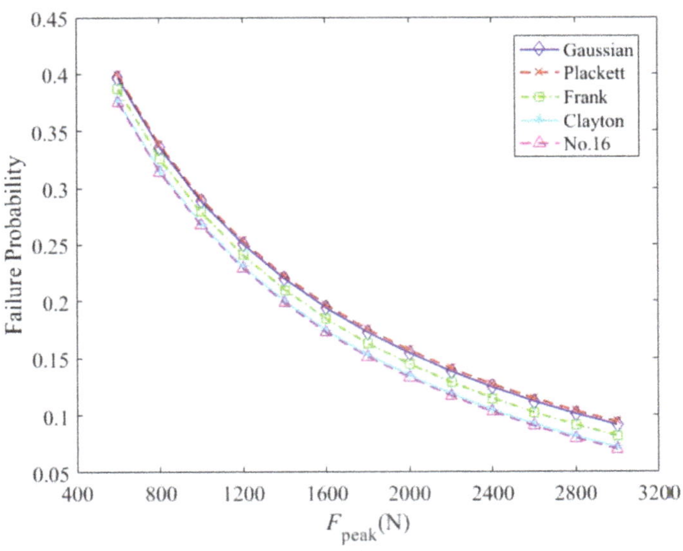

(a)

Figure 9. *Cont.*

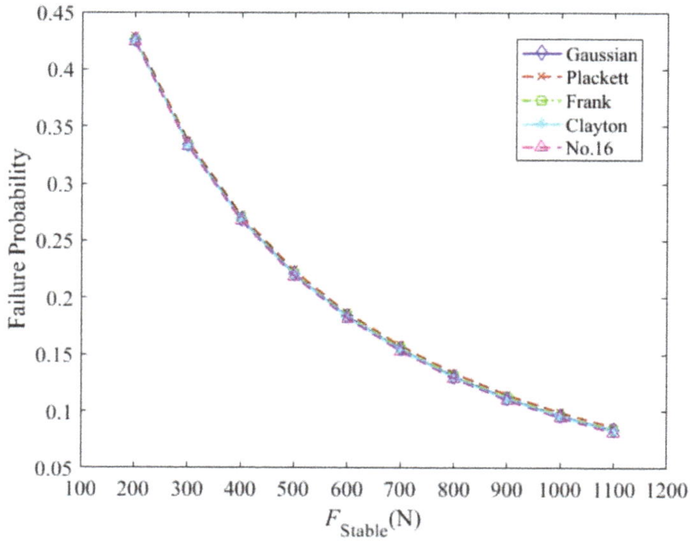

(**b**)

Figure 9. Failure probability under the different safety standards: (**a**) Vertical peak forces; (**b**) Stable value of vertical forces.

6. Conclusions

This paper proposed a probability analysis method of pipeline failure under impact damage due to submarine slides based on the copula function. Under incomplete probability information, reasonable characterization of the correlation among the risk variables, improvement in the calculation precision of the pipeline failure probability under impact damage due to submarine slides, and the accurate assessment of the pipeline reliability under impact damage due to submarine slides are of great importance. The main conclusions were as follows:

(1) Based on the copula function, a joint probability distribution model of the risk variables could be established given any marginal distribution function and related structure. The process of reliability analysis through a joint probability analysis model is as follows: a. the optimal marginal distribution function and optimal copula function were determined, and a joint probability distribution model was established and simulated in accordance with the distribution characteristics of the risk variables; b. based on the established joint probability distribution model, the MCS method was applied to generate a large number of random samples to calculate the failure probability of the pipeline impact damage attributed to submarine slides;

(2) Under the condition of incomplete probability information, the copula function could reasonably represent the relevant nonnormal distribution characteristics of the risk variables, effectively establish a corresponding joint probability distribution model, simulate data conforming to the distribution pattern of the risk variables, and provide reliable and statistically significant samples for the reliability evaluation of the submarine slide effect on pipeline damage;

(3) The traditional independent normal distribution model ignores the nonnormal distribution characteristics of the risk variables, and the calculated failure probability was very low, which could result in the serious overestimation of the reliability of submarine pipelines. Therefore, the correlation and nonnormal distribution characteristics of the risk variables should be comprehensively considered when evaluating the reliability of submarine pipelines.

The method proposed in this paper integrated a variety of uncertain factors, left the uncertain factors unrefined, established a joint probability distribution model of the distribution characteristics of the risk variables, and analysed the reliability of pipelines impacted by submarine slides. Based on existing research results, the first author will conduct a large number of model tests, study the correlation among dual-risk variables, and examine a large number of numerical simulations considering various uncertain factors of submarine slides to supplement the research results obtained in this paper.

Author Contributions: Conceptualization, L.S. and N.F.; methodology, N.F. and L.S.; software, N.F.; validation, X.D.; formal analysis, H.Y.; investigation, W.W.; data curation, L.S.; writing—review and editing, L.S. All authors have read and agreed to the published version of the manuscript.

Funding: This work was supported by the China National Natural Science Foundation (Grant Nos. 52009017 and 51979026), Open Fund Project of the State Key Laboratory of Water Resources and Hydropower Engineering Science (Grant No. 2021SGG04), China National Postdoctoral Program for Innovative Talents (Grant No. BX20190057), China National Postdoctoral Science Foundation (Grant No. 2020M680951), Liaoning Province Science Foundation (2020-BS-06), and Scientific Research Project of the Education Department of Zhejiang Province (Y202146367).

Institutional Review Board Statement: Not applicable.

Informed Consent Statement: Not applicable.

Data Availability Statement: All data, models, and code generated or used during the study appear in the submitted article.

Conflicts of Interest: The authors declare no conflict of interest.

References

1. Yonggang, J.; Chaoqi, Z.; Liping, L.; Dong, W. Marine geohazards: Review and future perspective. *Acta Geol. Sin. Engl. Ed.* **2016**, *90*, 1455–1470. [CrossRef]
2. Zhao, E.; Dong, Y.; Tang, Y.; Sun, J. Numerical investigation of hydrodynamic characteristics and local scour mechanism around submarine pipelines under joint effect of solitary waves and currents. *Ocean. Eng.* **2021**, *222*, 108553. [CrossRef]
3. Hance, J.J. Development of a database and assessment of seafloor slope stability based on published literature. Ph.D. Thesis, University of Texas at Austin, Austin, TX, USA, 2003.
4. Nodine, M.C.; Gilbert, R.B.; Cheon, J.Y.; Wright, S.G.; Ward, E.G. Risk Analysis for Hurricane-Wave Induced Submarine Mudslides. In *Submarine Mass Movements and Their Consequences*; Springer: Dordrecht, The Netherlands, 2010; pp. 335–351.
5. Dutta, S.; Hawlader, B. Pipeline-soil-water interaction modeling for submarine landslide impact on suspended offshore pipelines. *Geotechnique* **2018**, *69*, 29–41. [CrossRef]
6. Demars, K.R. Design of marine pipelines for areas of unstable sediment. *Transp. Eng. J.* **1978**, *104*, 109–112. [CrossRef]
7. Randolph, M.F.; Houlsby, G.T. Discussion: The limiting pressure on a circular pile loaded laterally in cohesive soil. *Geotechnique* **1986**, *36*, 457. [CrossRef]
8. Zakeri, A.; Hawlader, B. Drag forces caused by submarine glide block or out-runner block impact on suspended (free-span) pipelines-Numerical analysis. *Ocean. Eng.* **2013**, *67*, 89–99. [CrossRef]
9. Pazwash, H.; Robertson, J.M. Forces on bodies in Bingham fluids. *J. Hydraul. Res.* **1975**, *13*, 35–55. [CrossRef]
10. Chehata, D.; Zenit, R.; Wassgren, C.R. Dense granular flow around an immersed cylinder. *Phys. Fluids* **2003**, *15*, 1622–1631. [CrossRef]
11. Liu, J.; Tian, J.; Yi, P. Impact forces of submarine slides on offshore pipelines. *Ocean. Eng.* **2015**, *95*, 116–127. [CrossRef]
12. Randolph, M.F.; White, D.J. Interaction forces between pipelines and submarine slides-A geotechnical viewpoint. *Ocean. Eng.* **2012**, *48*, 32–37. [CrossRef]
13. Dong, Y.K. Runout of submarine slides and their impact to subsea infrastructure using material point method. Ph.D. Thesis, University of Western Australia, Perth, Australia, 2016.
14. Martin, C.M.; Randolph, M.F. Upper bound analysis of lateral pile capacity in cohesive soil. *Geotechnique* **2006**, *56*, 141–145. [CrossRef]
15. Fan, N.; Sahdi, F.; Zhang, W.; Nian, T.; Randolph, M.F. Effect of pipeline-seabed gaps on the vertical forces of a pipeline induced by submarine slide impact. *Ocean. Eng.* **2021**, *221*, 108506. [CrossRef]
16. Fan, N.; Jiang, J.; Dong, Y.; Guo, L.; Song, L. Approach for evaluating instantaneous impact forces during submarine slide-pipeline interaction considering the inertial action. *Ocean. Eng.* **2022**, *245*, 110466. [CrossRef]
17. Vu, K.A.T.; Stewart, M.G. Structural reliability of concrete bridges including improved chloride-induced corrosion models. *Struct. Saf.* **2000**, *22*, 313–333. [CrossRef]

18. Nowak, A.S.; Szerszen, M.M. Structural reliability as applied to highway bridges. *Prog. Struct. Eng. Mater.* **2015**, *2*, 218–224. [CrossRef]
19. Guo, X.; Thorsten, S.; Nian, T.; Jia, Y.; Liu, X. Effect of pipeline surface roughness on peak impact forces caused by submarine mudflow. *Ocean. Eng.* **2021**, *243*, 110184. [CrossRef]
20. Fan, N.; Nian, T.K.; Jiao, H.B.; Jia, Y.G. Interaction between submarine slides and suspended pipelines with a streamlined contour. *Mar. Georesour. Geotechnol.* **2018**, *36*, 652–662. [CrossRef]
21. Guo, X.; Nian, T.; Zhao, W.; Gu, Z.; Liu, C.; Liu, X.; Jia, Y. Centrifuge experiment on the penetration test for evaluating undrained strength of deep-sea surface soils. *Int. J. Min. Sci. Technol.* 2021; *in press*. [CrossRef]
22. Guo, X.S.; Nian, T.K.; Wang, D.; Gu, Z.D. Evaluation of undrained shear strength of surficial marine clays using ball penetration-based CFD modelling. *Acta Geotech.* **2021**, 1–17. [CrossRef]
23. Rosenberg, J.V.; Schuermann, T. A general approach to integrated risk management with skewed, fat-tailed risks. *J. Financ. Econ.* **2006**, *79*, 569–614. [CrossRef]
24. Wen, X.Q.; Wei, Y.; Huang, D.S. Measuring contagion between energy market and stock market during financial crisis: A Copula approach. *Energy Econ.* **2012**, *34*, 1435–1446. [CrossRef]
25. Ghosh, S. Modelling bivariate rainfall distribution and generating bivariate correlated rainfall data in neighbouring meteorological subdivisions using Copula. *Hydrol. Processes* **2010**, *24*, 3558–3567. [CrossRef]
26. Wang, Y.; Ma, H.; Sheng, D.; Wang, D. Assessing the interactions between Chlogophyll *a* and environmental variables using Copula method. *J. Hydrol. Eng.* **2012**, *17*, 495–506. [CrossRef]
27. Ghosh, S.; Sheppard, L.W.; Holder, M.T.; Loecke, T.D.; Reid, P.C.; Bever, J.D.; Reuman, D.C. Chapter Eleven-Copulas and their potential for ecology. *Adv. Ecol. Res.* **2020**, *62*, 409–468.
28. Liebscher, E.; Taubert, F.; Waltschew, D.; Hetzer, J. Modelling multivariate data using product copulas and minimum distance estimators: An exemplary application to ecological traits. *Environ. Ecol. Stat.* **2022**, 1–24. [CrossRef]
29. Zhou, Y.; Lu, Z.; Shi, Y.; Cheng, K. The copula-based method for statistical analysis of step-stress accelerated life test with dependent competing failure modes. *Proc. Inst. Mech. Eng. Part O J. Risk Reliab.* **2019**, *233*, 401–418. [CrossRef]
30. Wang, Y.-C.; Emura, T.; Fan, T.-H.; Lo, S.M.S.; Wilke, R.A. Likelihood-based inference for a frailty-copula model based on competing risks failure time data. *Qual. Reliab. Eng. Int.* **2020**, *36*, 1622–1638. [CrossRef]
31. Li, D.-Q.; Tang, X.-S.; Phoon, K.-K.; Chen, Y.-F.; Zhou, C.-B. Bivariate simulation using Copula and its application to probabilistic pile settlement analysis. *Int. J. Numer. Anal. Methods Geomech.* **2013**, *37*, 597–617. [CrossRef]
32. Li, D.-Q.; Zhang, L.; Tang, X.-S.; Zhou, W.; Li, J.-H.; Zhou, C.-B.; Phoon, K.-K. Bivariate distribution of shear strength parameters using Copulas and its impact on geotechnical system reliability. *Comput. Geotech.* **2015**, *68*, 184–195. [CrossRef]
33. Tang, X.-S.; Li, D.-Q.; Rong, G.; Phoon, K.-K.; Zhou, C.-B. Impact of Copula selection on geotechnical reliability under incomplete probability information. *Comput. Geotech.* **2013**, *49*, 264–278. [CrossRef]
34. Tang, X.-S.; Li, D.-Q.; Zhou, C.-B.; Phoon, K.-K. Copula-based approaches for evaluating slope reliability under incomplete probability information. *Struct. Saf.* **2015**, *52*, 90–99. [CrossRef]
35. Song, L.; Xu, B.; Kong, X.; Zou, D. Reliability Analysis of 3D Rockfill Dam Slope Stability Based on the Copula Function. *Int. J. Geomech.* **2021**, *21*, 04021001. [CrossRef]
36. Kong, X.J.; Song, L.F.; Xu, B.; Zou, D.G. Correlation and distribution model for nonlinear strength parameters of rockfill based on Copula function. *Chin. J. Geotech. Eng.* **2020**, *42*, 797–807. (In Chinese)
37. Sklar, A. Fonctions de repartition an dimensions et leurs marges. *Publ. L Inst. Stat. L Univ. Paris* **1959**, *8*, 229–231.
38. Joe, H. *Dependence Modelling with Copulas*; CRC Press: London, UK, 2014.
39. Durante, F.; Sempi, C. *Principles of Copula Theory*; CRC Press: Boca Raton, FL, USA, 2016.
40. Salvadori, G.; De Michele, C. On the use of copulas in hydrology: Theory and practice. *J. Hydrol. Eng.* **2007**, *12*, 369–380. [CrossRef]
41. Kaike, H.A. A new look at the statistical model identification. *IEEE Trans. Autom. Control.* **1974**, *19*, 716–723.
42. Schwarz, G. Estimating the dimension of a model. *Ann. Stat.* **1978**, *6*, 461–464. [CrossRef]
43. Nelsen, R.B. *An Introduction to Copulas*; Springer: New York, NY, USA, 2006.
44. Xing, J.; Tang, X.-S.; Li, D.-Q.; Zhao, Y.-F. Bivariate distribution of shear strength parameters for rock mass using Copulas method. *Rock Soil Mech.* **2016**, *37*, 783–792. (In Chinese)
45. Tang, X.-S.; Li, D.-Q.; Zhou, C.-B.; Zhang, L.-M. Bivariate distribution models using Copulas for reliability analysis. *J. Risk Reliab.* **2013**, *227*, 499–512. [CrossRef]
46. Pang, R.; Song, L. Stochastic Dynamic Response Analysis of the 3D Slopes of Rockfill Dams Based on the Coupling Randomness of Strength Parameters and Seismic Ground Motion. *Mathematics* **2021**, *9*, 3256. [CrossRef]
47. Jiang, S.-H.; Li, D.-Q.; Cao, Z.-J.; Zhou, C.-B. Efficient system reliability analysis of slope stability in spatially variable soils using monte carlo simulation. *J. Geotech. Geoenvironmental Eng.* **2014**, *141*, 04014096. [CrossRef]
48. Wang, Y.; Kulhawy, F.H. Reliability index for serviceability limit state of building foundations. *J. Geotech. Geoenvironmental Eng.* **2008**, *134*, 1587–1594. [CrossRef]
49. Aladejare, A.E.; Wang, Y. Influence of rock property correlation on reliability analysis of rock slope stability: From property characterization to reliability analysis. *Geosci. Front.* **2018**, *9*, 1639–1648. [CrossRef]

50. Pan, Y.; Zhang, L.; Wu, X.; Qin, W.; Skibniewski, M.J. Modeling face reliability in tunneling: A Copula approach. *Comput. Geotech.* **2019**, *109*, 272–286. [CrossRef]
51. De Blasio, F.V.; Elverhøi, A.; Issler, D.; Harbitz, C.B.; Bryn, P.; Lien, R. Flow models of natural debris flows originating from over consolidated clay materials. *Mar. Geol.* **2004**, *213*, 439–455. [CrossRef]
52. Murphy, B.P.; Chakravarti, I.M.; Laha, R.G.; Roy, J. Handbook of Methods of Applied Statistics, Vol. I: Techniques of Computation, Descriptive Methods and Statistical Inference. *Appl. Stat.* **1968**, *17*, 293. [CrossRef]
53. Song, L.; Pang, R.; Xu, B.; Zhou, Y. Dynamic Reliability Analysis of the Stability of the 3D Slope of a Rockfill Dam Based on the Copula Function and Generalized Probability Density Evolution Method. *Int. J. Geomech.* **2022**, *22*, 04022038. [CrossRef]
54. Bajeel, P.N.; Kumar, M. Reliability test plan for a series system with variable failure rates. *Int. J. Qual. Reliab. Manag.* **2017**, *34*, 849–861. [CrossRef]

mathematics

Article

Diffusion Mechanism of Slurry during Grouting in a Fractured Aquifer: A Case Study in Chensilou Coal Mine, China

Minglei Zhai [1], Dan Ma [2,3,*] and Haibo Bai [1]

1 State Key Laboratory of Geomechanics and Deep Underground Engineering, China University of Mining and Technology, Xuzhou 221116, China; minglzcumt@126.com (M.Z.); hbbaicumt@126.com (H.B.)
2 School of Mines, China University of Mining and Technology, Xuzhou 221116, China
3 State Key Laboratory of Coal Resources and Safe Mining, China University of Mining and Technology, Xuzhou 221116, China
* Correspondence: dan.ma@cumt.edu.cn; Tel.: +86-176-2650-0518

Abstract: Grouting is one of the main technical means to prevent water inrush hazards in coal seam floor aquifers. It is of great significance to elucidate the diffusion law of slurry in the process of grouting in fractured aquifers for safe mining in coal mines. In this paper, the mechanism of slurry diffusion in horizontal fractures of fractured aquifers was studied based on the Bingham slurry with time-varying characteristics; additionally, a one-dimensional seepage grouting theoretical model considering the temporal and spatial variation of slurry viscosity under constant grouting rate was established. In this model, the grouting pressure required by the predetermined slurry diffusion radius can be obtained by knowing the grouting hole pressure and injection flow. Slurry properties, fracture parameters, grouting parameters, and water pressure were the parameters affecting the slurry diffusion process. Looking at the problem of water disaster prevention of coal seam floor in the Working Face 2509 of the Chensilou Coal Mine, according to the aquifer parameters and model calculation results, a grouting scheme with a slurry diffusion radius of 20 m and grouting pressure of 12 MPa was proposed. Finally, with the comparative analysis of the transient electromagnetic method (TEM) and water inflow before and after grouting, it was verified that the design grouting pressure and the spacing of grouting holes were reasonable and the grouting effect was good.

Keywords: fractured aquifer; Bingham slurry; grout diffusion model; slurry diffusion distance; grouting effect

Citation: Zhai, M.; Ma, D.; Bai, H. Diffusion Mechanism of Slurry during Grouting in a Fractured Aquifer: A Case Study in Chensilou Coal Mine, China. *Mathematics* **2022**, *10*, 1345. https://doi.org/10.3390/math10081345

Academic Editor: Aleksandr Rakhmangulov

Received: 9 March 2022
Accepted: 15 April 2022
Published: 18 April 2022

1. Introduction

With the exploitation of coal resource extending deeper in China, the threat of mine water disasters to coal mining safety is becoming more and more obvious [1–5]. In the mining process of a Carboniferous–Permian coal seam in a North China coal field, the working face is seriously threatened by a high pressure limestone aquifer [6,7] in the coal seam floor (Figure 1). As a kind of coal mine geological guarantee technology, grouting technology is often applied in mining to control for water disasters, and the analysis of slurry diffusion rules and grouting effects in grouting engineering are urgent and difficult problems [8–11].

At present, many scientific researchers have been conducting significant research on slurry diffusion rules, and fruitful research results have been achieved [12–16]. Grouting theory is the basis for the study of slurry diffusion rules, which can provide guidance for the design and implementation of grouting engineering [17–22]. The existing grouting theories mainly include pore rock mass grouting theory [23], fractured rock mass grouting theory [24], fracturing grouting theory [25], compaction grouting theory [26], and dynamic water grouting theory [27]. Theoretical analysis is an effective means to study the rules of slurry diffusion, in which the rule of slurry flow in a single fracture is the basis of the study of slurry diffusion rules.

Figure 1. Disaster caused by floor water inrush in a mine working face.

Some scholars simulated the grout diffusion rule in hydrostatic and hydrodynamic conditions through artificial equipment, and put forward the flow equation of slurry in a single fracture [28–31]. A quasi-three-dimensional fracture grouting test system was developed for hydrodynamic conditions, the slurry diffusion rule was studied and the grouting plugging method of water inrush in fractured rock mass was put forward [32]. The grouting plugging mechanism of rock mass was studied by using a seepage grouting simulation test device with three-dimensional constant pressure [33].

The evaluation of grouting effect is an indispensable step in grouting engineering [34,35]. Liu et al., treated loess strata in the tunnel by curtain grouting and evaluated the grouting effect through ground penetrating radar (GPR) and numerical simulation method [36]. It was concluded that grouting can effectively block the inflow and seepage of groundwater, and effectively control disasters such as water and mud inflow in the heading face. Zhang et al., systematically classified the grouting effect evaluation methods and put forward the inspection methods and standards of various grouting technologies, which provided a reference for the grouting construction of similar projects [37]. In addition, many scholars studied the diffusion law of cement slurry in planar fractures by numerical simulation software and obtained the parameters such as diffusion radius and grouting pressure of slurry [8,9,38–42].

The above research mainly focused on grouting simulation experiments in the laboratory, nevertheless, most of the grouting projects for mine water disaster control are carried out in limestone or sandstone aquifers, and the parameters in the grouting process are often determined based on experience and lack of corresponding theoretical basis [43]. More importantly, as the most commonly used grouting in engineering, the viscosity of cement slurry is time-varying; that is, the viscosity tends to increase with time, and the slurry diffusion radius will be much smaller if time variability is considered. However, many grouting diffusion theories ignored this property, and the viscosity used in the establishment of the grouting diffusion model was fixed as the initial viscosity value [44]. The theoretical values used in the model were obviously much larger than the actual values, and the grouting hole distance designed was also unreasonable, which was difficult to ensure the grouting effect when used to guide the construction. In the light of existing problems, this paper aims to study the diffusion mechanism and grouting effect of slurry with time-dependent behavior of viscosity, and then provides guidance for the design and implementation of aquifer grouting engineering.

2. Methodology

2.1. Basic Assumptions of Slurry Flow Model

The following hypotheses are presented [32]:

1. The slurry is non-compressible and isotropic.
2. The influence of fracture roughness is not considered and the migration velocity of grouting slurry on the fracture walls is constant at 0.

3. Slurry does not enter the rock mass during the flow process and penetrate through the fracture walls.
4. Constant pressure and uniform speed grouting are adopted in the grouting method.
5. The fractures are horizontally distributed and evenly distributed, and the influence of gravity on the slurry diffusion process is not considered.
6. The right side of slurry under static water pressure is abrupt, and the additional stress caused by slurry movement and groundwater displacement is ignored.

2.2. Basic Equations of the Slurry Flow Model

To study the diffusion law of slurry under fluid–solid coupling, it is necessary to accurately describe all the details of slurry flow in fractures. Therefore, the Navier–Stokes (N-S) equation is used as the motion equation of slurry diffusion, which is based on momentum conservation, and its expression is [45]

$$\rho\frac{\partial v}{\partial t}+\rho(v\cdot\nabla)v=\nabla\cdot\left\{-p\cdot\boldsymbol{I}+\mu\left[\nabla v+(\nabla v)^T\right]-\frac{2}{3}\mu(\nabla\cdot v)\boldsymbol{I}\right\}+F \tag{1}$$

where ρ is slurry density, v is flow velocity, t is grouting time, ∇ is divergence operator, p is grout pressure, $\boldsymbol{I}$ is identity tensor, μ is slurry viscosity, and F is volume force.

The flow law of slurry can be expressed by a continuity equation because the flow process is continuous. The slurry is assumed incompressible in the flow process and the continuity equation is based on mass conservation. The continuity equation is expressed as [45]

$$\frac{\partial\rho}{\partial t}+\nabla\cdot(\rho v)=0 \tag{2}$$

In the actual grouting process, the compressibility of the slurry is negligible [1], therefore, Equations (1) and (2) can be rewritten as

$$\rho\frac{\partial v}{\partial t}+\rho(v\cdot\nabla)v=\nabla\cdot\left\{-p\cdot\boldsymbol{I}+\mu\left[\nabla v+(\nabla v)^T\right]\right\}+F \tag{3}$$

$$\rho\nabla\cdot v=0 \tag{4}$$

The inertia term in the equation can be ignored if the viscous deformation stress of slurry is ignored, Equation (3) can be further simplified as [45]

$$\rho\frac{\partial v}{\partial t}=\nabla\cdot\left\{-p\cdot\boldsymbol{I}+\mu\left[\nabla v+(\nabla v)^T\right]\right\}+F \tag{5}$$

The constitutive relation of slurry flow is expressed the viscosity of slurry, the general expressions is

$$\tau=\mu\dot{\gamma}-\frac{2}{3}\mu(\nabla v)\boldsymbol{I} \tag{6}$$

Ignoring the compressibility of the slurry, Equation (6) can be simplified as

$$\tau=\mu\dot{\gamma} \tag{7}$$

In Equation (7), $\dot{\gamma}$ is the engineering strain rate tensor, and the expression is

$$\dot{\gamma}=\left[\nabla v+(\nabla v)^T\right] \tag{8}$$

The viscosity of slurry has time-varying characteristics. The results show that slurry with low water cement ratio (w/c) is a power-law fluid (w/c = 0.5~0.7), the w/c of Bingham fluid slurry is 0.8–1.0, and slurry with w/c > 2.0 is Newtonian fluid [32]. The w/c used for grouting the limestone aquifer in coal measures is generally 1.0, so it is considered as Bingham fluid. The viscosity variation law is in the form of exponential function

$$\mu(t)=\mu_{t_0}e^{kt} \tag{9}$$

Therefore, the expression of Bingham fluid constitutive equation is

$$\tau = \tau_0 + \mu\dot{\gamma} \tag{10}$$

By substituting Equation (9) into Equation (10), the rheological equation of Bingham fluid with the curve which not passing through the origin is usually given as [32]

$$\tau = \tau_0 + \mu_{t_0} e^{kt} \dot{\gamma} \tag{11}$$

2.3. Detection Methods of Water Abundance of Working Face Floor

Due to the growing threat of water disasters, it has become particularly important to detect water abundance within the seam floor before stoping. On the basis of *The Detailed Rules for Water Disaster Prevention and Control of Coal Mines* (*NCMSA* 2018), geophysical and drilling exploration methods should be applied simultaneously when a high-pressure karst aquifer and good water abundance exist in the coal seam floor [46]. Therefore, the transient electromagnetic method (TEM) and drilling exploration were used simultaneously to detect the floor of working face to detect the water bearing properties of the limestone. TEM is based on the time domain electromagnetic induction method. By manually supplying current pulse square wave, the law of the secondary magnetic field is observed to determine the characteristics of the geological structure. In the TEM detection results, the areas where the attenuation rate of the secondary magnetic field slows down mainly indicate areas where the rock stratum is broken, the water-abundance is strong, or a fracture has developed.

3. Mathematical Modeling of the Suspension Diffusion Process

The negligence of the influence of gravity on slurry diffusion in a single plate fracture with equal opening makes it possible to simplify to a two-dimensional problem, and therefore the slurry diffusion form is axisymmetric diffusion. The rectangular coordinate system as shown in the Figure 2 with the symmetry axis and vertical direction of the fracture as the coordinate axis. We analyze the forces in view of the micro element of slurry, taking the fracture center as the symmetry axis.

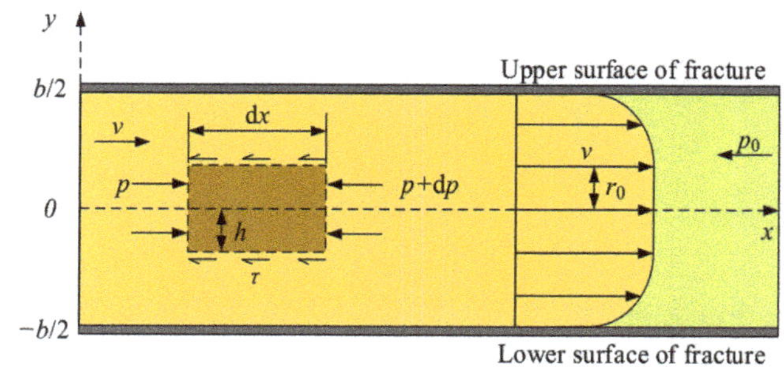

Figure 2. Force analysis of slurry motion [32].

As shown in Figure 2, b is fracture width, p_0 is hydrostatic pressure, dx is micro element length, dp is slurry pressure increment per unit volume, h is half the height of the micro element.

According to the sectional shear force distribution formula at any position in the fracture, the distribution law of shear stress along the fracture width direction can be obtained based on the stress analysis of the micro element

$$\tau = -y\frac{dp}{dx} \tag{12}$$

The order of the pressure gradient in x direction is

$$A = -\frac{dp}{dx} \tag{13}$$

Besides, the flow core zone is in the symmetric region of the fracture center when Bingham fluid flows in the fracture (see Figure 3).

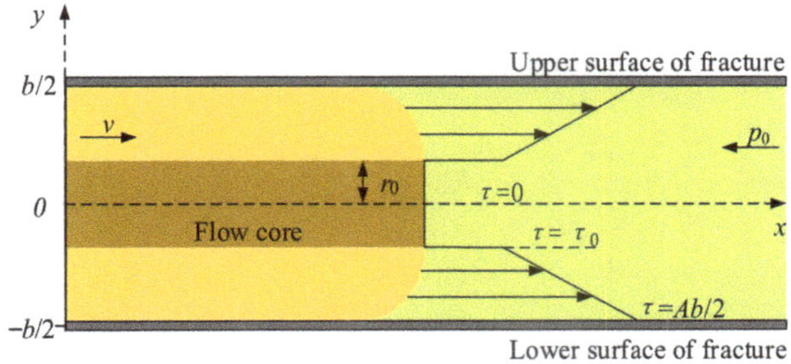

Figure 3. Sectional shear force distribution [32].

Assuming that the shear stress on the edge of the flow core zone is τ_0, the distribution of shear stress is [32]

$$\tau = \begin{cases} 0, -r_0 < y < r_0 \\ \tau_0, y = \pm r_0 \\ Ay, r_0 < |y| < \frac{b}{2} \\ \frac{Ab}{2}, y = \pm\frac{b}{2} \end{cases} \tag{14}$$

thus, we can obtain

$$\tau_0 = -r_0 \frac{dp}{dx} \tag{15}$$

That is, the radius of the flow core area is

$$r_0 = -\tau_0 \cdot A^{-1} \tag{16}$$

In addition, the flow core area is not greater than the fracture width

$$r_0 \leq \frac{b}{2} \tag{17}$$

thus, we can obtain [47]

$$-\frac{dp}{dx} \geq \frac{2\tau_0}{b} \tag{18}$$

Equation (18) shows that there is a starting pressure gradient when unsteady Bingham slurry flows

$$\lambda = \frac{2\tau_0}{b} \tag{19}$$

Combining Equations (11) and (12) results in

$$\frac{dv}{dy} = \frac{\tau_0}{\mu_{t_0} e^{kt}} + \frac{y}{\mu_{t_0} e^{kt}} \cdot \frac{dp}{dx} \tag{20}$$

The boundary conditions can be written as

$$\begin{cases} y = \pm\frac{b}{2}, v = 0 \\ y \leq r_0, v = v \\ r_0 \leq \frac{b}{2} \end{cases} \tag{21}$$

Substituting Equation (21) into Equation (20), the benchmark solution of velocity can be given as [47]

$$v = \begin{cases} -\frac{b^2 - 4y^2}{8\mu_{t_0} e^{kt}} \frac{dp}{dx} - \frac{\tau_0}{\mu_{t_0} e^{kt}} \left(\frac{b}{2} - |y|\right), r_0 \leq |y| \leq \frac{b}{2} \\ -\frac{b^2 - 4{r_0}^2}{8\mu_{t_0} e^{kt}} \frac{dp}{dx} - \frac{\tau_0}{\mu_{t_0} e^{kt}} \left(\frac{b}{2} - r_0\right), |y| \leq r_0 \end{cases} \tag{22}$$

By integrating and averaging the slurry velocity in the fracture width direction, the average slurry velocity in the fracture can be obtained as [47]

$$\overline{v} = \frac{-b^2}{12\mu_{t_0}e^{kt}}\left[\frac{dp}{dx} + \frac{3\tau_0}{b} + \frac{4\tau_0{}^3\left(\frac{dp}{dx}\right)^{-2}}{b^3}\right] \tag{23}$$

Assuming that the grouting flow is Q, then

$$Q = 2\pi x b\overline{v} = \frac{-b^3\pi x}{6\mu_{t_0}e^{kt}}\left[\frac{dp}{dx} + \frac{3\tau_0}{b} + \frac{4\tau_0{}^3\left(\frac{dp}{dx}\right)^{-2}}{b^3}\right] \tag{24}$$

Considering the pressure gradient of slurry is generally much greater than its own shear yield stress in grouting engineering, the high-order minor term in Equation (24) can be ignored and integrate x in Equation (24)

$$p = -\frac{3\tau_0}{b}x - \frac{6\mu_{t_0}e^{kt}Q}{\pi b^3} \cdot lnx + C \tag{25}$$

The injection amount of slurry is equal to the diffusion amount of slurry in the fracture according to the law of mass conservation, we have

$$Qt = b\pi\left(r_t{}^2 - r_c{}^2\right) \tag{26}$$

The radius of grouting hole, r_c, can be ignored since the grouting hole size is very small compared with the slurry diffusion area, Equation (26) can be written as

$$t = \frac{Q}{\pi b r_t{}^2} \tag{27}$$

With the diffusion of slurry, the slurry pressure in the fracture gradually decreases. When the farthest point of slurry diffusion is r_t, the slurry stops diffusion. Then we can obtain

$$\begin{cases} x = r_t, p = p_0 \\ x = r_c, t = 0 \end{cases} \tag{28}$$

By substituting Equation (28) into Equation (25), the relationship between grouting pressure and slurry diffusion distance can be obtained

$$p = p_0 + \frac{3\tau_0}{b}(r_t - r_c) + \frac{6\mu_{t_0}e^{\frac{kQ}{\pi b r_t{}^2}}Q}{\pi b^3} ln\frac{r_t}{r_c} \tag{29}$$

Equation (29) shows the grouting pressure required for predetermined slurry diffusion radius can be obtained by known grouting hole pressure and injection flow. Slurry properties, fracture parameters, grouting parameters, and water pressure are the parameters affecting the slurry diffusion process.

4. Validation with In Situ Engineering

4.1. Overview of the Chensilou Coal Mine

The Chensilou Coal Mine is located in the northeast of Yongcheng-Xiayi coal mining area, Henan Province in China (Figure 4) [48]. It covers an area of 62 square kilometers and has an annual production of 4.5 million metric tons. Working Face 2509 is mined in 2_2 coal seams, creating a complex structure and fold developed with an average thickness of 2.34 m; 22 normal faults were actually exposed during roadway excavation. According to the hydrogeological data collected during the roadway excavation, karst and fractures are relatively developed, which belong to the aquifer with medium water abundance, and the indirect water-filled source of Working Face 2509 is limestone water in the upper Taiyuan formation (L11–L8). The 2_2 coal seam floors are 41.55 m, 60.16 m, 70.16 m, and 75.19 m away from L11, L10, L9, and L8 limestone of upper Taiyuan formation, respectively (Figure 5).

According to observation data of water pressure, the floor of Working Face 2509 was subject to the highest hydrostatic pressure of limestone aquifer in the upper Taiyuan formation is 5.34 MPa. The calculation formula of safe head pressure in working face floor according to the *NCMSA* was [46]

$$P = T_S \cdot M \tag{30}$$

where P is the water pressure at the base of the floor aquitard (MPa), T_S is the water inrush coefficient (MPa/m), the *NCMSA* states that the water inrush coefficient should not exceed 0.06 MPa/m in an area where the coal seam floor has been fractured or 0.1 MPa/m in an unfractured area, M is the thickness of the aquitard (m).

As the floor of Working Face 2509 was complex a structure and fold developed, so the T_S was selected according to the seam floor has been fractured to calculate the thickness of the aquitard. Then [46]

$$M = \frac{P}{0.06} \tag{31}$$

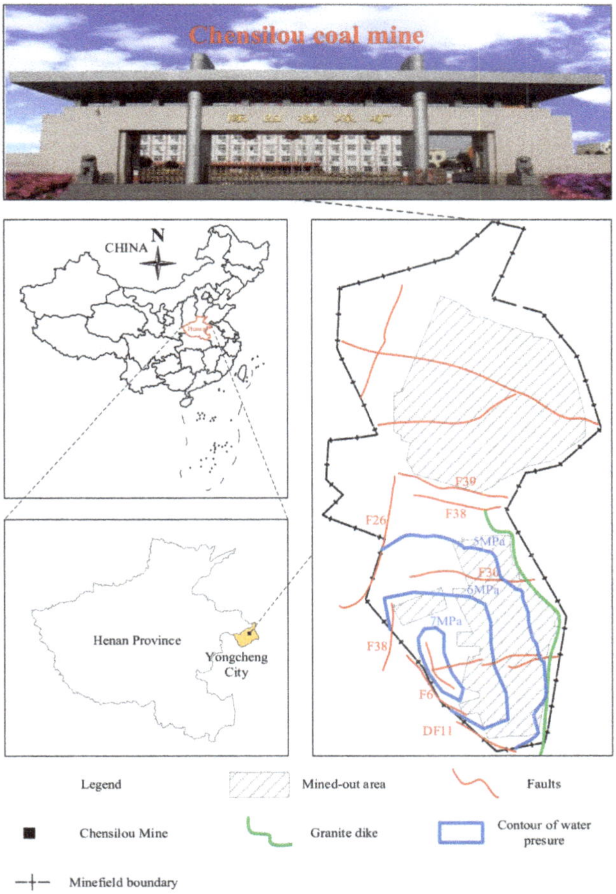

Figure 4. Location of the study area.

Therefore, to make the aquifuge thickness of Working Face 2509's floor reach the safety aquifuge (89 m) and realize the role of blocking water, it was finally determined to transform L8 to an aquitard, thereby ensuring the safe mining of Working Face 2509.

4.2. Detection Results of Water Abundance of Working Face Floor

From Figure 6, we can see that 11 areas showed an attenuated secondary magnetic field under TEM; i.e., zones A, B, . . . , K, were all located 50 m below the floor of Working

Face 2509, and the water-rich areas were located 70 m below working face floor. It was believed that L8 was rich in water in these zones.

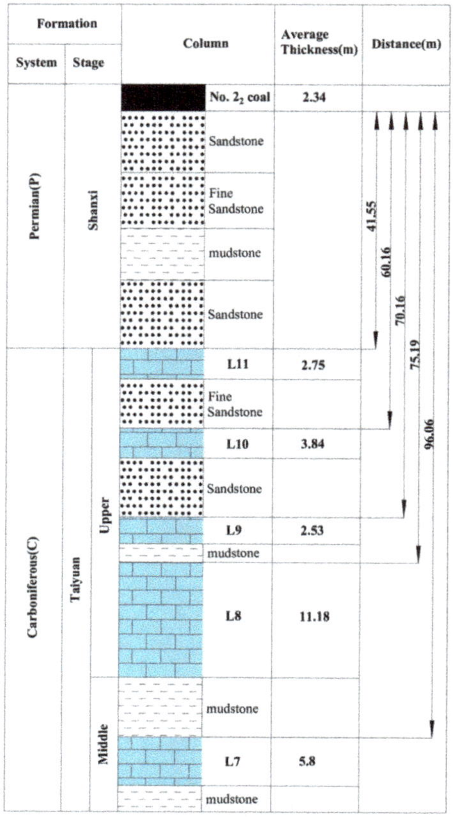

Figure 5. Histogram of the floor strata.

As the most direct and accurate method, drilling was often used to determine the distribution of water rich areas in the floor of working face. A total of 63 boreholes were drilled in the L8 aquifer and the water inflow was analyzed to verify the abnormal areas obtained by TEM. For the various boreholes, by water inflow group as shown in Figure 7, the numbers with water inflow above 5 m^3/h and exceeding 50 m^3/h were 52 and 10, respectively. The drilling results confirmed the existence of the water rich areas in the floor detected by TEM in Working Face 2509.

4.3. Determination of Grouting Pressure

Equation (29) shows the relationship between slurry diffusion distance and the grouting pump pressure. According to grouting material of the Chensilou Coal Mine, the following parameters were applied: the hydrostatic pressure p_0 was 5.34 MPa, the fracture width b was 0.5 mm, the radius of grouting hole r_c was 4.45×10^{-2} m^2, and the slurry injection flow Q was 150 L/min. According to the existing research results, and by taking the effects of the viscosity, separated water ratio, and compressive strength of stone body of slurry into account, the value of w/c of the slurry should be 0.8–1.0. The larger the w/c, the easier the slurry settles and the higher the stone rate [49]. To ensure the filling and blocking effect owing to the large water inflow of the floor of Working Face 2509, the w/c of slurry selected was 1.0, and hence the slurry can be regarded as Bingham fluid.

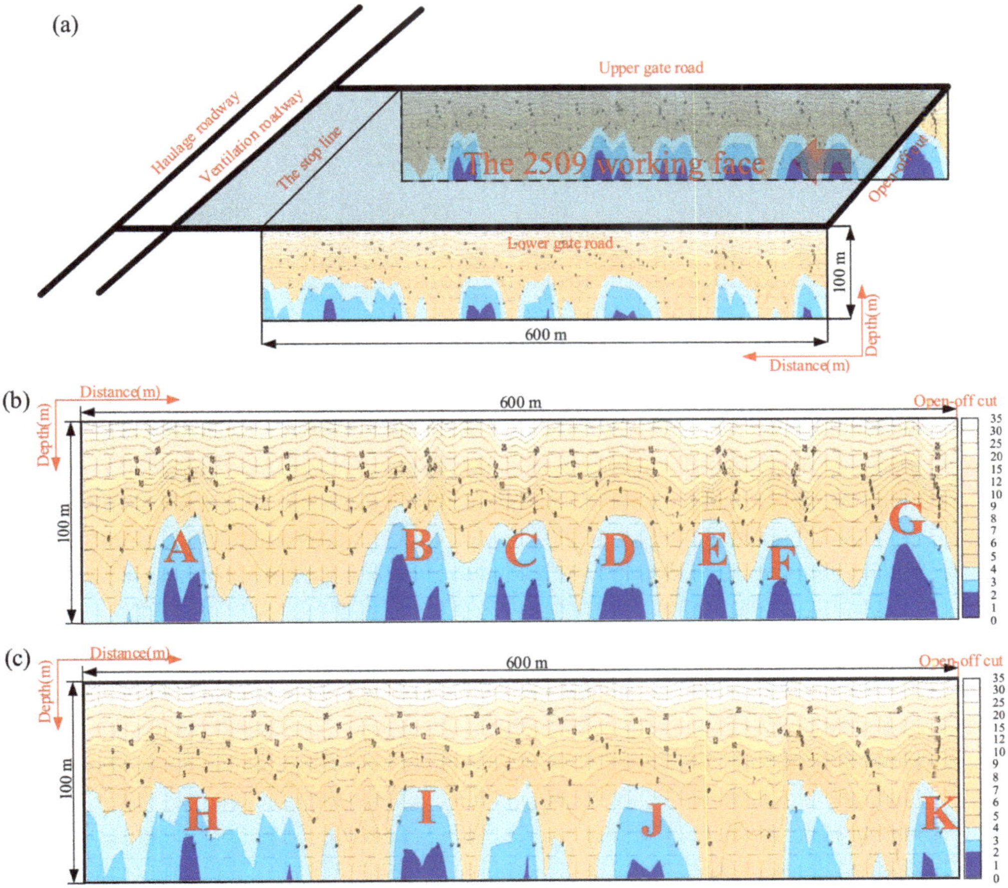

Figure 6. TEM results along the gate road of Working Face 2509 before grouting: (**a**) TEM detection profile; (**b**) the upper gate road; (**c**) the lower gate road.

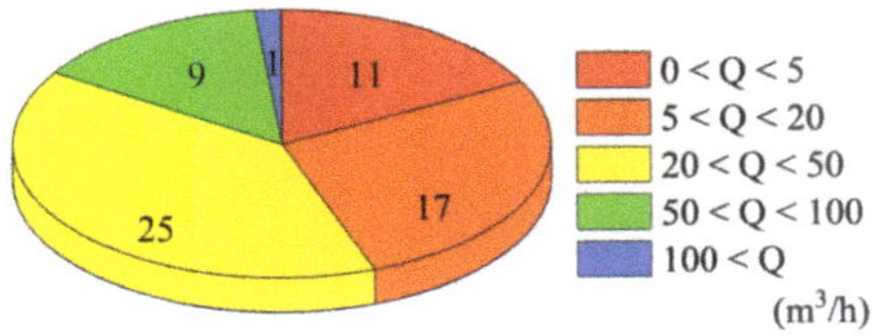

Figure 7. Number of boreholes with different water inflows (Q).

Substituting the above data into Equation (29), the relationship between the slurry diffusion distance with grouting time under different grouting pressure is shown in Figure 8. As can be seen from Figure 8, the grout diffusion distance has a prominent stage characteristic with the change of grouting pump pressure. The slurry diffusion distance increases with the increase in grouting time with different grouting pressure; however, the growth rate decreases gradually and the change rate of slurry diffusion distance becomes smaller and smaller. When the grouting time reaches a certain value, the slurry diffusion distance will tend to a stable value. If the grouting was continued at this time, the slurry diffusion range was limited and the grouting became more and more difficult.

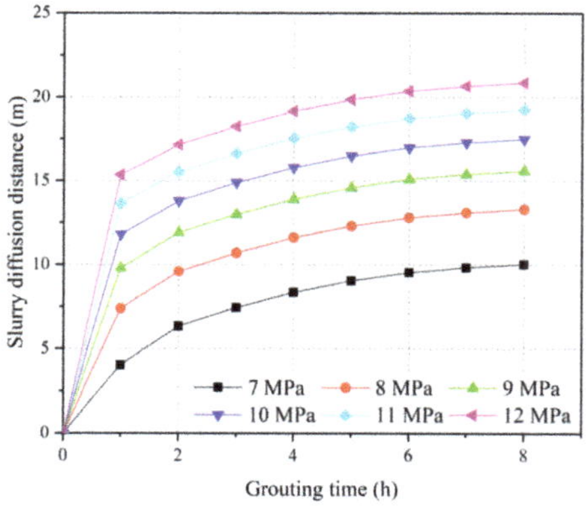

Figure 8. Variation of slurry diffusion distance with grouting time under different grouting pressures.

The general view was that the greater the grouting pressure, the greater the slurry diffusion distance. However, higher grouting pressure will expand the fracture and widen the flow channel, and the time required to reach the limit diffusion distance will also increase, even leading to roadway floor and wall heave, and the slurry will diffuse to the section that does not need reinforcement [1]. Therefore, the grouting pressure should be reasonably determined. The research showed that, after high-pressure grouting, the diffusion distance of Bingham slurry in the fracture development direction was designed as 20–30 m [7], and the conservative value of 20 m was adopted in this paper. According to these analysis results, the final pressure was 12 MPa when the designed slurry diffusion radius r_t was 20 m.

4.4. Testing the Effectiveness of Grouting Reinforcement

Grouting effectiveness test was carried out on floor limestone aquifers of Working Face 2509. The results showed that the areas and size of the water rich areas detected by TEM was significantly reduced, indicating that the grouting was effective (see Figure 9).

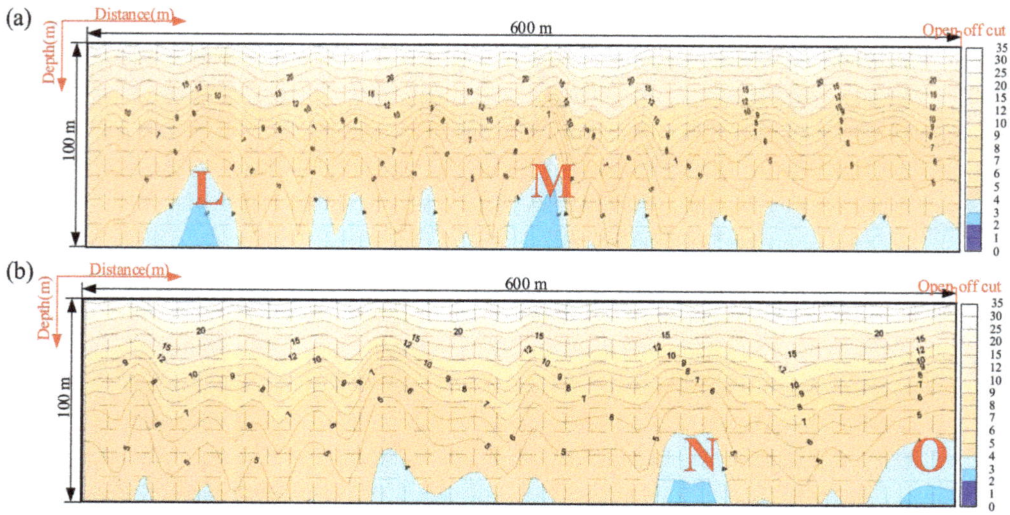

Figure 9. TEM results along the gate road of Working Face 2509 after grouting: (**a**) the upper gate road; (**b**) the lower gate road.

To determine the effectiveness of the grouting reinforcement, the water rich areas detected by TEM named L, M, N, and O were drilled, two or three test holes in each drilling site. As was already mentioned, it was considered as a safe condition if the water flow of a single test hole was less than 5 m^3/h. A total of nine testing holes were drilled and the test results are shown in Table 1, which displays that the smallest water inflow was 0.5 m^3/h (N1 hole), and the largest was 4.5 m^3/h (L2 hole). Water inflows at all test holes were under 5 m^3/h and reduced clearly compared with those listed in Figure 7. It was proved by practice that the grouting reinforcement was successful for no floor water inrush occurred during production of Working Face 2509.

Table 1. Water inflow statistics of the drilling test.

Zone No.	Hole No.	Water Inflow (m^3/h)
L	L1	3
	L2	4.5
M	M1	2
	M2	4
	M3	2
N	N1	0.5
	N2	3
O	O1	3.5
	O2	4

5. Conclusions

To prevent water-bursting disasters from occurring in the floor of a mine in the North China coalfield and to minimize deaths and economic loss, grouting transformation for limestone aquifers in coal floor is an effective means.

In this paper, by regarding slurry as a Bingham liquid of time-dependent behavior, the slurry diffusion mechanism in the horizontal fracture of a fractured aquifer was studied. A theoretical model of one-dimensional permeation grouting considering the temporal and spatial variation of slurry viscosity under constant grouting rate was developed. In this model, the grouting pressure required for predetermined slurry diffusion radius can be obtained by known grouting hole pressure and injection flow.

The mathematical modeling of the suspension diffusion process was verified in the Chensilou Coal Mine. To ensure the safe mining of Working Face 2509, the floor of the working face should be grouted to the bottom of the L8 limestone aquifer, according to the geological conditions of Working Face 2509 and the water inrush coefficient, so as to make the floor a safe water resisting layer of the working face that reaches 89 m. The grouting areas of fractured aquifer were determined according to TEM and drilling results, the grouting pressure with predetermined slurry diffusion distance was determined by using the slurry diffusion theoretical model, and it was found that the final pressure was 12 MPa when the designed slurry diffusion radius was 20 m. To guarantee the grouting quality under high water pressure, cement was regard as the main dry material of the grouting slurry and the grouting method of repeated pipe fixation was used. Finally, it has been proved by practice that the grouting reinforcement was successful in eliminating floor water inrush during production of Working Face 2509.

Author Contributions: M.Z. is responsible for the in situ investigation, data curation, and paper writing; D.M. is responsible for model investigation and funding acquisition; H.B. is responsible for the data processing and part of the paper writing. All authors have read and agreed to the published version of the manuscript.

Funding: The work presented in this paper was financially supported by the National Natural Science Foundation of China (grant no. 41977238 and 52122404) and the Graduate Innovation Program of China University of Mining and Technology.

Institutional Review Board Statement: Not applicable.

Informed Consent Statement: Not applicable.

Data Availability Statement: The study did not report data.

Acknowledgments: We would like to acknowledge the reviewers for their invaluable comments.

Conflicts of Interest: The authors declare no conflict of interest.

References

1. Hu, Y.; Liu, W.; Shen, Z.; Gao, K.; Liang, D.; Cheng, S. Diffusion mechanism and sensitivity analysis of slurry while grouting in fractured aquifer with horizontal injection hole. *Carbonate Evaporite* **2020**, *35*, 1–16. [CrossRef]
2. Zhang, J.; Liu, L.; Zhang, F.; Cao, J. Development and application of new composite grouting material for sealing groundwater inflow and reinforcing wall rock in deep mine. *Sci. Rep.* **2018**, *8*, 5642.
3. Ju, J.; Xu, J.; Yang, J. Experimental Study on the Flow Behavior of Grout Used in Horizontal Directional Drilling Borehole Grouting to Seal Mining-Induced Overburden Fractures. *Geofluids* **2021**, *2021*, 8823902. [CrossRef]
4. Wang, J.; Ma, D.; Li, Z.; Huang, Y.; Du, F. Experimental investigation of damage evolution and failure criterion on hollow cylindrical rock samples with different bore diameters. *Eng. Fract. Mech.* **2022**, *260*, 108182. [CrossRef]
5. Ma, D.; Zhang, J.; Duan, H.; Huang, Y.; Li, M.; Sun, Q.; Zhou, N. Reutilization of gangue wastes in underground backfilling mining: Overburden aquifer protection. *Chemosphere* **2021**, *264*, 128400. [CrossRef] [PubMed]
6. Han, C.; Zhang, W.; Zhou, W.; Guo, J.; Yang, F.; Man, X.; Jiang, J.; Zhang, C.; Li, Y.; Wang, Z.; et al. Experimental investigation of the fracture grouting efficiency with consideration of the viscosity variation under dynamic pressure conditions. *Carbonate Evaporite* **2020**, *35*, 30. [CrossRef]
7. Liu, S.; Yu, F.; Xu, Y.; Huang, L.; Guo, W. Full-floor Grouting Reinforcement for Working Faces with Large Mining Heights and High Water Pressure: A Case Study in China. *Mine Water Environ.* **2020**, *39*, 268–279. [CrossRef]
8. Zhang, C.; Chang, J.; Li, S.; Liu, C.; Qin, L.; Bao, R.; Liu, H.; Cheng, R. Experimental study comparing the microscopic properties of a new borehole sealing material with ordinary cement grout. *Environ. Earth Sci.* **2019**, *78*, 149.
9. Ma, D.; Duan, H.; Liu, W.; Ma, X.; Tao, M. Water–Sediment Two-Phase Flow Inrush Hazard in Rock Fractures of Overburden Strata During Coal Mining. *Mine Water Environ.* **2020**, *39*, 308–319. [CrossRef]
10. Gong, J.; Rossen, W.R. Modeling flow in naturally fractured reservoirs: Effect of fracture aperture distribution on dominant sub-network for flow. *Petrol. Sci.* **2017**, *14*, 138–154. [CrossRef]
11. Shimada, H.; Hamanaka, A.; Sasaoka, T.; Matsui, K. Behaviour of grouting material used for floor reinforcement in underground mines. *Int. J. Min. Reclam. Environ.* **2014**, *28*, 133–148. [CrossRef]
12. Chen, Y.; Zhou, C.; Sheng, Y. Formulation of strain-dependent hydraulic conductivity for a fractured rock mass. *Int. J. Rock Mech. Min.* **2007**, *44*, 981–996. [CrossRef]
13. Chien, S.; Ou, C. A novel technique of harmonic waves applied electro-osmotic chemical treatment for soil improvement. *Appl. Clay Sci.* **2011**, *52*, 235–244. [CrossRef]
14. Ma, D.; Kong, S.; Li, Z.; Zhang, Q.; Wang, Z.; Zhou, Z. Effect of wetting-drying cycle on hydraulic and mechanical properties of cemented paste backfill of the recycled solid wastes. *Chemosphere* **2021**, *282*, 131163. [CrossRef]
15. Miller, E.A.; Roycroft, G.A. Compaction Grouting Test Program for Liquefaction Control. *J. Geotech. Geoenviron.* **2004**, *130*, 355–361. [CrossRef]
16. Stoll, M.; Huber, F.M.; Trumm, M.; Enzmann, F.; Meinel, D.; Wenka, A.; Schill, E.; Schäfer, T. Experimental and numerical investigations on the effect of fracture geometry and fracture aperture distribution on flow and solute transport in natural fractures. *J. Contam. Hydrol.* **2019**, *221*, 82–97. [CrossRef]
17. Draganović, A.; Stille, H. Filtration and penetrability of cement-based grout: Study performed with a short slot. *Tunn. Undergr. Space Technol.* **2011**, *26*, 548–559. [CrossRef]
18. Draganović, A.; Stille, H. Filtration of cement-based grouts measured using a long slot. *Tunn. Undergr. Space Technol.* **2014**, *43*, 101–112. [CrossRef]
19. Eklund, D.; Stille, H. Penetrability due to filtration tendency of cement-based grouts. *Tunn. Undergr. Space Technol.* **2008**, *23*, 389–398. [CrossRef]
20. Funehag, J.; Gustafson, G. Design of grouting with silica sol in hard rock—New methods for calculation of penetration length, Part I. *Tunn. Undergr. Space Technol.* **2008**, *23*, 1–8. [CrossRef]
21. Gothäll, R.; Stille, H. Fracture dilation during grouting. *Tunn. Undergr. Space Technol.* **2009**, *24*, 126–135. [CrossRef]
22. Ma, D.; Duan, H.; Zhang, J.; Liu, X.; Li, Z. Numerical simulation of water-silt inrush hazard of fault rock: A three-phase flow model. *Rock Mech. Rock Eng.* 2022, *in press*.
23. Takano, S.; Hayashi, K.; Zen, K.; Rasouli, R. Controlled Curved Drilling Technique in the Permeation Grouting Method for Improvement Works of an Airport in Operation. *Proc. Eng.* **2016**, *143*, 539–547. [CrossRef]
24. Gothäll, R.; Stille, H. Fracture–fracture interaction during grouting. *Tunn. Undergr. Space Technol.* **2010**, *25*, 199–204. [CrossRef]
25. Zhang, Q.; Zhang, L.; Liu, R.; Wen, S.; Zheng, Z.; Wang, H.; Zhu, G. Split grouting theory based on slurry-soil coupling effects. *Chin. J. Geotech. Eng.* **2016**, *38*, 323–330.

26. Li, P.; Zhang, Q.; Zhang, X.; Li, S.; Zhang, W.; Li, M.; Wang, Q. Analysis of fracture grouting mechanism based on model test. *Rock Soil Mech.* **2014**, *35*, 3221–3230.
27. Pinto, A.; Tomásio, R.; Marques, G. Ground Improvement with Jet Grouting Solutions at the New Cruise Terminal in Lisbon, Portugal. *Proc. Eng.* **2016**, *143*, 1495–1502. [CrossRef]
28. Amadei, B.; Savage, W.Z. An analytical solution for transient flow of Bingham viscoplastic materials in rock fractures. *Int. J. Rock Mech. Min. Sci.* **2001**, *38*, 285–296. [CrossRef]
29. Zhan, K.; Sui, W.; Gao, Y. A model for grouting into single fracture with flowing water. *Rock Soil Mech.* **2011**, *32*, 1659–1663.
30. Zhang, G.; Zhan, K.; Sui, W. Experimental investigation of the impact of flow velocity on grout propagation during chemical grouting into a fracture with flowing water. *J. China Coal Soc.* **2011**, *36*, 403–406.
31. Ma, D.; Wang, J.; Cai, X.; Ma, X.; Zhang, J.; Zhou, Z.; Tao, M. Effects of height/diameter ratio on failure and damage properties of granite under coupled bending and splitting deformation. *Eng. Fract. Mech.* **2019**, *220*, 106640. [CrossRef]
32. Li, S.; Liu, R.; Zhang, Q.; Sun, Z.; Zhang, X.; Zhu, M. Research on C-S Slurry diffusion Mechanism with Time-Dependent Behavior of Viscosity. *Chin. J. Rock Mech. Eng.* **2013**, *32*, 2415–2421.
33. Wang, Q.; Feng, Z.; Wang, L.; Tang, D.; Feng, C.; Li, S. Numerical analysis of grouting radius and grout quantity in fractured rock mass. *J. China Coal Soc.* **2016**, *41*, 2588–2595.
34. Shen, S.; Wang, Z.; Horpibulsuk, S.; Kim, Y. Jet grouting with a newly developed technology: The Twin-Jet method. *Eng. Geol.* **2013**, *152*, 87–95. [CrossRef]
35. Zhang, W.; Zhu, X.; Xu, S.; Wang, Z.; Li, W. Experimental study on properties of a new type of grouting material for the reinforcement of fractured seam floor. *J. Mater. Res. Technol.* **2019**, *8*, 5271–5282. [CrossRef]
36. Liu, P.; Liang, S.; Zheng, L. Application of Curtain Grouting Reinforcement Technique in Tunnel with High Water Content Loess Stratum. *Chin. J. Undergr. Space Eng.* **2018**, *14*, 1137–1144.
37. Zhang, M.; Zhang, W.; Sun, G. Evalution technique of grouting effect and its application to engineering. *Chin. J. Rock Mech. Eng.* **2006**, *25* (Suppl. S2), 3909–3918.
38. Gustafson, G.; Claesson, J.; Fransson, Å. Steering Parameters for Rock Grouting. *J. Appl. Math.* **2013**, 269594. [CrossRef]
39. Masumoto, K.; Sugita, Y.; Fujita, T.; Martino, J.B.; Kozak, E.T.; Dixon, D.A. A clay grouting technique for granitic rock adjacent to clay bulkhead. *Phys. Chem. Earth Parts A B C* **2007**, *32*, 691–700. [CrossRef]
40. Wang, K.; Wang, L.; Ren, B.; Fan, H. Study on Seepage Simulation of High Pressure Grouting in Microfractured Rock Mass. *Geofluids* **2021**, 6696882. [CrossRef]
41. Watanabe, N.; Hirano, N.; Tsuchiya, N. Diversity of channeling flow in heterogeneous aperture distribution inferred from integrated experimental-numerical analysis on flow through shear fracture in granite. *J. Geophys. Res. Solid Earth* **2009**, *114*, B04208. [CrossRef]
42. Zhu, D.; Guo, Y.; Wang, W.; Guo, G.; An, T. Grouting Reinforcement Technique in Wind Oxidation Zone by Power Law Superfine Cement Slurry Considering the Time-Varying Rheological Parameters. *Adv. Civ. Eng.* **2019**, 2495850. [CrossRef]
43. Ma, D.; Duan, H.; Zhang, J. Solid grain migration on hydraulic properties of fault rocks in underground mining tunnel: Radial seepage experiments and verification of permeability prediction. *Tunn. Undergr. Space Technol* **2022**, *in press*.
44. Ruan, W. Research on diffusion of grouting and basic properties of grouts. *Chin. J. Geotech. Eng.* **2005**, *27*, 69–73.
45. Li, S.; Zheng, Z.; Liu, R.; Wang, X.; Zhang, L.; Wang, H. Analysis on fracture grouting mechanism considering grout-rock coupling effect. *Chin. J. Rock Mech. Eng.* **2017**, *36*, 812–820.
46. Zhai, M.; Bai, H.; Wu, L.; Wu, G.; Yan, X.; Ma, D. A reinforcement method of floor grouting in high-water pressure working face of coal mines: A case study in Luxi coal mine, North China. *Environ. Earth Sci.* **2022**, *81*, 28. [CrossRef]
47. Zhang, L.; Zhang, Q.; Liu, R.; Li, S.; Wang, H.; Li, W.; Zhang, S.; Zhu, G. Penetration grouting mechanism of quick setting slurry considering spatiotemporal variation of viscosity. *Rock Soil Mech.* **2017**, *38*, 443–452.
48. Li, H.; Bai, H.; Wu, J.; Wang, C.; Ma, Z.; Du, Y.; Ma, K. Mechanism of water inrush driven by grouting and control measures—A case study of Chensilou mine, China. *Arab. J. Geosci.* **2017**, *10*, 468. [CrossRef]
49. Shan, R.; Yang, H.; Zhang, L.; Guo, Z.; Liu, X. Research on proportion and applicable conditions of cement stable slurry. *Coal Eng.* **2014**, *46*, 97–100.

mathematics

Article

Mining Stress Evolution Law of Inclined Backfilled Stopes Considering the Brittle-Ductile Transition in Deep Mining

Yuan Zhao, Guoyan Zhao *, Jing Zhou *, Xin Cai and Ju Ma

School of Resources and Safety Engineering, Central South University, Changsha 410083, China; zhaoyuan92@csu.edu.cn (Y.Z.); xincai@csu.edu.cn (X.C.); majucsu@csu.edu.cn (J.M.)
* Correspondence: gyzhao@csu.edu.cn (G.Z.); zhoujing205@csu.edu.cn (J.Z.); Tel.: +86-135-0731-1842 (G.Z.); +86-152-7313-8014 (J.Z.)

Abstract: To study the mining stress evolution law of inclined backfilled stope in deep mining, this paper first proposes a method for determining the parameters of the brittle-ductile transition model corresponding to the Hoek–Brown criterion and Mohr-Coulomb criterion under high geostress. Then, a model composed of inclined backfilled stopes with different depths is established to simulate the sequential mining process of ore bodies with varying depths from shallow to deep. The numerical model's stratum displacement, rock mass stress distribution, and risk factors show that the mining-induced stress will move to the upper stopes and the stratum below the deepest stope. The transfer range and influence degree of mining-induced stress will increase with the increase of the deep mining, resulting in the most dangerous backfilled stope occurring one to two layers above the deepest stope and the apparent stress concentration area occurring below the deepest stope. To prevent disasters caused by mining stress, pillars in inclined deep stopes should have large safety factors. Replacing low-strength backfills with high-strength backfills can reduce the stress concentration in the stratum below the deepest stope.

Keywords: deep mining; mining disturbance; stress evolution; brittle-ductile transition; backfilled stopes

MSC: 74A10; 74L10; 74S20

Citation: Zhao, Y.; Zhao, G.; Zhou, J.; Cai, X.; Ma. J. Mining Stress Evolution Law of Inclined Backfilled Stopes Considering the Brittle-Ductile Transition in Deep Mining. *Mathematics* **2022**, *10*, 1308. https://doi.org/10.3390/math10081308

Academic Editor: David Greiner

Received: 4 March 2022
Accepted: 12 April 2022
Published: 14 April 2022

1. Introduction

Backfill mining has become a standard mining method for deep metal mines [1–3]. With the increase of metal mines' mining depth, the in situ stress of the formation increases [4,5]. A large mining-induced stress field will form near the excavation area with the high geostress environment, which will increase the risk of instability and rockburst in backfilled stopes [6,7]. Studying the mining stress evolution law of surrounding rock and backfill is helpful to better understand the stress state of deep backfilled stopes and to prevent disasters [8,9]. However, due to the influence of factors such as the in situ stress condition, orebody shape, stope size, and the mechanical properties of surrounding rock and backfill, it is difficult to obtain a reasonable evolution law of mining stress in backfilled stopes. Therefore, scholars have carried out a lot of research on this.

Li L. et al. studied the analytical solution of the stress state of vertical backfilled stopes [10], and the nonuniform distribution of vertical stress in the horizontal direction is considered [11]. Ting et al. proposed the analytical solution of the vertical stress of the inclined mine stope with non-parallel walls [12]. Li L. et al. studied the stress state of inclined backfilled stopes with FLAC2D [13]. They found that the cohesion, friction angle, Poisson's ratio, and expansion angle of backfills significantly impact the stress distribution when the geometry of the stope remains unchanged. The vertical stress decreases significantly along the hanging wall and central line of the stope when the stope inclination angle is increased. Still, the horizontal stress along the footwall decreases significantly with the rise in the stope inclination. Jahanbakhshzadeh et al. studied the stress distribution of the inclined backfilled stope with void space on the top by FLAC, obtained conclusions similar to Li L. et al., and further proposed the new analytical solution for the stress state in included backfilled mine stopes based on Marston-arch equation [14,15]. Yan et al. proposed a three-dimensional analytical solution for inclined backfilled stops based on the limit equilibrium theory and verified it by the numerical simulation [16]. Based on the above research about the single stope, Falaknaz et al. simulated the stress distribution of two adjacent vertical stopes. They found that the geometry and backfill characteristics of the second stope will affect the stress distribution of the first backfilled stope. The greater the buried depth of the two stopes, the higher the stress level of the stope [17].

The backfill and surrounding rock's mechanical properties determine the supporting force that the backfill can provide to the surrounding rock and then affect the stress distribution of them. Liu et al. studied the reasonable matches between and rock mass [18]. They found that the strength and stiffness of rock mass are the dominant factors determining the instability of the surrounding rock and backfill system. When the elastic modulus of rock mass is constant, the system's stability declines with the decrease of the cement-tailing ratio. To simulate the failure process of the surrounding rock and backfill system, some scholars carried out the conventional triaxial test of rock-backfill composite specimens [19,20]. They found that multiple peaks will appear in the failure process of combined specimens. Increasing the volume fraction of the backfill will reduce the peak stress of the combined sample but increase the ductility of the post-peak deformation. To explain the mechanism of long-term stress growth in the backed stope, Qi et al. carried out a numerical simulation considering the creep behaviour of rock mass and the time-dependent characteristics of backfill [21]. They found that the squeeze-induced stress effect is the reason for the long-term stress growth in the backfilled stope [22].

The current research has revealed the mining stress distribution and the interaction mechanism between surrounding rock and backfill to a certain extent. However, there are still two shortcomings in the simulation research on the mining stress evolution law in deep metal mines. On the one hand, the research on the mining stress evolution law mainly focused on the horizontal or vertical orebody [23,24], and a few studies on the inclined orebody only analysed the mining stress evolution law in a single stope. The ore body dips in metal mines are different from those in coal mines. Specifically, the coal mine orebodies are primarily horizontal, while most metal mine orebodies are inclined. The results of the in situ stress measurements conducted at deep levels in metal mines show that the growth rate of the horizontal stress is significantly larger than that of the vertical stress at the post-mining stage, and the mining stress in the surrounding rock is transferred from the top to bottom [25,26]. Therefore, the mining stress evolution law summarized from horizontal orebodies is not entirely applicable to the metal mines.

On the other hand, most of these simulation studies used conventional plastic models or strain-softening models [26], which are not affected by minimum principal stress, to simulate the surrounding rock and backfill of stopes, which cannot sufficiently simulate the deformation behaviour of the deep rock mass. Many triaxial compression tests of rock materials have shown that the rock materials will have the characteristics of brittle-ductile transition when their post-peak deformation is affected by the minimum principal stress under triaxial stress [27,28]. Some triaxial compression tests with cemented backfill samples show that the backfill also has a brittle-ductile transition behaviour similar to that of rock materials [29,30]. Under the high geostress, mining-induced stress will make the surrounding rock and backfill of the stope enter the post-peak stage. Therefore, the brittle-ductile transition behaviour of the rock mass and backfill will also impact the evolution law of the mining stress [31]. The in situ stress restoration test conducted by Heping Xie et al. also shows that under the in situ stress, the post-peak deformation characteristics of deep rock cores are more ductile than those of shallow rock cores [32].

To study the evolution law of mining stress in inclined backfilled stopes under high geostress in deep mining, a strain-softening model considering the brittle-ductile transition behaviour of rock mass and backfill is proposed in this paper. The method is first proposed for determining the parameters of the brittle-ductile transition model corresponding to the H-B criterion and the M-C criterion to simulate rock mass and backfill, respectively. Then the proposed model is verified with a conventional triaxial compression numerical test in Section 2.1. In Section 2.2, based on the geological data and stope parameters of the Sanshandao gold mine, a model composed of multiple inclined backfilled stopes with different depths is established to simulate the process of sequential mining and filling ore bodies in different depths from shallow to deep. Then, the simulation results of the strata displacement, the stress distribution of rock mass, and risk factor distribution is described in Section 3. Finally, the stress evolution law in the backfilled stope and the influence of stope depth and backfill mechanical properties is discussed in Section 4.

2. Materials and Methods

2.1. Brittle-Ductile Transition Model

2.1.1. Simulation Theory

In the conventional triaxial test of rock materials, rock materials have the post-peak deformation characteristics of changing from brittleness to ductility with the increase of the confining pressure [27,33]. Wu X. et al. proposed a numerical model of the brittle-ductile transition to calculate the stress distribution and deformation after tunnel excavation under high geostress [34]. This numerical model adds the brittle-ductile transition formula related to the confining pressure based on the strain-softening model, which can simulate the transformation process of post-peak deformation characteristics of rock materials from brittleness to ductility with the increase of confining pressure. According to the results of the rock triaxial compression test, there is a nonlinear relationship between the confining pressure, σ_3, and residual strength, σ_1^* [35], which can be expressed as:

$$\sigma_1^*(\sigma_3) = \sigma_1^c(\sigma_3) - \beta \cdot e^{-\gamma \cdot \sigma_3} \tag{1}$$

where $\sigma_1^c(\sigma_3)$ is the principal yield stress corresponding to the confining pressure σ_3; β is the difference between peak and residual strength under the uniaxial compression test; γ is an exponential parameter. The conventional triaxial test of rock materials can fit β and γ through Equation (1).

To simulate the brittle-ductile transition, it is also necessary to establish the relationship between the post-peak strength parameters and minimum principal stress according to the strength criterion based on Equation (1). In the numerical simulation, the realization method of the strain-softening model is to make the strength parameter of the model element in the plastic stage decrease with the increase of plastic strain. The yield stress of the model element finally drops to the residual stress level.

Based on the Mohr-Coulomb (M-C) strength criterion, the relationship between peak strength, σ_1^c, and confining pressure, σ_3, is shown as:

$$\sigma_1^c(\sigma_3) = N_p\sigma_3 + 2c\sqrt{N_p} \tag{2}$$

where c is the cohesion of the material; N_p is the confinement coefficient, which remains unchanged within the plastic region. The relationship between N_p and friction angle of the rock is as follows:

$$N_P = \frac{1+\sin\phi}{1-\sin\phi} \tag{3}$$

where ϕ is the friction angle of the material.

According to the principle of the strain-softening model, the strain-softening model based on the M-C criterion needs to gradually adjust the cohesion of the element from c to

c^* with the increase of plastic deformation after the element enters the plastic stage. Based on Equations (1) and (2), the relationship between c and c^* satisfies:

$$\sigma_3 N_p + 2c^* \sqrt{N_p} = \sigma_3 N_p + 2c\sqrt{N_p} - \beta \cdot e^{-\gamma \cdot \sigma_3} \tag{4}$$

After simplifying Equation (4), the following relationship can be obtained:

$$c^* = c - \frac{\beta \cdot e^{-\gamma \cdot \sigma_3}}{2\sqrt{N_p}} \tag{5}$$

Based on the Hoek–Brown (H–B) empirical strength criterion of rock [36,37], the relationship between peak strength, σ_1^c, and confining pressure, σ_3, can be obtained as follows:

$$\sigma_1^c(\sigma_3) = \sigma_3 + \sigma_{ci}\left(m_b \frac{\sigma_3}{\sigma_{ci}} + s\right)^a \tag{6}$$

where σ_{ci} is the unconfined compressive strength. m_b, s, and a are the rock mass material constants obtained by fitting from the results of the conventional triaxial test of rock through Equation (6). For intact rock, $s = 1$ and $a = 0.5$.

The strain-softening model based on the H–B criterion also needs to change the mechanical parameters σ_{ci} to σ_{ci}^* gradually with the increase of plastic deformation to achieve the strain-softening effect. Based on Equations (1) and (6), the relationship between σ_{ci} and σ_{ci}^* can be expressed as follows:

$$\sigma_{ci}^*\left(m_b \frac{\sigma_3}{\sigma_{ci}^*} + s\right)^a = \sigma_{ci}\left(m_b \frac{\sigma_3}{\sigma_{ci}} + s\right)^a - \beta \cdot e^{-\gamma \cdot \sigma_3} \tag{7}$$

Equation (7) cannot be simplified to the same expression as Equation (5). Although the numerical method can obtain the accurate value of σ_{ci}^*, the calculation speed of the model will be affected. Therefore, it is necessary to deduce the approximate relationship to calculate σ_{ci}^* with σ_3. When σ_3 is 0, Equation (7) can be expressed as follows:

$$\sigma_{ci}^* = \sigma_{ci} - \frac{\beta}{s^a} \tag{8}$$

When σ_3 is large enough, the post-peak deformation of the material will be ductility, at this time $\sigma_{ci}^* = \sigma_{ci}$. Then the equation is obtained as follows:

$$\sigma_{ci}^* = \sigma_{ci} - \frac{\beta}{s^a} e^{-\gamma' \cdot \sigma_3} \tag{9}$$

where γ' is the exponential parameter of the strain-softening model based on the H–B criterion.

Before the simulation calculation, Equation (7) is used to calculate the corresponding numerical solution of σ_{ci}^* under multiple σ_3 and then use Equation (9) to fit γ'. During the simulation, σ_{ci}^* is calculated directly by using Equation (9), and no numerical solution is required. This way can reduce the calculation time significantly.

2.1.2. Verification Simulation

The brittle-ductile transition model is used it to simulate the conventional triaxial compression test in FLAC3D software to verify its effectiveness. The mechanical parameters of simulated rock material are determined according to the engineering geological exploration report and the literature of the Sanshandao Gold Mine [38,39]. The Young's modulus of the simulated rock material is set to 8 GPa based on the modulus of the ore body. The peak strength and residual strength of the simulated rock material under different confining pressures are shown in Table 1. Referring to the latest triaxial rock test research conducted on the high-stiffness test machine [40], the residual strength of simulated rock material under uniaxial compression is corrected to 0 to ensure the mining safety.

Table 1. The peak strength and residual strength of the simulated rock material.

Confining Pressure (MPa)	Peak Strength (MPa)	Residual Strength (MPa)
0.00	64.54	0.00
5.00	87.28	43.52
10.00	106.82	73.23
15.00	124.44	101.64
20.00	140.75	124.30
25.00	156.09	144.21
30.00	170.67	162.10

As shown in Table 1, the relationship between the peak strength of the simulated rock material and the confining pressure is close to the Hoek–Brown failure criterion. Therefore, the relevant parameters of the Brittle-ductile transition model can be calculated according to Equations (6) and (9). The parameters of the Brittle-ductile transition model σ_{ci}, m_b, s, a, β, γ, and γ' for the simulated rock material are 72.16 MPa, 7.22, 0.8, 0.5, 60.62 MPa, 0.07, and 0.08, respectively. The results of the conventional triaxial compression test simulated by these parameters are shown in Figure 1.

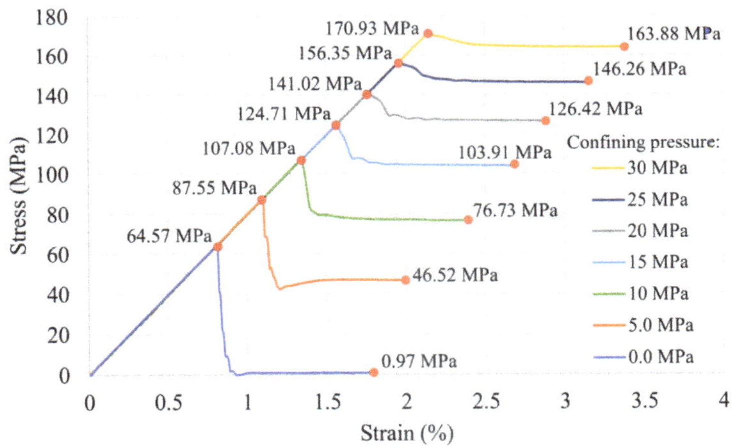

Figure 1. Stress-strain curves of the simulated rock material under different confining pressures.

From Table 1 and Figure 1, it is clear that the proposed brittle-ductile transition model can reasonably simulate the behaviour whereby the peak strength and residual strength of rock materials increase with the increase in the confining pressure. Therefore, the Brittle-ductile transition model can analyse inclined backfilled stopes' mining stress evolution law at different depths.

2.2. *Simulation of Backfill Mining*

2.2.1. Numerical Model of Stopes

The backfill mining process of orebody from −510 m to −960 m is simulated with FLAC3D. As shown in Figure 2, the height (H) of the model is 960 m, the top elevation is −300 m, the bottom elevation is −1260 m, the length (L_1) along the orebody strike is 250 m, and the length (L_2) perpendicular to the orebody strike is 1000 m.

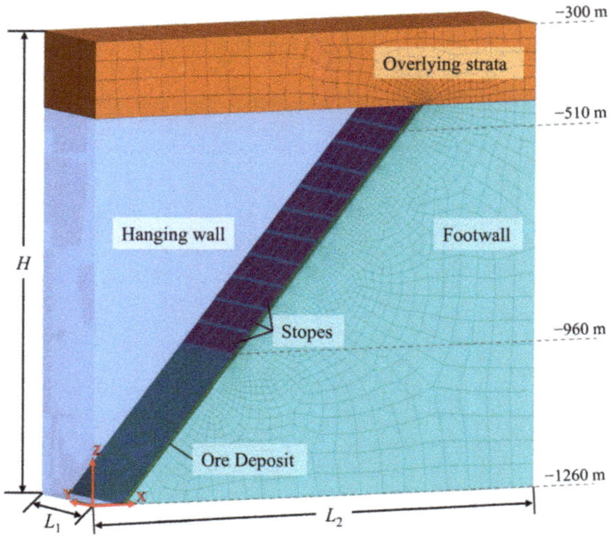

Figure 2. The simulation model of backfilled stopes.

The orebody is in the centre of the model, with a thickness of 10 m and an inclination of 40°. The width (l) of the stope is 50 m and the height (h) is 45 m. The width of barrier pillars and the height of bottom pillars are both 3 m. Five stopes at each depth in the model are symmetrical, and there are 11 floors of stopes from top to bottom.

Grids of the simulation model are hex-dominant meshes created by Itasca's automatic grid generator, Griddle. In order to balance the calculation accuracy and speed, the grid side length of the model boundary is set to 50 m, and the grid side length of stopes and pillars is set to 2 m. The grids of the transition area are automatically generated by Griddle.

2.2.2. Model Parameters

1. Parameters of the in situ stress field

To truly simulate the mining stress transfer process caused by backfill mining, obtaining the in situ stress in the case mine is necessary. The Sanshandao gold mine is located in the northwest of Jiaozhou. According to the research of Peng Li et al. [41], the maximum principal in situ stress in the northwest of Jiaozhou is the horizontal stress perpendicular to the strike of the orebody $\sigma_{h\ \max}$, the intermediate principal in situ stress is the vertical stress σ_v, and the minimum principal in situ stress is the horizontal stress along the strike of the orebody $\sigma_{h\ \min}$. The variation law of each stress with depth H is as follows:

$$\begin{aligned} \sigma_{h\ \max} &= 0.0510H + 1.9045 \\ \sigma_{h\ \min} &= 0.0276H - 0.2094 \\ \sigma_v &= 0.0303H - 0.4355 \end{aligned} \tag{10}$$

In the generation of the in situ stress model, vertical stress of 8.65 MPa and 37.74 MPa were applied to the top (−300 m) and bottom (−1260 m) of the numerical model shown in Figure 2 according to Equation (10). At the same time, $\sigma_{\mathrm{h\ max}}$ and $\sigma_{\mathrm{h\ min}}$ were calculated according to Equation (10) and applied to the two sides of the model in the XX direction and YY direction, respectively.

2. Parameters of rock masses

The Sanshandao gold mine, which mainly applies the backfill mining method, is a super large metal mine with a mining depth of over 1 km. According to the engineering geological survey reports and rock mechanical tests in the laboratory of the mine, the peak stress of rock masses from there applies to the H–B empirical strength criterion [38]. Thus, the physical and mechanical parameters of rock masses are determined, as shown in Table 2.

Table 2. The physical and mechanical parameters of rock masses (E is the Young's modulus; σ_{ci} is the unconfined compressive strength; m_b, s, and a are the rock mass material constants of H–B empirical strength criterion; β, γ, and γ' are residual strength parameters.).

Name	Density (kg/m^3)	Poisson's Ratio	E (GPa)	σ_{ci} (MPa)	m$_b$	s	a	β (MPa)	γ	γ'
Hanging wall	2700	0.2	11	110.40	8.60	0.8	0.5	96.27	0.04	0.048
Orebody	2700	0.2	8	71.89	7.24	0.8	0.5	63.53	0.07	0.081
Footwall	2700	0.25	16	160.71	11.28	0.8	0.5	139.15	0.03	0.038

Some studies have shown that the uniaxial compressive strength of rock increases with the buried depth in the coal mine [42], but the similar laws did not appear in the testing and research on the rock cores from the Sanshandao gold mine with the depth of 1 km [39]. Therefore, this paper makes a conservative assumption that the uniaxial compressive strength of the rock does not increase with the increase of depth.

3. Parameters of the backfill

Studies show that the mechanical parameters of the backfill will affect the stress distribution of the backfilled stope, and changing the materials of the backfill and the mixing proportion of the backfill can adjust these mechanical parameters. Therefore, it is necessary to compare the effects of different strength backfills to improve the stope stress state to determine the best filling scheme.

The common materials of the backfill in metal mines are non-cemented tailings, cemented tailings, and cemented block stones. The non-cemented backfill generally needs to be wrapped with cemented tailings to form a low-strength cemented backfill. Therefore, three different backfill materials were designed to analyse the stress evolution law of backfilled stope according to the mechanical parameters of the backfill commonly used in metal mines. The three materials of the backfill are: low strength cemented tailings, high strength cemented tailings, and cemented block stones. Their uniaxial compressive strengths (UCS) are 1 MPa, 3 MPa, and 5 MPa, respectively The experiments showed that the strength criterion of backfill is suitable for the M-C criterion [30]. The specific physical and mechanical parameters are shown in Table 3.

Table 3. The physical and mechanical parameters of backfill (E is the Young's modulus; c is the cohesion of the backfill; ϕ is the friction angle of the backfill; β and γ are residual strength parameters.).

Name	Density (kg/m^3)	Poisson's Ratio	E (GPa)	UCS (MPa)	c (MPa)	ϕ (°)	β (MPa)	γ
Low strength cemented tailings	2000	0.3	0.2	1	0.35	26	0.86	4.13
High strength cemented tailings	2000	0.3	0.3	3	0.90	30	2.58	3.74
Cemented block stones	2300	0.3	0.4	5	1.30	35	4.30	2.19

2.2.3. Numerical Simulation Procedure

To study the stress evolution law of backfilled stope, the numerical simulation was divided into the following three steps:

Step 1: Establishing the model. The linear elastic numerical model was established according to the model size in Figure 2 and the physical and mechanical parameters of rock masses in Table 2. The parameters of the overlying strata were the same as those of the hanging wall rock mass.

Step 2: Applying in situ stress. The boundary stress was applied to the model according to the in situ stress field parameters in Section 2.2.2. After the calculation to the equilibrium state, the linear elastic model was modified to the plastic model with the H–B criterion, and the displacement field was reset to 0.

Step 3: Mining the orebody within the stopes layer by layer. Starting from the −510 m, the orebody within the five stopes in the current layer is modified to the null model and one cycle is performed to update the model. Then, the null model within the stope was modified to the M-C model according to the physical and mechanical parameters in Table 3 to simulate the backfill. A 1 m high null model was reserved between the backfill and the bottom pillar at the top of the stope to simulate the effect of the backfill gap. Finally, the stopes in the next layer would be excavated after the model was calculated to the equilibrium state. Since the shallow part of the mine is generally filled with low-strength cemented tailings, the −825 m and above stopes were filled with low-strength cemented tailings. A comparative study of three backfilling methods of low-strength cemented tailings, high-strength cemented tailings, and cemented block stones was carried out only for the stopes in layers of −870 m, −915 m and −960 m.

2.2.4. Monitoring Method

To obtain the variation law of the deformation and stress of backfilled stopes with the mining, 33 stress monitoring points are set to monitor the stress state of the barrier pillars, bottom pillars, and backfilled stopes, as shown in Figure 3. Since the structure and stress state of the model are symmetrical along the y-axis, the simulation program only records the stress of the middle stope in each layer.

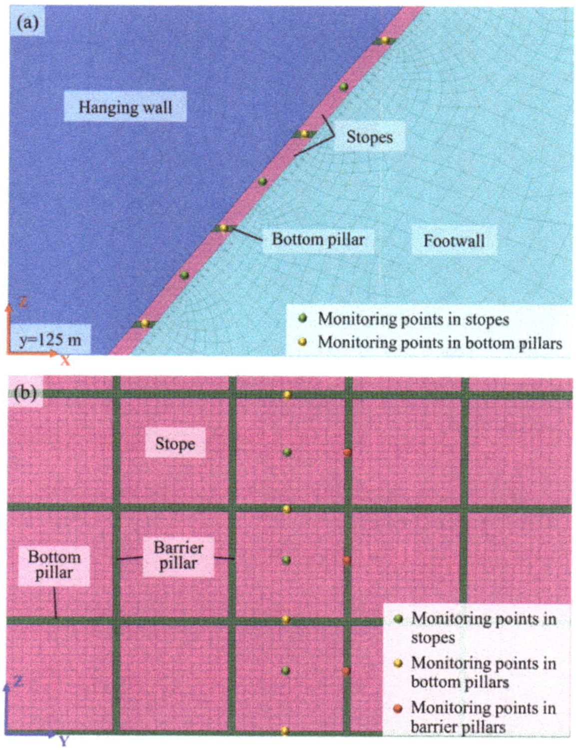

Figure 3. Schematic diagram of the location of stress monitoring points. (**a**) The view perpendicular to the Y axis; (**b**) The view perpendicular to the ore body.

3. Numerical Simulation Results

3.1. Formation Displacement

Figures 4 and 5 are cloud diagrams of the horizontal and vertical displacements of the ground, respectively. They are the model section at the middle position along the orebody strike (y = 125 m), and the number in the upper left corner of each figure represents the mining depth. The color ramps in the cloud diagrams of different mining depths adopt different scales to clearly show the distribution of displacement field near the ore body.

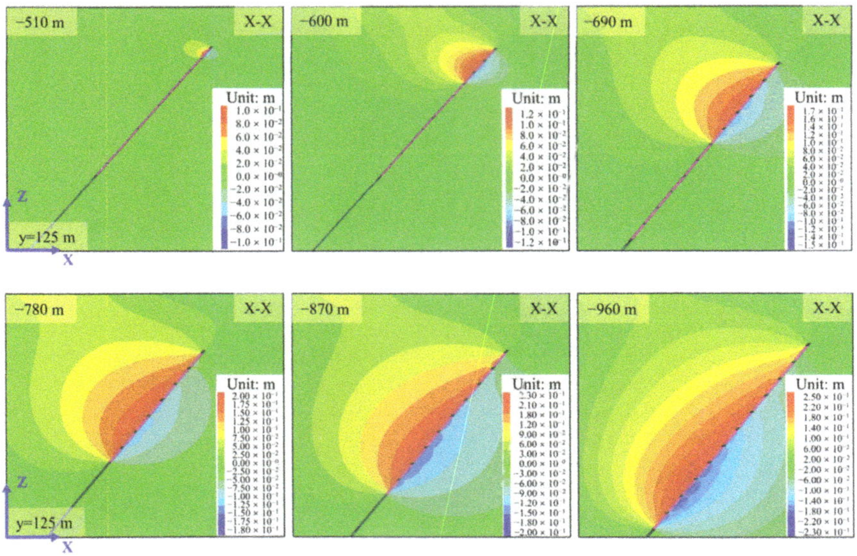

Figure 4. Horizontal displacement of ground at different depths of mining.

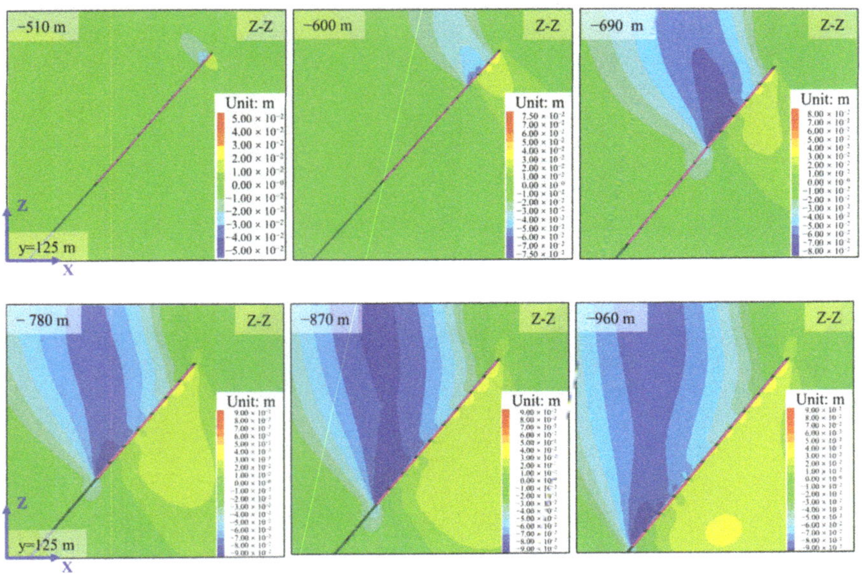

Figure 5. Vertical displacement of ground at different depths of mining.

After mining and filling the stopes, the horizontal displacement cloud diagram is in the shape of a water droplet, like a water droplet flowing along the inclined orebody. The horizontal displacement caused by cut and fill mining has two characteristics: the large influence range and the lag of the maximum displacement. In terms of the influence range, it can be seen from Figure 4 that even when mining the −960 m stope, the horizontal displacement of the −510 m stratum is still at a relatively large level, indicating that the −510 m stratum has been affected by deep mining. The specific performance of the lag of maximum displacement is that the maximum horizontal displacement of the ground always occurs in the upper stope during mining. As the mining depth increases, the stope where the maximum horizontal displacement occurs is farther from the deepest stope. The maximum horizontal displacement occurs in the −600 m and −645 m stopes when mining the −690 m stope, and the maximum horizontal displacement occurs at −825 m and −875 m stopes when mining the −960 m stope.

As shown in Figure 5, the vertical displacement cloud diagram formed by backfill mining of the inclined orebody is smoky, like the smoke rising from the hanging wall of the deepest stope. Compared with the horizontal displacement, the vertical displacement caused by backfill mining has a smaller influence range, and the maximum displacement of the surrounding rock in the hanging wall is always above the deepest stope. Starting from the mining of the −780 m stope, the vertical displacement of the surrounding rock in the hanging wall of the −510 m stope has been significantly smaller than the main deformation area, indicating that the influence of deep mining on shallow stopes has been weakened. In addition, the surrounding rock in the footwall of the shallow stope will show vertical displacement upward, the vertical displacement of the centre position in the vertical direction of each deep stope is more significant, and the vertical displacement near the bottom pillar is smaller.

3.2. Stress Distribution of Rock Mass

Figure 6 shows the distribution of stratum stress along the midline section of the stope (y = 125 m) when mining −600 m, −780 m, and −960 m stopes. The color ramps in the cloud diagrams of different mining depths use different scales to clearly show the stress concentration areas in the stratum. On the whole, there are low-stress areas in the surrounding rocks of the hanging wall and footwall near the mined stope. As the depth increases, the range of the low-stress areas decreases. The low-stress area appears in the bottom pillar because the in situ stress initially acting on the mined stope is transferred to the adjacent pillars.

Due to the impact of mining, a prominent high-stress area will appear below the deepest stope. The stress increase areas in the XX direction are symmetrically distributed along the vertical direction, while the stress increase areas in the ZZ direction are concentrated in the footwall.

After excavating the orebody in the stope, the in situ stress acting on the orebody will transfer to the bottom and barrier pillars of the stope. Figure 7 is a diagram of the stratum stress distribution along the centre line of the barrier pillar (y = 100 m). The X-X and Z-Z stresses of barrier pillars in backfilled stopes and adjacent strata have increased noticeably, forming a high-stress area. The area with the highest stress is not on the barrier pillar in the deepest stope but rather on the barrier pillars one or two stopes above the deepest stope. This phenomenon corresponds to Figure 4, where the maximum horizontal displacement appears above the deepest stope.

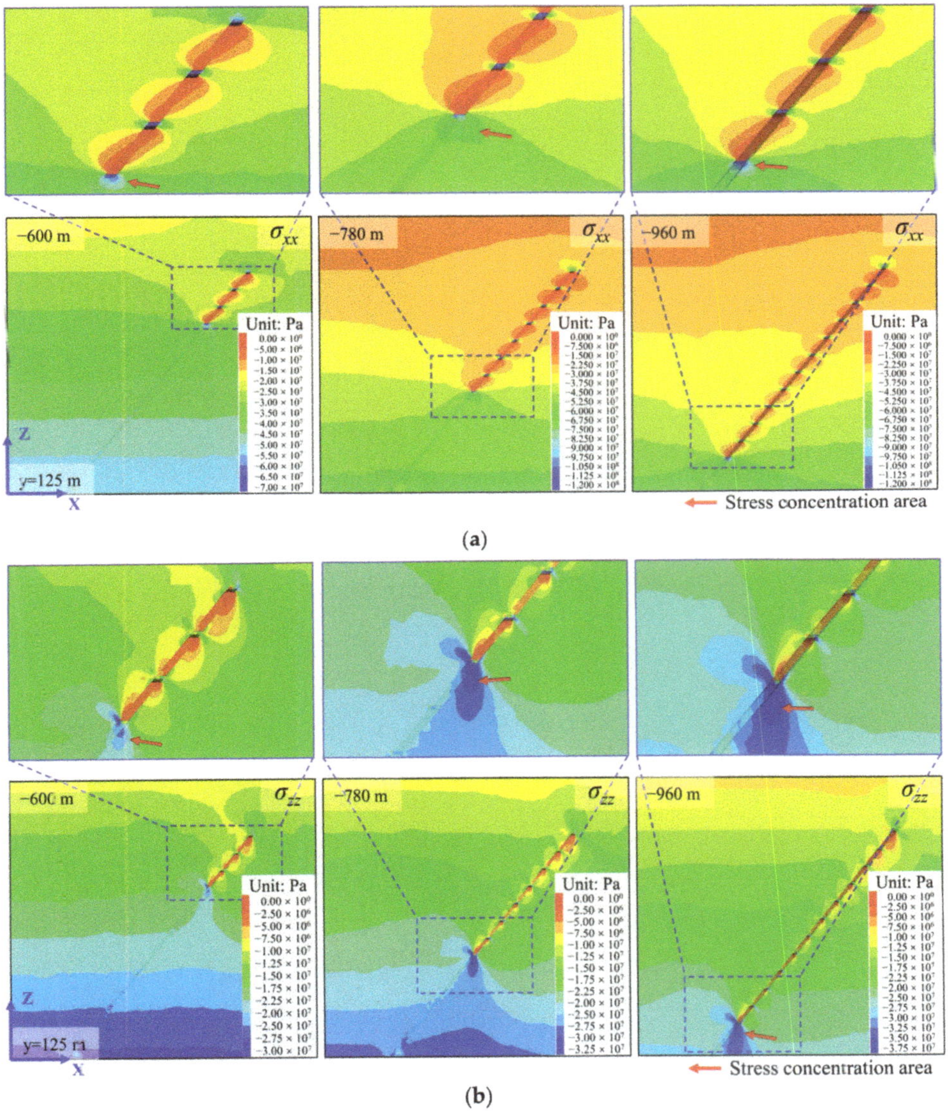

Figure 6. Stress distribution along the stope's midline (y = 125 m) (tensile stress is positive). (**a**) Horizontal stress distribution, σ_{xx}; (**b**) Vertical stress distribution, σ_{zz}.

3.3. Risk Factor

Since the maximum principal stress at the moment of rock failure will increase with the minimum principal stress under the triaxial compression test, the critical maximum principal stress, σ_1^c, is calculated by Equations (2) and (6) for filling and rock mass, respectively. The ratio of the current principal stress, σ_1, to the critical stress, σ_1^c, represents the risk factor, f. The cloud diagram of risk factor, f, is shown in Figure 8.

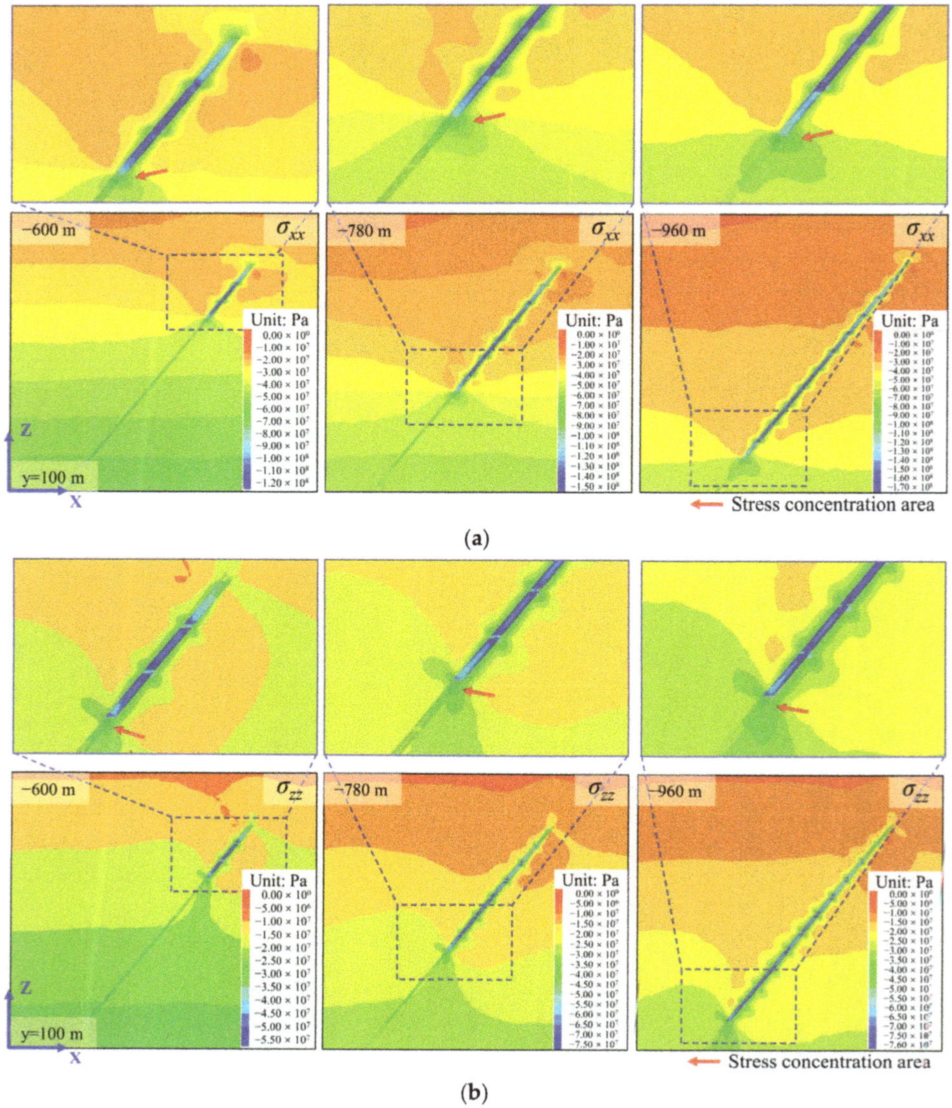

Figure 7. Stress distribution along the centre line of the barrier pillars (y = 100 m) (tensile stress is positive). (**a**) Horizontal stress distribution, σ_{xx}; (**b**) Vertical stress distribution, σ_{zz}.

Figure 8 is a cloud diagram of the stratum risk factor along the stope's midline (y = 125 m) and along the barrier pillars' midline (y = 100 m). Due to the low stress and high strength parameters, the risk coefficient of rock mass in the footwall is generally lower than that in the hanging wall. After backfill mining inclined ore bodies, mining stress transfer results in low-risk areas in the rock mass around the middle of each stope, high-risk areas in the bottom and barrier pillars, and the surrounding rock mass. As the mining depth increases, the range of the low-risk areas caused by mining stress transfer is reduced, and the range of the high-risk areas is increased.

From Figure 8, we can find that the location with the highest risk factor near the barrier pillars is not at the deepest stope, but rather at the position of one or two stopes above. As the mining depth increases, the distance between the position of the maximum risk factor and the deepest stope tends to increase. Therefore, it is necessary to discuss the influence of mining stress generated by deep mining on the upper stope structure in detail.

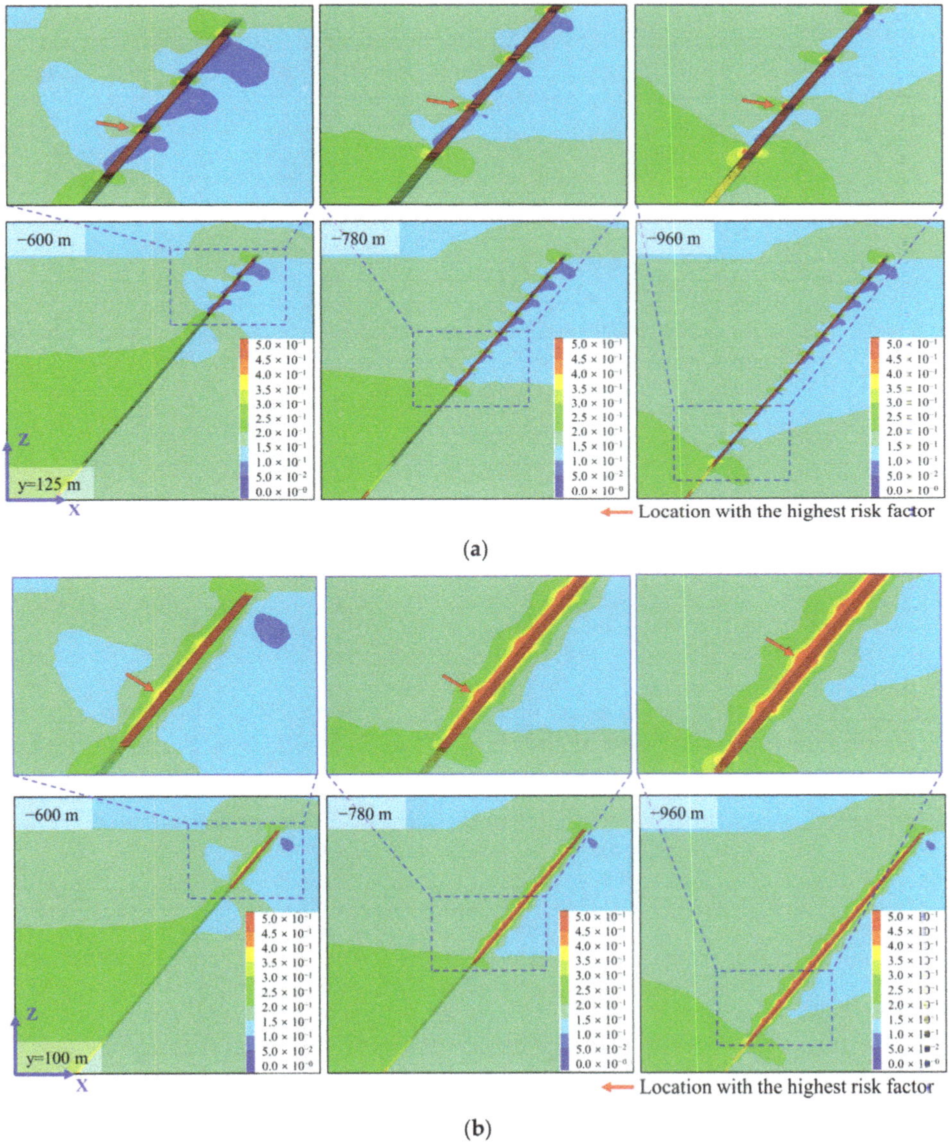

Figure 8. Cloud diagram of risk factor (σ_1/σ_1^c). (**a**) The risk factor along the stope's midline (y = 125 m); (**b**) The risk factor along the barrier pillars' midline (y = 100 m).

4. Discussion

4.1. Stress Evolution Law of Backfilled Stopes

From the results of Sections 3.2 and 3.3, it can be seen that after excavation of the orebody in the stope, the in situ stress mainly transfers to barrier pillars and bottom pillars around the stope. Thus, monitoring the stress changes at the midpoints of barrier pillars and bottom pillars can reflect the stress evolution law of all backfilled stopes.

Due to different in situ stresses to varying depths of underground mines, using the stress value to express the stress evolution law of stopes is not conducive to comparing the effects of the mining depth. To make the obtained stress evolution law of stopes more universal, the stress concentration ratio (SCR) is used to represent the stress growth of rock mass under mining-induced stress. The expression of SCR is as follows:

$$\mathrm{SCR} = \frac{\sigma_i}{\sigma_i^0}(i = x, y, z) \tag{11}$$

where, σ_i^0 is the in situ stress at the measuring point in the i direction before the orebody is excavated; σ_i is the real-time stress at the measuring point in the i direction.

The stress evolution laws of barrier and bottom pillars are discussed separately below. SCR-XX is the concentration of σ_{xx} perpendicular to the strike of the ore body, SCR-YY is the concentration of σ_{yy} parallel to the strike of the orebody, and SCR-ZZ is the concentration of σ_{zz} in the vertical direction.

4.1.1. Barrier Pillars

Figure 9 is the change of SCR of the barrier pillars at different depths. Figure 9a–c displays SCR-XX, SCR-YY, and SCR-ZZ of the barrier pillars, respectively.

The X-direction perpendicular to the strike of the orebody is the same as the direction of the maximum principal stress of the in situ stress. It is evident from Figure 9a that the SCR-XX values of all barrier pillars are continuously increasing with mining activities. When the orebody in the stope is excavated, the barrier pillars in the current stope will bear more significant transfer stress. Still, the SCR-XX increment of the barrier pillars caused by excavation decreases with the increase of stope depth. The mining stress transferred to the barrier pillars in the −555 stope during the excavation of the −555 stope is 2.35 times the in situ stress, and the SCR-XX increment of the barrier pillars in the −960 stope dropped to about 1.2 during the excavation of the −960 stope. Since the mine room and pillar at each depth are the same, the decrease in the SCR-XX increment is due to the change of mining stress transfer mode, not the difference in the stope structure.

Figure 9a shows that the transfer range of mining stress expands with increasing mining depth. After excavating the orebody, the increase of SCR-XX of the barrier pillars in the current stope is most apparent. Some stress is transferred to the barrier pillar in the upper stopes and the rock mass below the current stope, causing the increase of SCR-XX of barrier pillars in the upper stopes and the rock mass below. Judging from the effect of causing rock mass SCR-XX changes, the range of mining stress transferred to upper stopes is much more extensive than that to the lower rock mass. For example, when the −825 stope was excavated, the SCR-XX of the above stopes had varying degrees of growth, while the SCR-XX of the rock mass within the range of two stopes below increased significantly.

After excavating the orebody, the midpoint of the barrier pillar in the YY direction was in a temporary unloading state. After filling the stope, the stress of the midpoint of the barrier pillar in the YY direction will gradually increase and exceed the in situ stress. Therefore, in Figure 9b, the SCR-YY of each barrier pillar in the stope will first decrease to a negative value and then gradually increases to above 0. With the increase of mining depth, the SCR-YY of barrier pillars at different depths is all approaching 0.5, which indicates that the transfer law of mining stress in the Y direction of the barrier pillars at different depths is the same.

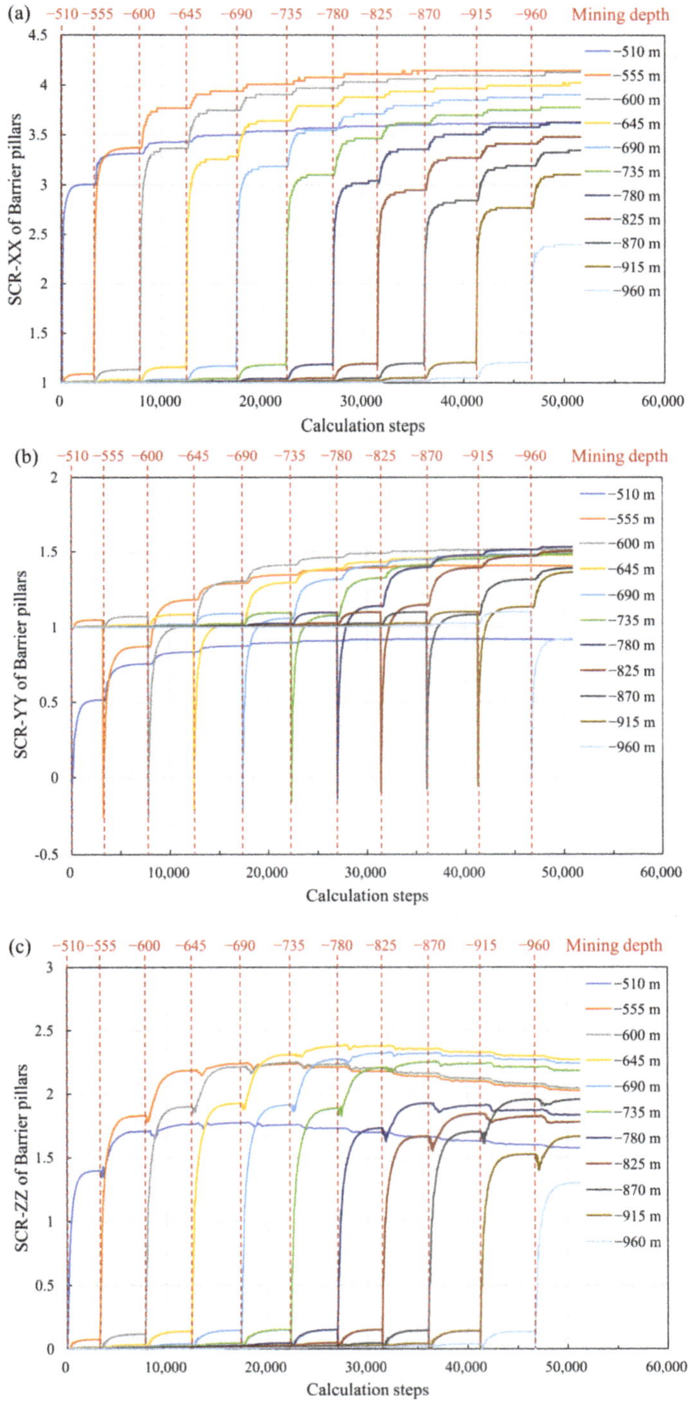

Figure 9. The change law of the stress concentration ratio (SCR) of the barrier pillars at different depths: (**a**) the SCR in the XX direction, (**b**) the SCR in the YY direction, and (**c**) the SCR in the ZZ direction.

The transfer law of the vertical stress on the barrier pillars is different from that of horizontal stress, showing a law of increasing first and then decreasing slightly (see Figure 9c). This phenomenon indicates that after mining deep ore bodies, the vertical stress had transferred to the shallow barrier pillars moved to the deep. The reason for this law may be that the rock mass in the hanging wall of the shallow stope will warp upward after mining the deep stope, as shown in Figure 5, thus reducing the vertical stress transferred from the rock mass in the hanging wall to the barrier pillars in the shallow stope.

4.1.2. Bottom Pillars

Figure 10 is the variation of the SCR of the bottom pillar at each depth. Figure 10a–c display SCR-XX, SCR-YY, and SCR-ZZ of the bottom pillars respectively.

As shown in Figure 10a, the SCR-XX of the bottom pillar will increase significantly after mining the stope above the bottom pillar, but it will decrease after mining the stope below the bottom pillar. As the depth increases, the increase of SCR-XX of the bottom pillar caused by mining the stope above the bottom pillar is greater, and the decrease of SCR-XX caused by the stope below the mining bottom pillar is also greater. When mining deep stopes, SCR-XX of the bottom pillar of the upper stope will gradually stabilize. The stable value of SCR-XX of the bottom pillars is between 1.0 and 2.0, which decreases as the depth of the bottom pillar increases.

The variation rule of SCR-YY of the bottom pillar is the same as that of SCR-XX. There is also a law that mining the stope above the bottom pillar causes the SCR-YY of the bottom pillar to increase, and mining the stope below the bottom pillar causes the SCR-YY of the bottom pillar to decrease, see Figure 10b. The final value of the SCR-YY of the bottom pillar in the shallow −510 stope is the largest, about 1.69. As the depth of the bottom pillar increases, its final value of SCR-YY gradually decreases to about 1.4.

Figure 10c shows that the SCR-ZZ of the bottom pillar decreases significantly after mining stopes above and below the bottom pillar due to the unloading effect of mining, and the SCR-ZZ increases slightly after filling the stope below the bottom pillar. The final value of the SCR-ZZ of the bottom pillar also gradually decreases as the depth of the bottom pillar increases. The SCR-ZZ of the bottom pillar at −510 m eventually drops 0.83, while the SCR-ZZ of the bottom pillar at −870 m eventually drops to 0.40.

4.2. Influence of Stope Depth on Stress Evolution

Figure 11 is a graph showing the variation of SCR-XX with the depth of the stope along the direction of maximum principal stress for all stopes. The curves of different colors represent different mining depths. Figure 11a,b are the SCR-XX of barrier pillars and bottom pillars. It can be seen from Figure 11 that the SCR curve of barrier pillars and bottom pillars shows a downward trend as a whole. After mining, the SCR of barrier pillars and bottom pillars in deeper stopes is smaller than that of the shallower stopes. Because all stopes in the model have the same size, the differences in SCR between pillars in different stopes is caused by the depth.

There are two reasons why the SCR of barrier pillars and bottom pillars in the deep stopes is lower than that in shallow stopes. One is that mining activities cause the horizontal principal stress to move to shallow stopes along the hanging wall and footwall. The other is that the shallow orebody's in situ stress is low, and its safety factor is large so that it can bear more mining-induced stress. Since the barrier pillar is a continuous whole in the vertical direction, it can reflect the process of transferring the mining-induced stress to shallow parts. For example, after mining the −780 m stope, the SCR-XX of the barrier pillar in the −555 m stope also increased significantly (see Figure 11a). Unlike the barrier pillar, the bottom pillars are separated by backfilled stopes vertically. The mining-induced stress generated by mining the deep stope is difficult to transfer to the bottom pillar in the shallow stope. As a result, the SCR-XX of the bottom pillars will only increase slightly, as shown in Figure 11b.

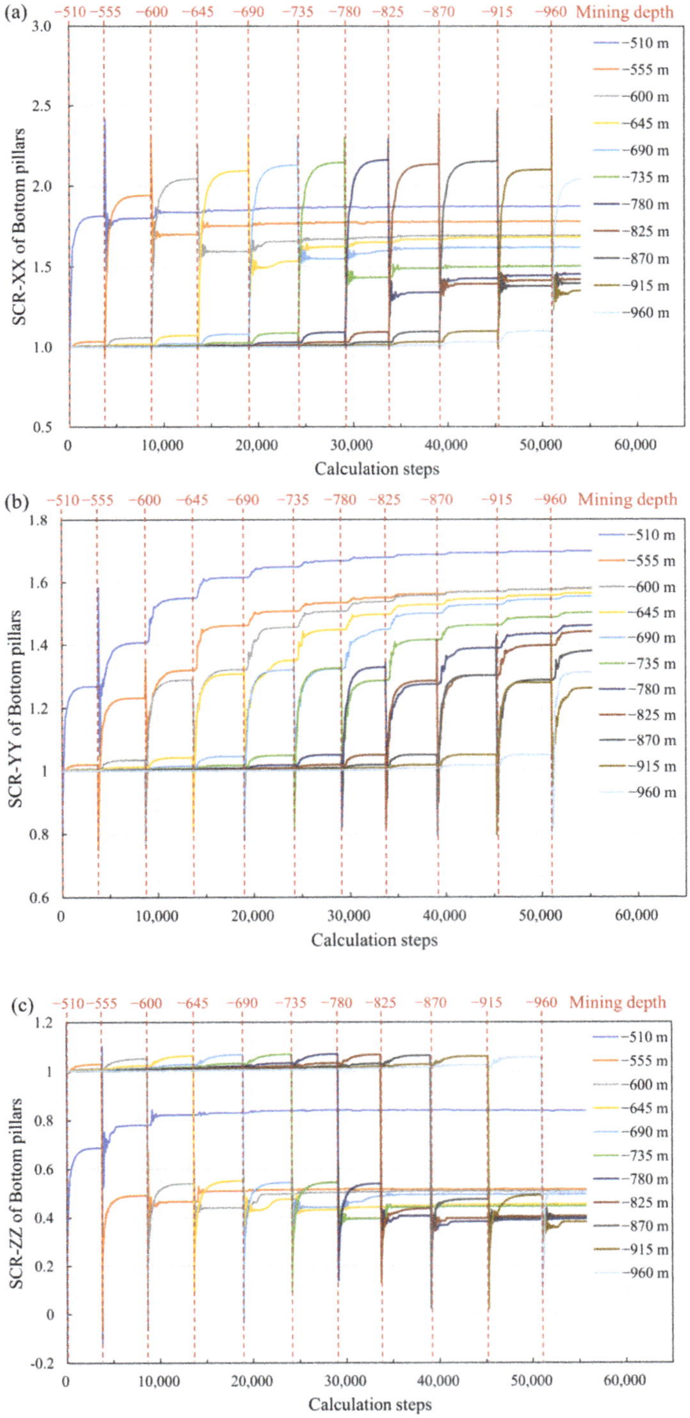

Figure 10. Variation regularity of stress ratio of bottom columns at various depths. (**a**) Stress ratio in XX direction, (**b**) Stress ratio in YY direction, and (c) Stress ratio in ZZ direction.

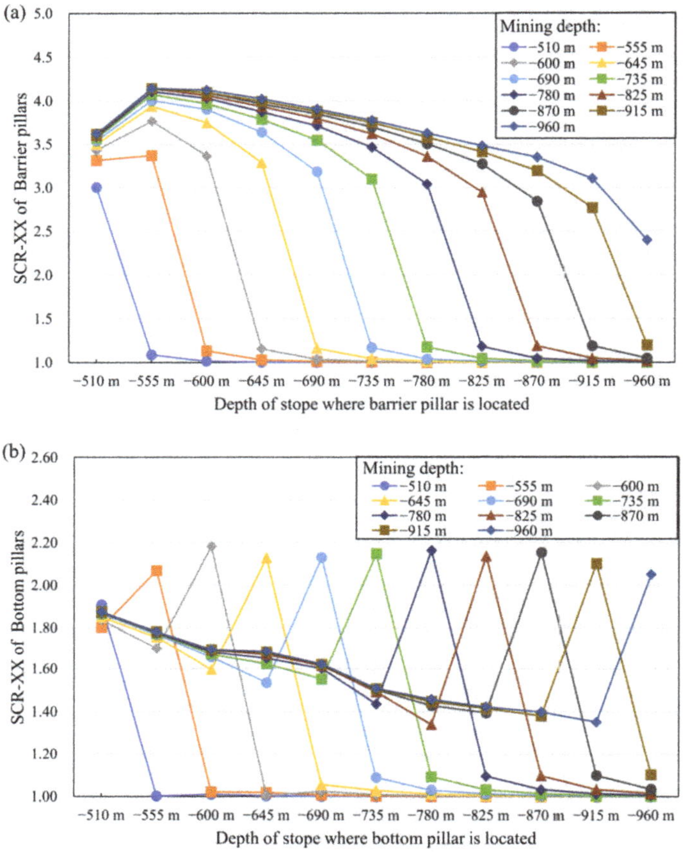

Figure 11. SCR-XX curves of (**a**) barrier pillars and (**b**) bottom pillars at different depths after excavation of stopes at different depths.

The SCR-XX of most barrier pillars in mined stopes is over 3.0, as shown in Figure 11, because the backfill well encapsulates the barrier pillars. On the contrary, the bottom pillar's SCR rapidly increases from around 1.0 to above 2.0 after mining the stope above and decreases significantly after mining the stope below. The main reason for this phenomenon is that the bottom pillars are not fully contacted with the backfill below. Therefore, when mining deep stopes, compared with the barrier pillars, the bottom pillars will have a significant stress drop, and it is difficult to withstand more mining-induced stress.

As the mining depth increases, barrier pillars and bottom pillars' SCR decrease after mining the stopes. More mining stress is transferred to the strata under the excavation stope, and a stress concentration area forms in the strata which will be excavated in the next step. It can be seen from Figures 6 and 7 that there is a stress concentration area below the deepest stope where the stress is greater than the in situ stress at the same level. The deeper the mining depth, the more significant the stress concentration area is. This phenomenon increases the risk of rockburst when excavating tunnels in deep strata.

The upper stopes need to bear more transfer loads to reduce the stress concentration in the strata below the deepest stope. The methods suitable for engineering are increasing the size and quantity of ore pillars and increasing the backfill strength. Increasing the size and amount of barrier pillars and bottom pillars will reduce the resource recovery rate of mining. Therefore, increasing the backfill strength is a more economical method to reduce the stress concentration of the strata below.

4.3. Influence of Mechanical Properties of the Backfill on the Stress Evolution

Studying the influence of mechanical properties on the stress evolution of the backfilled stope can explain the effect of increasing the strength of the backfill to reduce the stress concentration of the stratum below the deepest stope. In the numerical simulation for comparison, the stopes below −870 m are filled with low strength cemented tailings, high strength cemented tailings, and cemented block stones. These simulation models are used to compare the effects of different filling bodies on the stress concentration of the underlying stratum. The stress of the orebody in the next layer of the stope after filling the mining stope was recorded. Then the stress was used to calculate the SCR. The SCR of the orebody in the next layer of the excavated stope varies with the mining depth, as shown in Figure 12.

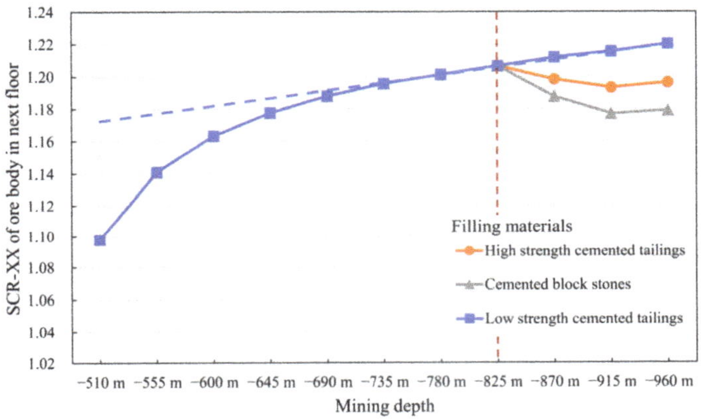

Figure 12. Variation of SCR with mining depth in the next layer of the excavated stope.

In the simulation, the stopes above −825 m are still filled with the low-strength cemented tailings. From Figure 12, the SCR of the orebody below the mining stope gradually rises with the increase of mining depth. Starting from mining −735 m, the SCR of the orebody under the mining stope increases linearly. When mining the stope above −735 m, the SCR of the orebody below the stope is lower because the overburden stratum can withstand mining stress greater than the barrier pillars and bottom pillars of stopes. When the mining depth exceeds −735 m, the vertical distance from the overlying stratum to the bottom of the stope is more than 270 m. After that, the mining stress is mainly borne by the pillars above the mining stope and the underlying stratum. Therefore, the SCR of the orebody below the stope increases linearly with the mining depth.

High strength cemented tailings and cemented block stones are used to replace low strength cemented tailings for filling from the −870 m. From the simulation results of Figure 12, the SCR curve showed a significant drop, indicating that the high strength filling method has indeed played a role in improving the stress of the orebody to be mined. The strength and elastic modulus of the backfill is higher, and the SCR of the underlying stratum is smaller.

5. Conclusions

Differences in mechanical properties between deep and shallow rock masses pose challenges in analysing the evolution law of mining stress in inclined backfilled stopes in deep mining. In this manuscript, a new method is proposed to determine the parameters of the brittle-ductile transition model corresponding to the H-B criterion and the M-C criterion, respectively. Next, the proposed method is verified with the conventional triaxial compression numerical test. Then, the model composed of multiple backfilled stopes with different depths is established. Finally, we discussed the formation displacement, the stress distribution of rock mass, the risk factor, and the general law of stress evolution in the backfilled stope according to the variation laws of SCR. The primary conclusions can be presented as follows:

(1) The mining-induced stress will transfer to the shallow stope after mining and filling the inclined orebody. Therefore, the largest horizontal displacement of the formation, the highest stress area of the stratum, and the largest risk factor area all appear in one to two layers above the deepest stope. The most dangerous place is farther from the deepest stope as the mining depth increases. The SCR-XX of the shallow barrier pillars will continue to grow when mining and filling the deep stopes. Under the action of new transfer stress, the backfilled stope subjected to higher stress may cause disasters such as rockburst. The excavations closer to the orebody such as haulage drifts are typically exhibiting varying degrees of squeezing [43]. Therefore, the strength design of the pillars and the backfill should have a large safety factor when mining inclined deep stopes.
(2) Barrier pillars and bottom pillars are the main support structures of the stope. Due to the brittle-ductile transition behaviour of rock mass, the backfill can increase the bearing capacity of pillars by increasing the minimum principal stress. After mining the deep stopes, the transfer stress of the barrier pillars in shallow stopes can always increase because the backfill wraps the barrier pillars nicely. On the contrary, the transfer stress that bottom pillars can withstand will decrease significantly after excavating the stope below because of the backfill gap.
(3) Mining deep ore bodies transfers more mining stress to shallow stopes and the underlying stratum. The barrier pillars and bottom pillars' SCR in the deep part is lower than in the shallow part since the mining-induced stress in the deep part has moved to the shallow part, which has a larger strength safety reserve. Similarly, more mining stress is transferred to the underlying stratum as the mining depth increases and forms a more obvious high-stress concentration area. The apparent stress of seismic events at the deep metal mine shows that the stress of the orebody below the excavated stope and the surrounding rock of the filled stope above will increase after the stope is excavated [44].
(4) Part of the mining stress transferred to the underlying stratum after mining and filling the inclined orebody will form a stress concentration area below the deepest stope. The SCR monitoring results of the underlying stratum shows that the ratio of the transfer stress to in situ stress of the underlying stratum increases as the mining depth increases. Replacing low-strength backfill with high-strength backfill can reduce the stress concentration of the underlying stratum and improve the stress environment in the following mining operation.

Author Contributions: Conceptualization, Y.Z. and J.Z.; methodology, Y.Z.; software, J.M.; validation, Y.Z., X.C. and J.Z.; resources, Y.Z.; data curation, J.Z.; writing—original draft preparation, Y.Z.; writing—review and editing, J.Z.; visualization, X.C.; supervision, G.Z.; project administration, G.Z. All authors have read and agreed to the published version of the manuscript.

Funding: This research is supported by financial grants from the National Key Research and Development Program of China (2018YFC0604606), the National Natural Science Foundation for Young Scientists of China (52104111), the Natural Science Foundation of Hunan Province (2021JJ30819), and the Fundamental Research Funds for the Central Universities of Central South University (2019zzts306).

Institutional Review Board Statement: Not applicable.

Informed Consent Statement: Not applicable.

Conflicts of Interest: The authors declare no conflict of interest.

References

1. Sobhi, M.A.; Li, L. Numerical investigation of the stresses in backfilled stopes overlying a sill mat. *J. Rock Mech. Geotech. Eng.* **2017**, *9*, 490–501. [CrossRef]
2. Raffaldi, M.J.; Seymour, J.B.; Richardson, J.; Zahl, E.; Board, M. Cemented Paste Backfill Geomechanics at a Narrow-Vein Underhand Cut-and-Fill Mine. *Rock Mech. Rock Eng.* **2019**, *52*, 4925–4940. [CrossRef] [PubMed]
3. Xinmin, W.; Desheng, G.; Qinli, Z. *Theory of Backfilling Activity and Pipeline Transportation Technology of Backfill in Deep Mines*; Central South University Press: Changsha, China, 2010; ISBN 978-7-5487-0152-1.
4. Chen, X.; Li, L.; Wang, L.; Qi, L. The current situation and prevention and control countermeasures for typical dynamic disasters in kilometer-deep mines in China. *Saf. Sci.* **2019**, *115*, 229–236. [CrossRef]
5. Wang, S.; Li, X.; Yao, J.; Gong, F.; Li, X.; Du, K.; Tao, M.; Huang, L.; Du, S. Experimental investigation of rock breakage by a conical pick and its application to non-explosive mechanized mining in deep hard rock. *Int. J. Rock Mech. Min. Sci.* **2019**, *122*, 104063. [CrossRef]
6. Wagner, H. Deep Mining: A Rock Engineering Challenge. *Rock Mech. Rock Eng.* **2019**, *52*, 1417–1446. [CrossRef]
7. Cai, X.; Cheng, C.; Zhao, Y.; Zhou, Z.; Wang, S. The role of water content in rate dependence of tensile strength of a fine-grained sandstone. *Arch. Civ. Mech. Eng.* **2022**, *22*, 58. [CrossRef]
8. Zhou, Z.L.; Zhao, Y.; Cao, W.Z.; Chen, L.; Zhou, J. Dynamic Response of Pillar Workings Induced by Sudden Pillar Recovery. *Rock Mech. Rock Eng.* **2018**, *51*, 3075–3090. [CrossRef]
9. Xiao, P.; Li, D.; Zhao, G.; Liu, H. New criterion for the spalling failure of deep rock engineering based on energy release. *Int. J. Rock Mech. Min. Sci.* **2021**, *148*, 104943. [CrossRef]
10. Li, L.; Aubertin, M. An improved analytical solution to estimate the stress state in subvertical backfilled stopes. *Can. Geotech. J.* **2008**, *45*, 1487–1496. [CrossRef]
11. Li, L.; Aubertin, M. An analytical solution for the nonlinear distribution of effective and total stresses in vertical backfilled stopes. *Geomech. Geoengin.* **2010**, *5*, 237–245. [CrossRef]
12. Ting, C.H.; Sivakugan, N.; Read, W.; Shukla, S.K. Analytical Expression for Vertical Stress within an Inclined Mine Stope with Non-parallel Walls. *Geotech. Geol. Eng.* **2014**, *32*, 577–586. [CrossRef]
13. Li, L.; Aubertin, M. Numerical Investigation of the Stress State in Inclined Backfilled Stopes. *Int. J. Geomech.* **2009**, *9*, 52–62. [CrossRef]
14. Jahanbakhshzadeh, A.; Aubertin, M.; Li, L. A New Analytical Solution for the Stress State in Inclined Backfilled Mine Stopes. *Geotech. Geol. Eng.* **2017**, *35*, 1151–1167. [CrossRef]
15. Jahanbakhshzadeh, A.; Aubertin, M.; Li, L. Analysis of the Stress Distribution in Inclined Backfilled Stopes Using Closed-form Solutions and Numerical Simulations. *Geotech. Geol. Eng.* **2018**, *36*, 1011–1036. [CrossRef]
16. Yan, B.; Zhu, W.; Hou, C.; Guan, K. A three-dimensional analytical solution to the arching effect in inclined backfilled stopes. *Geomech. Geoengin.* **2019**, *14*, 136–147. [CrossRef]
17. Falaknaz, N.; Aubertin, M.; Li, L. Numerical Analyses of the Stress State in Two Neighboring Stopes Excavated and Backfilled in Sequence. *Int. J. Geomech.* **2015**, *15*, 04015005. [CrossRef]
18. Liu, Z.X.; Lan, M.; Xiao, S.Y.; Guo, H.Q. Damage failure of cemented backfill and its reasonable match with rock mass. *Trans. Nonferrous Met. Soc. China (Engl. Ed.)* **2015**, *25*, 954–959. [CrossRef]
19. Cao, S.; Yilmaz, E.; Song, W.; Xue, G. Assessment of Acoustic Emission and Triaxial Mechanical Properties of Rock-Cemented Tailings Matrix Composites. *Adv. Mater. Sci. Eng.* **2019**, *2019*, 6742392. [CrossRef]
20. Yu, X.; Kemeny, J.; Tan, Y.; Song, W.; Huang, K. Mechanical properties and fracturing of rock-backfill composite specimens under triaxial compression. *Constr. Build. Mater.* **2021**, *304*, 124577. [CrossRef]
21. Thompson, B.D.; Grabinsky, M.W.; Bawden, W.F.; Counter, D.B. In-situ measurements of cemented paste backfill in long-hole stopes. In Proceedings of the 3rd CANUS Rock Mechanics Symposium, Toronto, ON, Canada, 9–15 May 2009; pp. 197–198.
22. Qi, C.; Fourie, A. Numerical Investigation of the Stress Distribution in Backfilled Stopes Considering Creep Behaviour of Rock Mass. *Rock Mech. Rock Eng.* **2019**, *52*, 3353–3371. [CrossRef]
23. Xue, D.; Wang, J.; Zhao, Y.; Zhou, H. Quantitative determination of mining-induced discontinuous stress drop in coal. *Int. J. Rock Mech. Min. Sci.* **2018**, *111*, 1–11. [CrossRef]
24. Wu, X.; Li, G.; Luo, F.; Duan, S. A Study on the Distribution and Evolution of Mining Induced Stress under the Condition of Multiple Mining. *Geotech. Geol. Eng.* **2021**, *39*, 1637–1648. [CrossRef]
25. Shankar, V.; Kumar, D.; Subrahmanyam, D. Impact and Severity of Deep Excavations on Stress Tensors in Mining. *J. Min. Sci.* **2019**, *55*, 213–218. [CrossRef]
26. Hu, Y.; Zhang, J.; Li, C.; Song, Z.; Xiao, Y.; Wang, Y. Characteristics and Time-Space Evolution of Mining Stress in High Stope. *Adv. Mater. Sci. Eng.* **2021**, *2021*, 2785933. [CrossRef]
27. Wawersik, W.R.; Fairhurst, C. A study of brittle rock fracture in laboratory compression experiments. *Int. J. Rock Mech. Min. Sci. Geomech. Abstr.* **1970**, *7*, 561–575. [CrossRef]
28. Li, D.; Gao, F.; Han, Z.; Zhu, Q. Experimental evaluation on rock failure mechanism with combined flaws in a connected geometry under coupled static-dynamic loads. *Soil Dyn. Earthq. Eng.* **2020**, *132*, 106088. [CrossRef]
29. Cui, L.; Fall, M. An evolutive elasto-plastic model for cemented paste backfill. *Comput. Geotech.* **2016**, *71*, 19–29. [CrossRef]

30. Yang, L.; Xu, W.; Yilmaz, E.; Wang, Q.; Qiu, J. A combined experimental and numerical study on the triaxial and dynamic compression behavior of cemented tailings backfill. *Eng. Struct.* **2020**, *219*, 110957. [CrossRef]
31. Wang, S.; Tang, Y.; Wang, S. Influence of brittleness and confining stress on rock cuttability based on rock indentation tests. *J. Cent. South Univ.* **2021**, *28*, 2786–2800. [CrossRef]
32. Xie, H.; Li, C.; He, Z.; Li, C.; Lu, Y.; Zhang, R.; Gao, M.; Gao, F. Experimental study on rock mechanical behavior retaining the in situ geological conditions at different depths. *Int. J. Rock Mech. Min. Sci.* **2021**, *138*, 104548. [CrossRef]
33. Li, D.; Sun, Z.; Xie, T.; Li, X.; Ranjith, P.G. Energy evolution characteristics of hard rock during triaxial failure with different loading and unloading paths. *Eng. Geol.* **2017**, *228*, 270–281. [CrossRef]
34. Wu, X.; Jiang, Y.; Guan, Z. A modified strain-softening model with multi-post-peak behaviours and its application in circular tunnel. *Eng. Geol.* **2018**, *240*, 21–33. [CrossRef]
35. Fang, Z.; Harrison, J.P. A mechanical degradation index for rock. *Int. J. Rock Mech. Min. Sci.* **2001**, *38*, 1193–1199. [CrossRef]
36. Hoek, E.; Brown, E.T. Practical estimates of rock mass strength. *Int. J. Rock Mech. Min. Sci.* **1997**, *34*, 1165–1186. [CrossRef]
37. Hoek, E.; Brown, E.T. The Hoek–Brown failure criterion and GSI – 2018 edition. *J. Rock Mech. Geotech. Eng.* **2019**, *11*, 445–463. [CrossRef]
38. Lu, C. Reaserch on the Energy System of Deep Granite Considering the Stored Energy and Dissipated Energy Characteristics. Ph.D. Thesis, University of Science and Technology Beijing, Beijing, China, 2019.
39. Kang, P.; Zhaopeng, L.; Quanle, Z.; Zhenyu, Z.; Jiaqi, Z. Static and Dynamic Mechanical Properties of Granite from Various Burial Depths. *Rock Mech. Rock Eng.* **2019**, *52*, 3545–3566. [CrossRef]
40. Hou, P.Y.; Cai, M.; Zhang, X.W.; Feng, X.T. Post-peak Stress–Strain Curves of Brittle Rocks Under Axial- and Lateral-Strain-Controlled Loadings. *Rock Mech. Rock Eng.* **2022**, *55*, 855–884. [CrossRef]
41. Li, P.; Cai, M.; Guo, Q.; Miao, S. In Situ Stress State of the Northwest Region of the Jiaodong Peninsula, China from Overcoring Stress Measurements in Three Gold Mines. *Rock Mech. Rock Eng.* **2019**, *52*, 4497–4507. [CrossRef]
42. Jiang, G.; Zuo, J.; Li, Y.; Wei, X. Experimental Investigation on Mechanical and Acoustic Parameters of Different Depth Shale Under The Effect of Confining Pressure. *Rock Mech. Rock Eng.* **2019**, *52*, 4273–4286. [CrossRef]
43. Mercier-Langevin, F. LaRonde Extension–mine design at three kilometres. *Min. Technol.* **2011**, *120*, 95–104. [CrossRef]
44. Brown, L.; Hudyma, M. Identification of Stress Change within a Rock Mass Through Apparent Stress of Local Seismic Events. *Rock Mech. Rock Eng.* **2017**, *50*, 81–88. [CrossRef]

mathematics

MDPI

Article

Optimization of Genetic Algorithm through Use of Back Propagation Neural Network in Forecasting Smooth Wall Blasting Parameters

Ying Chen [1,2,3], Shirui Chen [1], Zhengyu Wu [4,*], Bing Dai [1,*], Longhua Xv [3] and Guicai Wu [2]

1 School of Resource Environment and Safety Engineering, University of South China, Hengyang 421001, China; usc_chen@usc.edu.cn (Y.C.); 20212002110038@stu.usc.edu.cn (S.C.)
2 China Tin Group Co., Ltd., Liuzhou 545026, China; 13877206058@163.com
3 School of Environment and Resource, Southwest University of Science and Technology, Mianyang 621010, China; neuxulonghua@163.com
4 School of Engineering, Fujian Jiangxia University, Fuzhou 350108, China
* Correspondence: zhengyu_wu@fjjxu.edu.cn (Z.W.); daibingusc@usc.edu.cn (B.D.); Tel.: +86-159-8023-9390 (Z.W.); +86-178-7341-5528 (B.D.)

Abstract: With the continuous development in drilling and blasting technology, smooth wall blasting (SWB) has been widely applied in tunnel construction to ensure the smoothness of tunnel profile, diminish overbreak and underbreak, and preserve the tunnel's interior design shape. However, the complexity of the actual engineering environment and the deficiency of current optimization theories have posed certain challenges to the optimization of SWB parameters under arbitrary geological conditions, on the premise that certain control targets are satisfied. Against the above issue, a genetic algorithm (GA) and back propagation (BP) neural network-based computational model for SWB design parameter optimization is proposed. This computational model can comprehensively reflect the relation among geological conditions, design parameters, and results by training and testing the 285 collected sets of test data samples at different conditions. Moreover, it automatically searches optimal blasting design parameters through the control of SWB targets to acquire the optimal design parameters based on specific geological conditions of surrounding rocks and under the specified control targets. When the optimization algorithm is compared with other current optimization algorithms, it is shown that this algorithm has certain computational superiority over the existing models. When the optimized results are applied in practical engineering, it is shown that in overall consideration of the geological conditions, control targets, and other influencing factors, the proposed GA_BP-based model for SWB parameter optimization has high feasibility and reliability, and that its usage can be generalized to analogous tunneling works.

Keywords: genetic algorithm; BP neural network; smooth wall blasting; parameter optimization

MSC: 00

Citation: Chen, Y.; Chen, S.; Wu, Z.; Dai, B.; Xv, L.; Wu, G. Optimization of Genetic Algorithm through Use of Back Propagation Neural Network in Forecasting Smooth Wall Blasting Parameters. *Mathematics* **2022**, *10*, 1271. https://doi.org/10.3390/math10081271

Academic Editor: Mario Versaci

Received: 21 February 2022
Accepted: 6 April 2022
Published: 11 April 2022

1. Introduction

Drilling and blasting method is the main technique for tunnel construction [1]. With the progress in construction techniques, the SWB technique has been widely accepted in drilling and blasting operations. SWB can effectively control the blasting effect, and hence diminish overbreak and underbreak, keep the tunnel's rocky wall smooth, and further maintain the stability of surrounding rocks, lessen the supporting workload, reduce the support materials and engineering costs required, and shorten the construction period [2]. The core issue of SWB lies in the control of overbreak and underbreak [3]. Mahtab [4] believes that the combination of traditional blasting methods with simulation technology can assist in the further evaluation of overbreak and underbreak through the tunnel while the tunnel is drilled and blasted forward; the research outcomes can predict the total overbreak and underbreak through the tunnel and further define the confidence interval of the probability of the region where overbreak and underbreak appear. With further development in computer technology, the machine learning method has been widely accepted and used by more and more scholars to predict overbreak and underbreak in surface blasting operations [5,6]. Likewise, the theoretical and experimental studies on overbreak and underbreak have also achieved new breakthroughs [7–9] and explained macroscopic and microscopic reasons for overbreak and underbreak [10–12].

According to research findings, numerous parameters are accountable for the SWB effect, including the physical and mechanical properties of mineral rocks (such as *compressive and tensile strengths, joint development degree*, etc.), blasting design parameters (such as *hole pitch, array pitch, blast hole depth, charge concentration*, etc.), and evaluation indexes of blasting effect (such as *average linear overbreak, average linear underbreak*, etc.). Therefore, the selection of influencing factors for SWB is a multilevel, multifactor, multigoal complex decision-making process, with extremely convoluted uncertainty and nonlinear relations between blasting parameters and results [13,14]. However, in the current stage, SWB parameters are determined by the mere empirical method or in mere consideration of one or more simple factors; thus, blasting design parameters are extremely subjective and random [15]. The optimization of SWB parameters is always a challenge under any geological condition with certain control targets (blasting construction targets) [16].

In practical engineering, especially underground engineering, field test data are typically finite and discrete due to a myriad of limits of the field environment [17]. A mainstream solution to the global optimization problem with finite and discrete samples in underground engineering is the support vector machine (SVM) [18,19] and artificial neural network (ANN) [20,21]. The core of SVM is the minimum structural risk such that it has small sample demand and low fitting precision. Therefore, it is more applicable in parameter optimization problems with small numbers of parameters and samples. However, with the continuous development in blasting technology, more and more influencing factors need to be taken into consideration, and higher and higher requirement is raised on the precision of design parameters. Therefore, the minimum empirical risk-based neural network technology has received more attention and is more commonly used [22,23].

This study proposes an improved neural network algorithm: the GA_BP neural network algorithm, which has optimized the neural network's initial weights and thresholds, enhanced the fitting precision of BP neural network under small and medium sample sizes, and collected the SWB parameters from other engineering projects under different geological conditions. With the 145 sets of measured data of SWB including the above data as the training samples, and with the 20 sets of field test data in the East Tianshan tunneling project in Xinjiang as the test samples, the nonlinear mapping relation between blasting design parameters and blasting results is obtained through GA_BP neural network fitting. On this base, the blasting effect parameters and parts of the blasting design parameters are controlled as per the practical engineering requirements, and the optimal solutions of blasting design parameters under the control conditions are searched for automatically, so as to achieve the optimization of the design parameters.

In the work of this paper, the coupling algorithm for the GA and BP neural network has been described in Section 2. Section 3 introduces the computation flow of the GA_BP neural network algorithm-based SWB parameter optimization model. Section 4 analyzes, evaluates, and verifies the optimized results in combination with an engineering case. Section 5 is the conclusion.

2. Methods

The traditional BP neural network is prone to parameter underfitting due to improper selection of initial parameters while training with small and medium samples. To address this problem, the genetic algorithm (GA) is combined with the BP neural network. The preferable weights and thresholds of the initial network are obtained by GA, and thus the fitting precision of the BP neural network is improved [24–27].

2.1. Genetic Algorithm (GA)

The implementation of GA includes the following 5 steps: [28]

(a) Population initialization.

Individuals are encoded by the real coding method. Each individual is a real string composed of 4 components: weight of connection between the input layer and the hidden layer, threshold of the hidden layer, weight of connection between the hidden layer and the output layer, and threshold of the output layer. The individuals comprise all weights and thresholds of the neural network. Provided that the network structure is known, a network with a definite mapping structure, number of nodes, weights, and thresholds can be constructed.

(b) Fitness function.

According to the initial values of the BP neural network obtained by the individuals, the system output is predicted after training the BP neural network using the training data, and the absolute value of the error and the variance E between the predicted output and the expected output as individual fitness F is taken, as calculated by the Equation (1):

$$F = k(\sum_{i=1}^{n} abs(y_i - o_i)) \tag{1}$$

where n is the number of the network's output nodes; y_i is the expected output of the ith node of the BP neural network; o_i is the predicted output of the ith node; k is the coefficient for normalization; in this paper, $k = 1$.

(c) Selection operation.

The selection probability, p_i, of each individual i is:

The selection operation in GA is based on the selection strategy of fitness proportion.

$$f_i = k/F_i \tag{2}$$

$$p_i = \frac{f_i}{\sum\limits_{j=1}^{N} f_j} \tag{3}$$

where F_i is the fitness value of individual i. As it is preferred that fitness be as small as possible, the fitness value shall be inverted prior to individual selection. k is the coefficient with the same value as Formula (1); N is the number of individuals in the population.

(d) Crossover operation.

As individuals are encoded by real coding, the crossover operation is performed by a real number crossover method. The crossover operation on the kth chromosome a_k with the lth chromosome a_l at position j is performed by the following method:

$$\begin{aligned} a_{kj} &= a_{kj}(1-b) + a_{lj}b \\ a_{lj} &= a_{ij}(1-b) + a_{kj}b \end{aligned} \tag{4}$$

where: b is a random number within $(0, 1)$.

(f) Mutation operation.

The jth gene of the ith individual, a_{ij}, is selected to undergo mutation, as operated by the following method:

$$\begin{aligned} a_{ij} &= a_{ij} + (a_{ij} - a_{\max}) \times f(g) \qquad r > 0.5 \\ a_{ij} &= a_{ij} + (a_{\min} - a_{ij}) \times f(g) \qquad r \leq 0.5 \end{aligned} \tag{5}$$

where a_{max} is the upper bound to gene a_{ij}; a_{min} is the lower bound to gene a_{ij}; $f(g) = r_2(1 - g/G_{max})^2$; r_2 is a random number; g is the count up to the current iteration; G_{max} is the maximum evolution count; r is a random number used to judge the mutation operation within (0, 1), which is automatically generated when selecting a_{ij}.

2.2. BP Neural Network

The BP neural network is a multilayer feedforward neural network that can be regarded as a nonlinear function, whose independent and dependent variables are the network's input value and predicted value, respectively. When the number of input nodes is n and the number of output nodes is m, the BP neural network expresses the function mapping relation from the n independent variables to the m dependent variables. The BP neural network shall be trained prior to prediction so that it is endowed with associative memory and predictive ability. The training process of BP neural network includes the following 7 steps: [29]

(a) Network initialization.

According to the system's input and output sequences (X, Y), determine the number of nodes, n, at the network's input layer, the number of nodes, l, at the hidden layer, the number of nodes, m, at the output layer, initialize the weights w_{ij} and w_{jk} of connections between neurons at the input layer, hidden layer, and the output layer, respectively, and the threshold, a, of the hidden layer and the threshold, b, of the output layer, and give the learning rate and the neuron excitation function.

(b) Hidden layer output calculation.

According to the input variable X, weight w_{ij} of connection between the input layer and the hidden layer, and threshold a of the hidden layer, calculate the hidden layer output H.

$$H_j = f\left(\sum_{i=1}^{n} w_{ij}x_i - a_j\right) \qquad j = 1, 2, \ldots, l \tag{6}$$

where l is the number of nodes at the hidden layer; f is the excitation function of the hidden layer, which can be expressed in many ways. Here, the excitation function is selected as:

$$f(x) = \frac{1}{1 + e^{-x}} \tag{7}$$

(c) Output layer output calculation.

According to the hidden layer output H and the weight of connection w_{ij} and threshold b, calculate the predicted output O of the BP neural network.

$$O_k = \sum_{j=1}^{l} H_j w_{jk} - b_k \qquad k = 1, 2, \ldots, m \tag{8}$$

(d) Error calculation.

According to the network's predicted output O and expected output Y, calculate the network's predicted error e.

$$e_k = Y_k - O_k \qquad k = 1, 2, \ldots, m \tag{9}$$

(e) Weights update.

According to the network's predicted error e, update the weights w_{ij} and w_{jk} of network connections.

$$w_{ij} = w_{ij} + \eta H_j(1 - H_j)x(i)\sum_{k=1}^{m} w_{jk}e_k \qquad i = 1, 2, \ldots, n; j = 1, 2, \ldots, l \tag{10}$$

$$w_{jk} = w_{jk} + \eta H_j e_k \qquad j = 1,2,\ldots,l; k = 1,2,\ldots,m \tag{11}$$

where η is the learning rate.

(f) Thresholds update.

According to the network's predicted error e, update the thresholds a and b of network nodes.

$$a_j = a_j + \eta H_j(1 - H_j)x(i)\sum_{k=1}^{m} w_{jk}e_k \qquad j = 1,2,\ldots,l \tag{12}$$

$$b_k = b_k + e_k \qquad k = 1,2,\ldots,m \tag{13}$$

(g) Judge whether the algorithm iterations come to an end; if not, return to Step (b).

2.3. GA_BP Neural Network

The parameter fitting calculation of the GA_BP neural network falls into two components: the BP neural network and GA optimization. The calculation flowchart is shown in Figure 1.

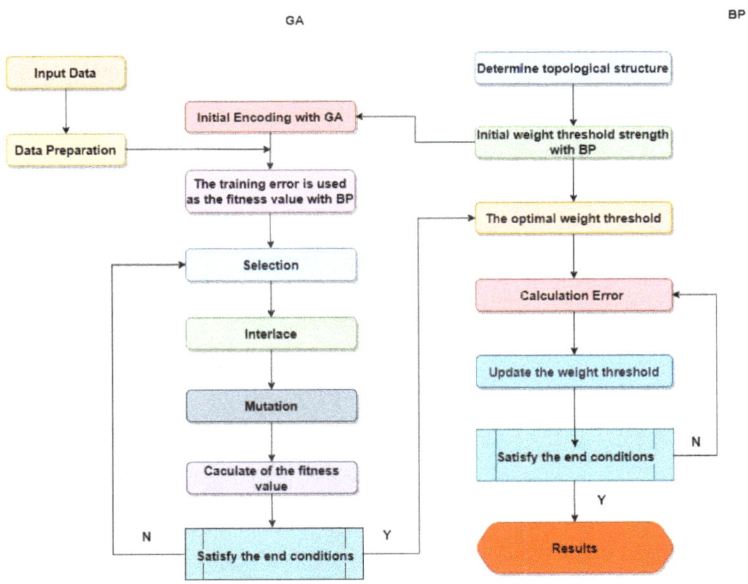

Figure 1. The algorithm structure of GA-BP neural network.

Where the BP neural network is composed of two parts: One is the part of the BP neural network structure determination, namely to determine the results of the BP neural network, and thereby the length of GA individuals, according to the number of input and output parameters of the fitting function; the other is the part of the BP neural network parameter fitting, which is responsible for fitting the neural network's input and output parameters, determining the nonlinear relations between parameters, and predicting the function output after network training. The GA optimization component uses the weights and thresholds of the genetic optimization algorithm of the BP neural network. Each individual in the population comprises all weights and thresholds of a network. Individuals' fitness is calculated via the fitness function, in order for the genetic algorithm to find the individual with the optimal fitness value through selection, crossover, and mutation operations, and thus to optimize the initial weights and thresholds of the BP neural network. The concrete procedure of GA_BP neural network goes as follows:

(a) Determine the input and output parameters of GA_BP neural network; (b) initialize the weights and thresholds between the initial parameters of the BP neural network; (c) optimize the above weights and thresholds by GA and select the optimal ones; (d) fit the input and output parameters of GA_BP neural network; (e) error test as to whether the requirement is met, if yes end the calculation, or else return to Step (c).

3. Optimization Model

3.1. Parameter Optimization

The calculation for the SWB parameter optimization is to determine the nonlinear relations between input parameters and output parameters by fitting all these parameters via the GA_BP neural network, and to implement the prediction of the output results under the condition of input parameters; next, through the control of the output results, it automates the optimization model to search for the optimal solutions among the input parameters. The calculation process falls into two parts: One is the fitting of GA_BP neural network parameters, and the other is the calculation for parameter optimization. The concrete calculation model is shown in Figure 2.

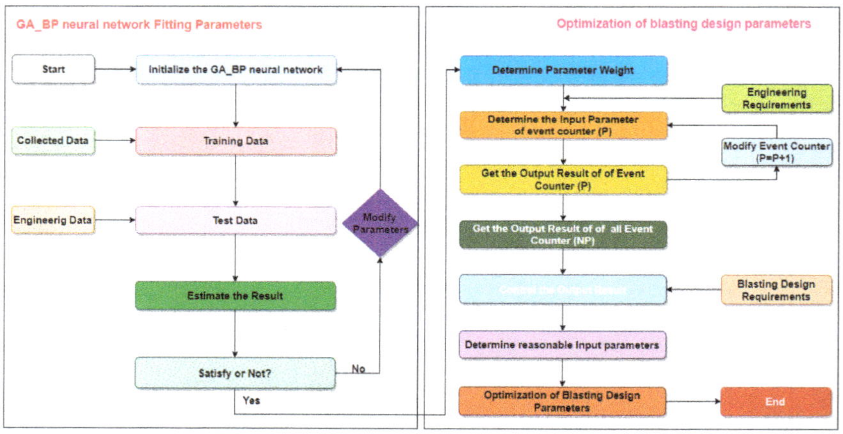

Figure 2. Schematic diagram of the calculation model for parameter optimization.

The concrete steps of calculation for the SWB parameter optimization go as follows:

- Initialize the GA_BP neural network and determine the optimal neural network parameters;
- Fit the input parameters and output parameters of SWB among the training samples of the GA_BP neural network;
- After fitting the parameters of the GA_BP neural network, conduct an error analysis into the GA_BP neural network using the test samples. If the requirement is met, go to the next step, else return to the first step and modify the basic parameters of the neural network;
- When the fitting results meet the requirement of test error, the nonlinear relations between the input and output parameters are reflected, and thus the prediction of results is implemented;
- According to the purpose of practical engineering and the parameter design requirement (input parameter control), determine the calculation parameters when the event counter P = 1;
- Try figuring out the SWB calculation results under the feasible condition when the event counter P = 1, using the nonlinear relations reflected in the GA_BP neural network.

- After going through the case when the event counter P = 1, determine the calculation parameters when the event counter P = 2, and return to the previous step for calculation until covering all event counters for P = P + 1;
- Control the calculation results and thus automatically search for the optimal values of them under this trial condition, so as to derive the optimal design parameters by inversion and implement the optimization of SWB design parameters.

3.2. Parameters

In practical tunneling, SWB involves numerous data. Taking the parameters into consideration of the calculation model for SWB parameter optimization is bound to the problems of data redundancy and computational complexity. Therefore, it is necessary to screen the important parameters in SWB and perform the corresponding simplification of the concrete fitting parameters. In general, the SWB effect is subject mainly to geological conditions and blasting parameters. Accordingly, the selected parameters should include factors in three aspects: geological conditions, blasting design parameters, and blasting result parameters. Considering the practical engineering application, and for the convenience of the uniform measurement of test data, the SWB parameters are simplified with overall consideration of the blasting design standard and scholars' research outcomes [8,30,31]. Figure 3 is a schematic diagram of the SWB design parameters. Figure 4 is a schematic diagram for the calculation of average linear overbreak and underbreak.

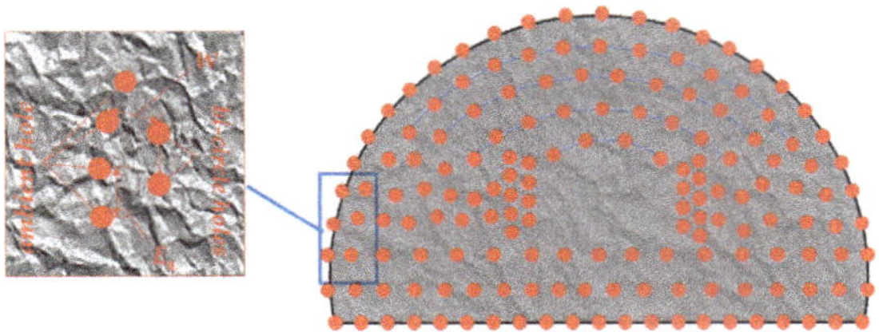

Figure 3. SWB design parameters.

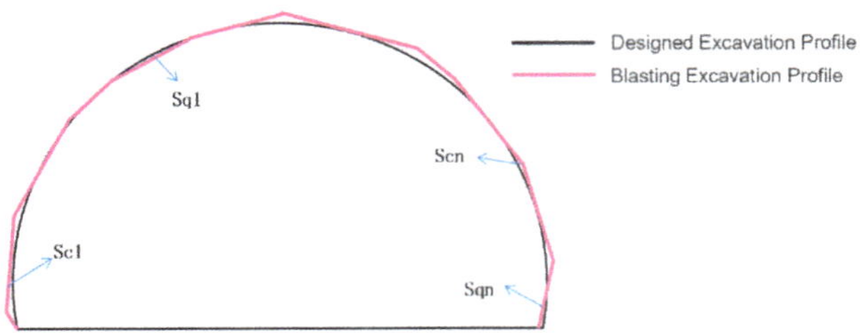

Figure 4. Schematic diagram for the calculation of average overbreak and underbreak.

Average overbreak and underbreak are calculated by the formulae:

$$S_c = \frac{S_{c1} + S_{c2} + \ldots + S_{cn}}{I_{c1} + I_{c2} + \ldots + I_{cn}}$$

for *average linear overbreak*, and:

$$S_q = \frac{S_{q1} + S_{q2} + \ldots + S_{qn}}{I_{q1} + I_{q2} + \ldots + I_{qn}}$$

for *average linear underbreak*, where I_c and I_q correspond, respectively, to the arclength corresponding to each overbreak area and underbreak area.

For optimization calculation, the input parameters are the ones that affect the blasting results, whereas the output parameters are the blasting results. Moreover, in practical engineering, due to the influences of geological conditions, boring equipment, and tunneling purpose, the parameters already have fixed values or designed values and cannot or need not be optimized further. Therefore, the core purpose of SWB parameter optimization is to search for the optimal solutions of *spacing between lines of least resistance W*, *auxiliary hole pitch* E_b, and *ambient hole pitch* E_a, under the condition of controlling the *average linear overbreak* and *average linear underbreak*.

3.3. Determination of GA_BP Neural Network Topology and Basic Parameters

Through the above description, the number of input parameters of the GA_BP neural network can be determined to be 11, whereas the number of output parameters is 2. The neural network adopts a three-layer topological structure. The excitation function is selected as a *sigmoid* function.

Generally, the number of nodes in the hidden layer is calculated by the following empirical formulas [29]:

$$l < \sqrt{mn} \tag{14}$$

$$l < \sqrt{(m+n)} + a \tag{15}$$

$$l = \log_2 n \tag{16}$$

where n is the number of input layer nodes, l is the number of hidden layer nodes, m is the number of output layer nodes, and a is a constant between 0 and 10.

In this paper, the number of nodes in the input layer is n = 11 and the number of nodes in the output layer is m = 2. Therefore, according to the above formula, it is considered that the value range of l is between 4 and 20. Figures 5 and 6 show the convergence speed during sample training and the error rate after sample fitting under different hidden layer nodes during trial calculation. Considering the results of Figures 1 and 2, it is most reasonable to set the number of nodes of the hidden layer as l = 13.

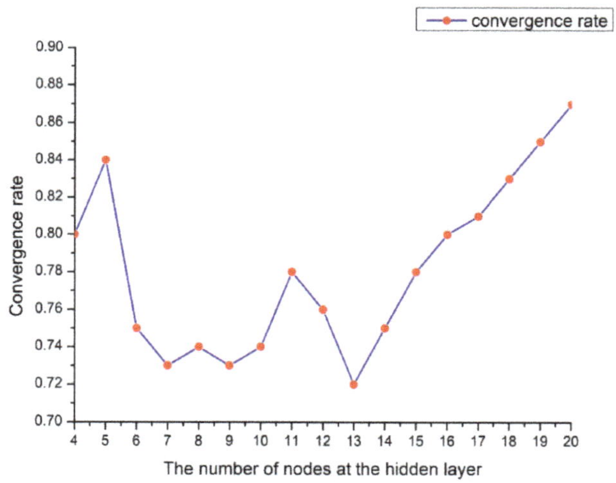

Figure 5. Relationship between sample convergence speed and the number of hidden layer nodes.

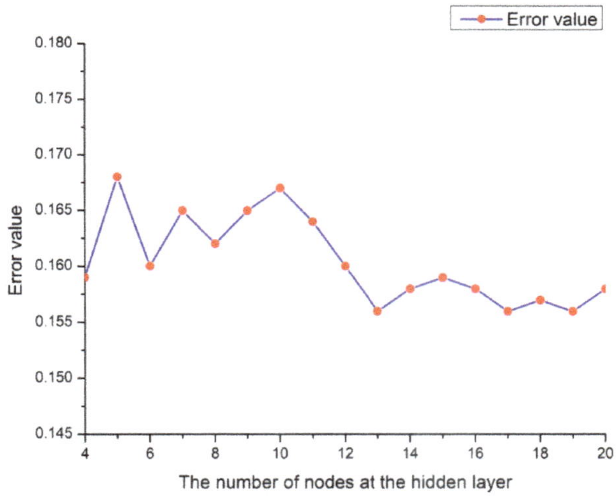

Figure 6. Relationship between prediction error rate and number of hidden layer nodes.

Relevant references [32,33], are selected to determine the network's initial basic parameters: The number of nodes at the hidden layer of GA_BP neural network is 13, with 11 × 13 + 2 × 13 = 169 weights and 13 + 2 = 15 thresholds; individual encoding length in GA is 169 + 15 = 184, population size is 20, evolution count is 100, crossover probability is 0.94, and mutation probability is 0.2. The final GA_BP neural network topology is shown in Figure 7.

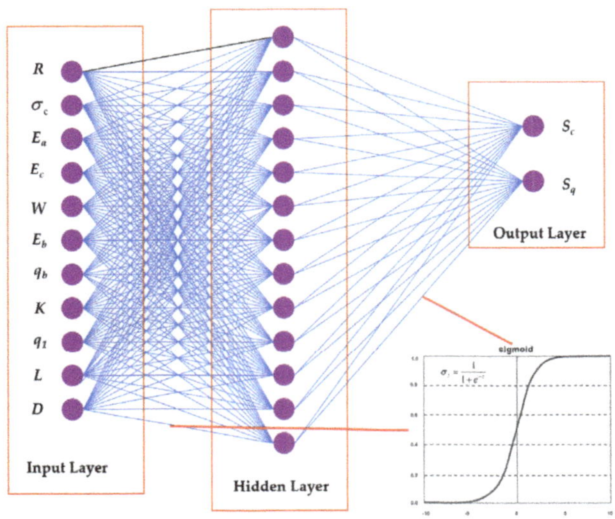

Figure 7. Schematic diagram of GA_BP neural network topology.

3.4. Control Targets of SWB Parameter Optimization

The purpose of parameter optimization is to search for the optimal results. SWB involves numerous parameters, and, without a unique evaluation index of the blasting results, the evaluation is a multitask and multipurpose problem. At the current stage, the multitask and multipurpose optimization using neural network is implemented mainly by two methods: One is to figure out the multiple goals into mutually independent target solutions by the method of *Pareto* solutions, and the other is to turn the multiple goals into single goals by some calculation model. Combining the practical engineering, the two target parameters, the average linear overbreak and average linear underbreak, are difficult to become mutually independent target solutions. Therefore, the latter approach can be more effective. Moreover, the optimal solutions for the results of SWB typically need to meet two requirements: One is that the tunnel profile shall be as smooth as possible; the other is that the contour line of the practically blasted tunnel shall be as designed as possible. Therefore, reflected in the calculation model for optimization, the control targets of the output parameters are:

- The minimum of "average linear overbreak S_c + average linear underbreak S_q";
- The minimum of average linear overbreak S_c.

4. Engineering Case

4.1. Project Overview

This paper relies on the East Tianshan tunneling project for the Barkol-Hami (first-class) highway G575, which is located in the area of Hami, Xinjiang. The two-way separated superlong tunnel covers a full length of 11.767 km of the sinistral tunnel and a full length of 11.776 km of the dextral tunnel; the maximum burial depth is 1225 m and the average burial depth is 706 m. Near the tunnel, one can find the typical western mountainous area with obvious underwater seepage. Affected by the burial depth and seepage, the overall strength of the tunnel's surrounding rocks is low, and some construction sections are even vulnerable to the gushing of a large amount of underground water after being tunneled. The schematic diagram of tunnel construction is shown in Figure 8, and the schematic diagram of the project overview is shown in Figure 9.

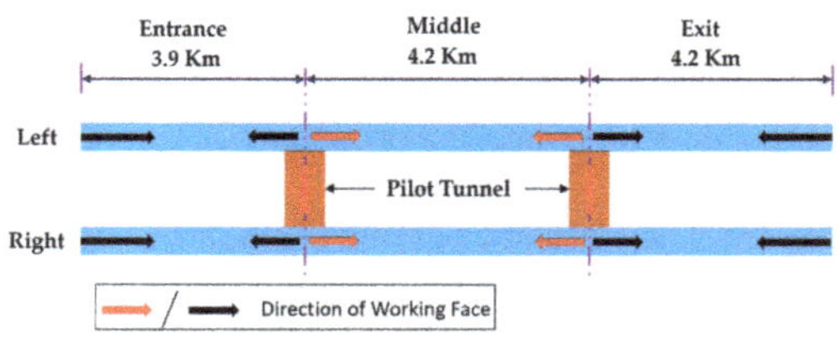

Figure 8. Schematic diagram of the tunneling construction scheme.

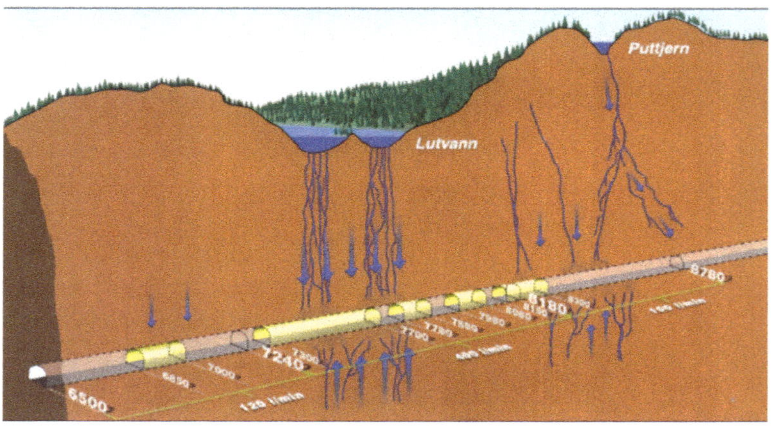

Figure 9. Schematic diagram of the tunneling overview.

4.2. *Fitting Parameter Training of GA_BP Neural Network*

4.2.1. Sample Collection

While 20 sets of SWB test data are collected from this project, 265 sets of the SWB test data are also collected from other tunneling projects. Figure 10 provides the sources of other SWB data. The input and output parameters of SWB are fitted with the 265 sets of data from other projects as the training samples of the GA_BP neural network and the 20 sets of field test data as the test samples.

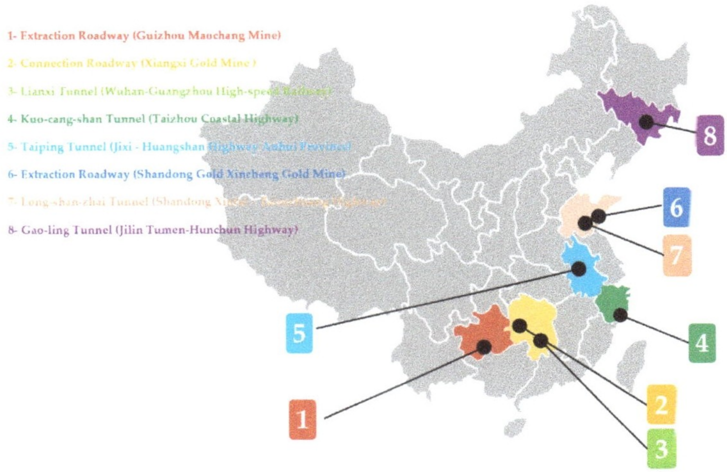

Figure 10. Tunneling sources of other SWB data.

According to the BQ method in the "Standard for Engineering Classification of Rock Masses" (GB 50218-94) of China and the BQ-RMR relation in (17), the parameter *surrounding rock rank* pertains to geological conditions among the sample data. Limited by space, this paper only displays the surrounding rock mass ranks under preliminary investigation in the Xinjiang-based project, as shown in Table 1.

$$\mathrm{RMR} = (\mathrm{BQ} - 80.786)/6.0943 \tag{17}$$

Table 1. Surrounding rock mass ranks under preliminary investigation.

Surrounding rock mass rank	BQ	V	IV
	RMR	V	IV
Length (km)/percentage (%)	Sinistral Tunnel	6.31/54	5.46/46
	Dextral Tunnel	6.88/58	4.89/42

Still limited by space, only the 20 sets of blasting test sample data in the Xinjiang-based project for testing the fitting precision of the GA_BP neural network are presented in Table 2, where the *average overbreak* is expressed in positive values and the *average underbreak* is expressed in negative values.

Table 2. Test sample parameter list.

Order	Input Parameters											Output Parameters	
No.	R	σ_c/MPa	E_c/mm	E_a/cm	E_b/cm	W/cm	q_b/kg	K	q_l	L/m	D/mm	S_c/cm	S_q/cm
1	4	37.93	12	50.1	73.3	45.4	1.8	1.5	0.25	4	48	3.6	−1.2
2	4	41.22	14	54.2	85	57	1.8	1.5	0.25	4	48	2.1	−7.8
3	4	38.36	12	53.9	79.4	86.7	1.8	1.5	0.25	4	48	0.9	−3.9
4	4	48.2	17	46.3	84.2	90.5	1.8	1.5	0.25	4	48	1.1	−4.2
5	4	44.6	16	56.7	71	68.2	1.8	1.5	0.25	4	48	6.9	−1.2
6	4	36.7	21	54.2	102	73.1	1.8	1.5	0.25	4	48	6.3	−1
7	5	26.6	41	65.1	145.1	55.6	1.8	1.5	0.2	4	48	1.2	−25.7
8	5	38.54	17	54.9	132.4	48.1	1.8	1.5	0.2	4	48	15.1	−1.4
9	5	36.78	34	59.2	102.5	52.5	1.8	1.5	0.2	4	48	1.7	−9.6
10	5	33.25	14	44.5	75.5	43.2	1.8	1.5	0.2	4	48	4.65	−1
11	5	38.34	42	60.1	90.6	70.1	1.8	1.5	0.2	4	48	15.2	−1.4
12	5	30.24	18	56.4	76.3	49.4	1.8	1.5	0.2	4	48	16.7	−1.7
13	5	28.71	11	56.5	86.4	50.1	1.8	1.5	0.2	4	48	19.4	−2.3
14	5	29.95	30	42.4	94.5	50.7	1.8	1.5	0.2	4	48	1.5	−15.3
15	5	25.75	26	57.3	123.1	62.8	1.8	1.5	0.2	4	48	1.9	−16.7
16	4	30.6	24	60.2	122.7	51	1.8	1.5	0.25	4	48	2.6	−17.5
17	4	31.4	23	76.4	148.3	47.3	1.8	1.5	0.25	4	48	2.7	−22.8
18	4	33.1	36	77.1	117.4	43.6	1.8	1.5	0.25	4	48	4.6	−21.5
19	4	34.3	38	78.9	120.1	46.3	1.8	1.5	0.25	4	48	3.2	−20.6
20	4	33.7	30	83.3	114.7	50.2	1.8	1.5	0.25	4	48	1.8	−18.2

4.2.2. Sample Training Test

GA_BP neural network is used to train and test the above-collected samples. Figure 11 presents the fitness curves of the GA-optimized initial parameters of the BP neural network. Figure 12 presents the overall prediction error curves at distinct iteration counts when GA_BP neural network is fitting the input and output parameters of SWB. From Figures 11 and 12, both the fitting precision and error can meet the practical requirements when the GA_BP neural network is fitting the input and output parameters of SWB, without the problem of underfitting or overfitting, and the nonlinear relations between input and output parameters of SWB can be achieved satisfactorily, hence the prediction of SWB results can be implemented.

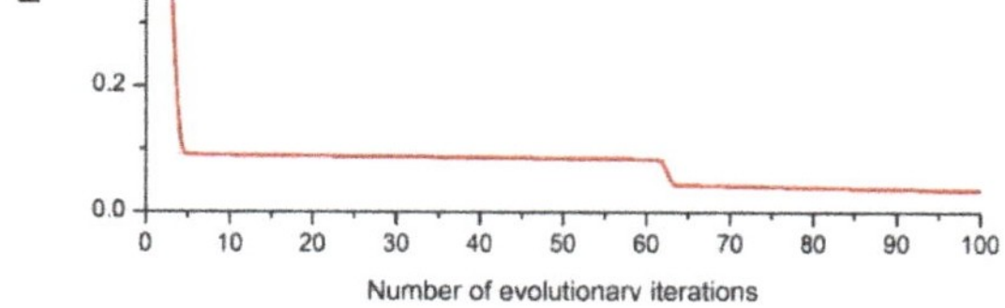

Figure 11. GA optimized fitness curves.

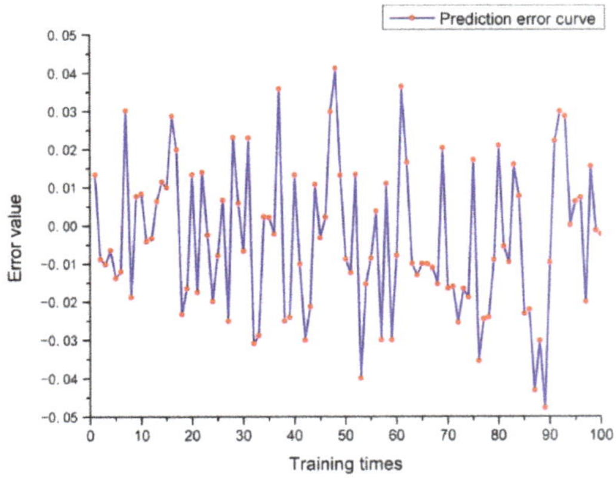

Figure 12. Overall error of training samples in GA_BP neural network.

4.2.3. The Accuracy of the Prediction in Both Training and Testing Data

Tables 3 and 4 list the measured value, predicted value, and error rates for 20 sets of training data and 20 sets of testing data. It can be considered that, whether it is training data or test data, the error between the predicted value and the measured value is within 10%. Considering drilling work, this 10% error can be neglected. In general, the accuracy of the prediction meets the actual engineering requirements.

Table 3. The accuracy of the prediction in training data.

No.	Overbreak/cm			Underbreak/cm		
	Measured Value	Predicted Value	Error Rate	Measured Value	Predicted Value	Error Rate
1	7.8	7.752	0.62%	−6.3	−6.115	2.94%
2	2.0	2.121	6.05%	−7.3	−7.725	5.82%
3	6.7	6.862	2.42%	−4.7	−4.823	2.62%
4	6.4	6.301	1.55%	−5.1	−5.341	4.73%
5	5.1	5.005	1.86%	−6.1	−6.198	1.61%
6	2.9	2.774	4.34%	−6.7	−7.033	4.97%
7	3.4	3.331	2.03%	−2.0	−1.887	5.65%
8	5.8	5.255	9.40%	−4.5	−4.411	1.98%
9	8.7	8.543	1.80%	−8.0	−8.431	5.39%
10	7.7	7.561	1.81%	−9.9	−10.323	4.27%
11	5.2	5.011	3.63%	−2.8	−2.704	3.43%
12	7.2	7.588	5.39%	−5.6	−5.719	2.13%
13	9.9	9.562	3.41%	−5.1	−5.373	5.35%
14	4.7	4.471	4.87%	−2.8	−2.567	8.32%
15	8.2	8.658	5.59%	−4.3	−4.511	4.91%
16	5.3	5.523	4.21%	−7.2	−7.846	8.97%
17	3.9	3.658	6.21%	−5.7	−5.246	7.96%
18	7.3	7.521	3.03%	−4.2	−4.299	2.36%
19	3.8	3.93	3.42%	−2.2	−2.353	6.95%
20	8.8	8.995	2.22%	−6.1	−6.002	1.61%

Table 4. The accuracy of the prediction in testing data.

No.	Overbreak/cm			Underbreak/cm		
	Measured Value	Predicted Value	Error Rate	Measured Value	Predicted Value	Error Rate
1	3.6	3.727	3.53%	−1.2	−1.265	5.42%
2	2.1	2.247	7.00%	−7.8	−7.621	2.29%
3	0.9	0.891	1.00%	−3.9	−3.844	1.44%
4	1.1	1.205	9.55%	−4.2	−4.141	1.40%
5	6.9	6.841	0.86%	−1.2	−1.114	7.17%
6	6.3	6.424	1.97%	−1	−1.045	4.50%
7	1.2	1.147	4.42%	−25.7	−24.876	3.21%
8	15.1	15.237	0.91%	−1.4	−1.412	0.86%
9	1.7	1.553	8.65%	−9.6	−9.613	0.14%
10	4.6	4.631	0.67%	−1	−1.023	2.30%
11	15.2	15.041	1.05%	−1.4	−1.404	0.29%
12	16.7	16.524	1.05%	−1.7	−1.712	0.71%
13	19.4	19.502	0.53%	−2.3	−2.323	1.00%
14	1.5	1.375	8.33%	−15.3	−15.421	0.79%
15	1.9	1.854	2.42%	−16.7	−16.521	1.07%
16	2.6	2.429	6.58%	−17.5	−17.323	1.01%
17	2.7	2.601	3.67%	−22.8	−22.651	0.65%
18	4.6	4.509	1.98%	−21.5	−21.212	1.34%
19	3.2	3.331	4.09%	−20.6	−20.412	0.91%
20	1.8	1.851	2.83%	−18.2	−18.023	0.97%

4.3. Results and Analysis of SWB Parameter Optimization Calculation

4.3.1. Parameter Input for Optimization Calculation

After the prediction of blasting results is implemented using the GA_BP neural network to complete SWB parameter fitting, the SWB design parameters can be optimized by controlling the predicted blasting results. Relying on the practical engineering project, this paper has designed five sets of different input parameters, among which the fixed parameters are shown in Table 5, and the ranges of values of the parameters to be optimized are shown in Table 6. As the values of W, E_b, and E_a are of little significance below the mm level in practical engineering, the parameters are optimized only at the cm level.

Table 5. SWB involved fixed parameters in practical engineering.

No.	R	σ_c/MPa	E_c/mm	q_b/kg	K	q_1	L/m	D/mm
1	4	42.34	20	1.8	1.5	0.25	4	48
2	4	55.27	20	1.8	1.5	0.25	4	48
3	4	48.45	20	1.8	1.5	0.25	4	48
4	5	38.12	20	1.8	1.5	0.2	4	48
5	5	36.94	20	1.8	1.5	0.2	4	48

Table 6. Ranges of the parameters to be optimized.

Parameter	E_a/cm	W/cm	E_b/cm
Range	40~90	40~90	70~120

4.3.2. Optimization Calculation Results and Engineering Application

Table 7 displays the optimization calculation results at different design numbers and the predicted overbreak and underbreak values at the corresponding optimal values.

Table 7. Optimization calculation results.

No.	E_a/cm	W/cm	E_b/cm	Predicted Overbreak/cm	Predicted Underbreak/cm
1	50	77	80	2.142	−0.964
2	51	78	82	3.512	−2.274
3	52	82	88	3.375	−1.731
4	54	77	84	1.972	−3.212
5	55	87	80	2.021	−1.397

The optimization calculation results in Table 6 are applied in the engineering practice, and the overbreak and underbreak values in real blasting are compared with the predicted overbreak and underbreak values with the results shown in Table 8. The actual effects of engineering blasting are shown in Figures 13 and 14.

Table 8. Comparison between predicted values and actual values.

No.	Predicted Overbreak/cm	Actual Value/cm	Predicted Underbreak/cm	Actual Value/cm
1	2.142	2.2	−0.964	−0.9
2	3.512	3.4	−2.274	−2.1
3	3.375	3.3	−1.731	−1.8
4	1.972	3.5	−3.212	−4.1
5	2.021	2.0	−1.397	−1.2

Figure 13. Schematic diagram of practical SWB construction.

Figure 14. Schematic diagram of practical SWB effect.

From the comparison between the predicted results of optimization calculation and the practical engineering results in Table 8, it can be found that the relative errors between the predicted overbreak/underbreak values and the actual values in engineering are small and within 10%, demonstrating that the method of optimization calculation is scientific and effective and that the predicted results of optimization are realistic and scientific. From Figures 13 and 14, it can be found that applying the optimized results of parameters in practical engineering can effectively enhance the accuracy of SWB during blasting tunneling and further ensure the security and stability of the surrounding rocks.

The accomplishment in engineering application demonstrates that the GA_BP neural network-based calculation model for SWB parameter optimization can be applied to not only effectively implement the prediction of SWB effect but also upgrade the SWB effect and the overall safety in tunnel construction.

5. Discussion

5.1. Discussion of the Superiority of GA_BP Neural Network Algorithm

To verify the superiority of the GA_BP neural network algorithm in data processing, a comparison is made between the predicted results under the test samples after the fitting calculation of its training samples and the actual values of blasting, the predicted values of the BP neural network, and the predicted values by the GA_ISVR algorithm. The predicted values of the three are shown in Table 9, and the algorithm average training errors are shown in Figures 15 and 16.

Table 9. Comparison between actual overbreak/underbreak values and predicted values by various algorithms.

No.	Measured Overbreak Value/cm	GA_BP Neural Network/cm	BP Neural Network/cm	GA_ISVR Algorithm/cm	Measured Underbreak Value/cm	GA_BP Neural Network/cm	BP Neural Network/cm	GA_ISVR Algorithm/cm
1	3.6	3.727	3.832	3.4	−1.2	−1.265	−1.321	−1.3
2	2.1	2.247	2.304	2.3	−7.8	−7.621	−7.934	−7.7
3	0.9	0.891	0.721	0.7	−3.9	−3.844	−4.021	−4.1
4	1.1	1.205	1.301	0.9	−4.2	−4.141	−4.052	−4
5	6.9	6.841	6.741	6.8	−1.2	−1.114	−1.105	−1.1
6	6.3	6.424	6.553	6.1	−1	−1.045	−0.848	−1.1
7	1.2	1.147	1.022	1.4	−25.7	−24.876	−24.125	−25.5
8	15.1	15.237	15.321	15.4	−1.4	−1.412	−1.501	−1.6
9	1.7	1.553	1.445	1.5	−9.6	−9.613	−9.501	−9.5
10	4.6	4.631	4.751	4.5	−1	−1.023	−0.826	−1.1
11	15.2	15.041	15.485	15.2	−1.4	−1.404	−1.297	−1.5
12	16.7	16.524	16.942	16.9	−1.7	−1.712	−1.622	−1.6
13	19.4	19.502	19.612	19.1	−2.3	−2.323	−2.145	−2.4
14	1.5	1.375	1.641	1.7	−15.3	−15.421	−15.997	−15.6
15	1.9	1.854	1.924	1.2	−16.7	−16.521	−15.981	−16.9
16	2.6	2.429	2.441	2.5	−17.5	−17.323	−16.248	−17.2
17	2.7	2.601	2.587	2.9	−22.8	−22.651	−23.684	−22.3
18	4.6	4.509	4.812	4.4	−21.5	−21.212	−20.014	−21.2
19	3.2	3.331	3.441	3.4	−20.6	−20.412	−22.121	−20.3
20	1.8	1.851	1.992	1.9	−18.2	−18.023	−19.01	−18.5

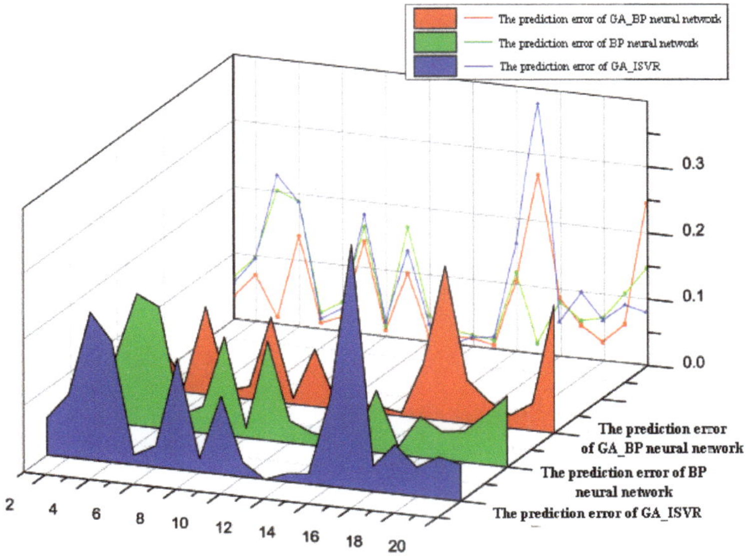

Figure 15. Schematic diagram of comparison between overbreak errors.

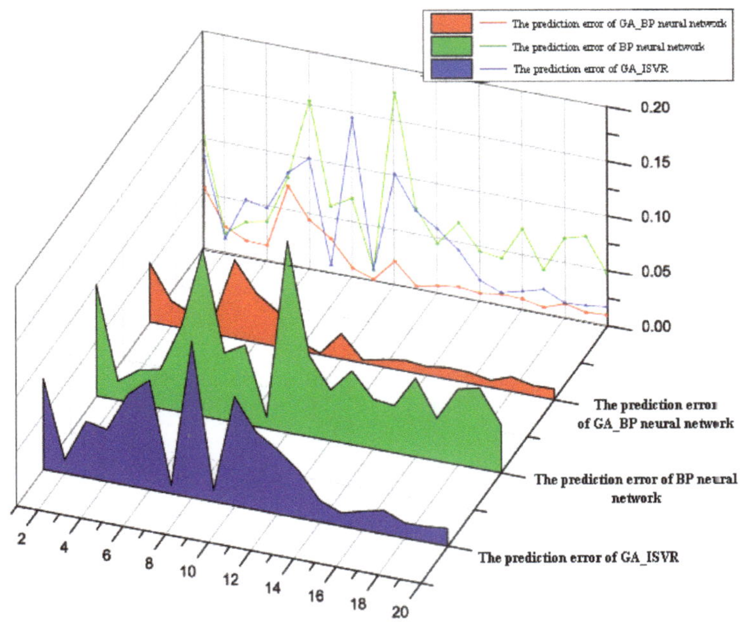

Figure 16. Schematic diagram of comparison between underbreak errors.

From Figures 15 and 16 and Table 9, one can find that the GA_BP neural network has a significant upgrade over the BP neural network in the accuracy of overbreak or underbreak prediction after the GA is introduced, and that the error of SWB parameter fitting by the GA_BP neural network is significantly smaller than by the GA_ISVR algorithm after the sample size is enlarged. Evidently, the GA_BP neural network can be applied for excellent fitting and prediction of SWB parameters. In practical engineering, the GA_BP neural network outperforms the BP neural network and GA_ISVR algorithm in predicting SWB results.

In addition, when discussing the superiority of the GA-BP algorithm, there is still a key problem, that is, the inaccuracy and uncertainty of the data. Therefore, it is also necessary to verify and analyze the accuracy and certainty of the data, so as to avoid calculation errors or analysis errors. In future work, based on References [34–36], the accuracy and certainty of the data will be investigated through the "prediction system based on fuzzy logic".

5.2. Discussion of Optimization Results of SWB Parameters

First, we define an indicator K called relative change rate:

$$K = \left| \frac{a-b}{a} \right| \times 100\%$$

where a and b represent the upper and lower bounds of the change interval, respectively. K represents a percentage.

According to Table 6, Scheme 1, Scheme 2 and Scheme 4 are selected. The sensitivity calculation of the optimization parameter index is shown in Table 10, and the analysis results are shown in Table 11.

Table 10. Sensitivity calculation of indices for optimization of SWB.

No.	Index	Change Interval/cm	Relative Change Rate K (%)	Predicted Overbreak			Predicted Underbreak		
				Change Interval/cm	Relative Change Rate K	Relative Change Rate K after Normalization (10%)	Change Interval/cm	Relative Change Rate K	Relative Change Rate K after Normalization (10%)
1	E_a/cm	50→45	10%	2.142→1.721	19.65%	19.65%	−0.964→−1.431	48.44%	48.44%
		50→55	10%	2.142→3.005	40.29%	40.29%	−0.964→−0.833	13.59%	13.59%
	W/cm	77→69	10.4%	2.142→1.859	13.21%	12.70%	−0.964→−1.343	39.32%	37.81%
		77→85	10.4%	2.142→2.851	33.10%	31.83%	−0.964→−0.854	11.41%	10.97%
	E_b/cm	80→72	10%	2.142→1.863	13.03%	13.03%	−0.964→−1.254	30.08%	30.08%
		80→88	10%	2.142→2.893	35.06%	35.06%	−0.964→−0.834	13.49%	13.49%
2	E_a/cm	51→46	9.8%	3.512→2.824	19.59%	19.99%	−2.274→−3.002	32.01%	32.66%
		51→56	9.8%	3.512→4.351	23.89%	24.38%	−2.274→−1.954	14.07%	14.36%
	W/cm	78→70	10.2%	3.512→3.031	13.70%	13.43%	−2.274→−2.851	25.37%	24.87%
		78→86	10.2%	3.512→4.012	14.24%	13.96%	−2.274→−1.886	17.06%	16.73%
	E_b/cm	82→74	9.7%	3.512→3.045	13.30%	13.71%	−2.274→−2.905	27.75%	28.61%
		82→90	9.7%	3.512→3.999	13.87%	14.30%	−2.274→−1.907	16.14%	16.64%
4	E_a/cm	54→49	9.3%	1.972→1.683	14.66%	15.76%	−3.212→−4.763	48.29%	51.92%
		54→59	9.3%	1.972→3.134	58.92%	63.35%	−3.212→−2.274	29.20%	31.40%
	W/cm	77→69	10.3%	1.972→1.763	10.60%	10.29%	−3.212→−4.554	41.78%	40.56%
		77→85	10.3%	1.972→2.865	45.28%	43.96%	−3.212→−2.587	19.46%	18.89%
	E_b/cm	84→76	9.5%	1.972→1.702	13.69%	14.41%	−3.212→−4.476	39.35%	41.42%
		84→92	9.5%	1.972→2.789	41.43%	43.61%	−3.212→−2.436	24.16%	25.43%

Table 11. Sensitivity analysis of indices for optimization of SWB.

No.	Relative Change Rate K of E_b		Relative Change Rate K of W		Relative Change Rate K of E_a	
	Maximum	Average	Maximum	Average	Maximum	Average
1	48.44%	30.49%	37.81%	23.33%	35.06%	22.92%
2	32.66%	22.85%	24.87%	17.25%	28.61%	18.32%
3	63.35%	40.61%	43.96%	28.43%	43.61%	31.22%
Total	63.35%	31.32%	43.96%	23.00%	43.61%	23.67%

After normalizing the optimization parameters "*auxiliary hole pitch* E_b*, spacing between lines of least resistance* W and *ambient hole pitch* E_a" according to the "change rate of 10%", it can be found that in each scheme, the relative change of over and underbreak caused by the change of "*auxiliary hole pitch E_b*" is the largest, its maximum value reaches 63.35%, and the average change rate of the three schemes is 31.32%. In contrast, the maximum relative change rate of overbreak and underbreak caused by the change of "*spacing between lines of least resistance W*" is 43.96%, and the average change rate of the three schemes is 23.00%. The maximum relative change rate of overbreak and underbreak caused by the change of "*ambient hole pitch E_a*" is 43.61%, and the average change rate of the three schemes is 23.67%. It can be observed that for the optimization parameter index, "*auxiliary hole pitch E_b*" is the most sensitive factor affecting the value of overbreak and underbreak.

In addition, according to Table 11, the overall relative change rate of Scheme 1 is 25.58%, that of Scheme 2 is 19.47%, and that of Scheme 4 is 33.42% when all optimization parameters change by 10%. Referring to Table 6, the σ_c of rock in Scheme 2 is the largest and its relative change rate is the smallest, while the σ_c of rock in Scheme 4 is the smallest and its relative change rate is the largest. Therefore, it can be considered that this change law may be related to rock properties, which need further study and analysis.

6. Conclusions

The combination of genetic algorithm and the BP neural network can improve the generalization degree and calculation accuracy of the prediction model, so as to solve the problem of insufficient accuracy caused by insufficient data quantity. Based on this, in this work, a new algorithm based on the GA_BP neural network is proposed, which is applied to the prediction and optimization of the tunnel SWB parameters. By training the input data (geological conditions and SWB parameters) and output data (overbreak/underbreak), the algorithm model builds the nonlinear relationship and realizes the prediction of the SWB effect. Moreover, based on the control of the prediction, the optimization of the design parameters of the SWB is realized.

In addition, through the analysis of the optimization of tunnel SWB parameters, it is believed that the "*E_b (auxiliary hole pitch)*" has the greatest impact on overbreak and underbreak. Therefore, this index should be given priority when determining parameters.

The application results demonstrate that the algorithm model is effective and feasible to predict and optimize the parameters of SWB, and the results can meet the requirements of practical engineering.

Author Contributions: Conceptualization, L.X.; methodology, G.W.; software, writing—original draft preparation, Y.C. and B.D.; writing—review and editing, S.C.; supervision, Z.W. and L.X.; funding acquisition, Y.C. and B.D. All authors have read and agreed to the published version of the manuscript.

Funding: This research is supported by financial grants from the National Natural Science Foundation of China (51804163, 52004130, 52104122), and the China Postdoctoral Science Foundation (2021M693799). The authors are very grateful to the financial contribution and convey their appreciation of the organization for supporting this basic research.

Institutional Review Board Statement: Not applicable.

Informed Consent Statement: Not applicable.

Data Availability Statement: Not applicable.

Conflicts of Interest: The authors declare no conflict of interest.

References

1. Wang, S.; Tang, Y.; Li, X.; Du, K. Analyses and Predictions of Rock Cuttabilities under Different Confining Stresses and Rock Properties Based on Rock Indentation Tests by Conical Pick. *Trans. Nonferrous Met. Soc. China* **2021**, *31*, 1766–1783. [CrossRef]
2. Wang, S.; Li, X.; Yao, J.; Gong, F.; Li, X.; Du, K.; Tao, M.; Huang, L.; Du, S. Experimental Investigation of Rock Breakage by a Conical Pick and Its Application to Non-Explosive Mechanized Mining in Deep Hard Rock. *Int. J. Rock Mech. Min. Sci.* **2019**, *122*, 104063. [CrossRef]

3. Mandal, S.K.; Singh, M.M.; Dasgupta, S. Theoretical Concept to Understand Plan and Design Smooth Blasting Pattern. *Geotech. Geol. Eng.* **2008**, *26*, 399–416. [CrossRef]
4. Mahtab, M.A.; Rossler, K.; Kalamaras, G.S.; Grasso, P. Assessment of Geological Overbreak for Tunnel Design and Contractual Claims. *Int. J. Rock Mech. Min. Sci.* **1997**, *34*, 185.e1–185.e13. [CrossRef]
5. Khandelwal, M.; Monjezi, M. Prediction of Backbreak in Open-Pit Blasting Operations Using the Machine Learning Method. *Rock Mech Rock Eng.* **2013**, *46*, 389–396. [CrossRef]
6. Jang, H.; Topal, E. Optimizing Overbreak Prediction Based on Geological Parameters Comparing Multiple Regression Analysis and Artificial Neural Network. *Tunn. Undergr. Space Technol.* **2013**, *38*, 161–169. [CrossRef]
7. Hamdi, E.; Romdhane, N.B.; Le Cléac'h, J.M. A Tensile Damage Model for Rocks: Application to Blast Induced Damage Assessment. *Comput. Geotech.* **2011**, *38*, 133–141. [CrossRef]
8. García Bastante, F.; Alejano, L.; González-Cao, J. Predicting the Extent of Blast-Induced Damage in Rock Masses. *Int. J. Rock Mech. Min. Sci.* **2012**, *56*, 44–53. [CrossRef]
9. Xia, X.; Li, H.B.; Li, J.C.; Liu, B.; Yu, C. A Case Study on Rock Damage Prediction and Control Method for Underground Tunnels Subjected to Adjacent Excavation Blasting. *Tunn. Undergr. Space Technol.* **2013**, *35*, 1–7. [CrossRef]
10. Hu, Y.; Lu, W.; Chen, M.; Yan, P.; Yang, J. Comparison of Blast-Induced Damage Between Presplit and Smooth Blasting of High Rock Slope. *Rock Mech Rock Eng.* **2014**, *47*, 1307–1320. [CrossRef]
11. Yan, P.; Lu, W.; Chen, M.; Hu, Y.; Zhou, C.; Wu, X. Contributions of In-Situ Stress Transient Redistribution to Blasting Excavation Damage Zone of Deep Tunnels. *Rock Mech Rock Eng.* **2015**, *48*, 715–726. [CrossRef]
12. Wang, S.; Sun, L.; Li, X.; Wang, S.; Du, K.; Li, X.; Feng, F. Experimental Investigation of Cuttability Improvement for Hard Rock Fragmentation Using Conical Cutter. *Int. J. Geomech.* **2021**, *21*, 06020039. [CrossRef]
13. Ryu, C.-H.; Sunwoo, C.; Lee, S.-D.; Choi, H.-M. Suggestions of Rock Classification Methods for Blast Design and Application to Tunnel Blasting. *Tunn. Undergr. Space Technol.* **2006**, *21*, 401–402. [CrossRef]
14. Yilmaz, O.; Unlu, T. An Application of the Modified Holmberg–Persson Approach for Tunnel Blasting Design. *Tunn. Undergr. Space Technol.* **2014**, *43*, 113–122. [CrossRef]
15. Monjezi, M.; Dehghani, H. Evaluation of Effect of Blasting Pattern Parameters on Back Break Using Neural Networks. *Int. J. Rock Mech. Min. Sci.* **2008**, *45*, 1446–1453. [CrossRef]
16. Wang, S.; Tang, Y.; Wang, S. Influence of Brittleness and Confining Stress on Rock Cuttability Based on Rock Indentation Tests. *J. Cent. South Univ.* **2021**, *28*, 2786–2800. [CrossRef]
17. Dai, B.; Chen, Y.; Zhao, G.; Liang, W.; Wu, H. A Numerical Study on the Crack Development Behavior of Rock-Like Material Containing Two Intersecting Flaws. *Mathematics* **2019**, *7*, 1223. [CrossRef]
18. Smola, A.J.; Schölkopf, B. A Tutorial on Support Vector Regression. *Stat. Comput.* **2004**, *14*, 199–222. [CrossRef]
19. Teymen, A.; Mengüç, E.C. Comparative Evaluation of Different Statistical Tools for the Prediction of Uniaxial Compressive Strength of Rocks. *Int. J. Min. Sci. Technol.* **2020**, *30*, 785–797. [CrossRef]
20. Hasanipanah, M.; Noorian-Bidgoli, M.; Jahed Armaghani, D.; Khamesi, H. Feasibility of PSO-ANN Model for Predicting Surface Settlement Caused by Tunneling. *Eng. Comput.* **2016**, *32*, 705–715. [CrossRef]
21. Tawadrous, A.S.; Katsabanis, P.D. Prediction of Surface Crown Pillar Stability Using Artificial Neural Networks. *Int. J. Numer. Anal. Methods Geomech.* **2007**, *31*, 917–931. [CrossRef]
22. Zhou, J.; Qiu, Y.; Armaghani, D.J.; Zhang, W.; Li, C.; Zhu, S.; Tarinejad, R. Predicting TBM Penetration Rate in Hard Rock Condition: A Comparative Study among Six XGB-Based Metaheuristic Techniques. *Geosci. Front.* **2021**, *12*, 101091. [CrossRef]
23. Dai, B.; Shan, Q.; Chen, Y.; Luo, X. Mechanical and Energy Dissipation Characteristics of Granite under Cyclic Impact Loading. *J. Cent. South Univ.* **2022**, *29*, 116–128. [CrossRef]
24. Li, Q.; Xu, Z.; Shen, X.; Zhong, J. Predicting Business Risks of Commercial Banks Based on BP-GA Optimized Model. *Comput. Econ.* **2021**. [CrossRef]
25. Zhu, C.; Zhang, J.; Liu, Y.; Ma, D.; Li, M.; Xiang, B. Comparison of GA-BP and PSO-BP Neural Network Models with Initial BP Model for Rainfall-Induced Landslides Risk Assessment in Regional Scale: A Case Study in Sichuan, China. *Nat. Hazards* **2020**, *100*, 173–204. [CrossRef]
26. Wang, W.; Tang, R.; Li, C.; Liu, P.; Luo, L. A BP Neural Network Model Optimized by Mind Evolutionary Algorithm for Predicting the Ocean Wave Heights. *Ocean Eng.* **2018**, *162*, 98–107. [CrossRef]
27. Li, B.; Ge, W.; Li, Q.; Li, Y.; Tan, C. Gearshift Sensorless Control for Direct-Drive-Type AMT Based on Improved GA-BP Neural Network Algorithm. *Math. Probl. Eng.* **2020**, *2020*, e6456410. [CrossRef]
28. Goldberg, D.E. *Genetic Algorithms in Search, Optimization and Machine Learning*; Addison-Wesley: Boston, MA, USA, 1989.
29. Pineda, F.J. Generalization of Back-Propagation to recurrent neural networks. *Phys. Rev. Lett.* **1987**, *59*, 2229–2232. [CrossRef] [PubMed]
30. Mohamadnejad, M.; Gholami, R.; Ataei, M. Comparison of Intelligence Science Techniques and Empirical Methods for Prediction of Blasting Vibrations. *Tunn. Undergr. Space Technol.* **2012**, *28*, 238–244. [CrossRef]
31. Li, J.C.; Li, H.B.; Ma, G.W.; Zhou, Y.X. Assessment of Underground Tunnel Stability to Adjacent Tunnel Explosion. *Tunn. Undergr. Space Technol.* **2013**, *35*, 227–234. [CrossRef]
32. Cheng, P.; Chen, D.; Wang, J. Clustering of the Body Shape of the Adult Male by Using Principal Component Analysis and Genetic Algorithm–BP Neural Network. *Soft Comput.* **2020**, *24*, 13219–13237. [CrossRef]

33. Zheng, D.; Qian, Z.; Liu, Y.; Liu, C. Prediction and Sensitivity Analysis of Long-Term Skid Resistance of Epoxy Asphalt Mixture Based on GA-BP Neural Network. *Constr. Build. Mater.* **2018**, *158*, 614–623. [CrossRef]
34. Kavousi-Fard, A.; Khosravi, A.; Nahavandi, S. A New Fuzzy-Based Combined Prediction Interval for Wind Power Forecasting. *IEEE Trans. Power Syst.* **2016**, *31*, 18–26. [CrossRef]
35. Barzani, M.M.; Zalnezhad, E.; Sarhan, A.A.; Farahany, S.; Ramesh, S. Fuzzy logic based model for predicting surface roughness of machined Al–Si–Cu–Fe die casting alloy using different additives-turning. *Measurement* **2015**, *61*, 150–161. [CrossRef]
36. Cacciola, M.; Pellicanò, D.; Megali, G.; Lay-Ekuakille, A.; Versaci, M.; Morabito, F.C. Aspects about air pollution prediction on urban environment. In Proceedings of the 4th IMEKO TC19 Symposium on Environmental Instrumentation and Measurements, Protection Environment, Climate Changes and Pollution Control, Lecce, Italy, 3–4 June 2013; pp. 15–20.

mathematics MDPI

Article

Research on Zonal Disintegration Characteristics and Failure Mechanisms of Deep Tunnel in Jointed Rock Mass with Strength Reduction Method

Baoping Chen [1], Bin Gong [2,*], Shanyong Wang [3] and Chun'an Tang [1]

[1] State Key Laboratory of Coastal & Offshore Engineering, Faculty of Infrastructure Engineering, Dalian University of Technology, Dalian 116024, China; 326398381@mail.dlut.edu.cn (B.C.); tca@mail.neu.edu.cn (C.T.)
[2] Department of Civil and Environmental Engineering, Brunel University London, London UB8 3PH, UK
[3] Discipline of Civil, Surveying & Environmental Engineering, School of Engineering, The University of Newcastle, Callaghan, NSW 2308, Australia; shanyong.wang@newcastle.edu.au
* Correspondence: bin.gong@brunel.ac.uk

Abstract: To understand the fracture features of zonal disintegration and reveal the failure mechanisms of circle tunnels excavated in deep jointed rock masses, a series of three-dimensional heterogeneous models considering varying joint dip angles are established. The strength reduction method is embedded in the RFPA method to achieve the gradual fracture process, macro failure mode and safety factor, and to reproduce the characteristic fracture phenomenon of deep rock masses, i.e., zonal disintegration. The mechanical mechanisms and acoustic emission energy of surrounding rocks during the different stages of the whole formation process of zonal disintegration affected by different-dip-angle joints and randomly distributed joints are further discussed. The results demonstrate that the zonal disintegration process is induced by the stress redistribution, which is significantly different from the formation mechanism of traditional surrounding rock loose zone; the dip angle of joint set has a great influence on the stress buildup, stress shadow and stress transfer as well as the failure mode of surrounding rock mass; the existence of parallel and random joints lead the newly formed cracks near the tunnel surface to developing along their strikes; the random joints make the zonal disintegration pattern much more complex and affected by the regional joint composition. These will greatly improve our understanding of the zonal disintegration in deep engineering.

Keywords: zonal disintegration; jointed rock mass; stress redistribution; strength reduction; numerical simulation

MSC: 74L10; 74Rxx

Citation: Chen, B.; Gong, B.; Wang, S.; Tang, C. Research on Zonal Disintegration Characteristics and Failure Mechanisms of Deep Tunnel in Jointed Rock Mass with Strength Reduction Method. *Mathematics* **2022**, *10*, 922. https://doi.org/10.3390/math10060922

Academic Editor: Andrey Jivkov

Received: 28 January 2022
Accepted: 5 March 2022
Published: 14 March 2022

1. Introduction

With the increasing development of global economy, the shallow resources and space are getting harder and harder to satisfy the growing demands of human beings for a better life. In recent decades, the deep-buried underground resources and transport infrastructure in mountain areas, especially those that are difficult to excavate or construct in the past, have become new growth points in energy exploitation, civil engineering and related fields. For example, the maximum depth of several tunnels on the Sichuan-Tibet Railway, China are over 1000 m, such as the Baxu Tunnel and Sangzhuling Tunnel [1,2], the maximum depth of the diversion tunnel at Jinping II Hydropower Station, China is around 2500 m [3] and the maximum depths of the Taimyrskii Mine in Russia [4] and Witwatersrand Mine in South Africa [5] are over 1000 m and 2000 m, respectively. In deep engineering, the complex geological conditions involving high ground temperature, high ground stress, high permeability and strong excavation disturbance leads to many characteristic geological problems, which are quite different from shallow buried underground engineering [6–10]. As one unconventional fracture phenomenon, the zonal disintegration phenomenon of deep surrounding rock has drawn a lot of experts' attention [11–13]. In 1970, this phenomenon was firstly discovered in the roof of a gold mine in South Africa [5], which is characterized by the interval distribution of fractured zone and intact zone of roof rocks. In 2008, the zonal disintegration phenomenon in Dingji mine—910 m of Huainan mining area in China [12,14,15] was monitored by the borehole TV imager, as shown in Figure 1.

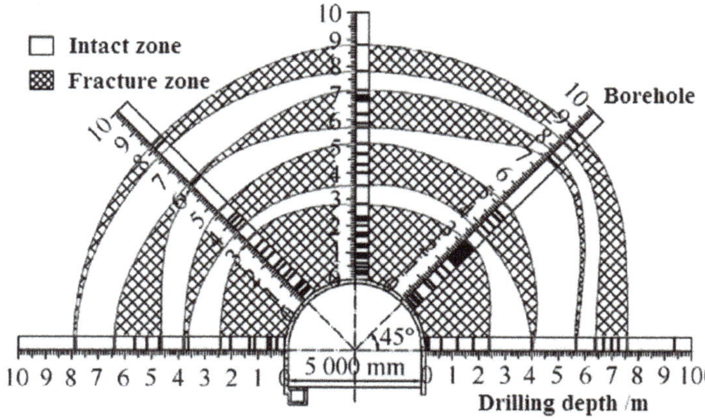

Figure 1. Sketch of the zonal disintegration phenomena in Huainan mine of China (Reproduced with permission from [12], Elsevier, 2017).

The formation mechanisms of this nonlinear deformation phenomenon are quite different from the traditional loosened zone of surrounding rock, which results in the difficulties of stability assessment and support design of deep tunnel. Therefore, it is necessary to reveal the mechanical mechanism and fracture characteristics of the zonal disintegration phenomenon. Through the theoretical analysis, numerical simulations and laboratory tests, some researchers [12,16–19] have discussed the formation conditions of zonal disintegration in deep engineering and suggested that when the tunnel axis is parallel to the direction of the maximum principal stress, the zonal disintegration is prone to occur. Based on this understanding, Zhang et al. [20] successfully reproduced the formation process of zonal disintegration through laboratory test using a true three-dimensional high-stress loading system, monitored the radial strain and displacement development and summarized the alternation distribution law of their peaks and troughs. Zhang et al. [21] conducted the zonal disintegration laboratory test using the soft-layered joint model with different spacing distances, and pointed out that the zonal disintegration is more obvious due to the weak interlayers considering the greater radial displacement and strain. Simultaneously, when the spacing of interlayers is denser, a greater number of fractured zones and larger damaged area will be observed. Pu and Xu [22] monitored the three-dimensional geological model with different uniaxial compressive strengths using the strain gauges and failure wires and found that during the loading process, both the radial tensile strain and the tangential compressive strain increased. According to their measurement results, the radius scale factor of the fracture area was also summarized. Factually, as a unique phenomenon in deep engineering [23], the zonal disintegration phenomenon generally appears in the complex environment of "high ground stress, high ground temperature, high permeability" [6–8], which directly leads to many difficulties in field and laboratorial tests, such as long-period, high-cost and hazardous operating environment and so on. Fortunately, the numerical simulation technology can effectively avoid these problems. By comparing the results of field monitoring, laboratory test and numerical simulation, the feasibility of the numerical simulation technology in studying zonal disintegration phenomenon has been verified [12]. For instance, Zhu et al. [24]

used the finite element method to analyze the influence of rock physical and mechanical parameters, including internal friction angle, cohesion and Poisson's ratio, on the fracture area and failure patterns of zonal disintegration phenomenon. Zhang et al. [25] studied the effect of confining pressure on the crack occurrence order and final failure range by computing the formation of zonal fractures using the 2D rock failure process analysis (RFPA) method. Wu et al. [26] applied the particle flow code to simulate the zonal fracture phenomenon of brittle rock mass, and carried out the sensitivity analysis of initial stress, lateral pressure coefficient and joint friction coefficient.

However, the zonal disintegration is a 3D failure phenomenon involving the gradual process of stress buildup, stress shadow and stress transfer in the 3D space and the initiation, propagation and coalescence of 3D cracks. Currently, most of the simulated results are obtained by 2D numerical methods and cannot effectively reflect the influence of mechanical interaction of cracks in the 3D space on the formation of zonal disintegration by taking the rock heterogeneity and discontinuity into account. Therefore, in this study, to reveal the fracture characteristics and failure mechanisms of deep jointed rock masses, a series of 3D heterogeneous numerical models considering varying joint dip angles and random spatial distribution of joints are built up and the strength reduction method (SRM) is embedded to reproduce the zonal disintegration phenomenon of deep rock masses. The discontinuous deformation field, evolution rule of background stress field, alternate regional destructions and safety factor are therefore discussed, which provides theoretical basis of stability assessment, support design and safe construction in terms of zonal disintegration in deep engineering.

2. Materials and Methods

2.1. Basic Principles

The current numerical methods applied in rock mechanics can be categorized into three groups [27], i.e., continuum methods, discontinuum methods and hybrid continuum/discontinuum methods. The finite difference method (FDM) [28]. finite element method (FEM) [29], boundary element method (BEM) [30], etc. are classical continuum methods. However, they generally cannot satisfactorily deal with the discontinuities widely existing in rock mass. The discrete element method (DEM) [31] and discontinuous deformation analysis (DDA) [32] are two mainstream discontinuum methods. However, the model block division for these methods often requires fine engineering geological survey and has a great influence on simulated results; the divided blocks generally cannot be broken during calculation and the much lower computing capacity and efficiency than continuum methods put limitations on the application of this kind of methods. The hybrid methods provide a feasible way of combining the strengths of both continuum methods and discontinuum methods. However, they still cannot handle the interaction of multiple joint sets in the 3D space. As a FEM-based method, RFPA is able to take the heterogeneity, nonlinearity and anisotropy of rock materials into account. Simultaneously, to simulate the evolution process of progressive failure of deep tunnels, the fundamental principles of strength reduction method are embedded into the RFPA3D code. Namely, the failure modes and safety factor can be gained without any assumption on the failure surface in a complex rock mass model.

To establish a 3D heterogeneous numerical model, a rock sample or structure will be discretized into hexahedral elements at first, which can be regarded as the basic mechanical units of the rock medium in the process of numerical simulation depending on the premise that the element size can reasonably reflect the characteristics of the rock medium.

In addition, with the aim of appropriately capturing the mechanical behaviors of rock mass, the non-uniformity of rock mass cannot be ignored. To fully reflect the heterogeneity of mesoscopic elements after discretization of rock mass model, the material properties of these elements, such as uniaxial-compressive strength and elastic modulus, are assumed to obey a statistical distribution function. As a common probability density function used in the field of rock mechanics, the Weibull distribution [33–35] is as shown in Equation (1).

$$\phi(\alpha) = \frac{m}{\alpha_0} \cdot \left(\frac{\alpha}{\alpha_0}\right)^{m-1} \cdot e^{-\left(\frac{\alpha}{\alpha_0}\right)^m} \tag{1}$$

where α is a mechanical parameter of rock medium, such as elastic modulus, strength, Poisson's ratio, weight, etc.; α_0 is the mean value of the mechanical parameter; m is termed the heterogeneity coefficient reflecting the uniformity degree of rock material.

The strength reduction method is introduced into the solution process. At each reduction step after all loads are applied, the initial strength of one element will be reduced according to the following criterion [36]:

$$f_s^{\text{trial}} = \frac{f_0}{f_0^{\text{trial}}} \tag{2}$$

where f_0^{trial} and f_0 are the test strength and the initial strength, respectively; f_s^{trial} is the test safety factor.

When the number of failure elements reaches the maximum at a step, it means that the model loses stability, and the corresponding test safety factor f_s^{trial} is the safety factor F_S of the model. F_S can be calculated using Equation (3):

$$F_S = \frac{1}{1-(k-1)\times\Delta} \tag{3}$$

where k is the number of calculation steps until model failure; Δ is the reduction coefficient of material strength, which should be set before simulation. When the instability of model occurs, $f_0^{\text{trial}} = f_0 - (k-1)\times\Delta\times f_0$.

According to the deformation and damage state, mesoscopic elements in RFPA can be classified into three phases, i.e., the matrix phase, air phase and contact phase. The RFPA code is able to model the crack initiation, propagation and coalescence through the transformation of the three element phases. The Mohr-Coulomb criterion, with a tensile-off [33], is used as the strength criterion, as shown in Equation (4). Clearly, the tension failure mode is judged by the maximum tensile stress criterion and the shear failure mode is judged by the Mohr-Coulomb criterion.

$$\begin{cases} \sigma_1 - \frac{(1+\sin\theta)}{(1-\sin\theta)}\sigma_3 \geq \sigma_c, \sigma_1 \geq \sigma_c\left(1-\frac{1+\sin\theta}{1-\sin\theta}\cdot\frac{1}{\lambda}\right) \\ \sigma_3 \leq -\sigma_t, \sigma_1 \leq \sigma_c\left(1-\frac{1+\sin\theta}{1-\sin\theta}\cdot\frac{1}{\lambda}\right) \end{cases} \tag{4}$$

where σ_c is the uniaxial-compressive strength, σ_t is the tensile strength, and θ is the internal friction angle; λ is the ratio of the compressive strength to tensile strength; σ_1 and σ_3 are the maximum and minimum principal stresses, respectively.

A matrix element represents the solid medium. Under uniaxial stress state, its mechanical behavior can be described by the constitutive relation of rock material shown in Figure 2. When the deformation of the element exceeds the ultimate tensile strain $\varepsilon_{\max\_t}$ under uniaxial tensile state, the matrix element will be converted into an air element and its elastic modulus will decrease to a very low value so that it does not transfer stress any more. Simultaneously, when the deformation of the element exceeds the ultimate compressive strain $\varepsilon_{\max\_c}$ under uniaxial compression state, it will enter the extrusion phase and be transformed into a contact element with a gradually increasing stiffness. Considering that most of the failure of rock materials is in the tensile mode at the mesoscopic scale, the priority is given to the maximum tensile stress criterion. Namely, it will be firstly judged if the stress state of an element satisfies the maximum tensile stress criterion. If not, the Mohr-Coulomb criterion will be checked next.

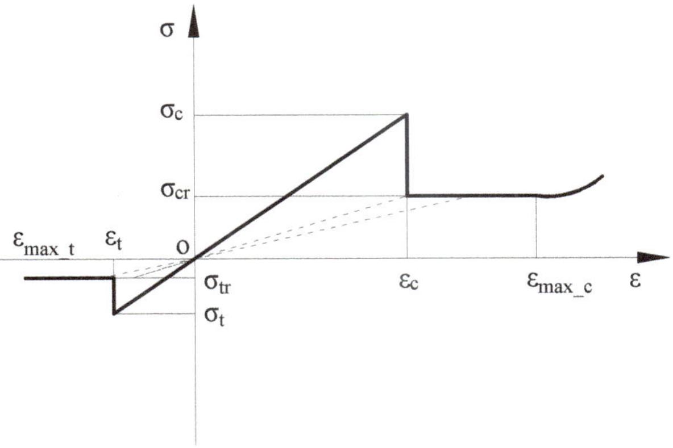

Figure 2. Elastic-brittle constitutive relation of an element under uniaxial stress state.

The strength reduction method is introduced into the RFPA method to realize the progressive failure progress simulation of the tunnel models. The uniaxial compressive strength and tensile strength of the mesoscopic elements will be reduced linearly in a certain proportion during calculation [36]. The stress, strain and failure states of mesoscopic elements will be analyzed after each strength reduction until the external and internal forces reach new equilibrium. Meanwhile, considering rock medium is a kind of quasi-brittle material and the strain energy will be released in the form of acoustic emission when new cracks occur, and the acoustic emissions caused by element failure will be monitored. Actually, when the stress state of an element reaches the failure criteria, the failure of the element will occur, and then the corresponding acoustic emission will be considered to happen. Namely, if one element gets damaged, an acoustic emission event will be generated. The location of the acoustic emission is the center of the element, and the energy magnitude of acoustic emission is calculated according to the stress–strain curve at the moment of failure. Therefore, it is reasonable to evaluate the stability of the model using the released count and energy of acoustic emissions. The reduction step with the largest number of acoustic emissions corresponds to the macro destruction of the numerical models [33,35]. Hence, it will be regarded as a criterion for judging model instability in this study, and the safety factor of the models can be therefore determined for comparing the stability of the tunnels with different joint sets.

2.2. Numerical Model Setting Up

In this paper, the feasibility and correctness of the combination of the strength reduction method and RFPA in studying the mechanical mechanisms of zonal disintegration are verified by comparing the simulated results with the experimental results of Gao et al. (2018) [37]. The prototype of the indoor test is a 910 m deep roadway tunnel in Dingji Coal Mine, Huainan, China. The similar materials are used to prefabricate a 0.6 m × 0.6 m × 0.6 m cube model and a circular tunnel is excavated at the center with a diameter of 0.1 m. The physical model is compressed along the tunnel axis with a load as high as 2.0 times of the material uniaxial tensile strength. Simultaneously, 1.5 times of the in situ stress load is applied in the horizontal direction, and 1.2 times of the material uniaxial compressive strength is applied in the vertical direction. A series of three-dimensional heterogeneous models are built up, whose material properties including elastic modulus, uniaxial compressive strength and tensile strength are assumed to be subject to the Weibull distribution. At the same time, according to Cai and He (2013) [38], the numerical models with the size of 8700 mm × 8700 mm × 4200 mm in the X, Y and Z directions, respectively, are established, and a circular tunnel with a radius of 700 mm is excavated at the center. The models whose axial direction is parallel to the Z axis are discretized into 2,500,000 hexahedron elements, as shown in Figure 3a. Besides, to reduce the end effect caused by the stiffness mismatch between the loading end and the model specimen [39], a backing plate with a thickness of 400 mm is set on every loading face of the tunnel models. Note that the elastic modulus of the backing plates is equal to the tunnel models but is considered to be homogeneous. Meanwhile, their strength is much greater than the tunnel models, as shown in the red parts of Figure 3b. The physical and mechanical parameters of surrounding rocks, mainly composed of hard rock granite, are listed in Table 1.

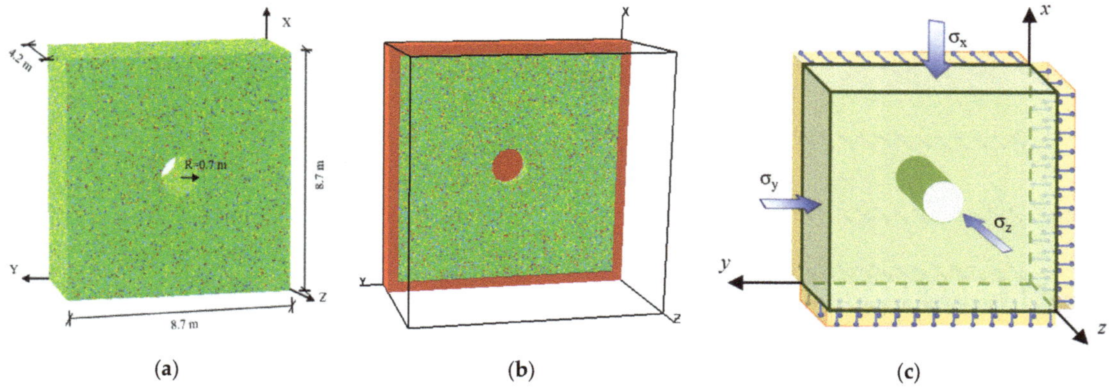

Figure 3. Heterogeneous numerical model containing circular tunnel without joints: (**a**) Model size; (**b**) layout of backing plates; (**c**) loading conditions.

Table 1. Physical and mechanical parameters of model material.

Parameter	Value
Elastic modulus (E)/GPa	80
Compressive strength (σ_c)/MPa	150
Poisson ratio (μ)	0.25
Friction angle (ϕ)/°	25
C/T coefficient	10
Residual strength coefficient	0.1
Heterogeneity coefficient	4
Reduction coefficient for strength	0.01

For the boundary conditions, the surfaces of x = 0, y = 0 and z = 0 are fixed along the normal direction and uniformly distributed loads are applied on the surfaces of x = 9.5 m, y = 9.5 m and z = 5.0 m, respectively. According to the previous research [13], the stress load along the axial direction of the tunnel is set as 2 times the confining pressure. Clearly, the confining pressure of 5 MPa is loaded on the surfaces of x = 9.5 m and y = 9.5 m, and the 10 MPa stress is loaded on the surface of z = 5.0 m in five steps before strength reduction, as shown in Figure 3c.

3. Results

3.1. Comparison of Numerical Simulation and Laboratory Test

In order to verify the correctness of the developed model in studying the zonal disintegration phenomenon, the comparison of the indoor physical test [37] and numerical simulation results of Model A is shown in Figure 4. For examining the numerical simulation results accurately, the different colors represent the different damage states of the elements in Figure 4a. Clearly, the elastic modulus of the elements corresponding to the red color is the smallest, which means the elastic modulus is reduced to the minimum value after those elements get fully failed; meanwhile, the elastic modulus of green elements is about 80 GPa, which has little change compared with the initial value. Namely, no failure and phase transformation occur to these elements.

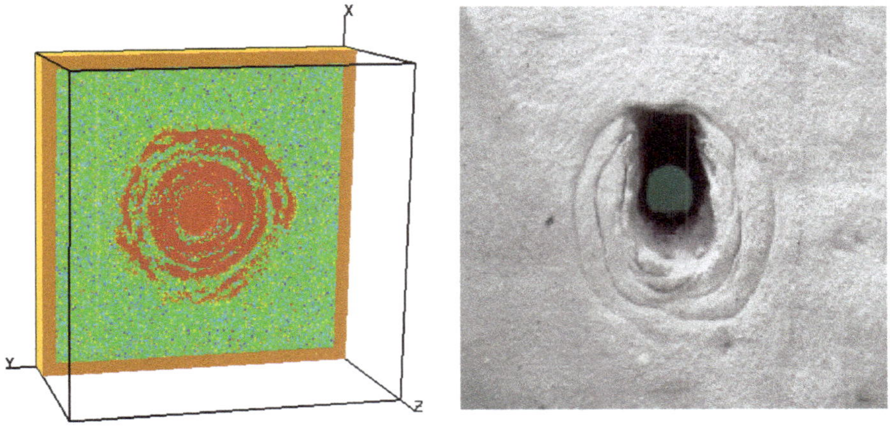

(a) (b)

Figure 4. Comparison of model fracture patterns: (**a**) Numerical simulation result; (**b**) laboratory model test result (Gao et al. [37]).

The existence of scattered yellow and blue elements in the surrounding rock is caused by the different degrees of damage and the discreteness of the strength parameters of the elements in the heterogeneity model. Under the increasing external load, the stress and damage states of each element are closely related to its own properties, such as strength, elastic modulus, etc. Those elements with low phase transition threshold are easily damaged under a low load, such as yellow elements, but they may not grow and form cracks because of the unconnected distribution. However, the elements with high phase transition threshold, such as blue elements, may not get damaged even under high stress. From Figure 4a, it can be observed that the 3D RFPA method combined with the strength reduction method can successfully reproduce the zonal disintegration phenomenon characterized by the interval distribution of fractured zones and intact zones in surrounding rock. Furthermore, it can be found that the fractured ring farther away from the tunnel wall has thinner failure thickness, lower integrity and fuzzier boundary.

The elastic modulus evolution process of zonal disintegration obtained by RFPA numerical simulation is shown in Figure 5. It can be observed from Figure 5a that when the strength reduction is performed to the 94th step, the first fracture area appears near the tunnel wall, and the peak value of maximum principal stress σ_1 of surrounding rock is about 15 MPa, as shown in Figure 6. The failure area can be regarded as the traditional loose zone produced by the excavation of the tunnel. After that, when the model is calculated to the 98-2nd reduction step, the stress redistributes, the mean value of σ_1 rises up to 19.5 MPa, and the second complete circular fracture zone of surrounding rock is generated in an interval way, as shown in Figure 5b. Similarly, with the further reduction of the surrounding rock strength, the stress transfers to the deeper part of the surrounding rock again, and the third circular fracture area appears. The maximum value of σ_1 reaches 24.5 MPa now. In the same way, there are six fracture zones appearing in the surrounding rock of the model. Finally, when the sixth fracture zone located 2 m away from the tunnel wall is formed, the peak stress decreases to 27 MPa, and the stress caused by the excavation of the tunnel is basically released, which cannot cause more damage.

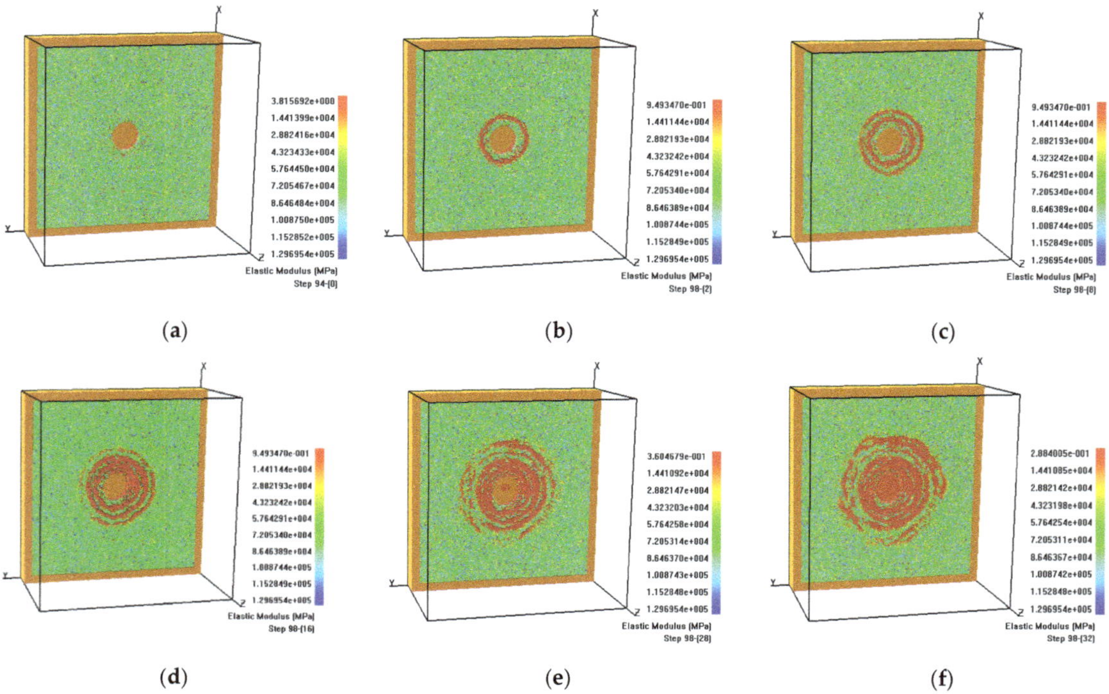

Figure 5. Distribution of elastic modulus during the zonal disintegration process of heterogeneous model around a circular tunnel: (**a**) Step 94; (**b**) Step 98-2; (**c**) Step 98-8; (**d**) Step 98-16; (**e**) Step 98-28; (**f**) Step 98-32.

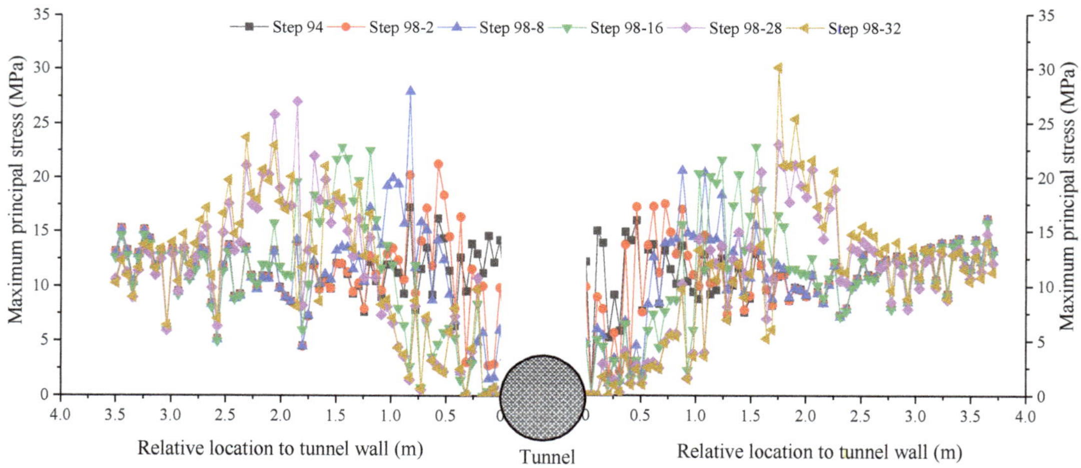

Figure 6. The maximum principal stress-transfer process of surrounding rock during zonal disintegration formation of Model A.

3.2. Numerical Simulation of Zonal Disintegration Phenomenon in Jointed Rock Mass with Different Inclinations

To discuss the zonal disintegration behavior of circular tunnel in jointed rock mass with different dip angles, the related models are established as shown in Figure 7, including three inclinations of 0°, 30° and 45°, respectively, i.e., Models B0, B30 and B45. The physical and mechanical parameters of these joints, composed of soft sandstone and other materials, are shown in Table 2. The size and boundary conditions of the models are the same as the jointless model A.

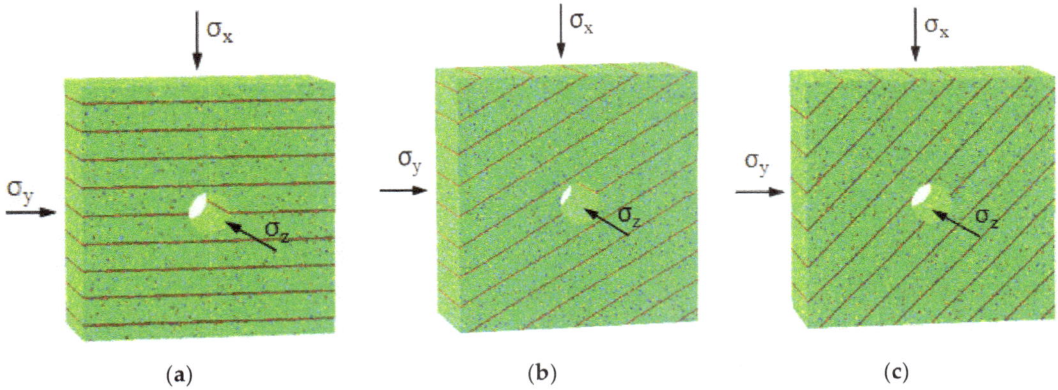

Figure 7. Numerical models and boundary conditions of surrounding rock with three kinds of dip angle joints (0°, 30° and 45°): (**a**) Model B0; (**b**) Model B30; (**c**) Model B45.

Table 2. Physical and mechanical parameters of joint material.

Parameter	Value
Elasticity modulus (E)/GPa	16
Compressive strength (σ_c)/MPa	30
Poisson ratio (μ)	0.25
Friction angle (ϕ)/°	25
C/T coefficient	10
Residual strength coefficient	0.1
Heterogeneity coefficient	200

The elastic modulus diagrams shown in Figures 8–10 depict the reduction steps corresponding to the formation of each fracture ring in Models B0, B30 and B45, through which we can intuitively understand the whole evolution process of the zonal disintegration phenomenon of jointed rock mass with different dip angles, from crack initiation, propagation to final coalescence.

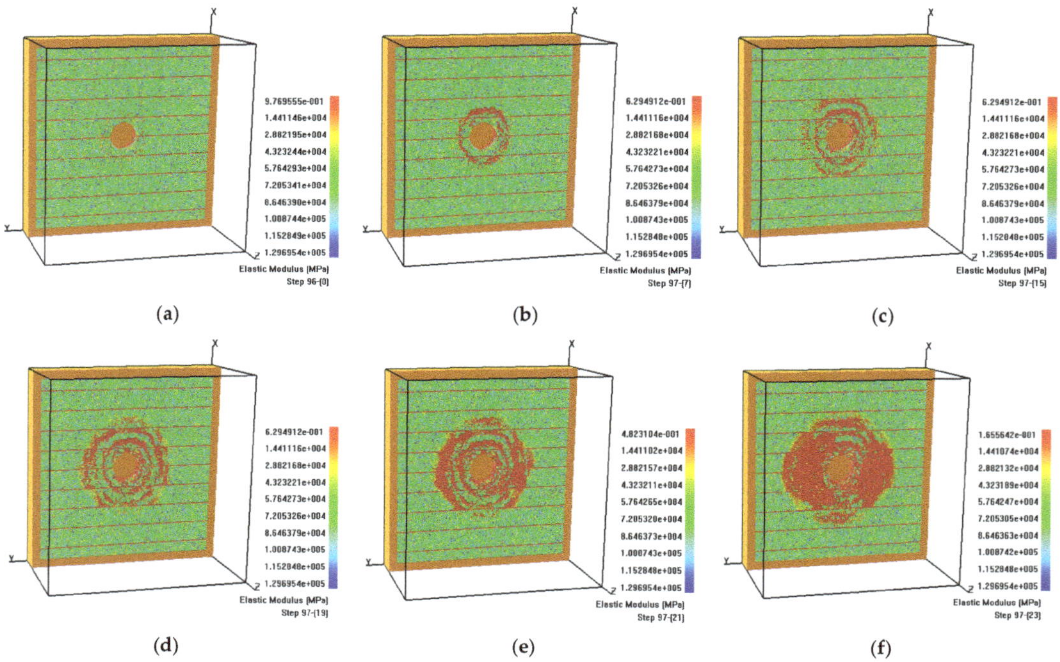

Figure 8. Distribution of elastic modulus during zonal disintegration process of surrounding rock with Model B0: (**a**) Step 96-0; (**b**) Step 97-7; (**c**) Step 97-15; (**d**) Step 97-19; (**e**) Step 97-21; (**f**) Step 97-23.

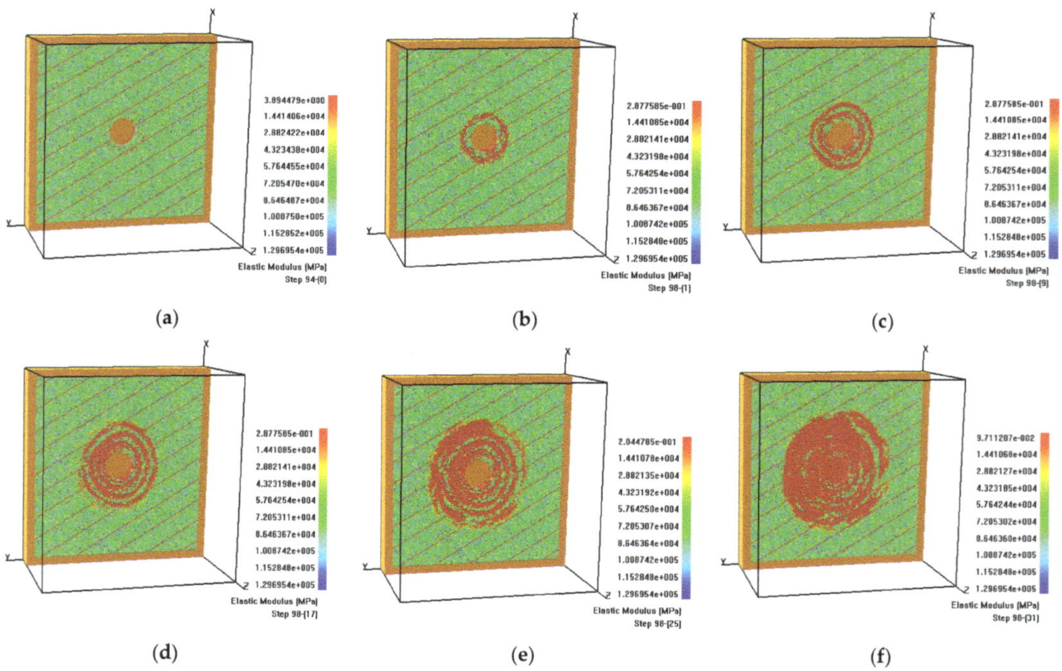

Figure 9. Distribution of elastic modulus during zonal disintegration process of surrounding rock with Model B30:(**a**) Step 94-0; (**b**) Step 98-1; (**c**) Step 98-9; (**d**) Step 98-17; (**e**) Step 98-25; (**f**) Step 98-31.

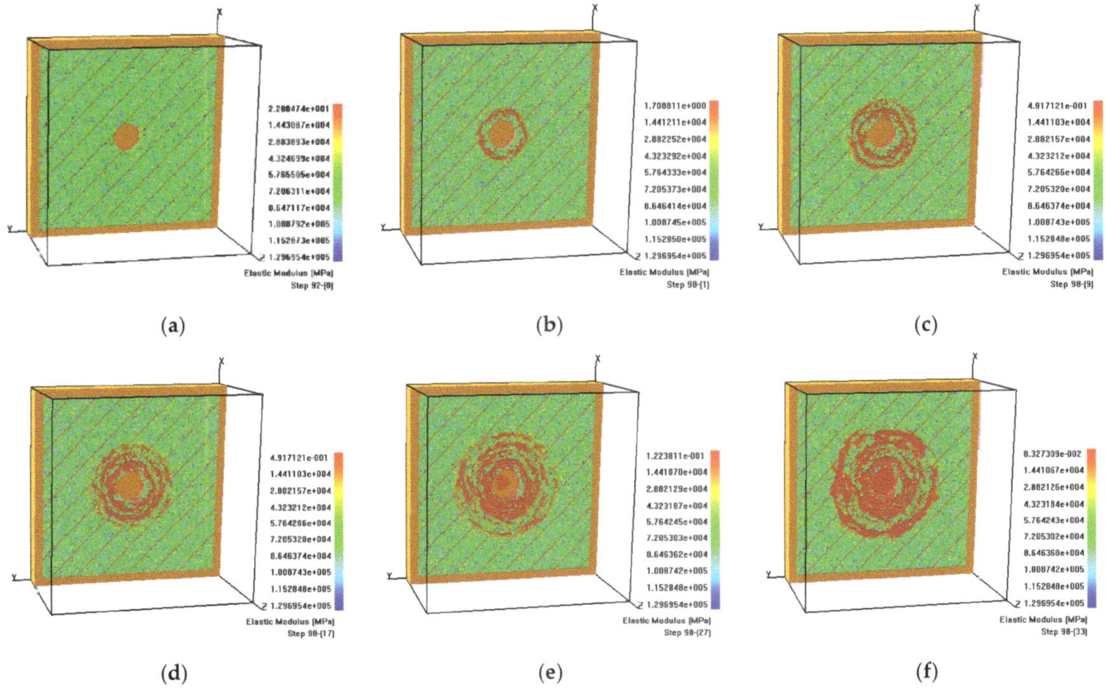

Figure 10. Distribution of elastic modulus during zonal disintegration process of surrounding rock with Model B45: (**a**) Step 92-0; (**b**) Step 98-1; (**c**) Step 98-9; (**d**) Step 98-17; (**e**) Step 98-27; (**f**) Step 98-33.

Figure 8 shows the continuous damage evolution of the horizontal joint model B0 under the triaxial stress state, from which can be observed that with the rock mass strength weakening, the first fractured zone adjacent to the tunnel wall appears at the 96th reduction step. The elements in the corresponding region are damaged due to tensile failure and their elastic moduli gradually decrease with the developing damage. When the compressive/tensile strength continues reducing to the 97-7th step, the rock mass at a certain distance away from the first fracture ring forms an intermittent second fracture ring. Then, at the 97-15th step, a third fracture ring is observed at the interval of thicker intact rock mass. However, it is obvious that the new fracture ring owns a lower degree of integrity comparing with the two previous fracture rings. The rupture area of the fracture ring located above the tunnel develops along the horizontal direction due to the existence of the horizontal soft joint. After that, when the strength of the model drops to the 97-19th step, the fourth fracture ring appears. However, this fracture ring shows a strong intermittent feature and develops along the horizontal joint under the bottom of the tunnel. At the 97-21st and 97-23rd steps, the last two fracture rings with fuzzy boundary appears, and some of the damaged elements are even disorderly connected with previous one.

As shown in Figures 9 and 10, compared with the horizontal joint circular tunnel Model B0, the zonal fracture modes of Models B0, B30 and B45 show similarities to a certain extent. For example, the number of fracture rings is the same, and the fracture rings are not as complete as the jointless Model A with a lower degree of integrity and fuzzier boundaries. Meanwhile, the differences between them are significant.

Clearly, the first fracture ring of Model B0 appears at the 96th reduction step, and the last fracture ring forms at the 97-23rd step. Under the same boundary condition, the first fracture ring of Model B30 advances to the 94-0th step, and the last fracture ring appearance postpones to the 98-31st step. As for Model B45, its first fracture ring appears earliest, being advanced to the 92nd step. However, its last failure ring is not complete until the 98-33rd step. Consequently, the simulated results demonstrate that the larger the angle between the joints and the horizontal direction is, the earlier the fracture ring appears. However, the last fracture ring displays the opposite formation rule, i.e., a model with larger joint dip angle needs more reduction steps to form the final failure mode, meaning several fracture rings distribute alternately, which also means that the corresponding surrounding rock is more stable.

Figure 11 displays the contrast diagrams of the accumulated acoustic emission energy and accumulated acoustic emission counts of each reduction step during the deformation and failure process of the different jointed Models B0, B30 and B45, respectively. The model with larger angle between the joints and the horizontal plane generally releases higher acoustic emission energy and more accumulated acoustic emission counts, which means a larger destruction area and more serious damage occurs before the final collapse. These results indicate that the model with larger inclination between the joints and the horizontal plane has higher ultimate load-bearing capacity and therefore higher stability, but the intensity of failure is higher before the final collapse. This conclusion also agrees with the above analysis of Figures 8–10 from the perspective of energy.

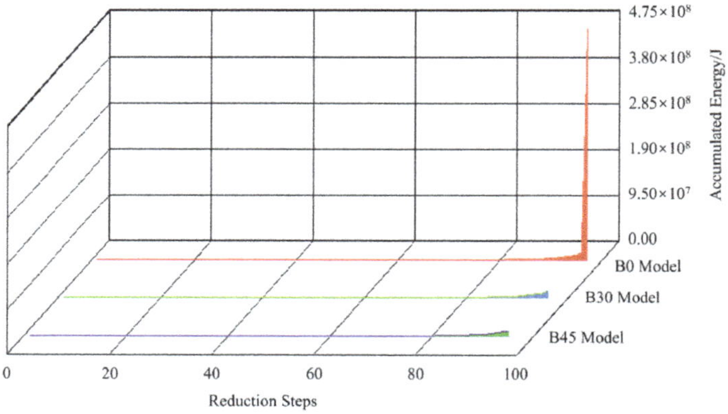

(a)

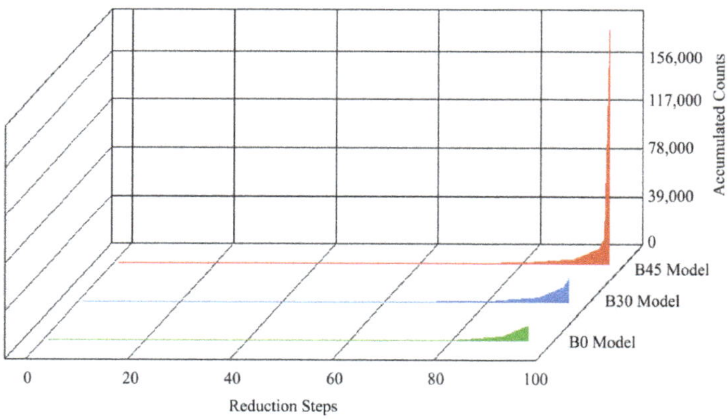

(b)

Figure 11. Comparison of acoustic emission in different models: (**a**) accumulated acoustic emission energy; (**b**) accumulated acoustic emission counts.

In order to further analyze the zonal disintegration law of the jointed circular tunnels and reveal the influence of joints with different dip angles on the zonal disintegration phenomenon, the maximum principal stresses σ_1 of the elements on the x = 4750 mm and z = 850 mm line on both sides of the tunnels related to the occurred fracture rings are studied comprehensively. The detailed maximum principal stress σ_1 distribution along the broken line when the fracture rings occur in each model are shown in Figure 12. It can be observed from Figure 12a that the peak values of σ_1 in Model B0 are about 17~30 MPa. As illustrated by the black broken line with rectangular dots, when the first fracture zone is formed, the peak value is about 17.5 MPa. However, this value rises to 21 MPa at the 97-7th step when the second fracture ring is formed, and the spatial position of the peak value moves to the inner rock mass along the radial direction. The subsequent peak values related to each new fracture ring are 23.75 MPa, 25 MPa, 28 MPa and 28.75 MPa, respectively, and the peak positions of each curve are distributed outward, in turn. With the continuous process of stress buildup, stress shadow and stress transfer, the maximum principal stresses of the surrounding rocks gradually become stable at a relatively low level in the far-field rock masses, except for some slight fluctuations caused by the inhomogeneity of rock mass. The corresponding far-field rock masses are basically undamaged under the low excavation disturbance.

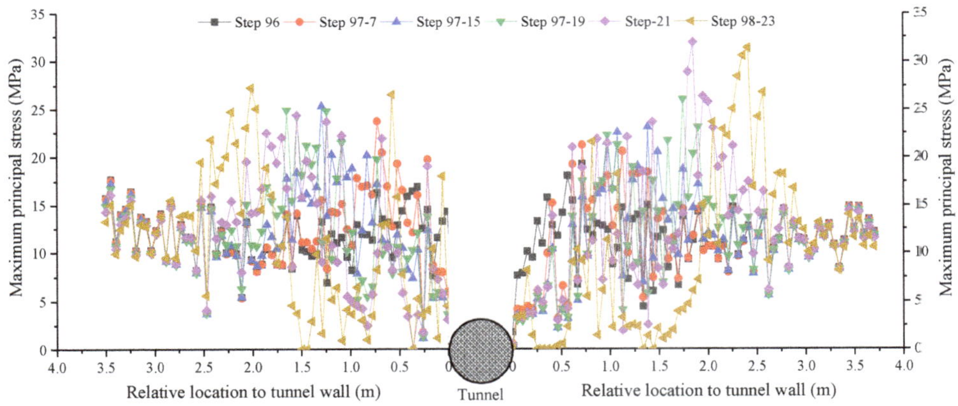

(a)

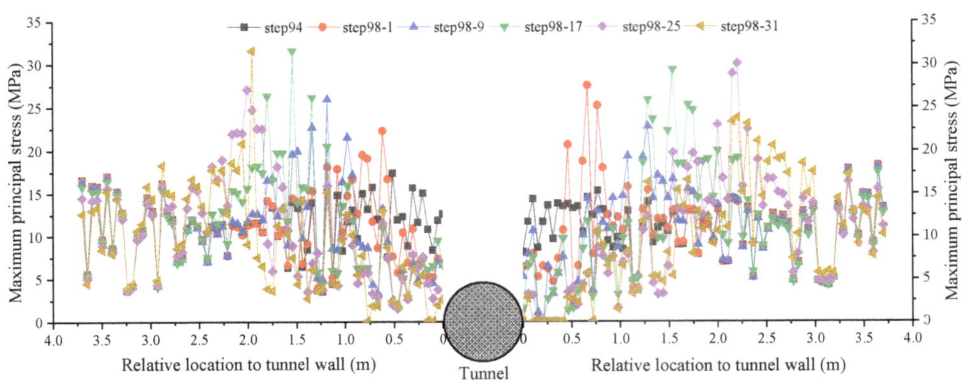

(b)

Figure 12. *Cont.*

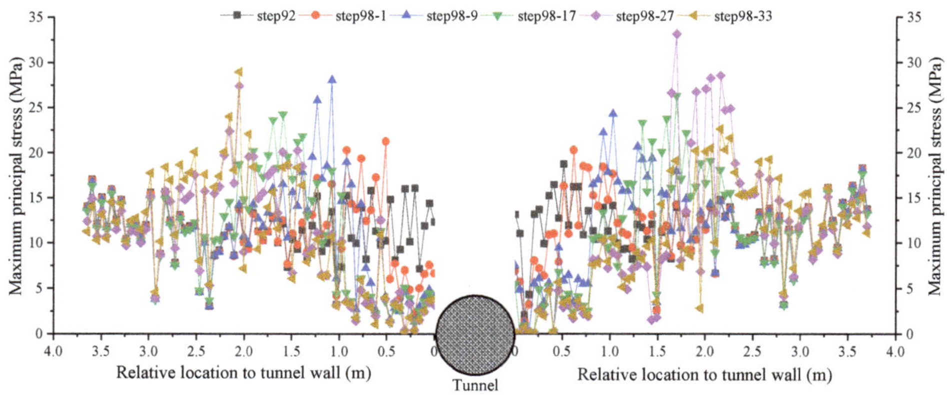

(c)

Figure 12. The maximum principal stress transfer process of surrounding rock during zonal disintegration formation in different models: (**a**) Model B0; (**b**) Model B30; (**c**) Model B45.

Obviously, the failure features of the zonal disintegration phenomenon show significant differences from the traditional excavation loose zone around shallow tunnels. The above analysis of the changes of surrounding rock stresses during the formation of each fracture ring at the key reduction steps suggests that the continuous process of stress buildup, stress shadow and stress transfer plays a critical role in promoting the surrounding rock masses of the circular tunnel to fail in the pattern of alternate regional destructions. Namely, the mechanical process of stress transferring and re-concentrating to the deeper part of the surrounding rock masses keeps the integrity of the rocks between two fracture rings and lead to zonal disintegration phenomenon in the view of mechanical mechanism. Moreover, when comparing the stress distributions of Models B0, B30 and B45 along the broken line, it can be concluded that the stress values show a downward tendency with the increasing of dip angle. In addition, the radius of the outermost fracture ring expands out gradually, i.e., 1.75 m, 1.9 m and 2.1 m away from the tunnel walls, respectively, which means that the destroyed range of surrounding rock masses extend wider if the joint dip angle increases.

3.3. *Zonal Disintegration Evolution of Random Joint Model*

The circular tunnel Model C with random joints distributed in surrounding rocks is shown in Figure 13. The physical and mechanical parameters of these joints are listed in Table 2 and the boundary conditions and strength reduction process are the same as Model A. The random joint model generates zonal disintegration phenomenon as shown in Figure 14, from which it can be observed that the basic fracture characteristics of Model C are similar to the previous models, i.e., there are also several fracture rings distributed at intervals near the tunnel wall. However, the fracture mode of the random joint model is significantly affected by weak joints and shows specific characteristics.

As shown in Figure 14, the first fracture ring appears at the 92nd strength reduction step, and then the second fracture ring is formed when the strength of the rock materials is reduced to the 96th step. Note that the rupture area D are not damaged along the radial direction as predicted, but forms a linear crack by connecting the upper end of the joint A and the lower end of the joint B. Similarly, when the rock strength is reduced to the 97-2nd step, the lower end of the joint B connects with the lower end of the joint C, forming a straight crack. Then, as shown in Figure 14d, the third fracture ring with low integrity degree is formed at the 97-8th step. After that, two new fracture rings with fuzzy boundaries and discontinuous shape occur at the 97-16th and 97-18th steps, respectively, as shown in Figure 14e,f. Furthermore, it can be observed from Figure 14 that during the failure process of the random joint Model C, the linear cracks formed by connecting close joints have priority over the damage path along the radial direction because of the stress concentration at joint ends and dense joint distribution. It means that the zonal disintegration phenomenon affected by random joints in surrounding rock masses has a low possibility to form the closed complete fracture rings because the existed joints may break the integrity of the potential fracture ring.

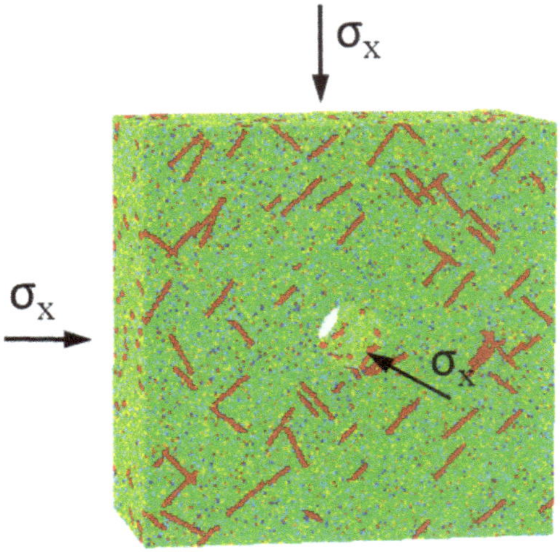

Figure 13. Random jointed surrounding rock numerical Model C and boundary conditions.

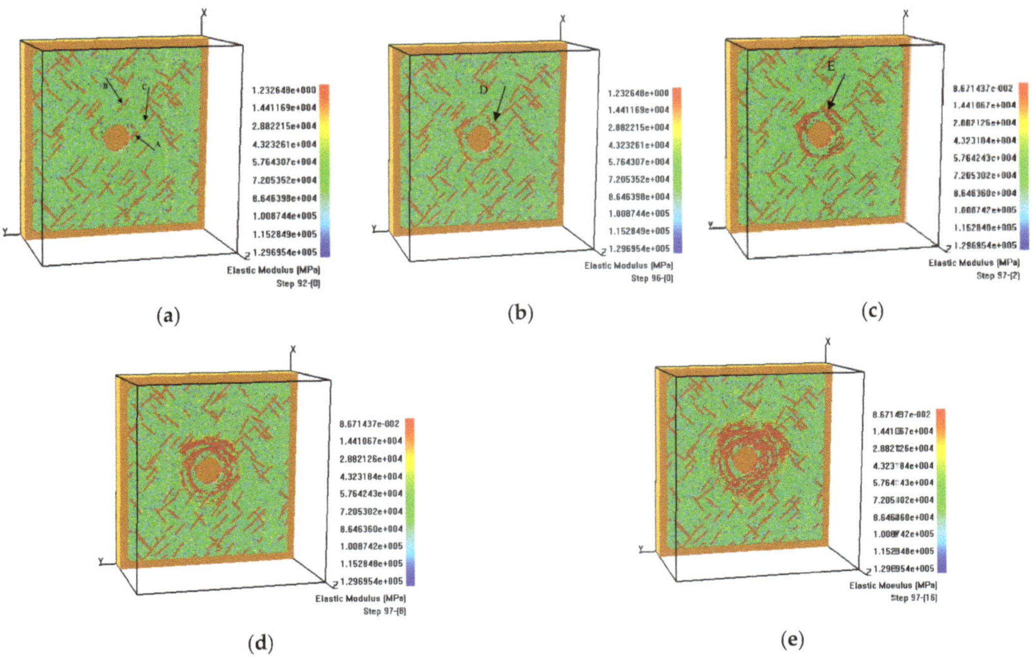

Figure 14. Elastic modulus distribution during zonal disintegration process of randomly jointed surrounding rock: (**a**) Step 92; (**b**) Step 96; (**c**) Step 97-2; (**d**) Step 97-8; (**e**) Step 97-16.

As shown in Figure 15, the maximum principal stresses of the randomly jointed rock around the circular tunnel are approximately 15~20 MPa, which are lower than the previous models. Additionally, the last fracture ring is located only 1.2 m away from the tunnel wall, which indicates the overall damage area of the surrounding rock is obviously smaller than the models discussed above.

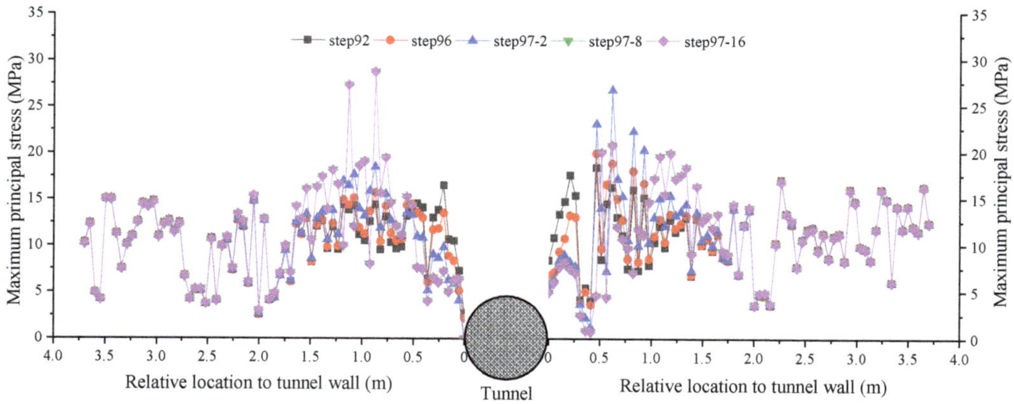

Figure 15. The maximum principal stress-transfer process of surrounding rock during zonal disintegration formation in Model C.

As shown in Figure 16, as many as 20,488 acoustic emission events occur before macro instability, and there is a significant upward trend before fully researching the overall failure of the tunnel, demonstrating a salient failure precursory characteristic of the inhomogeneous surrounding rock. Besides, around 1.58 × 10^7 J, acoustic emission energy is released by the damaged elements, which is also much lower than the above jointed rock models and jointless model because of its smaller overall damage area.

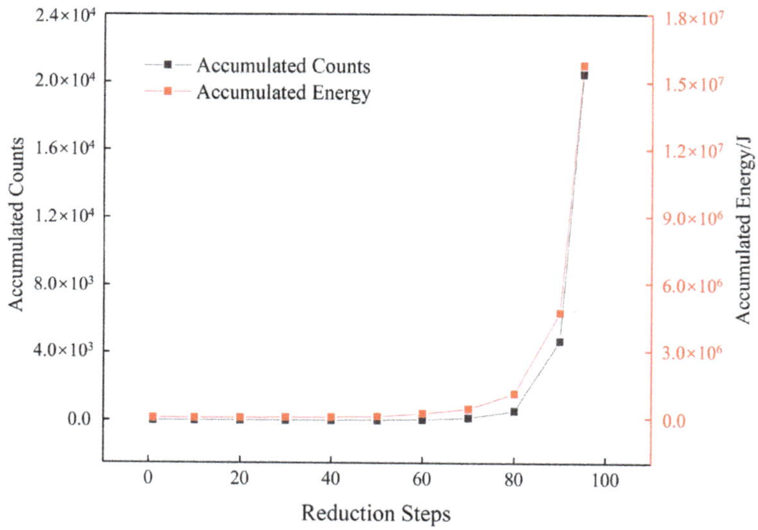

Figure 16. Accumulated acoustic emission counts and energy.

4. Discussion

Table 3 lists the values of strength reduction times, critical compressive and tensile strengths and tunnel safety factors F_S of the studied models obtained by the strength reduction method. It can be observed that the F_S of Model A, i.e., the jointless model, is larger than the other models mainly because the jointless surrounding rock of the tunnel is the most stable, and the relatively closed complete fracture rings are produced in this model, showing the distinct zonal disintegration phenomenon. Moreover, after comparing the accumulated acoustic emission energy and the F_S values of the jointed Models B0, B30 and B45, it can be concluded that with the dip angle between the joint set and the horizontal direction increasing, the surrounding rock of the tunnel will become more and more stable under the boundary conditions and material parameters in this study.

Table 3. The safety factor of the circular tunnels with different surrounding rock masses.

Model	Strength Reduction Step at Failure	Critical Compressive/Tensile Strength (MPa)	Safety Factor
A	100	7.5/0.75	20.00
B0	96	13.5/1.35	11.11
B30	97	12/1.20	12.50
B45	98	10.5/1.05	14.29
C	96	13.5/1.35	11.11

Furthermore, there is a positive correlation between the accumulated acoustic emission energy before final collapse and the safety factor of the tunnel, as shown in Figure 17. It is worth noting that the accumulated acoustic emission energy of each model compressed by initial stresses is very small, and the accumulated acoustic emission counts are basically negligible, as shown in Figures 11 and 16. Therefore, it can be proven that the rock masses surrounding the tunnel are not damaged for the adopted parameters without reducing the strength. Besides, the safety factor of the random joint Model B0 is coincidentally equal to the horizontal joint Model C. However, the horizontal joints and randomly distributed joints play a completely different role in forming the zonal disintegration in terms of background stress evolution; i.e., the gradual process of stress buildup, stress shadow and stress transfer induced by the existed joints would result in different fracture patterns. Further comparison of the accumulated acoustic emission energy released in the two models demonstrates that at the same reduction steps, the latter release more energy than the former because of its highly fractured surrounding rock masses. Reasonably, Model B0 has a better stability than Model C under the influence of the distributed joints. Hence, the acoustic emission events provide more accurate evaluation information rather than the tunnel safety factor.

According to the above analysis, we know that in deep geological environments, the fracture characteristics of the zonal disintegration of surrounding rocks are very different from the traditional loose zone of surrounding rocks. Therefore, the support measures adopted for the two phenomena should also be different. Firstly, in terms of the failure mode, the traditional loose circle only produces fractures within a certain range away from the tunnel wall, while the zonal disintegration produces spaced failure areas within the interior of surrounding rocks, which leads to the different specific positions needing to be supported. Secondly, the primary process of stress redistribution inducing the loose zone of surrounding rocks can be finished immediately after tunnel excavation. However, the surrounding rocks will go through several times of stress redistribution when forming zonal disintegration. Hence, compared with the traditional support measures, the occurrence sequence of fracture rings should be considered for handling zonal disintegration, and the supporting schemes have to be determined. Finally, considering that the zonal fractures of surrounding rocks generally happen in high-stress environments in deep engineering, which may result in unimaginable accidents, it is necessary to further study more effective support methods.

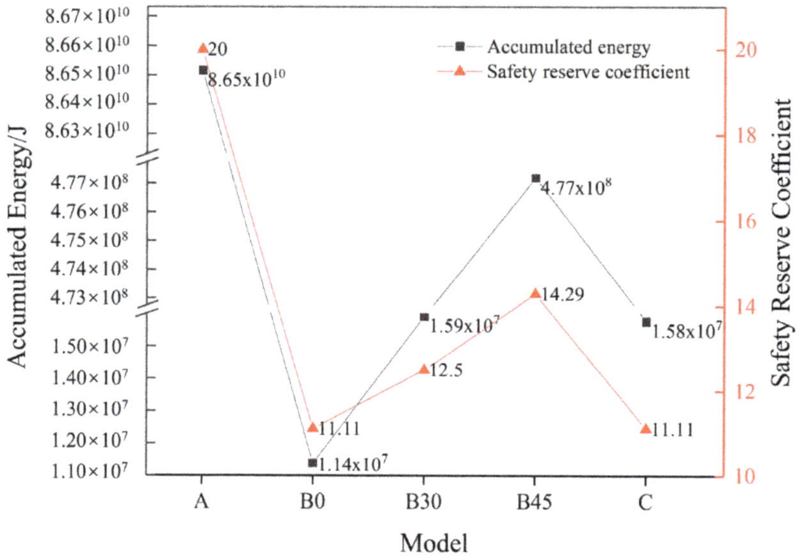

Figure 17. Relationship between the accumulated acoustic emission energy and the safety factor of the different tunnel models.

5. Conclusions

With the aim of understanding the zonal disintegration phenomenon and revealing the intrinsic mechanical mechanisms, a series of 3D heterogeneous tunnel models considering varying joint dip angles are built up and tested by the rock failure process analysis method on the basis of the continuum mechanics, meso-damage mechanics and statistical strength theory. The strength reduction method is embedded to achieve the gradual fracture process, final failure mode and safety factor and to reproduce the characteristic fracture phenomenon of deep tunnels under high geo-stress level, i.e., zonal disintegration. The following conclusions can be reached:

The combined approach has been proven effective in capturing the zonal disintegration characteristics by the indoor physical test [37]. The detailed failure process of the jointed rock models demonstrates that the zonal disintegration is induced by the stress redistribution of surrounding rock masses. Namely, the continuous process of stress buildup, stress shadow and stress transfer play a critical role in promoting the surrounding rock masses to fail in the pattern of alternate regional destructions. It is the mechanical process of stress transferring and re-concentrating to the deeper part of the surrounding rock masses to keep the integrity of the rocks between two fracture rings and lead to zonal disintegration phenomenon in the view of mechanical mechanism, which is obviously different from the formation mechanism of the traditional surrounding rock loose zone.

The dip angle of the existed joint set shows great influence on the stress evolution and inner crack propagation of the surrounding rock mass. On the one hand, the fracture ring of the 45° joint model first appears at the 92-0th step, but the last ring forms at the 98-33rd step. Actually, a model with larger joint dip angle needs more reduction steps to generate the final failure mode. On the other hand, the acoustic emission energy released by Model B45 is the most, up to 4.77×10^8 J, while Model B0 only releases 1.58×10^7 J energy. This means that the model with larger inclination angle will be damaged more seriously before the final collapse. Besides, as the joint inclination increases from 0° to 45°, the safety factor of the model rises from 11.11 to 14.29. In addition, the model with larger joint inclination has higher ultimate load-bearing capacity and therefore higher stability. However, the intensity of failure is also higher before the final collapse.

Furthermore, the existence of random joints might significantly affect the integrity and regularity of the zonal disintegration pattern because the high stress concentrations at the ends of the existed joints could change the crack growth path and prevent the damage development along the radial direction. Namely, the zonal disintegration affected by randomly distributed joints has a low possibility to form the closed complete fracture rings. Meanwhile, the upward trend of acoustic emission energy released before nearly researching the overall failure of the tunnel could be regarded as a macro-failure precursor. Models C and B0 have the same safety factor of 11.11. The acoustic emission count of the former is 20,488, which is much higher than 12,801 of the latter. Hence, the acoustic emission events are able to provide more accurate evaluation basis rather than the tunnel safety factor.

These achievements will improve our understanding of the mechanical mechanism of zonal disintegration formation, provide insights into the characteristic fracture phenomenon of deep-buried tunnels, and establish the basis for design, construction and treatment of tunnels in deep engineering.

Author Contributions: Conceptualization, B.G.; data curation, B.C.; formal analysis, B.C.; funding acquisition, B.G.; investigation, B.C. and B.G.; software, C.T.; supervision, B.G. and C.T.; writing—original draft, B.C.; writing—review and editing, B.G. and S.W. All authors have read and agreed to the published version of the manuscript.

Funding: This work was funded by the National Natural Science Foundation of China (Grant No. 42102314) and the China Postdoctoral Science Foundation (Grant No. 2020M680950).

Data Availability Statement: The datasets generated and/or analyzed during the current study are available from the corresponding author upon reasonable request.

Conflicts of Interest: The authors declare no conflict of interest.

References

1. He, S.; Lai, J.; Zhong, Y.; Wang, K.; Xu, W.; Wang, L.; Liu, T.; Zhang, C. Damage behaviors, prediction methods and prevention methods of rockburst in 13 deep traffic tunnels in China. *Eng. Fail. Anal.* **2021**, *121*, 105178. [CrossRef]
2. Hu, Y.; Wang, Q.; Wang, M.; Liu, D. A study on the thermo-mechanical properties of shotcrete structure in a tunnel, excavated in granite at nearly 90 °C temperature. *Tunn. Undergr. Space Technol.* **2021**, *110*, 103830. [CrossRef]
3. Lin, P.; Liu, H.; Zhou, W. Experimental study on failure behaviour of deep tunnels under high in-situ stresses. *Tunn. Undergr. Space Technol.* **2015**, *46*, 28–45. [CrossRef]
4. Shemyakin, E.I.; Fisenko, G.L.; Kurlenya, M.V.; Oparin, V.N.; Reva, V.N.; Glushikhin, F.P.; Rozenbaum, M.A.; Tropp, E.A.; Kuznetsov, Y.S. Zonal disintegration of rocks around underground workings, Part 1: Data of in situ observations. *J. Min. Sci.* **1986**, *22*, 157–168. [CrossRef]
5. Adams, G.R.; Jager, A.J. Petroscopic observations of rock fracturing ahead of stope faces in deep-level gold mines. *J. South Afr. Inst. Min. Metall.* **1980**, *80*, 204–209.
6. Wang, R.; Bai, J.B.; Yan, S.; Chang, Z.G.; Song, Y.B.; Zhang, W.G.; Xu, J. The elastoplastic solutions of deep buried roadway based on the generalized 3D Hoek-Brown strength criterion considering strain-softening properties. *Geofluids* **2021**, *2021*, 1–15. [CrossRef]
7. Kang, F.; Li, Y.; Tang, C. Numerical study on airflow temperature field in a high-temperature tunnel with insulation layer. *Appl. Therm. Eng.* **2020**, *179*, 115654. [CrossRef]
8. Gong, B.; Wang, Y.; Zhao, T.; Tang, C.; Yang, X.; Chen, T. AE energy evolution during CJB fracture affected by rock heterogeneity and column irregularity under lateral pressure. *Geomat. Nat. Haz. Risk* **2022**, *13*, 877–907. [CrossRef]
9. Wang, S.F.; Tang, Y.; Li, X.B.; Du, K. Analyses and predictions of rock cuttabilities under different confining stresses and rock properties based on rock indentation tests by conical pick. *Trans. Nonferrous Met. Soc. China* **2021**, *31*, 1766–1783. [CrossRef]
10. Wang, S.F.; Tang, Y.; Wang, S.Y. Influence of brittleness and confining stress on rock cuttability based on rock indentation tests. *J. Cent. South Univ.* **2021**, *28*, 2786–2800. [CrossRef]
11. Qian, Q.; Zhou, X. Failure behaviors and rock deformation during excavation of underground cavern group for Jinping I Hydropower Station. *Rock Mech. Rock Eng.* **2018**, *51*, 2639–2651. [CrossRef]
12. Zhang, Q.; Zhang, X.; Wang, Z.; Xiang, W.; Xue, J. Failure mechanism and numerical simulation of zonal disintegration around a deep tunnel under high stress. *Int. J. Rock Mech. Min. Sci.* **2017**, *93*, 344–355. [CrossRef]
13. Jia, P.; Zhu, W. Mechanism of zonal disintegration around deep underground excavations under triaxial stress—Insight from numerical test. *Tunn. Undergr. Space Technol.* **2015**, *48*, 1–10. [CrossRef]

14. Li, S.C.; Wang, H.P.; Qian, Q.H.; Li, S.C.; Fan, Q.Z.; Yuan, L.; Xue, J.H.; Zhang, Q.S. In-situ monitoring research on zonal disintegration of surrounding rock mass in deep mine roadways. *Chin. J. Rock Mech. Eng.* **2008**, *27*, 1545–1553.
15. Chen, X.G.; Wang, Y.; Zhang, Q.Y.; Li, S.C.; Nordlund, E. Analogical model test and theoretical analysis on zonal disintegration based on filed monitoring in deep tunnel. *Eur. J. Environ. Civ. Eng.* **2013**, *17*, s33–s52. [CrossRef]
16. Gong, B.; Liang, Z.Z.; Liu, X.X. Nonlinear deformation and failure characteristics of horseshoe-shaped tunnel under varying principal stress direction. *Arab. J. Geosci.* **2022**, *15*, 475. [CrossRef]
17. Ma, X.; Wei, J.; Liu, J.; Song, Z.; Bai, Y. Study on the generation mechanism and development law of the zonal disintegration in deep burial tunnels. *Shock Vib.* **2020**, *2020*, 1–16. [CrossRef]
18. Tang, Q.; Xie, W.; Wang, X.; Su, Z.; Xu, J. Numerical study on zonal disintegration of deep rock mass using three-dimensional bonded block model. *Adv. Civ. Eng.* **2019**, *2019*, 1–12. [CrossRef]
19. Chen, X.G.; Zhang, Q.Y.; Lin, B.; Zhang, N.; Liu, D.J.; Wang, M.B. Experimental and numerical study on zonal disintegration of deep tunnel. In Proceedings of the Controlling Seismic Hazard and Sustainable Development of Deep Mines: 7th International Symposium on Rockburst and Seismicity in Mines, Dalian, China, 21–23 August 2009.
20. Zhang, X.; Xue, J.; Duan, C. Research on formation process of zonal disintegration in deep rock mass based on field monitoring and geomechanical model test. *Geotech. Geol. Eng.* **2018**, *36*, 2725–2733.
21. Zhang, X.T.; Zhang, Q.Y.; Xiang, W.; Gao, Q.; Yuan, S.B.; Wang, C. Model test study of zonal disintegration in deep layered jointed rock mass. *Rock Soil Mech.* **2014**, *35*, 2247–2254.
22. Yuan, P.; Xu, Y. Analyses on deformation and fracture evolution of zonal disintegration during axial overloading in 3D geomechanical model tests. *J. Vibroeng.* **2019**, *21*, 1163–1174. [CrossRef]
23. Qian, Q.H. Research status and some thoughts on zonal disintegration. In Proceedings of the Activities of the 21st Academic Salon of New Ideas and Theories, Chinese Society for Rock Mechanics & Engineering, Beijing, China, 23 June 2008.
24. Zhu, X.; Wang, Y.; Ren, Y. Numerical simulation to research on the fracture mechanism of the surrounding rock in deep tunnels. *Geotech. Geol. Eng.* **2020**, *38*, 319–327. [CrossRef]
25. Zhang, H.; Shi, H.; Jing, H.; Wu, Y.; Pu, H. Numerical study of remote fracturing around a circular opening in rock. *Eur. J. Environ. Civ. Eng.* **2018**, *24*, 1032–1050. [CrossRef]
26. Wu, S.; Chen, L.; Cheng, Z. Macro and meso research on the zonal disintegration phenomenon and the mechanism of deep brittle rock mass. *Eng. Fract. Mech.* **2019**, *211*, 254–268. [CrossRef]
27. Jing, L. A review of techniques, advances and outstanding issues in numerical modelling for rock mechanics and rock engineering. *Int. J. Rock Mech. Min. Sci.* **2003**, *40*, 283–353. [CrossRef]
28. Fallah, N.; Bailey, C.; Cross, M.; Taylor, G. Comparison of finite element and finite volume methods application in geometrically nonlinear stress analysis. *Appl. Math. Model.* **2000**, *24*, 439–455. [CrossRef]
29. Feng, X.; Gong, B.; Tang, C.; Zhao, T. Study on the non-linear deformation and failure characteristics of EPS concrete based on CT-scanned structure modelling and cloud computing. *Eng. Fract. Mech.* **2022**, *261*, 108214. [CrossRef]
30. Brady, B.H.G.; Bray, J.W. Boundary element method for determining stresses and displacements around long openings in a triaxial stress-field. *Int. J. Rock Mech. Min. Sci.* **1978**, *15*, 21–28. [CrossRef]
31. Cundall, P.A. A computer model for simulating progressive large scale movements in blocky rock systems. In Proceedings of the Symposium of the International Society for Rock Mechanics (ISRM), Nancy, France, 4–6 October 1971.
32. Shi, G.H. Discontinuous Deformation Analysis: A New Numerical Model for the Statics and Dynamics of Block Systems. Ph.D. Thesis, University of California, Berkeley, CA, USA, 1988.
33. Wang, Y.; Gong, B.; Tang, C.; Zhao, T. Numerical study on size effect and anisotropy of columnar jointed basalts under uniaxial compression. *Bull. Eng. Geol. Environ.* **2021**, *81*, 41. [CrossRef]
34. Tang, C.; Liu, H.; Lee, P.; Tsui, Y.; Tham, L. Numerical studies of the influence of microstructure on rock failure in uniaxial compression—Part I: Effect of heterogeneity. *Int. J. Rock Mech. Min. Sci.* **2000**, *37*, 555–569. [CrossRef]
35. Tang, C. Numerical simulation of progressive rock failure and associated seismicity. *Int. J. Rock Mech. Min. Sci.* **1997**, *34*, 249–261. [CrossRef]
36. Tang, C.A.; Li, L.C.; Li, C.W.; Ma, T.H. RFPA strength reduction method for stability analysis of geotechnical engineering. *Chin. J. Rock Mech. Eng.* **2006**, *25*, 1522–1530.
37. Gao, Q.; Zhang, Q.; Zhang, X.; Zhang, L. Geomechanical model test and energy mechanism analysis of zonal disintegration in deep surrounding rock. *Geosciences* **2018**, *8*, 237. [CrossRef]
38. Cai, M.F.; He, M.C. *Rock Mechanics and Engineering*; Science Press: Beijing, China, 2013; p. 298.
39. Liang, Z.Z. Three-Dimensional Failure Process Analysis of Rock and Associated Numerical Tests. Ph.D. Thesis, Northeastern University, Shenyang, China, 2005.

mathematics

Article

Novel Ensemble Tree Solution for Rockburst Prediction Using Deep Forest

Diyuan Li [1,*], Zida Liu [1], Danial Jahed Armaghani [2], Peng Xiao [1] and Jian Zhou [1]

[1] School of Resources and Safety Engineering, Central South University, Changsha 410083, China; liuzida@csu.edu.cn (Z.L.); xiaopengaizhanghuimin@csu.edu.cn (P.X.); j.zhou@csu.edu.cn (J.Z.)

[2] Department of Urban Planning, Engineering Networks and Systems, Institute of Architecture and Construction, South Ural State University, 76, Lenin Prospect, 454080 Chelyabinsk, Russia; danialarmaghani@susu.ru

* Correspondence: diyuan.li@csu.edu.cn

Abstract: The occurrence of rockburst can cause significant disasters in underground rock engineering. It is crucial to predict and prevent rockburst in deep tunnels and mines. In this paper, the deficiencies of ensemble learning algorithms in rockburst prediction were investigated. Aiming at these shortages, a novel machine learning model, deep forest, was proposed to predict rockburst risk. The deep forest combines the characteristics of deep learning and ensemble models, which can solve complex problems. To develop the deep forest model for rockburst prediction, 329 real rockburst cases were collected to build a comprehensive database for intelligent analysis. Bayesian optimization was proposed to tune the hyperparameters of the deep forest. As a result, the deep forest model achieved 100% training accuracy and 92.4% testing accuracy, and it has more outstanding capability to forecast rockburst disasters compared to other widely used models (i.e., random forest, boosting tree models, neural network, support vector machine, etc.). The results of sensitivity analysis revealed the impact of variables on rockburst levels and the applicability of deep forest with a few input parameters. Eventually, real cases of rockburst in two gold mines, China, were used for validation purposes while the needed data sets were prepared by field observations and laboratory tests. The promoting results of the developed model during the validation phase confirm that it can be used with a high level of accuracy by practicing engineers for predicting rockburst occurrences.

Keywords: rockburst prediction; deep forest; bayesian optimization; ensemble model

MSC: 68Txx

Citation: Li, D.; Liu, Z.; Armaghani, D.J.; Xiao, P.; Zhou, J. Novel Ensemble Tree Solution for Rockburst Prediction Using Deep Forest. *Mathematics* **2022**, *10*, 787. https://doi.org/10.3390/math10050787

Academic Editor: Manuel Pastor

Received: 9 February 2022
Accepted: 25 February 2022
Published: 1 March 2022

1. Introduction

Rockburst is a geological catastrophe induced by the sudden release of strain energy stored in rock mass during or after the excavation of underground engineering in high in-situ stress areas. Rockburst occurs in many countries around the world [1]. It is generally believed that intensity and frequentness of rockburst increase as depth increases. The occurrence of rockburst damages underground tunnels and facilities and poses a severe threat to the safety of the operators on site. The gold mines in South Africa have greater mining depths than those in other countries. Meanwhile, most rockburst disasters occur in South African gold mines. In 1975, 73 laborers died due to 680 rockburst incidents in 31 gold mines. From 1984 to 1993, 3275 laborers lost their lives in mining geological disasters due to the lack of mining technology to cope with the rockburst below 2000 m [2]. Rockburst is a complicated problem restricting the progress of underground engineering. It is necessary for researchers to study how to prevent and control rockburst.

Many scholars have taken various measures to evaluate rockburst risk. These methods contain empirical indicators, numerical modeling, rock mechanics tests, intelligent techniques, etc. [1,3]. Xue et al. [4] adopted the empirical method to estimate the rockburst grade at the Jiangbian hydropower station, China. The empirical method is simple and easy to implement, but its effectiveness is poor. Zhai et al. [5] carried out rockburst tests with six hard brittle rocks subjected to one-free-face true triaxial mechanical tests. Their test results revealed that strength, fracturing, fragmentation characteristics, and failure modes had a remarkable impact on rockburst proneness. Due to scale effect considerations, experimental methods are suitable for investigating the failure process and mechanism of rockburst rather than predicting rockburst [6–8]. Moreover, the field condition is challenging to be reproduced in the laboratory. Wang et al. [9] summarized the numerical simulations, including the continuum, discontinued, and hybrid techniques for rockburst evaluation. The numerical simulation is economical, secure, and time-saving [1,3,10]. Nevertheless, choosing an appropriate constitutive model and simulation method is very important according to specific problems. With the blossom of artificial intelligence and big data, intelligent algorithms are increasingly used to predict rockbursts. Compared to empirical, numerical, and experimental methods, the intelligent model has high efficiency, good practicability and can foretell and prevent rockbursts in time. However, it requires high-quality data.

The machine learning (ML) algorithm is an essential part of the intelligent algorithm [11,12]. The ML algorithms for rockburst classification mainly include linear models (LM), decision trees (DT), artificial neural networks (ANN), *k*-nearest neighbor (KNN), Bayes classifiers, support vector machines (SVM), ensemble models, etc. Each ML model has its own supremacy and drawback, and no model can perform best for every practical engineering based on the 'No Free Lunch theorem'. Table 1 compiles the ML techniques for rockburst estimation recently and compares their advantages and disadvantages.

Table 1. The superiority and drawback of ML techniques for rockburst estimation recently.

Algorithm	Superiority	Drawback
LDA [13] LR [14]	Fast training and prediction speed, and simple and easy to interpret.	Unsuitable for high-dimensional data.
C5.0 DT [15] DT [16]	Suitable for data with missing values, and can process continuous variables and discrete variables simultaneously.	Tend to produce an overly complex model to reduce its generalization.
KNN [13]	Simple and easy to implement.	Unsuitable for unbalanced samples.
Naïve Bayes [13] BN [17]	Simple and fast. Perform very well under the assumption that distribution is independent.	The assumption of independent distribution is difficult to meet in practical projects.
ANN [13,18–20]	Strong mapping ability and can deal with complex nonlinear problems.	Many hyperparameters to turn and easy to overfit.
SVM [13,21]	Solid theoretical basis and can be applied to complex nonlinear data.	Difficult to deal with multiple classification problems.
RF [13,22] Bagging [23]	Suitable for high dimension data and good generalization ability.	Overfitting appears when dealing with classification with high noise.
GBM [13] XGB [24]	Suitable for continuous values and discrete values. Robust to outliers using robust loss functions	Difficult to train data in parallel.

Note: LDA = linear discriminant analysis; LR = logistic regression; BN = Bayesian network; GBM = gradient boosting machines; RF = random forest; XGB = extreme gradient boosting;.

To overcome the limitations of single ML models, some researchers have combined multiple intelligent techniques to develop an ensemble model for rockburst estimation recently. Zhang et al. [25] combined seven extensively applied ML techniques using a voting strategy to construct an ensemble model, which had better performance than individual classifiers in rockburst prediction. Liang et al. [26] compared five ensemble models based on DT for forecasting short-term rockburst. They found that the RF was the optimal model. Liang et al. [27] utilized weighting voting to combine six intelligent techniques to forecast short-term rockburst. The capacity of the comprehensive combined model was better than that of the base classifiers. Yin et al. [28] used stacking to integrate KNN, SVM, deep neural networks, and recurrent neural networks for rockburst prediction. The ensemble model that adopted KNN and RNN performed best in all ensemble models. Although these ensemble models show a high level of prediction accuracy, some practical problems [27,28] prevent them from being widely used.

(1) Ensemble models are easy to overfit. For obtaining higher prediction accuracy, ensemble models, such as RF, become complex, which makes generalization of the model poor [13].
(2) The selection of a base classifier and combination strategy is difficult [25,27–29]. Different problems require different combined strategies, there should be a difference between the base classifiers, and it is necessary to choose the combination strategy and base classifier. Otherwise, performance cannot improve.
(3) Many hyperparameters need to be tuned. The parameter setting of the ensemble model has a significant effect on the capacity.

To address the above limitations, this study proposed a novel methodology, the deep forest model, for rockburst prediction. Motivated by the theory behind deep neural networks and ensemble models, Zhou et al. [30] proposed the deep forest (DF), which combines the characteristics of deep learning models and ensemble models. DF can deal with more complex problems such as deep learning models. However, it has fewer parameters to work with than deep learning models. The DF model is easy to use, and its complexity can be determined according to the data, which can effectively prevent overfitting. The DF model is suitable for small data sets. Through validation by different data in different fields, the DF model still performs better even if it adopts the default parameter configuration. It is meaningful to build the more powerful and robust intelligent model by DF model to predict and prevent rockburst.

Additionally, Bayesian optimization (BO) is applied for optimizing hyperparameters of DF. BO has been widely used in hyperparametric optimization in different ML studies (e.g., [31–33]). BO differs from other optimization algorithms, and it is a useful model for problems that are expensive to conduct. BO constructs a probabilistic model of the objective function to be optimized and then applies the probabilistic model to determine the next point to be evaluated [34]. BO has been increasingly applied to geotechnical engineering [31,35–37].

The structure of this study is as follows: the 'Methodology' section introduces the theory and composition of DF and BO. The 'Data' section presents the source and statistical description of the rockburst database. The 'Simulation' exhibits how to construct and optimize the DF model for rockburst prediction. In the 'Discussion' section, the capability of DF to predict rockburst is evaluated. Furthermore, the influence of variables on rockburst intensities is analyzed by sensitivity analysis. Finally, the DF is applied to forecast the rockburst in practical engineering.

2. Methodology

2.1. DF

Zhou et al. [38] presented the gcForest to build the DF model, which consists of the cascade forest and multi-grained scanning. When the gcForest addresses sequence or image-style data, it needs multi-grained scanning. The rockburst data does not have spatial or sequential relationships in this study, so the multi-grained scanning structure in gcForest is abandoned.

RF and complete random forest (CRF) are the base classifiers in the DF model. The RF is an ensemble model composed of K decision trees $\{h(X, \theta_k), k = 1, \ldots, K\}$, where θ_k is a random vector that satisfies independent identically distributed [39]. Figure 1 shows flowchart to build RF.

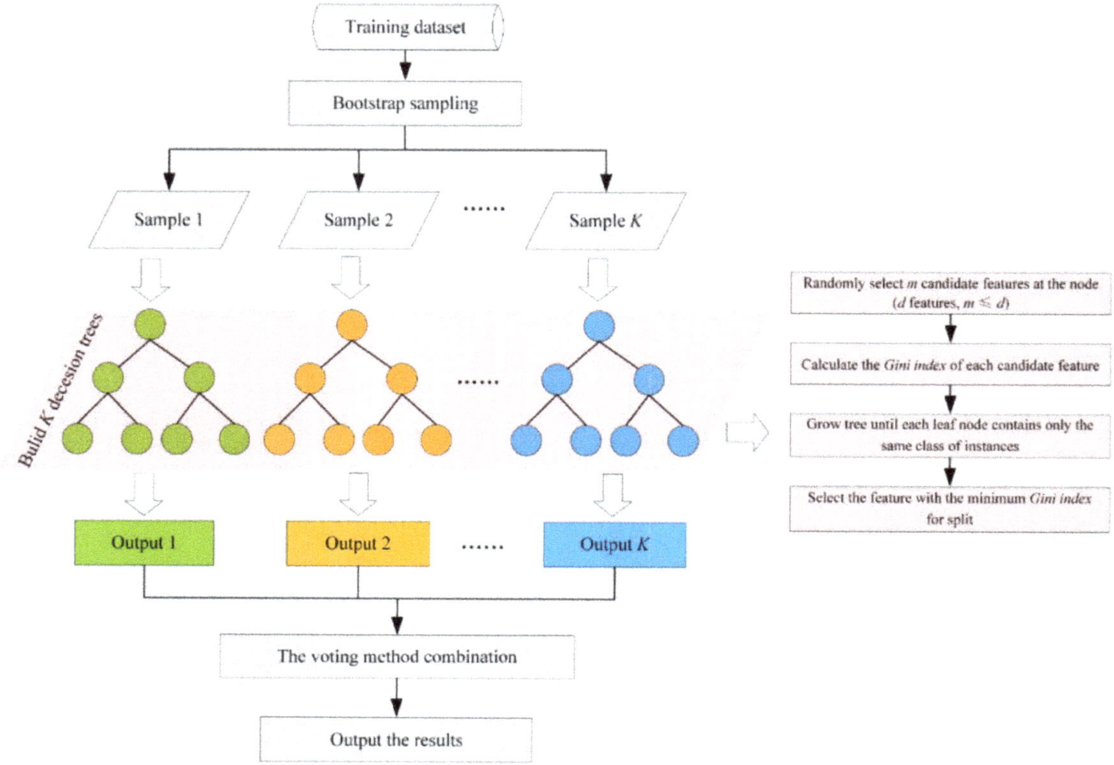

Figure 1. The flowchart to establish RF model.

Similarly, CRF randomly selects a feature for split, and the others are the same as the process of RF. As shown in Figure 2, the RF and CRF are implemented to construct cascade layers. The input features are input to the first cascade layer. After all this, the output of the previous cascade layer and the input features are input to the next layer. At this step, the training set is divided into a growing set and an estimating set. When the cascade forest increases by one layer, the estimating set tests the whole generated DF model. If the performance of the estimating set is lower than that of the previous layer, the DF model stops growing, and the cascade layer does not increase. In the last layer, the average of all the output probability vectors is calculated, and the label with the maximum probability is output as the prediction result. When the cascade layer does not increase, the DF model is retained based on the whole training set. The structure of the DF model is automatically determined, which reduces the risk of overfitting.

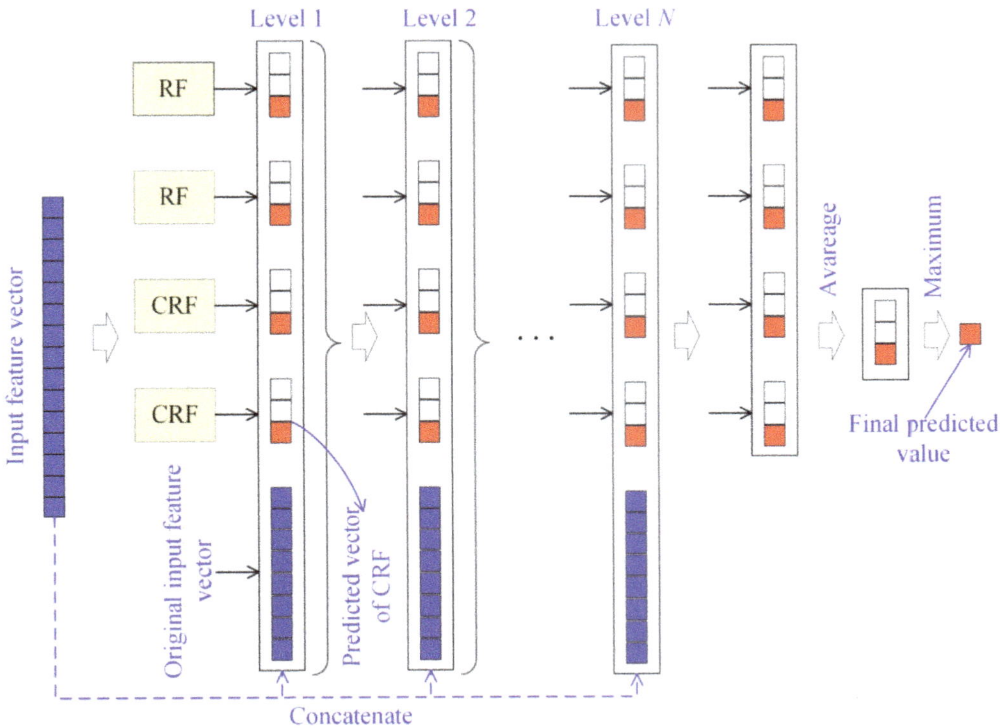

Figure 2. Schematic diagram of cascade forest.

2.2. BO

BO is appropriate for tasks with expensive evaluation costs [40]. BO consists of two parts, the surrogate model and the acquisition function [41]. The Gaussian process (GP) [42] is the most extensively applied surrogate model in BO due to its flexibility and tractability [40]. BO has three basic acquisition functions (AF), which are the probability of improvement [43], expected improvement [44,45], and upper/lower confidence bound. It is vital to choose an appropriate acquisition function to match the surrogate model. GP-Hedge is proposed to select an appropriate AF in each iteration, and detailed information about GP-Hedge can be found in previous studies [46].

3. Data

3.1. Data Collection and Description

The database, including 329 real rockburst cases worldwide, was established, as shown in Table 2. According to the criteria for the classification of rockburst in Table 3, rockburst levels can be grouped into four categories: none (53 cases), light (101 cases), moderate (119 cases), and strong (56 cases). In the collected database (Supplementary Materials), the number of light and moderate rockbursts is greater than that of none and strong rockbursts.

Rockburst often occurs on the excavation face in deep underground construction, it is induced by the sudden release of strain energy stored in the rock mass, and the most common phenomenon is strain burst. Rockburst mechanisms are complicated [47], and it is connected with stress in the earth's crust, rock property, rock mass structure, groundwater, and so on [48]. In this study, seven factors, including maximum tangential stress (σ_θ), uniaxial compressive strength (σ_c), tensile strength (σ_t), elastic strain energy index (W_{et}), stress concentration factor (SCF or σ_θ/σ_c), rock brittleness index B_1 ($B_1 = \sigma_c/\sigma_t$), and rock brittleness index B_2 ($B_2 = (\sigma_c - \sigma_t)/(\sigma_c + \sigma_t)$), are adopted as the input variables in the DF model [13,28]. Table 4 displays the statistical description of the four rockburst intensities. Pearson correlation coefficients (Equation (1)) between variables are calculated, as shown in Figure 3. Figure 4 exhibits the boxplots and histograms of the seven input variables of four rockburst intensities. The boxplots are not symmetrical, there are many points outside the upper and lower whiskers of boxplots, and the collected database does not satisfy a normal distribution.

$$r = \frac{\sum_{i=1}^{n}(X_i - \overline{X})(Y_i - \overline{Y})}{\sqrt{\sum_{i=1}^{n}(X_i - \overline{X})^2}\sqrt{\sum_{i=1}^{n}(Y_i - \overline{Y})^2}} \quad (1)$$

Table 2. The database source.

No.	Number of Cases	Reference
1	N (43 cases), L (78 cases), M (81 cases), S (44 cases)	Zhou et al. [13]
2	L (1 case), M (11 cases)	Pu et al. [49]
3	N (3 cases), L (4 cases), M (8 cases), S (1 case)	Liu et al. [50]
4	N (3 cases), L (7 cases), M (7 cases), S (3 cases)	Xue et al. [51]
5	L (1 case), M (5 cases), Strong (1 case)	Wu et al. [52]
6	N (1 case), L (2 cases), S (4 cases)	Du et al. [53]
7	L (3 cases), M (3 cases)	Jia et al. [54]
8	N (3 cases), L (5 cases), M (4 cases), S (3 cases)	Xue et al. [55]
Sum	N (53 cases), L (101 cases), M (119 cases), S (56 cases)	329 cases

Table 3. Standard of classification for four intensities of rockburst [48].

Rockburst Label	Failure Characteristics
None	No sound of rockburst and rockburst activities.
Light	The surrounding rock is spalled, cracked, or striped, and there is no ejection phenomenon and a weak sound.
Moderate	The surrounding rock is deformed and fractured, and there is aconsiderable number of rock chip ejection, loose and sudden destruction, accompanied by crisp crackling, and often presented in the local cavern of surrounding rock.
Strong	The surrounding rocks are severely bursted and suddenly thrown or shot into the tunnel, accompanied by strong bursts and roaring sounds, air jets, the continuity of storm phenomena, and the rapid expansion into deep surrounding rocks.

Table 4. Statistical description of the input parameters.

Grade	Statistical Indicators	σ_θ	σ_c	σ_t	SCF	B_1	B_2	W_{et}
None	Mean value	25.27	101.96	5.98	0.30	21.08	0.87	2.78
	Standard deviation	16.32	49.39	3.90	0.25	12.72	0.07	1.94
	Min value	2.60	20.00	0.40	0.05	5.38	0.69	0.81
	25th percentiles	12.30	67.40	3.00	0.13	10.75	0.83	1.50
	50th percentiles	21.50	96.41	5.00	0.21	18.75	0.90	2.04
	75th percentiles	31.20	123.60	7.60	0.31	29.40	0.93	3.60
	Max value	77.69	241.00	17.66	1.05	47.93	1.00	7.80

Table 4. *Cont.*

Grade	Statistical Indicators	σ_θ	σ_c	σ_t	SCF	B_1	B_2	W_{et}
Light	Mean value	44.42	116.64	6.68	0.41	21.53	0.89	3.72
	Standard deviation	20.63	39.56	3.91	0.19	10.12	0.07	1.54
	Min value	13.50	30.00	1.90	0.10	2.52	0.43	0.85
	25th percentiles	29.70	88.00	3.60	0.26	12.70	0.85	2.53
	50th percentiles	43.21	117.00	5.90	0.38	23.60	0.92	3.20
	75th percentiles	57.97	142.00	8.95	0.56	28.10	0.93	4.61
	Max value	126.72	263.00	22.60	0.90	69.69	0.97	9.00
	Mean value	44.42	116.64	6.68	0.41	21.53	0.89	3.72
Moderate	Mean value	51.50	116.58	6.12	0.47	25.20	0.90	5.06
	Standard deviation	22.91	43.03	3.80	0.20	16.34	0.05	2.69
	Min value	13.02	30.00	1.30	0.10	0.15	0.69	1.20
	25th percentiles	37.15	84.30	2.98	0.34	15.02	0.87	3.66
	50th percentiles	51.50	112.50	5.26	0.47	21.69	0.91	5.00
	75th percentiles	65.84	147.53	8.30	0.59	27.76	0.93	5.91
	Max value	118.77	237.20	17.66	1.27	80.00	0.98	21.00
Strong	Mean value	119.65	129.08	10.34	1.18	14.12	0.85	8.91
	Standard deviation	83.11	52.37	4.67	1.17	5.94	0.06	6.02
	Min value	16.43	30.00	2.50	0.10	5.53	0.59	2.03
	25th percentiles	61.98	91.30	7.04	0.53	11.16	0.84	5.86
	50th percentiles	91.37	127.09	10.27	0.72	13.27	0.86	7.20
	75th percentiles	126.71	158.60	13.86	0.97	16.84	0.89	9.03
	Max value	297.80	304.20	22.60	4.87	32.20	0.94	30.00

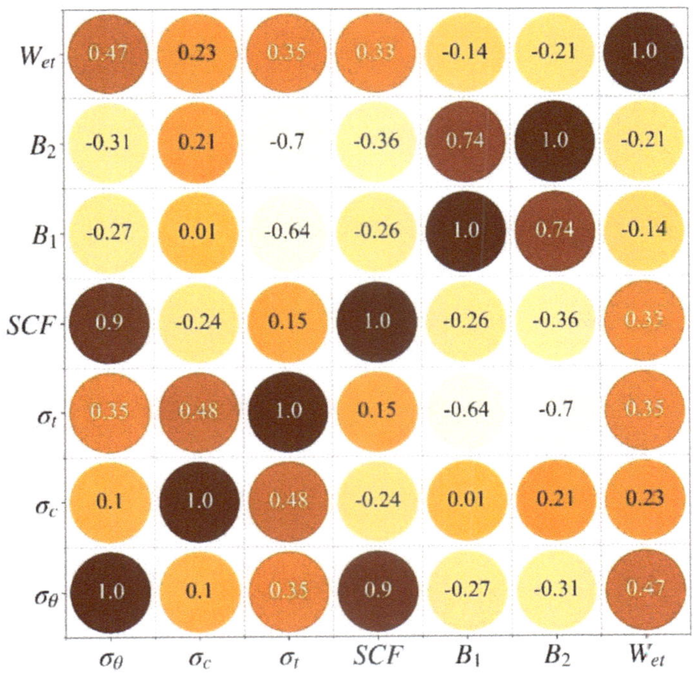

Figure 3. The heatmap of correlations between different variables.

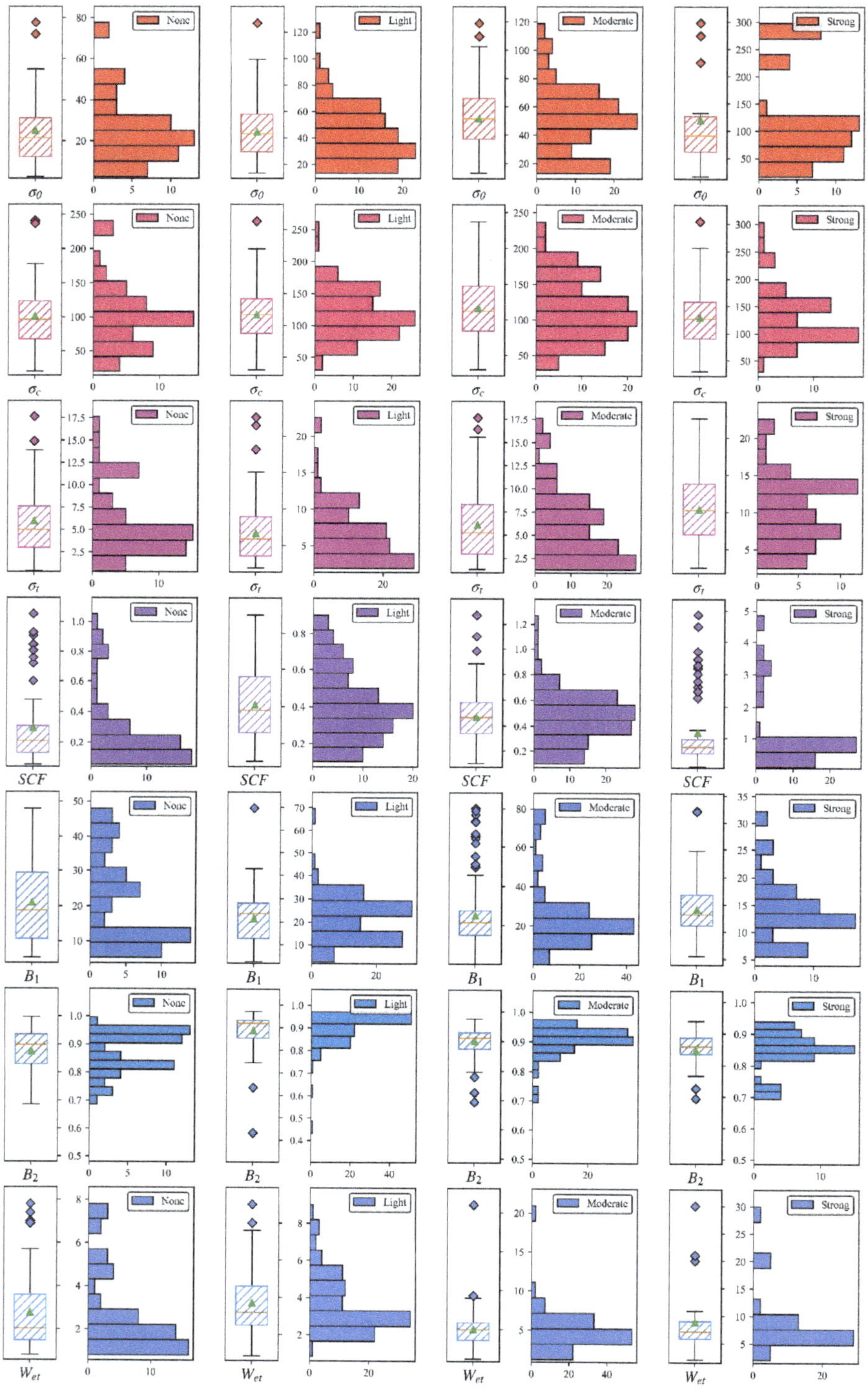

Figure 4. Boxplots and histograms of the seven variables of four rockburst intensities.

3.2. Step-by-Step Study Flowchart

The database is divided into a training set (Tr) and a testing set (Te) according to three split ratios, Tr (75%)-Te (25%), Tr (80%)-Te (20%), and Tr (85%)-Te (15%). Z-score (Equation (2)) is utilized to process the input parameters. As shown in Figure 5, the training part is applied to build the DF model for rockburst estimation. BO is implemented to tune the hyperparameters of the DF. 5 fold cross-validation is implemented to choose the optimized DF model. The permutation feature importance algorithm and partial dependence plots are introduced to interpret the DF model. A sensitivity analysis is employed to analyze effective variables on rockburst intensities and the robustness of the DF model. Finally, the intelligent model is applied to foretell rockburst in practical engineering.

$$X' = \frac{X - \overline{X}}{\sigma} \tag{2}$$

In Equation (2), $\overline{X}$ depicts the average value and σ represents the standard deviation.

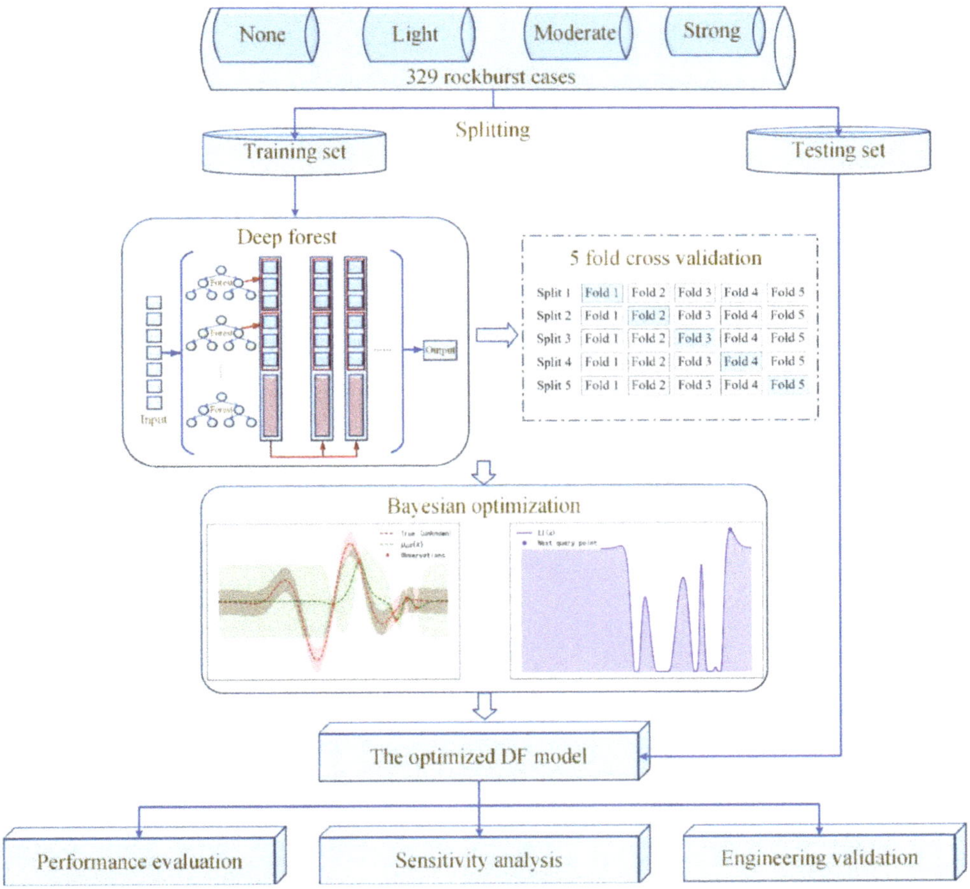

Figure 5. The flowchart to develop DF model for rockburst prediction.

4. Simulation

4.1. Model Metrics

Accuracy is often used as the metric index in the ML classification problem. Equation (3) shows the equation to calculate accuracy. According to the actual label and the predicted label, the sample can be divided into true positives (TP), false positives (FP), true negatives (TN), and false negatives (FN). The sum of the samples = TP + FP + TN + FN. According to these, the precision and recall can be calculated, as shown in Equations (4) and (5). In precision and recall, usually, when one is higher, the other is lower. In order to integrate precision and recall, their harmonically average value, i.e., f_1, is usually taken (Equation (6)). Additionally, a receiver operating characteristic (ROC) curve is introduced to evaluate the capability of single rockburst types. The area under the ROC curve is between 0 and 1. A larger area of the ROC curve indicates a better prediction effect of the model.

$$Accuracy = \frac{n'}{n} \tag{3}$$

$$precision = \frac{TP}{TP + FP} \tag{4}$$

$$recall = \frac{TP}{TP + FN} \tag{5}$$

$$f_1 = \frac{2 \times precision \times recall}{precision + recall} \tag{6}$$

In Equation (3), n represents the number of all datasets, and n' stands for the number of instances of which the predicted labels are equal to actual labels.

4.2. Cross-Validation

When the hyperparameters of ML models are optimized, the generalization of the model needs to be evaluated to select the optimal model. K fold cross-validation is often adopted in model evaluation. In cross-validation, k is usually set to 10 or 5 [56]. In this study, k was set to 5 by referring to a previous study [25] to reduce the running time. As shown in Figure 6, the training set is divided into five pieces of data equal to each other, four pieces of data are selected to train each time, and the remaining piece of data is used for validation. The process is repeated five times, and finally, the average value of the five validation scores is taken.

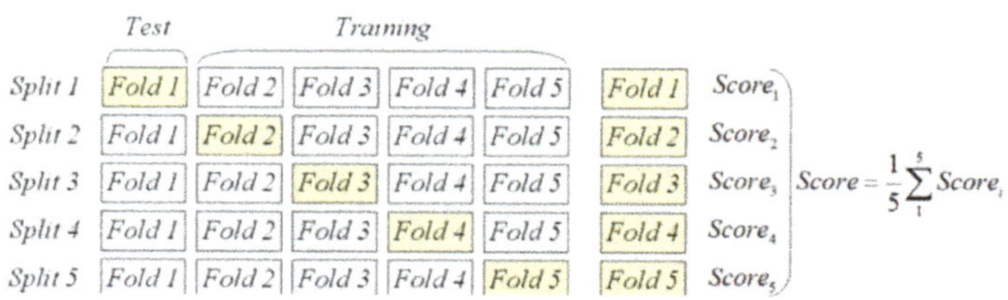

Figure 6. The steps to perform 5 fold CV.

4.3. DF Optimization

σ_θ, σ_c, σ_t, SCF, B_1, B_2, and W_{et} were input to the DF to develop a rockburst prediction model. The cascade layers in the DF can be automatically determined according to the training set. The number of forests, the trees in each forest, and the maximum number of cascade layers are the key parameters that influence the performance of DF. Referring to previous studies by Zhou et al. [30], the optimization range of these hyperparameters was determined, as shown in Table 5.

Table 5. The optimization range of hyperparameters in the DF model.

Hyperparameters	Optimization Range
The number of forests	(1, 4)
The trees in the forest	(10, 100)
The maximum number of cascade layers	(10, 30)

BO was implemented to optimize the hyperparameters of the DF to choose an optimal model. Before performing BO, the objective function needed to be defined, and BO was utilized to optimize this objective function. The cross-entropy loss function is commonly used for classification problems in ML areas, as shown in Equation (7). The smaller the value is, the better the capability of the model. To improve the generalization of the DF model, the cross-entropy loss function in 5 fold cross-validation was chosen as the objective function, as shown in Equation (8).

$$loss = -\frac{1}{n}\sum_{i=1}^{n} \log p_{model}[y_i \in C_{y_i}] \tag{7}$$

$$Objective\_function = \frac{1}{5}\sum_{1}^{5} loss_i \tag{8}$$

In Equation (7), $p_{model}[y_i \in C_{y_i}]$ is the prediction probability in the actual label. Equation (8) means that the training set is split into five folds, four folds are applied to train the DF model, and the cross-entropy loss function of the DF on the remaining one fold is calculated. Repeating the process five times, the average value of the cross-entropy loss function is chosen as the objective function.

BO was performed using *Scikit-Optimize*, an open-source *Python* library [57]. The parameters of BO utilized the default value of *Scikit-Optimize*. Table 6 lists the values of hyperparameters in BO in this study. Gaussian process was chosen as the surrogate model, and GP-Hedge was selected as the acquisition function. Figure 7 exhibits the flowchart in which BO tunes the hyperparameters of the DF model. In the BO process, GP-Hedge determines the following points that need to be evaluated. The DF model trains according to the hyperparameter value recommended by the GP-Hedge. After the DF is built, the objective function is calculated, and the GP model is updated. By repeating this process N times, the optimal hyperparameters can be obtained. Figure 8 exhibits the convergence of the objective function with the process of BO. BO can efficiently minimize the objective function to find the optimal DF model. Table 7 shows the optimal hyperparameters of the DF model with different training sets at the end of BO.

Table 6. The parameters of BO.

Surrogate Model	Acquisition Function	Surrogate Model Hyperparameters
GP	GP-Hedge	1. Kernel function: Matern Kernel and White Kernel 2. Noise: Gaussian distribution

Table 7. The optimal hyperparameters in the DF model.

Datasets	Hyperparameters	Value
Tr (75%)-Te (25%)	The number of forests	4
	The trees in the forest	36
	The maximum number of cascade layers	30
Tr (80%)-Te (20%)	The number of forests	4
	The trees in the forest	18
	The maximum number of cascade layers	19
Tr (85%)-Te (15%)	The number of forests	4
	The trees in the forest	25
	The maximum number of cascade layers	25

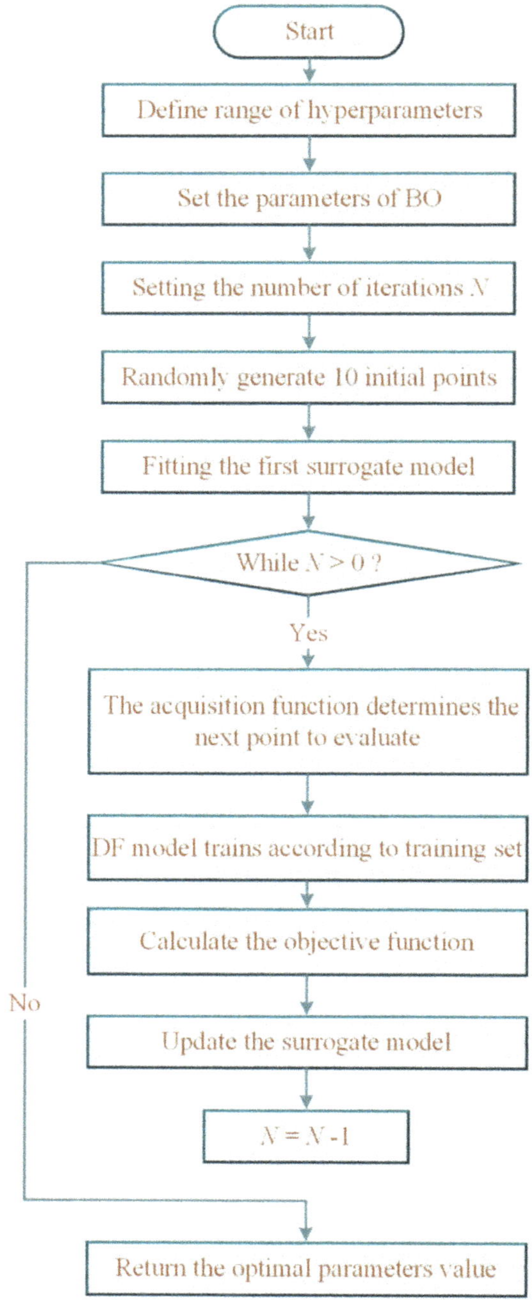

Figure 7. The flowchart of BO.

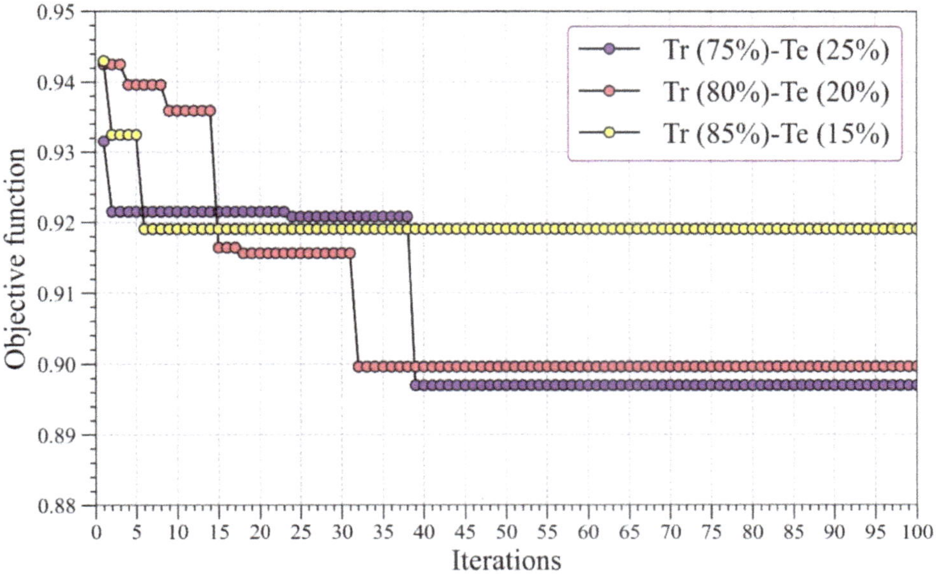

Figure 8. Convergence of objective function during BO.

4.4. Results

The optimized DF models with different training sets were obtained according to Table 7, and in three different training sets, the training accuracies of DF models were 100%. The remaining three testing sets were adopted to evaluate the capability of DF models. Tables 8–10 present the performances of DF models on the different testing sets. The three DF models developed by different training datasets had the same capabilities in predicting strong rockburst. In terms of testing accuracy, the DF model with Tr (80%)-Te (20%) has the best capacity. Accordingly, it is appropriate to develop DF models to predict rockburst by 80% training set and 20% testing set.

Table 8. The testing performance of DF models with Tr (75%)-Te (25%).

Rockburst Type	Precision	Recall	f_1	Number
None	0.86	0.92	0.89	13
Light	0.95	0.76	0.84	25
Moderate	0.88	1.00	0.94	30
Strong	1.00	1.00	1.00	14
Accuracy		91.5%		82

Table 9. The testing performance of DF models with Tr (80%)-Te (20%).

Rockburst Type	Precision	Recall	f_1	Number
None	0.91	0.91	0.91	11
Light	0.94	0.80	0.86	20
Moderate	0.89	1.00	0.94	24
Strong	1.00	1.00	1.00	11
Accuracy		92.4%		66

Table 10. The testing performance of DF models with Tr (85%)-Te (15%).

Rockburst Type	Precision	Recall	f_1	Number
None	1.00	0.88	0.93	8
Light	0.92	0.80	0.86	15
Moderate	0.86	1.00	0.92	18
Strong	1.00	1.00	1.00	8
Accuracy		91.8%		49

5. Discussion

5.1. Model Performance Comparison

DF model consists of RF and CRF, and to analyze the advantages of DF compared to its base classifiers, RF and CRF are also independently built with the same hyperparameters (Tr (80%)-Te (20%)) in Table 7. The training accuracy in RF and CRF was 100% and 99.6%, respectively, and Tables 11 and 12 display the testing performance of RF and CRF, respectively. According to Table 9, the DF has higher testing accuracy than RF and CRF, which reveals that model combination can improve the capability for predicting rockburst. The ROC curve is introduced to compare the performance of each rockburst intensity in DF, RF, and CRF. Figure 9 exhibits the ROC curves of the four rockburst intensities in DF, RF, and CRF. The larger area of the ROC curve is associated with better model performance. DF, RF, and CRF perform similarly in terms of strong rockburst prediction, but DF outperforms RF and CRF in terms of none, light, and moderate rockburst prediction.

Table 11. The testing performance of the RF model.

Rockburst Type	Precision	Recall	f_1	Number
None	0.75	0.82	0.78	11
Light	0.74	0.85	0.79	20
Moderate	0.95	0.83	0.89	24
Strong	1.00	0.91	0.95	11
Accuracy		84.8%		66

Table 12. The testing performance of the CRF model.

Rockburst Type	Precision	Recall	f_1	Number
None	0.75	0.82	0.78	11
Light	0.75	0.75	0.75	20
Moderate	0.84	0.88	0.86	24
Strong	1.00	0.82	0.90	11
Accuracy		81.8%		66

Additionally, to explore the power of the DF model, widely used ML models were also developed in this study, and they included LR, Naive Bayes, KNN, SVM, DT, adaptive boosting (AdaBoost), ANN, XGB, and GBM. XGB was built using the default parameters in *XGBoost* (a *Python* library) [51], and other models used the *Scikit-learn* [52] default parameters. Training set (80%) was applied to develop these models, and testing set (20%) was adopted to evaluate these models. Figure 10 displays the training and testing accuracy of these models. The DT suffers from serious overfitting, and its performance differs markedly between the training and testing sets. Many ensemble tree models, i.e., GBM, XGB, and RF, perform better than other ML models. Taylor diagrams [58,59] were introduced to determine the strength of the DF model compared to other models. In this study, Taylor diagrams combine the Matthews correlation coefficient (MCC), centered root mean square error (green dotted lines in Figure 11), and standard deviation into a polar diagram. Equation (9) shows the equation to calculate MCC. The reference points with black star shapes depict the actual rockburst, and when other points are closer to the reference points, the corresponding models have better performance in predicting rockburst. It is worth noting that the DF model outperforms other commonly used tree models in the training and testing sets, according to Figure 11.

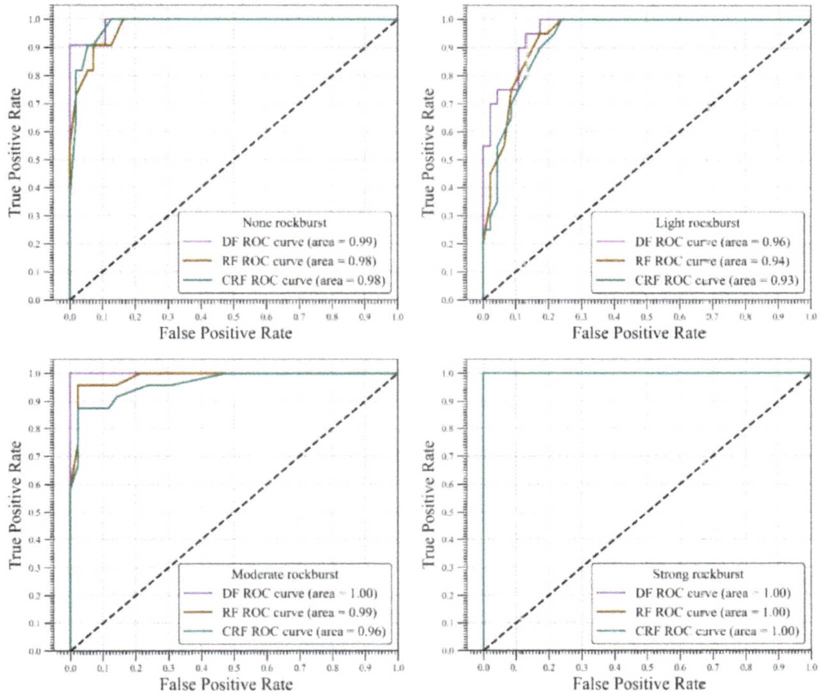

Figure 9. Model performance comparison in each rockburst by ROC curve.

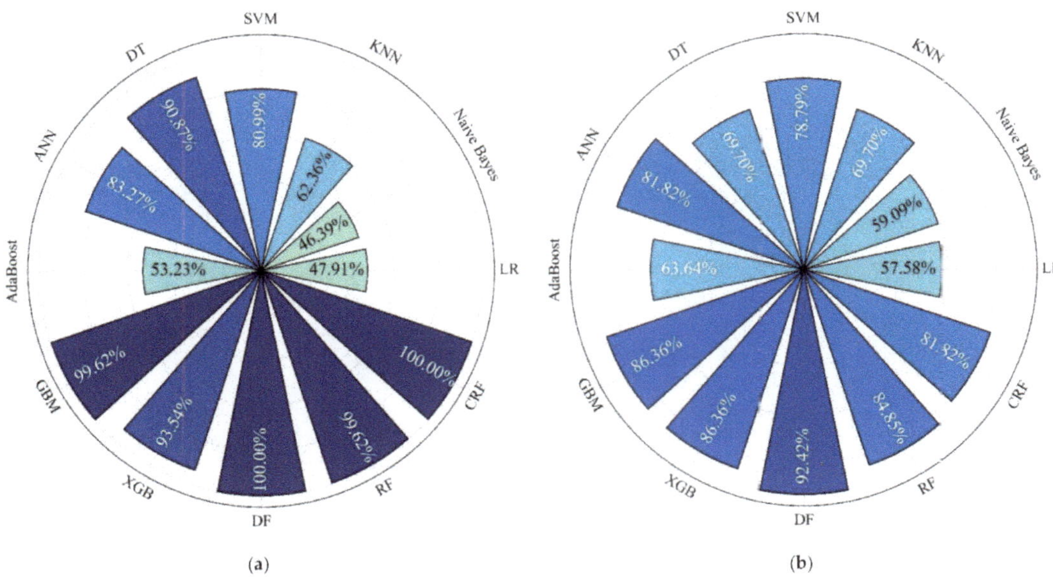

Figure 10. Training and testing results of the developed models (**a**) Training accuracy; (**b**) Testing accuracy.

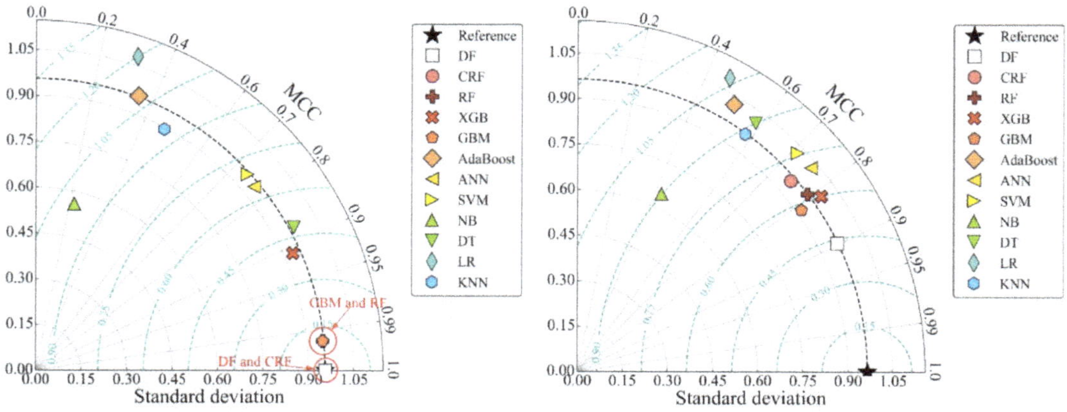

Figure 11. Taylor diagrams for comparing the performance of developed models. (a) Training results; (b) Testing results.

Additionally, Table 13 compares the capabilities of DF and some intelligent models proposed by other scholars in recent years, and the DF model has better performance than other models during training and testing phases. The results reveal that the DF model is a powerful technique to forecast and prevent rockburst.

$$MCC = \frac{TP \times TN - FP \times FN}{\sqrt{(TP + FP)(TP + FN)(TN + FP)(TN + FN)}} \tag{9}$$

Table 13. Comparison of DF and other ML models proposed in recent years.

Algorithm/Model	Input Parameters	Data Size	Accuracy	
			Training	Testing
Voting model [25]	$H, \sigma_\theta, \sigma_c, \sigma_t, W_{et}$	188	94%	80%
PSO-ELM [55]	$\sigma_\theta, \sigma_c, \sigma_t, SCF, B_1, W_{et}$	344	98.99%	88.89%
Bagging [23]	$\sigma_\theta, \sigma_c, \sigma_t, SCF, B_1, W_{et}$	102	100%	88.24%
Boosting [23]	$\sigma_\theta, \sigma_c, \sigma_t, SCF, B_1, W_{et}$	102	100%	91.18%
Stacking model [28]	$\sigma_\theta, \sigma_c, \sigma_t, SCF, B_1, B_2, W_{et}$	246		88.52%
DF	$\sigma_\theta, \sigma_c, \sigma_t, SCF, B_1, B_2, W_{et}$	329	100%	92.40%

Note: Voting model is the combination of back propagation neural network, KNN, SVM, LR, linear model, DT, and Naive Bayes; H = depth; POS = particle swarm optimization; ELM = extreme learning machine; stacking model is the combination of KNN, SVM, deep neural network, and recurrent neural network.

5.2. Sensitivity Analysis

The permutation feature importance was implemented to determine crucial variables that affected rockburst in the DF model. The permutation feature importance was beneficial for analyzing the relative importance of input variables in nonlinear or opaque ML models [60], and it is introduced to determine the importance score of input parameters. Figure 12 shows the importance score of the input parameters in the DF model. According to this Figure 12, σ_θ, W_{et}, and SCF are the vital parameters that influence the performance of rockburst prediction.

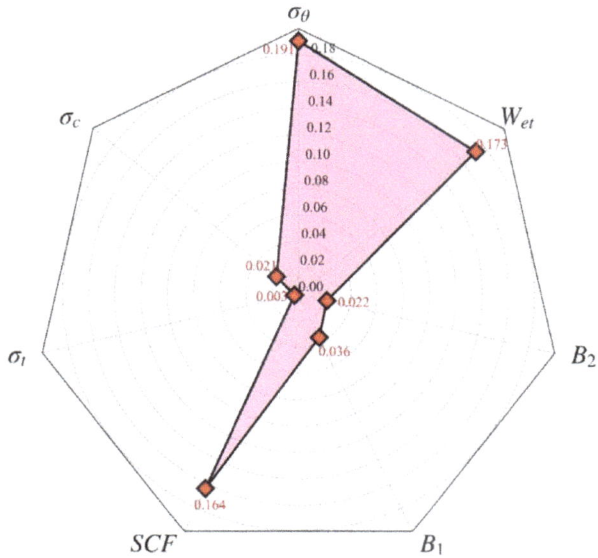

Figure 12. The relative importance of input variables in the DF model.

To determine the impact of parameters on the rockburst levels, partial dependence plots (PDP) [61] were introduced to analyze the relationship between variables and predicted results of the DF model, as shown in Figure 13. The PDP displays the dependence of the predicted probability of different rockburst levels on the variable of interest when other variables are fixed. With the increase of σ_θ, W_{et}, and SCF, the predicted probability of a strong rockburst increases, and the predicted probability of a none rockburst decreases. There is no apparent relationship between the other variables and the predicted probability of rockburst in the DF model. These results indicate that larger σ_θ, W_{et}, and SCF values are accompanied by more serious rockburst. Accordingly, it is vital to reduce the σ_θ, W_{et}, and SCF of surrounding rock in underground engineering to mitigate the rockburst risk. Some measures, such as smooth blasting, pressure relief blasting, and deformable bolts and mesh, can be applied to prevent rockburst on-site [62].

The key parameters affecting rockburst intensities are determined, which makes it possible to analyze the performance variation of the DF model under different influential variables. According to the relative importance of the input variables, seven models that adopted different input parameters were developed, as shown in Table 14. These seven models adopted the same hyperparameters (Tr (80%)-Te (20%)) in Table 7. Figure 14 displays the variations of training and testing accuracy with the input variable number varying. With the decrease of input parameters, DF has low accuracy, which suggests that considering more factors is beneficial to improving the generalization of DF and predicting and preventing rockburst due to the complexity of rockburst. Additionally, with only three input parameters, the testing accuracy of the DF model is still 81.82%, which indicates that the DF model has good robustness and is suitable for fewer input parameters.

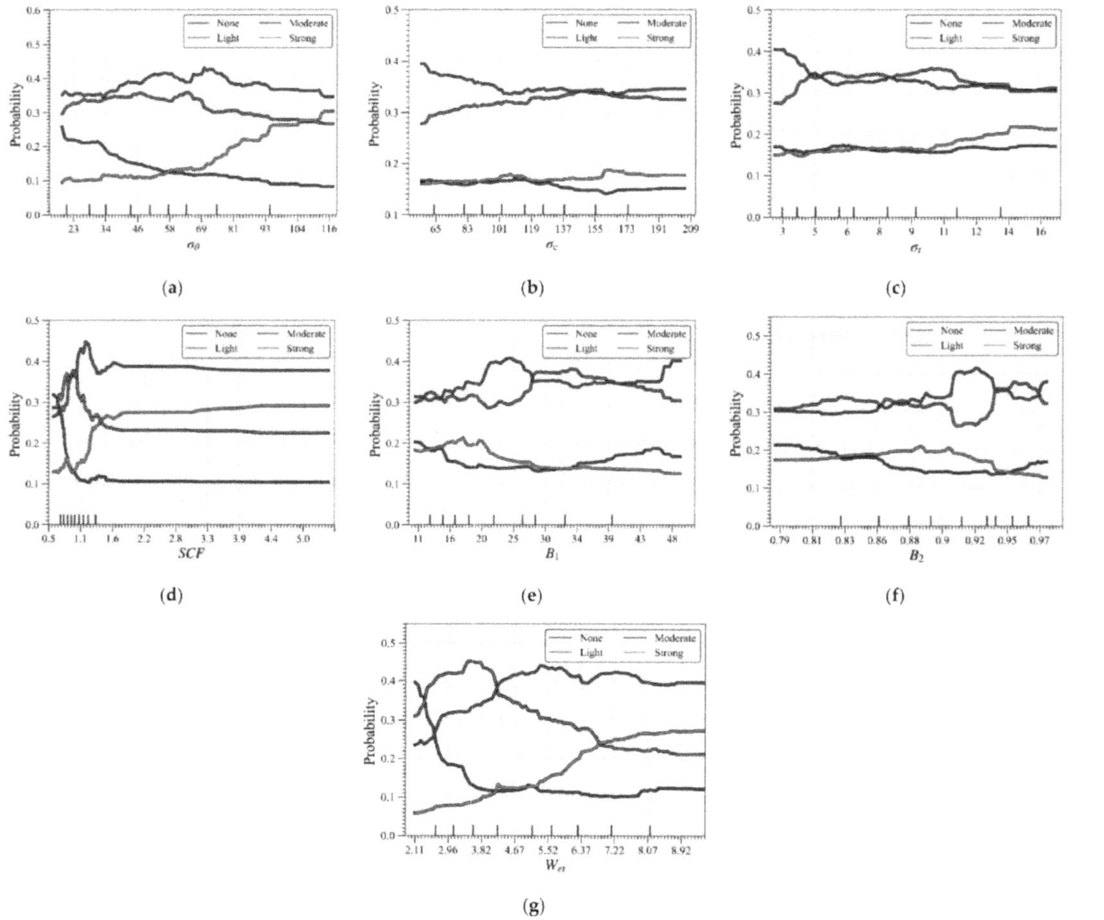

Figure 13. PDP to analyze the influence of variables on the rockburst predicted probability. (**a**) σ_θ; (**b**) σ_c; (**c**) σ_t; (**d**) SCF; (**e**) B_1; (**f**) B_2; (**g**) W_{et}.

Table 14. Seven models and their input parameters.

Model	Input Parameters	Input Parameters Number
M 1	$\sigma_\theta, \sigma_c, \sigma_t, SCF, B_1, B_2, W_{et}$	7
M 2	$\sigma_\theta, \sigma_c, SCF, B_1, B_2, W_{et}$	6
M 3	$\sigma_\theta, SCF, B_1, B_2, W_{et}$	5
M 4	$\sigma_\theta, SCF, B_1, W_{et}$	4
M 5	$\sigma_\theta, SCF, W_{et}$	3
M 6	σ_θ, W_{et}	2
M 7	σ_θ	1

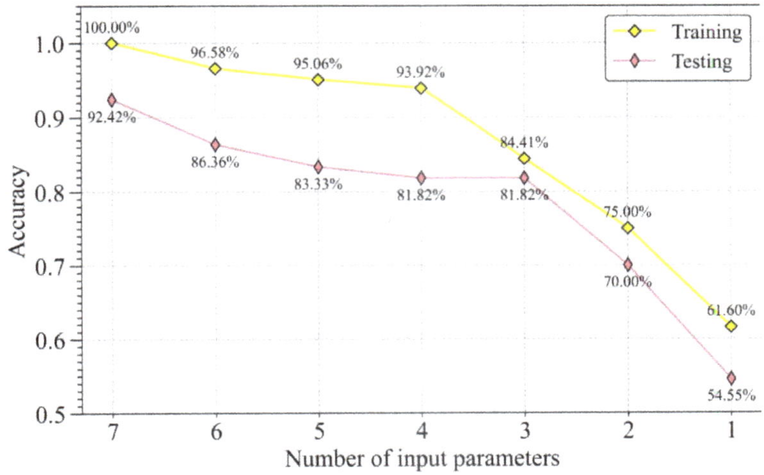

Figure 14. Model performance variation with the input parameters variation.

5.3. Engineering Validation

Xincheng Gold Mine and Sanshangao Gold Mine are located in Yantai, Shandong, China. Figure 15 presents their locations. After years of mining production, the two gold mines have become mines with a mining depth of more than 1000 m in China. To meet production needs, it is necessary to excavate ores in deeper strata. Due to the complexity of deep high stress and geological conditions, many engineering problems are inevitably faced in deep mining operations, among which rockburst poses a severe threat to the safety of facilities and workers. A series of field investigations and rock mechanics tests were carried out in the two gold mines to avoid the threat of rockburst to shaft construction. Seven rock blocks were taken in different locations in the Xincheng Gold Mine, and five rock blocks were taken in the Sanshandao Gold Mine. These rock blocks were processed into standard specimens for rock mechanics tests and in-situ stress analysis [63]. Table 15 presents the rock parameters used to foretell rockburst. According to the standard of classification for intensities of rockburst in Table 3, the rockburst level of each site was determined. The DF model was used to evaluate the rockburst, and Table 15 shows the predicted results. The predicted results of the DF model are consistent with the actual rockburst situation. These results suggest that the DF model has superior engineering practicability.

Table 15. Application of DF model in practical engineering.

No.	Engineering	σ_θ/MPa	σ_c/MPa	σ_t/MPa	*SCF*	B_1	B_2	W_{et}	Actual Grade	Predicted Grade
1	Xincheng Gold Mine	87.60	139.07	10.63	0.63	13.08	0.86	5.56	S	S
2		108.31	149.99	11.97	0.72	12.53	0.85	6.88	S	S
3		89.46	155.45	12.05	0.58	12.90	0.86	3.98	S	S
4		100.00	137.52	13.73	0.73	10.01	0.82	5.27	S	S
5		107.25	182.67	12.11	0.59	15.08	0.88	5.48	S	S
6		107.87	140.38	12.06	0.77	11.64	0.84	8.50	S	S
7		109.57	174.34	12.07	0.63	14.44	0.87	7.69	S	S
8	Sanshandao Gold Mine	94.64	160.94	9.74	0.59	16.52	0.89	4.94	M	M
9		32.45	138.25	9.04	0.23	15.29	0.88	3.73	L	L
10		23.13	146.29	19.6	0.16	7.46	0.76	6.45	N	N
11		34.12	154.28	13.98	0.22	11.04	0.83	4.61	L	L
12		34.07	128.5	11.71	0.27	10.97	0.83	1.92	N	N

Figure 15. The location of Xincheng Gold Mine and Sanshandao Gold Mine.

6. Limitations and Future Studies

Rockburst is a complex geological disaster that is related to many factors, such as geological structure, in-situ stress conditions, rock strength, excavation method, excavation size, etc. However, in this study, only seven related parameters were considered to build the intelligent model for rockburst prediction. Other parameters, such as rock quality index, rock integrity coefficient, and the geometric size of the cross-section of the excavation, are also essential to determine rockburst level. In the future, more influential factors should be considered to add to the database. Additionally, increasing the size of the database contributes to building a more powerful intelligent rockburst prediction model. As for the DF model, to apply DF to predict rockburst in different engineering problems, it is necessary to tune the hyperparameters of DF according to the size and complexity of data.

7. Conclusions

(1) Deep forest, a novel tree-based ensemble model, was proposed to build the rockburst prediction model based on 329 collected real rockburst cases. Bayesian optimization was used to turn the hyperparameters of the DF. The DF had 100% accuracy in the training set and 92.4% accuracy in the testing set, and it performed better than other ML models and can forecast massive rockburst disasters.

(2) σ_θ and W_{et} are the essential parameters that affect the performance of the DF model for rockburst prediction. Sensitivity analysis reveals that more factors can be taken into account to build a more accurate rockburst prediction model for the complexity of rockburst. Moreover, it also confirms that the proposed DF model has good performance with fewer input parameters.

(3) A field investigation was carried out in the Xincheng Gold Mine and Sanshandao Gold mine, Shandong, China, and the collected rock blocks were tested in the laboratory. The obtained parameters were input into the trained DF model, and the predicted results matched the rockburst situation on site. The validation datasets from gold mines can expand the rockburst database to establish more powerful models.

(4) The DF model is trained by the datasets from mines, tunnels, large underground chambers, traffic tunnels, etc., and it is validated by the cases from two gold mines.

Accordingly, it is worth noting that the proposed DF model is not only applied to gold mines with a high level of accuracy but also is suitable for other deep mine and underground excavation engineering.

Supplementary Materials: The collected rockburst database can be found be downloaded at: https://www.mdpi.com/article/10.3390/math10050787/s1, Table S1: Collected rockburst database.

Author Contributions: Conceptualization, D.L. and Z.L.; methodology, Z.L.; software, Z.L.; validation, P.X.; investigation, P.X.; resources, D.L.; writing—original draft preparation, Z.L. and D.L.; writing—review and editing, D.L., D.J.A. and J.Z.; visualization, Z.L.; supervision, D.L.; project administration, D.L.; funding acquisition, D.L. All authors have read and agreed to the published version of the manuscript.

Funding: This research was supported by the National Natural Science Foundation of China (Grant No.:52074349).

Institutional Review Board Statement: Not applicable.

Informed Consent Statement: Not applicable.

Data Availability Statement: Not applicable.

Conflicts of Interest: The authors declare no conflict of interest.

References

1. Zhou, J.; Li, X.; Mitri, H.S. Evaluation method of rockburst: State-of-the-art literature review. *Tunn. Undergr. Space Technol.* **2018**, *81*, 632–659. [CrossRef]
2. Cai, M. Prediction and prevention of rockburst in metal mines–A case study of Sanshandao gold mine. *J. Rock Mech. Geotech. Eng.* **2016**, *8*, 204–211. [CrossRef]
3. Afraei, S.; Shahriar, K.; Madani, S.H. Developing intelligent classification models for rock burst prediction after recognizing significant predictor variables, Section 1: Literature review and data preprocessing procedure. *Tunn. Undergr. Space Technol.* **2019**, *83*, 324–353. [CrossRef]
4. Xue, Y.; Bai, C.; Kong, F.; Qiu, D.; Li, L.; Su, M.; Zhao, Y. A two-step comprehensive evaluation model for rockburst prediction based on multiple empirical criteria. *Eng. Geol.* **2020**, *268*, 105515. [CrossRef]
5. Zhai, S.; Su, G.; Yin, S.; Zhao, B.; Yan, L. Rockburst characteristics of several hard brittle rocks: A true triaxial experimental study. *J. Rock Mech. Geotech. Eng.* **2020**, *12*, 279–296. [CrossRef]
6. Khan, N.M.; Ahmad, M.; Cao, K.; Ali, I.; Liu, W.; Rehman, H.; Hussain, S.; Rehman, F.U.; Ahmed, T. Developing a new bursting liability index based on energy evolution for coal under different loading rates. *Sustainability* **2022**, *14*, 1572. [CrossRef]
7. Wang, S.; Li, X.; Yao, J.; Gong, F.; Li, X.; Du, K.; Tao, M.; Huang, L.; Du, S. Experimental investigation of rock breakage by a conical pick and its application to non-explosive mechanized mining in deep hard rock. *Int. J. Rock Mech. Min. Sci.* **2019**, *122*, 104063. [CrossRef]
8. Wang, S.; Sun, L.; Li, X.; Wang, S.; Du, K.; Li, X.; Feng, F. Experimental investigation of cuttability improvement for hard rock fragmentation using conical cutter. *Int. J. Geomech.* **2021**, *21*, 06020039. [CrossRef]
9. Wang, J.; Apel, D.B.; Pu, Y.; Hall, R.; Wei, C.; Sepehri, M. Numerical modeling for rockbursts: A state-of-the-art review. *J. Rock Mech. Geotech. Eng.* **2021**, *13*, 457–478. [CrossRef]
10. Pu, Y.; Apel, D.B.; Liu, V.; Mitri, H. Machine learning methods for rockburst prediction-state-of-the-art review. *Int. J. Min. Sci. Technol.* **2019**, *29*, 565–570. [CrossRef]
11. Ma, L.; Khan, N.M.; Cao, K.; Rehman, H.; Salman, S.; Rehman, F.U. Prediction of sandstone dilatancy point in different water contents using infrared radiation characteristic: Experimental and machine learning approaches. *Lithosphere* **2022**, *2021*, 3243070. [CrossRef]
12. Khan, N.M.; Ma, L.; Cao, K.; Hussain, S.; Liu, W.; Xu, Y.; Yuan, Q.; Gu, J. Prediction of an early failure point using infrared radiation characteristics and energy evolution for sandstone with different water contents. *Bull. Eng. Geol. Environ.* **2021**, *80*, 6913–6936. [CrossRef]
13. Zhou, J.; Li, X.; Mitri, H.S. Classification of rockburst in underground projects: Comparison of ten supervised learning methods. *J. Comput. Civ. Eng.* **2016**, *30*, 04016003. [CrossRef]
14. Li, N.; Jimenez, R. A logistic regression classifier for long-term probabilistic prediction of rock burst hazard. *Nat. Hazards* **2018**, *90*, 197–215. [CrossRef]
15. Ghasemi, E.; Gholizadeh, H.; Adoko, A.C. Evaluation of rockburst occurrence and intensity in underground structures using decision tree approach. *Eng. Comput.* **2020**, *36*, 213–225. [CrossRef]
16. Pu, Y.; Apel, D.B.; Lingga, B. Rockburst prediction in kimberlite using decision tree with incomplete data. *J. Sustain. Min.* **2018**, *17*, 158–165. [CrossRef]

17. Li, N.; Feng, X.; Jimenez, R. Predicting rock burst hazard with incomplete data using Bayesian networks. *Tunn. Undergr. Space Technol.* **2017**, *61*, 61–70. [CrossRef]
18. Zhou, J.; Guo, H.; Koopialipoor, M.; Jahed Armaghani, D.; Tahir, M. Investigating the effective parameters on the risk levels of rockburst phenomena by developing a hybrid heuristic algorithm. *Eng. Comput.* **2021**, *37*, 1679–1694. [CrossRef]
19. Zhou, J.; Koopialipoor, M.; Li, E.; Armaghani, D.J. Prediction of rockburst risk in underground projects developing a neuro-bee intelligent system. *Bull. Eng. Geol. Environ.* **2020**, *79*, 4265–4279. [CrossRef]
20. Li, D.; Liu, Z.; Xiao, P.; Zhou, J.; Jahed Armaghani, D. Intelligent rockburst prediction model with sample category balance using feedforward neural network and Bayesian optimization. *Undergr. Space* **2022**, in press. [CrossRef]
21. Pu, Y.; Apel, D.B.; Wang, C.; Wilson, B. Evaluation of burst liability in kimberlite using support vector machine. *Acta Geophys.* **2018**, *66*, 973–982. [CrossRef]
22. Lin, Y.; Zhou, K.; Li, J. Application of Cloud Model in Rock Burst Prediction and Performance Comparison with Three Machine Learning Algorithms. *IEEE Access* **2018**, *6*, 30958–30968. [CrossRef]
23. Wang, S.-M.; Zhou, J.; Li, C.-Q.; Armaghani, D.J.; Li, X.-B.; Mitri, H.S. Rockburst prediction in hard rock mines developing bagging and boosting tree-based ensemble techniques. *J. Cent. South Univ.* **2021**, *28*, 527–542. [CrossRef]
24. Xie, X.; Jiang, W.; Guo, J. Research on rockburst prediction classification based on GA-XGB model. *IEEE Access* **2021**, *6*, 83993–84020. [CrossRef]
25. Zhang, J.; Wang, Y.; Sun, Y.; Li, G. Strength of ensemble learning in multiclass classification of rockburst intensity. *Int. J. Numer. Anal. Methods Geomech.* **2020**, *44*, 1833–1853. [CrossRef]
26. Liang, W.; Sari, A.; Zhao, G.; McKinnon, S.D.; Wu, H. Short-term rockburst risk prediction using ensemble learning methods. *Nat. Hazards* **2020**, *104*, 1923–1946. [CrossRef]
27. Liang, W.; Sari, Y.A.; Zhao, G.; McKinnon, S.D.; Wu, H. Probability estimates of short-term rockburst risk with ensemble classifiers. *Rock Mech. Rock Eng.* **2021**, *54*, 1799–1814. [CrossRef]
28. Yin, X.; Liu, Q.; Pan, Y.; Huang, X.; Wu, J.; Wang, X. Strength of stacking technique of ensemble learning in rockburst prediction with imbalanced data: Comparison of eight single and ensemble models. *Nat. Resour. Res.* **2021**, *30*, 1795–1815. [CrossRef]
29. Li, D.; Liu, Z.; Armaghani, D.J.; Xiao, P.; Zhou, J. Novel ensemble intelligence methodologies for rockburst assessment in complex and variable environments. *Sci. Rep.* **2022**, *12*, 1844. [CrossRef]
30. Zhou, Z.-H.; Feng, J. Deep forest. *arXiv* **2017**, arXiv:1702.08835. [CrossRef]
31. Zhou, J.; Qiu, Y.; Zhu, S.; Armaghani, D.J.; Khandelwal, M.; Mohamad, E.T. Estimation of the TBM advance rate under hard rock conditions using XGBoost and Bayesian optimization. *Undergr. Space* **2021**, *6*, 506–515. [CrossRef]
32. Zhou, J.; Asteris, P.G.; Armaghani, D.J.; Pham, B.T. Prediction of ground vibration induced by blasting operations through the use of the Bayesian Network and random forest models. *Soil Dyn. Earthq. Eng.* **2020**, *139*, 106390. [CrossRef]
33. Han, H.; Armaghani, D.J.; Tarinejad, R.; Zhou, J.; Tahir, M. Random forest and bayesian network techniques for probabilistic prediction of flyrock induced by blasting in quarry sites. *Nat. Resour. Res.* **2020**, *29*, 655–667. [CrossRef]
34. Shahriari, B.; Swersky, K.; Wang, Z.; Adams, R.P.; Freitas, N.D. Taking the human out of the loop: A review of bayesian optimization. *Proc. IEEE* **2015**, *104*, 148–175. [CrossRef]
35. Liang, X. Image-based post-disaster inspection of reinforced concrete bridge systems using deep learning with Bayesian optimization. *Comput.-Aided Civ. Infrastruct. Eng.* **2019**, *34*, 415–430. [CrossRef]
36. Zhang, Q.; Hu, W.; Liu, Z.; Tan, J. TBM performance prediction with Bayesian optimization and automated machine learning. *Tunn. Undergr. Space Technol.* **2020**, *103*, 103493. [CrossRef]
37. Sameen, M.I.; Pradhan, B.; Lee, S. Application of convolutional neural networks featuring Bayesian optimization for landslide susceptibility assessment. *Catena* **2020**, *186*, 104249. [CrossRef]
38. Lu, X.; Duan, Z.; Qian, Y.; Zhou, W. A malicious code classification method based on deep forest. *J. Softw.* **2020**, *31*, 1454–1464.
39. Breiman, L. Random Forests. *MLear* **2001**, *45*, 5–32.
40. Snoek, J.; Larochelle, H.; Adams, R.P. Practical bayesian optimization of machine learning algorithms. In Proceedings of the Advances in Neural Information Processing Systems 25: 26th Annual Conference on Neural Information Processing Systems 2012, Lake Tahoe, NV, USA, 3–6 December 2012; Volume 25.
41. Yang, L.; Shami, A. On hyperparameter optimization of machine learning algorithms: Theory and practice. *Neurocomputing* **2020**, *415*, 295–316. [CrossRef]
42. Seeger, M. Gaussian processes for machine learning. *Int. J. Neural Syst.* **2008**, *14*, 69–106. [CrossRef] [PubMed]
43. Kushner, H.J. A new method of locating the maximum point of an arbitrary multipeak curve in the presence of noise. *J. Basic Eng.* **1964**, *86*, 97–106. [CrossRef]
44. Mockus, J.; Tiesis, V.; Zilinskas, A. The application of Bayesian methods for seeking the extremum. *Towards Glob. Optim.* **1978**, *2*, 2.
45. Jones, D.R.; Schonlau, M.; Welch, W.J. Efficient global optimization of expensive black-box functions. *J. Glob. Optim.* **1998**, *13*, 455–492. [CrossRef]
46. Brochu, E.; Hoffman, M.W.; De Freitas, N. Portfolio allocation for bayesian optimization. *arXiv* **2010**, arXiv:1009.5419.
47. He, M.; Ren, F.; Liu, D. Rockburst mechanism research and its control. *Int. J. Min. Sci. Technol.* **2018**, *28*, 829–837. [CrossRef]
48. Zhou, J.; Li, X.; Shi, X. Long-term prediction model of rockburst in underground openings using heuristic algorithms and support vector machines. *Saf. Sci.* **2012**, *50*, 629–644. [CrossRef]

49. Pu, Y.; Apel, D.B.; Xu, H. Rockburst prediction in kimberlite with unsupervised learning method and support vector classifier. *Tunn. Undergr. Space Technol.* **2019**, *90*, 12–18. [CrossRef]
50. Ran, L.; Ye, Y.; Hu, N.; Hu, C.; Wang, X. Classified prediction model of rockburst using rough sets-normal cloud. *Neural Comput. Appl.* **2019**, *31*, 8185–8193.
51. Xue, Y.; Li, Z.; Li, S.; Qiu, D.; Tao, Y.; Wang, L.; Yang, W.; Zhang, K. Prediction of rock burst in underground caverns based on rough set and extensible comprehensive evaluation. *Bull. Eng. Geol. Environ.* **2019**, *78*, 417–429. [CrossRef]
52. Wu, S.; Wu, Z.; Zhang, C. Rock burst prediction probability model based on case analysis. *Tunn. Undergr. Space Technol.* **2019**, *93*, 103069. [CrossRef]
53. Du, Z.; Xu, M.; Liu, Z.; Xuan, W. Laboratory integrated evaluation method for engineering wall rock rock-burst. *Gold* **2006**, *27*, 26–30.
54. Jia, Q.; Wu, L.; Li, B.; Chen, C.; Peng, Y. The comprehensive prediction model of rockburst tendency in tunnel based on optimized unascertained measure theory. *Geotech. Geol. Eng.* **2019**, *37*, 3399–3411. [CrossRef]
55. Xue, Y.; Bai, C.; Qiu, D.; Kong, F.; Li, Z. Predicting rockburst with database using particle swarm optimization and extreme learning machine. *Tunn. Undergr. Space Technol.* **2020**, *98*, 103287. [CrossRef]
56. Zhou, J.; Li, E.; Wang, M.; Chen, X.; Shi, X.; Jiang, L. Feasibility of stochastic gradient boosting approach for evaluating seismic liquefaction potential based on SPT and CPT case histories. *J. Perform. Constr. Facil.* **2019**, *33*, 04019024. [CrossRef]
57. Head, T.; MechCoder, G.L.; Shcherbatyi, I. Scikit-optimize/scikit-optimize: v0.5.2. *Zenodo* **2018**, 1207017.
58. Taylor, K.E. Taylor Diagram Primer. Working Paper. 2005, pp. 1–4. Available online: http://www.atmos.albany.edu/daes/atmclasses/atm401/spring_2016/ppts_pdfs/Taylor_diagram_primer.pdf (accessed on 8 February 2022).
59. Zhou, J.; Zhu, S.; Qiu, Y.; Armaghani, D.J.; Zhou, A.; Yong, W. Predicting tunnel squeezing using support vector machine optimized by whale optimization algorithm. *Acta Geotech.* **2022**. [CrossRef]
60. Pedregosa, F.; Varoquaux, G.; Gramfort, A.; Michel, V.; Thirion, B.; Grisel, O.; Blondel, M.; Prettenhofer, P.; Weiss, R.; Dubourg, V. Scikit-learn: Machine learning in Python. *J. Mach. Learn. Res.* **2011**, *12*, 2825–2830.
61. Friedman, J.H. Greedy function approximation: A gradient boosting machine. *Ann. Stat.* **2001**, *29*, 1189–1232. [CrossRef]
62. Xiao, P.; Li, D.; Zhao, G.; Liu, H. New criterion for the spalling failure of deep rock engineering based on energy release. *Int. J. Rock Mech. Min. Sci.* **2021**, *148*, 104943. [CrossRef]
63. Xiao, P.; Li, D.; Zhao, G.; Liu, M. Experimental and Numerical Analysis of Mode I Fracture Process of Rock by Semi-Circular Bend Specimen. *Mathematics* **2021**, *9*, 1769. [CrossRef]

mathematics

Article

Effect of Lithology on Mechanical and Damage Behaviors of Concrete in Concrete-Rock Combined Specimen

Kewei Liu, Shaobo Jin, Yichao Rui *, Jin Huang and Zhanxing Zhou

School of Resources and Safety Engineering, Central South University, Changsha 410083, China; kewei_liu@csu.edu.cn (K.L.); jinshaobo@csu.edu.cn (S.J.); hj_changsha@csu.edu.cn (J.H.); zzx1230@csu.edu.cn (Z.Z.)
* Correspondence: ruiyichao@csu.edu.cn

Abstract: A concrete structure built on rock foundation works together with the connected rock mass, which has a significant effect on the mechanical behaviors of the concrete structure. To study the effect of lithology on the mechanical and damage behaviors of concrete in a concrete-rock combined specimen (CRCS), first, a test method for measuring the concrete part (concrete in CRCS) is adopted, then, uniaxial compression tests on seven types of specimens are performed and acoustic emission (AE) events are simultaneously monitored. Test results show that the low-strength concrete part plays a major role in the fracture behavior of CRCS. When the CRCS is failed, a sudden stress drop happens in CRCS, and the rock part (rock in CRCS) experiences a rapid axial strain recovery and intensifies the failure of the concrete part. The load-bearing and deformation capacities of the concrete part increase with the strength of the rock part, but the rock part shows the opposite behaviors under the influence of the concrete part. Furthermore, the damage of CRCS is mainly formed in the concrete part, and the damage extent of the concrete part is positively correlated with the strength of the rock part. Finally, a damage constitutive model of the concrete part is established and validated. This model can be used to accurately describe the effect of lithology on the mechanical response of the concrete part under uniaxial compression loading.

Keywords: lithology; concrete part; mechanical and damage behaviors; damage constitutive model

MSC: 74A45; 74A20; 74Q15; 74L10; 92F05

Citation: Liu, K.; Jin, S.; Rui, Y.; Huang, J.; Zhou, Z. Effect of Lithology on Mechanical and Damage Behaviors of Concrete in Concrete-Rock Combined Specimen. *Mathematics* **2022**, *10*, 727. https://doi.org/10.3390/math10050727

Academic Editor: Rami Ahmad El-Nabulsi

Received: 24 January 2022
Accepted: 22 February 2022
Published: 25 February 2022

1. Introduction

Concrete structures are widely built on rock foundations in the fields of hydraulic, civil and mining engineering, for instance, dams and retaining walls on a rock foundation, concrete support in underground excavation and underground nuclear waste repositories [1–11]. In the past, extensive studies have been carried out on pure concrete materials, and their deformation and failure evolution can be reasonably predicted. However, under the action of loading, in situ stress or temperature variation, the concrete structure and the connected rock foundation work and respond together. In this situation, the mechanical behaviors of concrete structures are significantly different from those of pure concrete structures, i.e., there are obvious shortcomings in revealing the mechanical performance of concrete structures built on rock foundations based on the experimental results of pure concrete specimens. In response to this situation, the concrete–rock structure should be considered as a combined body, and it is worthwhile to perform research on the concrete–rock combined specimen and further forecast the mechanical behaviors of concrete structures in practical applications.

In recent years, some efforts have been made to study the mechanical properties of CRCS subjected to various types of loads, such as compressive [12–16], shear [17–20], tensile [21–25], etc., in laboratory experiments. For example, Selçuk et al. [12] implemented uniaxial compressive tests on the concrete–rock bi-materials and found that with the increase of interface inclination angle, the strength of the combined specimen first decreases and then increases. Shen et al. [17] reported that the peak shear strength of the concrete–rock interface increases with the joint roughness coefficient, but the increase rate gradually decreases, and the peak strength tends to be stable. Chang et al. [21] observed that the load–bearing capacity of concrete–rock bi–material discs increases with the interface inclination angle, and the fracture pattern gradually changes from shear failure to a combined mode of shear fracture and tensile fracture, and finally it turns into tensile failure. Dong et al. [26] stated that under three–point bending, the failure of concrete–rock interface is mode I dominated fracture, and under four–point shearing, the mode II component may increase in the case of a small notched crack length–to–depth ratio. Although the above experimental studies mainly focus on the interface fracture and the influence of interface conditions on strength, it is implied that compared to pure concrete, the mechanical properties of CRCS are significantly affected when the rock material is involved.

Besides experimental studies, theoretical and numerical works have also been undertaken to analyze the mechanical behaviors of concrete–rock combined bodies. Javanmardi et al. [27] propose a theoretical model for transient water pressure variation along the concrete–rock crack during the earthquake, and the model is validated and further implemented in a finite element program for dynamic stability analysis of a concrete gravity dam resting on a rock foundation. Andjelkovic et al. [28] derived the mathematical models of shear deformability and shear strength for the contact of concrete and rock mass, and these models can serve as a basis for the structural analysis of concrete dams in the preliminary design. In addition, Tian et al. [29] developed a cohesive interface model for cemented concrete–rock joints to simulate the whole shear stress–shear displacement curve, including the elastic, bond failure and friction sliding stags. For seismic safety evaluation, Chavez et al. [30] used the Coulomb friction model to quantify the relative movement of earthquake–induced sliding at the interface between the concrete dam base and rock foundation. Furthermore, through experimental investigation, Dong et al. [23] propose an interfacial crack propagation criterion for the composite rock–concrete specimen, and by applying this criterion into the numerical simulation, the complete fracture process of a gravity dam is numerically analyzed, and the potential crack propagation paths are predicted.

Considering the difference between concrete and rock materials, these above–reviewed works emphasize the importance of concrete–rock contact and interface conditions, and some mechanical behaviors of the whole concrete–rock body under shear loads are also investigated. However, under the conditions of response to loading together, especially for compression, the concrete and rock in the combined body interact with each other and the mechanical behaviors of concrete structure are significantly affected by the rock mass. When designing and evaluating the concrete structure built on the rock foundation, this issue needs to be essentially concerned, while by far, few papers can be found to study the mechanical properties of concrete in the combined body. Moreover, in practical engineering, the concrete structures are built on bedrock with different lithology, and the mechanical performance of the concrete structures is closely related to the lithology of the rock foundation. In this paper, firstly an experimental method to obtain the stress and strain of concrete part in CRCS is adopted. Then uniaxial compression tests on seven types of specimens, including three types of pure rock specimens with different lithology, one type of pure concrete specimen and three types of CRCSs, are performed and acoustic emission (AE) events happening in the specimen are simultaneously monitored. The effect of rock lithology on the mechanical behaviors of the concrete part is analyzed. In addition, a statistical damage constitutive model of the concrete part is established and validated. Research in this study can provide fundamental insights into the mechanical and damage behaviors of concrete structures built on rock foundations with different lithology.

2. Specimen and Experiment Setup

2.1. Specimen

To study the effect of lithology on the mechanical and damage behaviors of the concrete part, in the present study, three types of rock, i.e., red sandstone, purple sandstone and granite, have been considered for the rock part. These three types of rock materials are processed into cylinders of 75 mm in diameter, and the heights of the cylinders are 75 mm and 150 mm. For the concrete, alluvial sands from a natural deposit with a fineness modulus of 2.9 have been used as fine aggregates. Clean gravels with a maximum size of 7 mm are used as coarse aggregates. Portland 42.5R cement, which is often adopted in practice, has been selected as cementitious material and the corresponding water/cement ratio is 0.67. The detailed material composition of current concrete is listed in Table 1.

Table 1. Concrete mix proportions.

Constituent	Amount (kg/m^3)
Cement (Portland 42.5R)	336
Water	226
Fine aggregate	791
Coarse aggregate	1036

Before casting the pure concrete specimens and combined specimens, the internal surfaces of designed cylindrical metal moulds with an inner diameter of 75 mm have been fully smeared with lubricating oil to avoid adhesion. When casting the combined specimens, 75 mm high rock parts are first placed at the bottom of the metal moulds. Considering the practical engineering operation that the concrete structures are built on a geological body, the cement–aggregate–water mixture is then directly poured onto the upper surfaces of rock parts and compacted by placing the moulds on a vibrating machine. For a combined specimen, the heights of concrete and rock part are both 75 mm, i.e., the concrete–rock height ratio in combined specimens is equal to 1, and for a pure material specimen, the height of cylindrical concrete or rock is 150 mm. The specimens are taken out from the moulds 24 h after casting and then moist–cured in an environment of 20 ± 2 °C and 100% relative humidity for 2 months. Before conducting the uniaxial compression tests, the specimens are ground by using a polisher to make both ends of the cylindrical specimens parallel and smooth, and the errors of specimen dimensions are within ± 1 mm. Then the specimens are coated with grease to reduce the friction between the specimen and the rigid loading platens. For the current study, one type of pure concrete specimen, three types of pure rock specimens and three types of CRCSs are prepared as shown in Figure 1, and at least five specimens for each type are processed for testing. The specimen labels and combination forms are listed in Table 2.

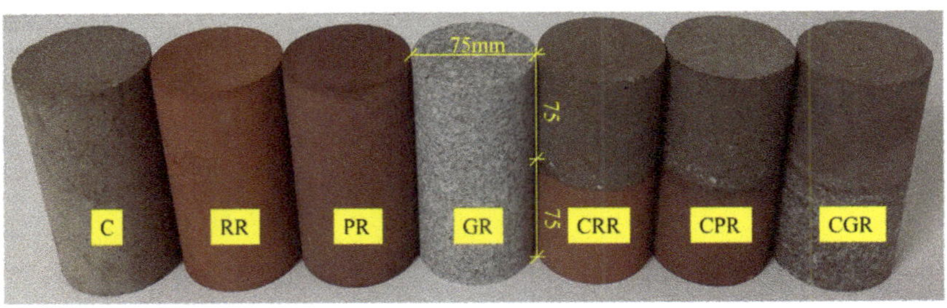

Figure 1. Seven types of specimens.

Table 2. Combination forms and concrete–rock height ratio.

Specimen Label	Lithology	Concrete–Rock Height Ratio
C	Concrete	1:0
RR	Red sandstone	0:1
PR	Purple sandstone	0:1
GR	Granite	0:1
CRR	Concrete–Red sandstone	1:1
CPR	Concrete–Purple sandstone	1:1
CGR	Purple–Granite	1:1

2.2. Test Method

Liu et al. [31] proposed an experimental method to obtain the stress and strain of coal in coal–rock combined specimens by gluing strain gauges on the rock surface. For the present study, similarly, the same experimental method is adopted. In this method, when the combined specimen is subjected to static uniaxial compression loading, the stresses in the concrete part, rock part and the whole combined specimen are equal to each other during the loading process which can be written as:

$$\sigma_C = \sigma_R = \sigma_W \tag{1}$$

where σ_C is the stress in the concrete part, σ_R is the stress in the rock part and σ_W is the stress of the whole combined specimen.

The total axial deformation of the combined specimen due to loading is equal to the sum of axial deformations of the concrete and rock parts, which can be expressed as:

$$\Delta L_W = \Delta L_C + \Delta L_R \tag{2}$$

where ΔL_W is the axial deformation of the whole combined specimen, ΔL_C is the axial deformation of the concrete part, and ΔL_R is the axial deformation of the rock part.

When a CRCS is loaded on a test device such as servo–controlled material testing machine, the axial stress and strain of the combined specimen, σ_W and ε_W, can be measured directly. According to Equation (1), the stresses in the concrete and rock parts, σ_C and σ_R, can be easily obtained. For the CRCS, generally, its failure is initiated by the break of the concrete part due to the fact that the strength and rigidity of rock are much higher than those of concrete. However, since concrete contains a large number of coarse aggregates, local deformations are significantly different at the positions of coarse aggregate and cement during the loading process. As a result, the strain of the concrete part, ε_C, cannot be directly and accurately measured by simply attaching strain gauges to the surface of the concrete part. While on the other side, the rock part is much more homogeneous. Its deformation is uniformly distributed along the height and its axial strain, ε_R, can be measured by gluing strain gauges on its surface. Therefore, according to the relationship between deformation and strain, and substituting the measured strains into Equation (2), the axial strain of the concrete part can also be obtained as follows:

$$\varepsilon_C = \frac{h_W \varepsilon_W - h_R \varepsilon_R}{h_C} \tag{3}$$

where h_W, h_R and h_C are the heights of the whole combined specimen, the rock and concrete parts, respectively.

2.3. Equipment Setup

The whole experimental equipment mainly consists of three sub–systems, i.e., the loading system, AE monitoring system and the strain testing system as shown in Figure 2. The INSTRON 1346 servo–controlled material testing machine with a capacity of 2000 KN is used as the uniaxial compression loading system. The specimens are performed uniaxial compression testing on this testing machine with a displacement–controlled loading mode, and the loading rate is maintained at 0.15 mm/min. In current testing, the lower rigid loading platen is loaded upwards to exert pressure on the specimen while the upper rigid loading platen keeps stationary. A linear variable differential transformer (LVDT), which is directly connected to Computer No. 1 for signal processing, is used to measure the axial strain of the whole specimen. The PCI–2 AE detector, developed by PAC, is used to monitor AE signals during the loading process. The AE sensor is attached to the bottom surface of the upper rigid loading platen with a magnetic fixing device. To ensure the signal reception, Vaseline taken as a coupling agent is smeared on the interfaces and the AE sensor is placed as close as possible to the specimen. The AE signals are received and then amplified and transmitted to Computer No. 2. The main parameters for the AE detector are set as pre–amplifier gain 40 dB, threshold 40 dB and sampling rate 1 MHz, respectively. The DH3817 static strain testing system, consisting of strain gauges, a data–collection device and a computer, is used to measure the axial strain of the rock part. The strain gauges glued on the cylindrical surface of the rock part are directly connected to the data–collection device and the data–collection device is subsequently connected to Computer No. 3. To ensure the accuracy of measured strain data, three strain gauges separated by 120° are glued on the surface of the rock part for each combined specimen, as shown in Figure 2b. During the experiment, three specimens for each type are first tested, and if sharp contrast existed between the results, more specimens of this type are further tested until three similar results are obtained.

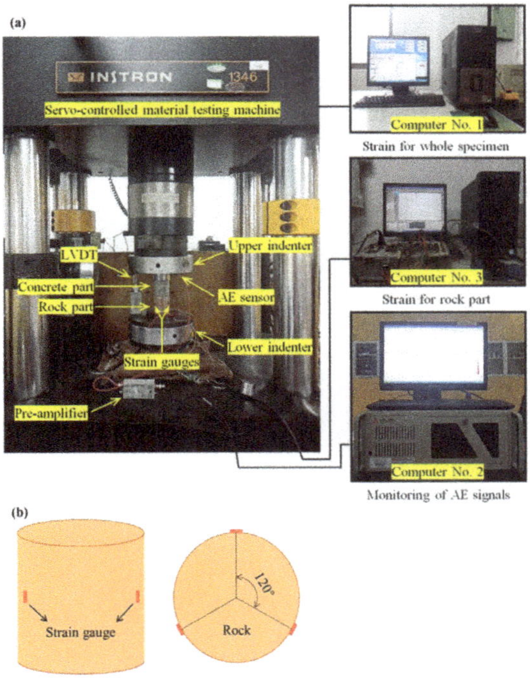

Figure 2. Test system: (**a**) Uniaxial compression testing system; (**b**) Schematic diagram of strain gauge sticking method.

3. Experimental Results and Analysis

3.1. Strength and Failure Pattern of Seven Types of Specimen

The uniaxial compressive strengths of the seven types of specimens are listed in Table 3. During the experiment, further tests are aborted when three similar results are obtained for one type of specimen, thus additional data for this type are not available and the average strength is determined based on the three similar values of all the obtained results for this type to avoid scattering and get a representative strength. As can be seen from the table, the average uniaxial compressive strengths of three types of pure rock specimens, i.e., red sandstone, purple sandstone and granite, are 76.50 MPa, 109.68 MPa and 166.27 MPa, respectively. After 2 months of curing, the average uniaxial compressive strength of pure concrete specimens is 33.71 MPa, and those of CRCSs are 36.99 MPa, 43.42 MPa, and 46.29 MPa, respectively. Herein, it can be noted that although the strengths of combined specimens are much smaller than those of pure rock specimens, they are obviously higher than that of pure concrete specimens, about 9.73–37.32% higher than the latter, and they increase with the increment of the strength of the rock part, which means the load–bearing capacity of the combined specimen is significantly affected by the lithology of rock part.

Table 3. Uniaxial compressive strength (MPa).

Specimen Label	No. 1	No. 2	No. 3	No. 4	No. 5	Average
C	34.24	33.21	33.68	/	/	33.71
RR	77.73	77.43	74.34	/	/	76.50
PR	110.45	111.56	107.03	/	/	109.68
GR	167.50	168.60	162.71	/	/	166.27
CRR	37.35	30.52	35.51	38.11	/	36.99
CPR	43.21	42.73	33.58	37.29	44.31	43.42
CGR	44.71	46.87	30.78	47.28	/	46.29

Figure 3 shows the ultimate failure patterns of concrete, rock and combined specimens in the testing. As can be observed, typical axial splitting fractures form in the pure concrete specimen, while shear–induced fractures occur in the three types of pure rock specimens. In the cases of CRCSs, although rock materials are also present, the specimens are all failed by axial splitting fractures along the loading direction. The axial cracks initiate and mostly propagate in the concrete parts and the crack density in concrete increases as the strength of rock increases. This is mainly because the axial cracks in concrete generate horizontal tensile stresses acting on the rock part, and higher strength rock needs more tension created in this way to overcome its tensile strength. With the growth of new cracks, the higher strength rock part is finally fractured in tension. From these observations, it can be found that the low–strength concrete plays a major role in the fracture behavior of CRCSs under uniaxial compression.

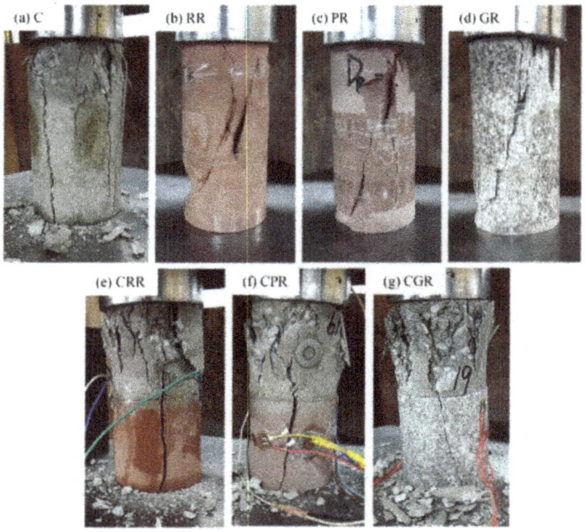

Figure 3. Failure patterns of different types of specimens.

3.2. Stress–Strain Curve of Seven Types of Specimen

In the uniaxial compression testing, the axial stress and strain of the whole specimen are directly measured and recorded by using the loading system. Figure 4 shows the stress–strain curves of the whole specimens and these stress–strain responses are closely related to the damage development of specimens. From the curves it can be seen that the whole loading process of concrete–rock specimens generally can be subdivided into five successive stages, i.e., compaction hardening, elastic deformation, plastic deformation, post–peak strain softening and residual friction.

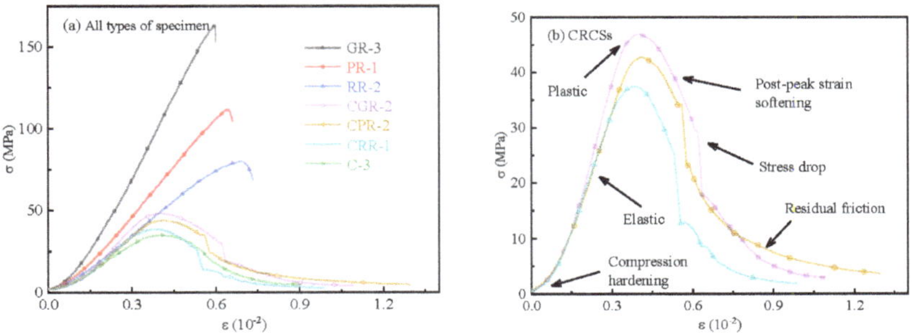

Figure 4. Stress—strain curves of the whole specimens.

The concrete usually contains lots of pores, and many micro cracks exist in the rock as well. In the initial compaction hardening stage, these existing flaws in the combined specimen are quickly crushed and closed under the pressure. Through self–adjustment, the density and stiffness of the combined specimen gradually increase, thus a concave curve is presented during this period. When the maximum compaction pressure is reached, the materials of concrete and rock tend to be completely compact and even, and the combined specimen begins to behave almost in a linear–elastic manner, that is to say, the combined specimen enters into an elastic deformation stage. At this stage, the stiffness of the specimen approximately keeps constant. Under the pressure, the concrete and rock parts work together, and the stress increases rapidly. With the further increase of stress, low–strength concrete first turns into a plastic state and the development of original randomly distributed micro cracks in the rock is also initiated. The stiffness of the combined specimen gradually decreases, and the plastic deformation increases nonlinearly. At the end of this stage, the combined specimen reaches its ultimate compressive strength. After that, the cracks in concrete quickly spread and converge, and the stress inside the specimen redistributes and rapidly decreases. In this situation, the specimen exhibits strain-softening behavior and undergoes a rapid decrease in load–bearing capacity. During this stage, the concrete changes into failure, and the cracks in the concrete form some major axial fractures and spread downwards to the interface. At the interface, the axial splitting fractures in concrete generate horizontal tensile stresses on the rock part, and with the growth of new cracks in concrete, the rock is finally fractured. The rock part is instantaneously fractured, at the same time, a sudden and steep stress drop happens and this phenomenon will be discussed in detail in Section 3.3 where the strain variation of the rock part is introduced. At last, the entire specimen is destroyed, but due to the friction between crack and fracture surfaces, it still maintains some residual resistance strength, and with the continuous increase of deformation, the stress in the combined specimen slowly decreases.

The main difference between the stress–strain curves of pure rock, pure concrete and combined specimens in the loading process is the post–peak strain–softening stage. Because of the brittle nature, the stress–strain curves of pure rock specimens can only be recorded a little later after the peaks. In contrast, due to being more ductile, the stress–strain curve of pure concrete specimen has good continuity and its post–peak strain–softening stage is relatively smooth. While in this regard, the CRCSs integrate both of the characteristics of the above two kinds of pure materials. Their stress–strain curves also present good continuity, but sudden stress drops appear at the ends of the post–peak strain softening stages when the rock in specimens are thoroughly fractured. Besides that, it is worth noting here that, with the increasing of rock strength in the combined specimen, the corresponding stress drop in the post–peak strain–softening stage delays. As mentioned in Section 3.1, this is again because of the horizontal tensile stresses generated by the axial cracks in concrete. When the higher strength rock is split, it needs more axial cracks of concrete to create enough horizontal tension to overcome its tensile strength, which means the corresponding combined specimen allows its concrete part to deform persistently along the loading direction with more axial strain.

3.3. Strain Variation of Rock Part

The axial strain of the rock part in the combined specimen is directly measured by using the strain testing system. Since the response behaviors of different rock parts in combined specimens during the loading are similar, here the stress–strain data of the rock part in CRR–1 (see Figure 5) is taken as an example for the analysis of strain variation of rock part. As can be seen, the axial strain variation of the rock part is similar to the stress variation of the corresponding combined specimen. Due to that, the strength and stiffness of rock are much higher than those of concrete, the rock part is almost intact before the fracture in concrete extends into it, during this period, according to the stress–strain curve in Figure 4a, the axial strain of rock in CRR–1 is still in the elastic stage, and a large amount of elastic energy is accumulated in rock part under the compressive loading. At the end of the post–peak strain–softening stage, the rock part is instantaneously fractured and a rapid axial strain recovery happens in the rock part. It is worth noting that the rapid axial strain recovery of the rock part and the sudden stress drop of CRCS almost happen at the same time. This phenomenon can be explained by the fact that when the rock part experiences a rapid axial strain recovery, the elastic energy accumulated in the rock part releases quickly and the rock part plays an axial loading role on the concrete part, which intensifies the failure of the concrete part, and as a result, the sudden stress drop happens in the CRCS. In addition, it can also be found from the figure that when the rapid axial strain recovery happens, the axial strain in the rock part and the stress in CRCS do not recover and decrease to zero, which indicates that the axial loading effect of the rock part cannot completely crush the concrete part, and the concrete part still has some residual strength.

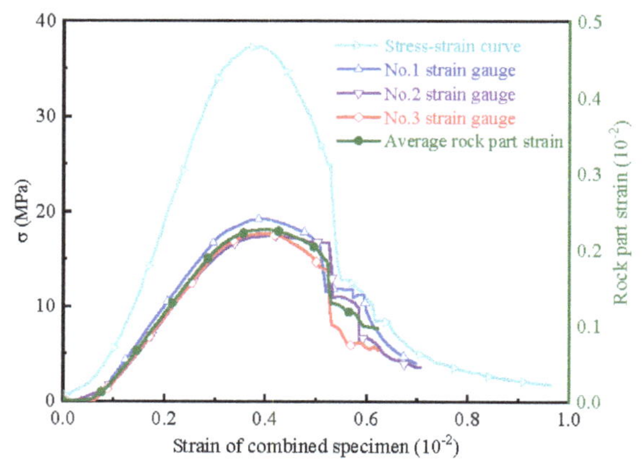

Figure 5. Strain variation of rock part in CRR–1.

Furthermore, it can be found that the strain of the rock part in the combined specimen at peak stress is much smaller than that of the corresponding combined specimen. For instance, the average strains of rock parts (rock in CRR–1, CPR–2 and CGR–2) at peak stresses are 2.28×10^{-3}, 1.90×10^{-3} and 1.47×10^{-3}, respectively. The average strains of corresponding combined specimens at peak stresses are 3.91×10^{-3}, 4.16×10^{-3} and 4.06×10^{-3}, respectively. In other words, the axial deformations of rock parts at the peak stresses account for 29.16%, 22.84% and 18.10% of the corresponding combined specimens, respectively. Compared to the strains of pure rock specimens at the peak stresses (for example, 6.92×10^{-3}, 6.46×10^{-3} and 5.55×10^{-3} for RR–2, PR–1 and GR–3), the axial strains of rock in combined specimens are only 32.95%, 29.41% and 26.49%. This is mainly due to the fact that the failure pattern of rock in the combined specimen is different from that of pure rock specimen under compressive loading. As mentioned in Section 3.1, the failure pattern of rock changes from a shear–induced fracture in a pure rock specimen to an axial tensile fracture in the combined specimen. Thus, it can be concluded that when the concrete–rock combined body is subjected to ultimate load, the concrete structure has a weakening effect on the axial load–bearing and deformation capacity of the connected rock foundation, and the weakening effect becomes more obvious with the increase of rock strength.

3.4. Stress–Strain Curve of Concrete Part

By substituting the strains of the rock part and corresponding combined specimen into Equation (3), the strain of the concrete part can be obtained. The stress of the concrete can be obtained by Equation (1). Because the whole combined specimen is near failed when the sudden stress drop happens, we only calculate the stress and strain of the concrete part before the stress drop happens. Figure 6 shows the stress–strain curve of the pure concrete specimen (C–3) and the derived stress–strain curves of concrete parts (concrete in CRR–1, CPR–2 and CGR–2). As can be seen, although the stress–strain curves of the pure concrete specimen and concrete parts show a similar evolution trend, several apparent differences can be observed from them. For instance, the strengths of concrete parts range from 37.35 MPa to 46.87 Mpa, and they are obviously higher than that of the pure concrete specimen (33.68 MPa). Moreover, the strains of concrete parts at peak stresses vary from 5.54×10^{-3} to 6.65×10^{-3}, which are also much larger than that of the pure concrete specimen (4.12×10^{-3}). Therefore, compared with the pure concrete specimen, the load-bearing and deformation capacities of the concrete part are both enhanced, and the strength and strain at peak stress of concrete parts also increase with the strength of the rock part. To illustrate these more clearly, the variations of the strength and the strain at peak stress of concrete and concrete parts are shown in Figure 7.

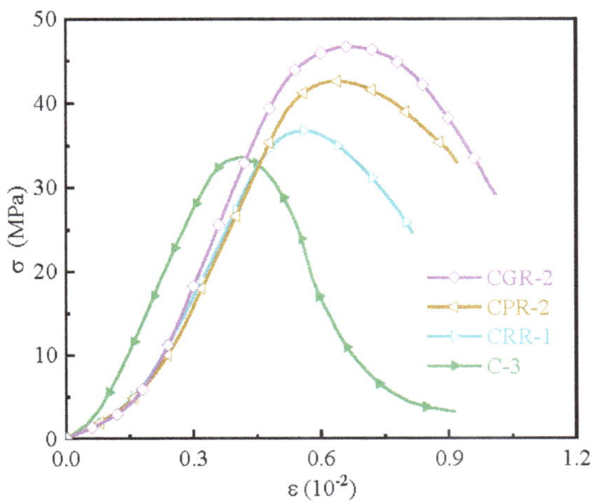

Figure 6. Stress−strain curves of the pure concrete specimen and concrete parts.

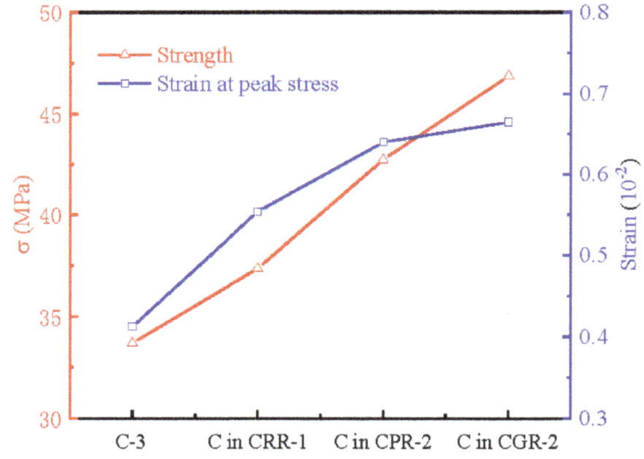

Figure 7. Variations of the strength and the strain at peak stress of concrete and concrete parts.

3.5. Damage Behaviors of Concrete Part

The AE signals (AE counts and cumulative AE counts) are monitored by using the PCI–2 AE detector to investigate the damage behaviors of the tested specimen. The AE detector runs simultaneously with the load–testing machine. Figure 8 shows AE counts and cumulative AE counts versus the stress–strain curve of each tested specimen. As can be seen from the figures, the variation of AE signals has a good correlation with the stress–strain curve. For all the specimens, at the beginning of loading, i.e., at the compaction hardening stage, due to the closure of micro pores and cracks, a small number of AE counts appeared, and then the AE activity was relatively quiet at the elastic deformation stage. When the plastic deformation stage is reached, the AE counts increase sharply, at the same time, the cumulative AE counts increase exponentially. The main differences of AE signals among these various types of specimens are that before the plastic deformation stage, the AE activities of pure rock specimens can be neglected, while due to the fact that concrete contains more pores, a number of AE events in pure concrete specimen and CRCSs are recorded. Moreover, because of the brittle nature, the AE signals of pure rock specimens can only be recorded a little later after the peaks, and due to being more ductile than rock, the AE events in the pure concrete specimen and CRCSs are active throughout the whole post–peak part. The peak values of AE counts of pure rock and concrete specimens appear near the peak load, while those of the CRCSs appear at the end of the post–peak soften stages where the sudden stress drop happens, and both mean the corresponding specimen is close to failing or failed.

During the loading process, the crack propagation and extension cause damage to the specimen, and the AE events happen in the specimen. According to the principle of AE technology, the damage extent of the specimen after the test can be inferred by the final cumulative AE counts. The final cumulative AE counts of pure rock specimens (RR–2, PR–1 and GR–3) are 9.23×10^4, 9.45×10^4, and 3.67×10^4, respectively, and those of pure concrete specimen (C–3) and CRCSs (CRR–1, CPR–2 and CGR–2) are 5.52×10^5, 7.61×10^5, 8.44×10^5, and 9.13×10^5, respectively. The values of pure concrete specimen and CRCSs are much higher than those of pure rock specimens, and this indicates that the internal damage of pure concrete and combined specimens are more severe than those of pure rock specimens. Furthermore, by comparing the values of cumulative AE counts of pure rock and concrete specimens, it can be concluded that the AE events of the CRCS mainly happen in the concrete part, and this illustrates that the damage of the CRCS is mainly formed in the concrete part. Furthermore, it can also be found that the final cumulative AE counts increase with the rock strength in CRCS, and this means that the damage extent of the concrete part has a positive correlation with the strength of the rock part.

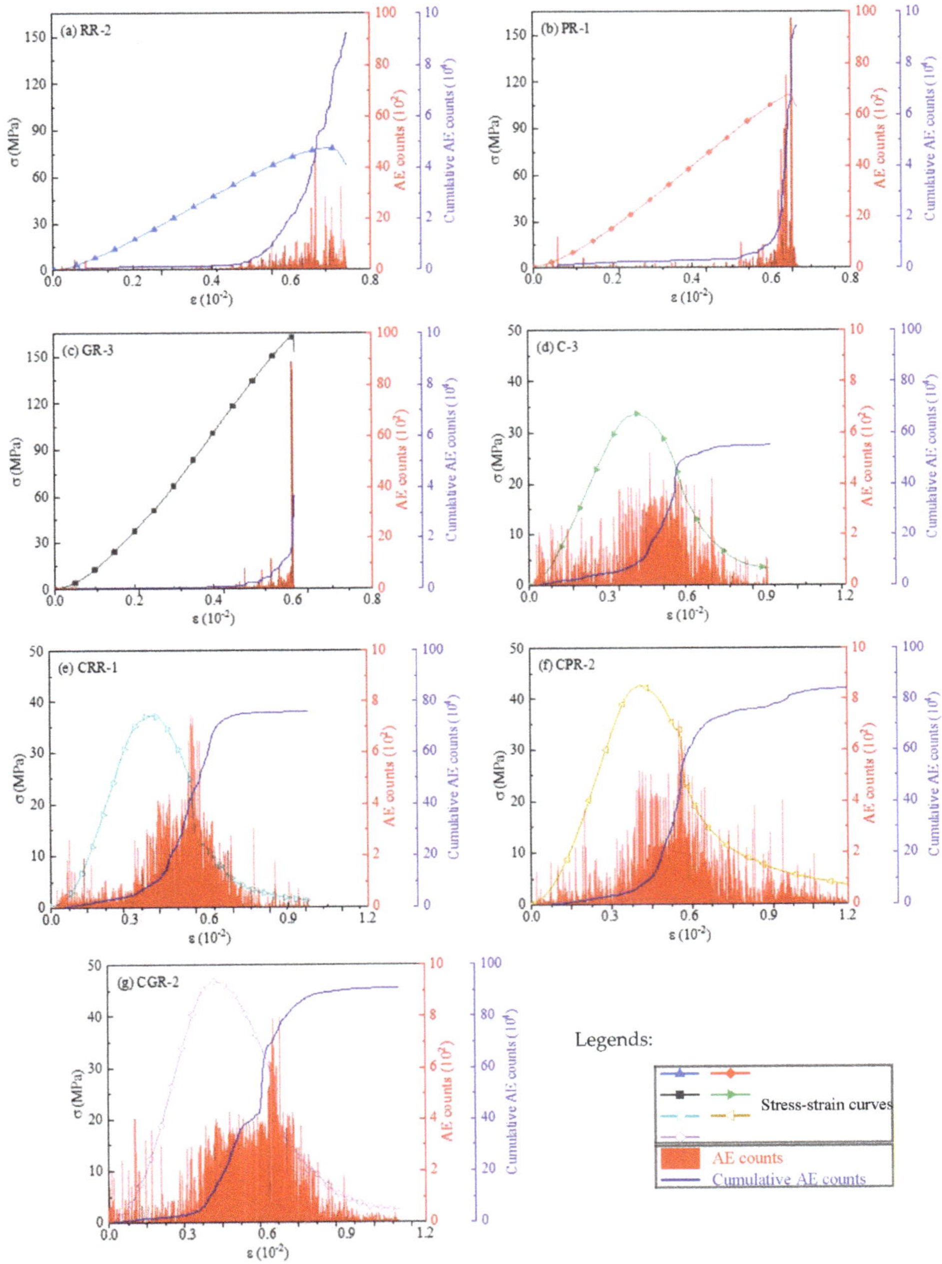

Figure 8. AE counts and cumulative AE counts of tested specimen versus strain.

4. Damage Constitutive Model of Concrete Part

4.1. Derivation of Constitutive Equation

There are a large number of defects in concrete in terms of micro pores and cracks, and these defects gradually develop and extend under external load and eventually lead to material damage. However, the properties of these micro defects, such as position, dimension, strength and stiffness, etc., are impossible to be known exactly. For the sake of simplification, it is generally thought that the micro–unit strength of concrete obeys some statistical distributions when trying to describe the mechanical behavior of concrete mathematically. In the same manner, for the present study of a constitutive model for the concrete part, it is assumed that the micro–unit strength of concrete satisfies Weibull distribution, and the probability density function of micro–unit strength can be written in the form of:

$$\varphi\left(\overline{f}\right) = \frac{m}{a}\left(\frac{\overline{f}}{a}\right)^{m-1} \exp\left[-\left(\frac{\overline{f}}{a}\right)^{m}\right] \tag{4}$$

where $\overline{f}$ is the micro–unit strength variable, and m and a are the shape parameter and the scale parameter, respectively.

The quantity of failure unit in interval $[0, \overline{f}]$ can be calculated as:

$$N_f\left(\overline{f}\right) = \int_0^{\overline{f}} N\varphi(x)dx = N\left\{1 - \exp\left[-\left(\frac{\overline{f}}{a}\right)^{m}\right]\right\} \tag{5}$$

The loading damage variable is defined as the ratio of the quantity of failure unit N_f to that of total unit N:

$$D = \frac{N_f}{N} \tag{6}$$

Substituting Equation (5) into Equation (6), then the loading damage variable is obtained as:

$$D = 1 - \exp\left[-\left(\frac{\overline{f}}{a}\right)^{m}\right] \tag{7}$$

When further assuming that the micro–unit strength of concrete satisfies the maximum–tensile strain yield criterion, the micro–unit strength variable can be written as:

$$\overline{f} = f(\varepsilon) = \varepsilon \tag{8}$$

where ε is the strain of the concrete part. By substituting Equation (8) into Equation (7), the loading damage variable becomes:

$$D = 1 - \exp\left[-\left(\frac{\varepsilon}{a}\right)^{m}\right] \tag{9}$$

In 1971, Lemaitre [32] puts forward a strain equivalence hypothesis. According to his hypothesis, the strain behavior of a damage material can be modified by damage only through the effective stress and represented by constitutive equations of the virgin material (without any damage) in which the stress is simply replaced by the effective stress. This hypothesis has been widely used in the researches for concrete under various loading conditions [33,34]. Based on this strain equivalence hypothesis, in 1985, Lemaitre [35] develops a damage constitutive model under uniaxial loading as follows:

$$\sigma = (1 - D)E_C\varepsilon \tag{10}$$

where E_C is the average Young's modulus of pure concrete specimens.

As mentioned in the foregoing analysis, the strength of concrete parts increases as the rock part strength increases, and the mechanical response of concrete parts is quite different from that of pure concrete. This is because rocks with different lithology have varying degrees of influence on the mechanical response of concrete parts. From this point of view, the above Equation (10), a traditional constitutive model for pure concrete, obviously cannot be used to accurately describe the behavior of concrete parts under uniaxial loading, and the influence of rocks with different lithology should be taken into consideration. At the beginning of loading, the concrete and rock parts are both in the compaction hardening process, but only a small amount of damage appeared in the rock, and the rock is almost in an elastic state before the sudden stress drop happens, thus the effect of rock part on the mechanical behavior of concrete part can be regarded as a non–ideal elastic body. In this way, the concrete in the combined specimen can be viewed as a cascade system of a damage body and a non–ideal elastic body (as shown in Figure 9) before the stress drop happens, and the constitutive equation for the non–ideal elastic body can be temporarily written as Equation (11):

$$\sigma = E_R \varepsilon \tag{11}$$

where σ and ε are the stress and strain of the non–ideal elastic body, respectively, and E_R is the average Young's modulus of pure rock specimens.

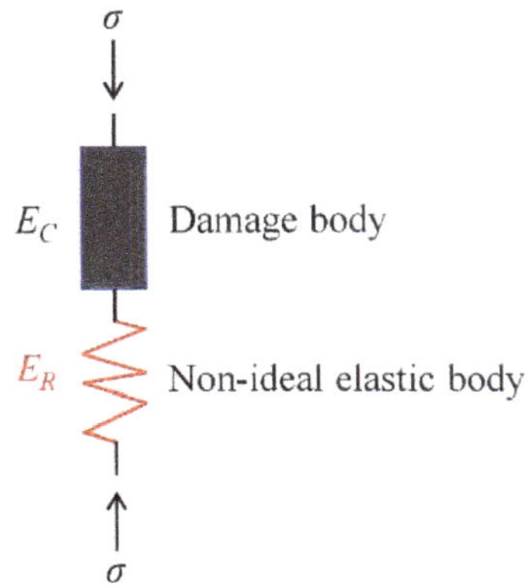

Figure 9. Damage constitutive model for concrete part.

Since the stress of the damage body is equal to that of the non–ideal elastic body, and the strain of the concrete part consists of those of the damage body and the non–ideal elastic body, then they can be written as

$$\begin{cases} \sigma = \sigma_D = \sigma_N \\ \varepsilon = \varepsilon_D + \varepsilon_N \end{cases} \tag{12}$$

where σ_D and ε_D are the stress and strain of damage body, respectively, and σ_N and ε_N are the stress and strain of non–ideal elastic body, respectively.

By substituting Equations (10) and (11) into Equation (12), the statistical damage constitutive equation of concrete part can be obtained as:

$$\sigma = \frac{(1-D)E_C E_R \varepsilon}{(1-D)E_C + E_R} \tag{13}$$

However, in the above establishment of damage constitutive equation, the compaction hardening process is not taken into account. As mentioned in Section 3.2, many micro pores and cracks exist within the concrete and rock, in the early stage of loading, these initial flaws in the combined specimen are rapidly compacted, and thus the stress–strain curve first starts with an obvious concave increase. To accurately express the characteristic of compaction hardening of the concrete part, a compaction hardening coefficient, α, is proposed to quantify the extent of compaction hardening:

$$\alpha = \begin{cases} \log_n \left[\frac{(n-1)\varepsilon}{\varepsilon_e} + 1 \right] & \varepsilon < \varepsilon_e \\ 1 & \varepsilon \geq \varepsilon_e \end{cases} \tag{14}$$

where n is a fitting constant obtained by fitting the experimental data, and ε_e is the strain of the concrete part corresponding to the stress at the starting point of the elastic deformation stage.

This compaction hardening coefficient is defined as the ratio of the slope of stress–strain curve to the Young's modulus of concrete part and it increases logarithmically with strain. By introducing this coefficient into Equation (14), the statistical damage constitutive model is modified as:

$$\sigma = \begin{cases} \frac{\alpha(1-D)E_C E_R \varepsilon}{(1-D)E_C + E_R} & \varepsilon < \varepsilon_e \\ \frac{(1-D)E_C E_R \varepsilon}{(1-D)E_C + E_R} & \varepsilon \geq \varepsilon_e \end{cases} \tag{15}$$

where m and a in "D" can be determined by fitting the experimental data.

4.2. Model Verification

Figure 10 shows the comparison of the stress–strain curves calculated using the above derived statistical damage constitutive model and the corresponding experimental data of concrete part under uniaxial compression loading. The curves are only calculated to the point of stress drop since the whole combined specimens are failed when the sudden stress drop happens. As can be seen, the calculated curves agree well with the experimental data, which indicates that the established statistical damage constitutive model can accurately describe the mechanical response of concrete parts under uniaxial compression loading.

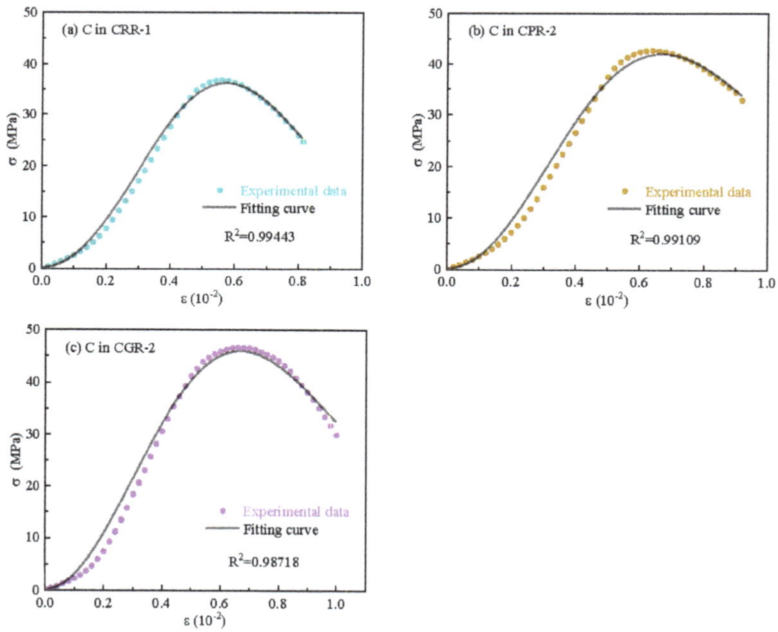

Figure 10. Comparison of experimental data and calculated stress–strain curves by using established statistical damage constitutive model.

Table 4 gives the fitting parameters and the related physical and mechanical parameters for the established damage constitutive model, and by introducing the fitting parameters a and m into Equation (9), the damage evolution curves (loading damage variable D versus strain) of the concrete part in combined specimens with different rock lithology are obtained as shown in Figure 11. As can be seen from the figure, the damage evolution curve reflects the mechanical performance of the concrete part under uniaxial compressive loading. When the concrete in different CRCSs suffers from the same damage at a higher level, the strain increases with the increment of the strength of the rock part. This indicates that the deformation capacity of the concrete part enhances with the rock part strength, and adds that the values of loading damage variable range from zero to one, they are concordant with the foregoing experimental results, and indirectly means that the statistical damage constitutive model and the corresponding fitting process are reliable.

Table 4. Fitting and related physical mechanical parameters for damage constitutive model.

Specimen (C in)	n	a	m	ε_e (10^{-3})	E_C (GPa)
C	34.24	33.21	33.68	/	/
RR	77.73	77.43	74.34	/	/
PR	110.45	111.56	107.03	/	/
GR	167.50	168.60	162.71	/	/
CRR	37.35	30.52	35.51	38.11	/
CPR	43.21	42.73	33.58	37.29	44.31
CGR	44.71	46.87	30.78	47.28	/

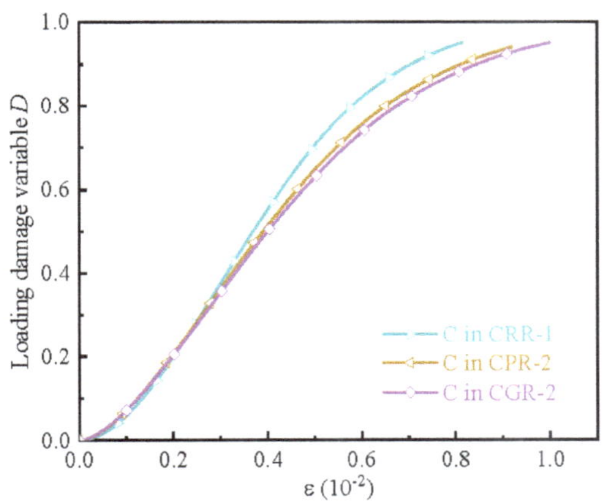

Figure 11. Variation of loading damage variable D of concrete parts versus strain.

5. Conclusions

In this study, the effect of lithology on the mechanical and damage behaviors of the concrete part in CRCS is investigated, and a statistical damage constitutive model for the concrete part is established and validated. The major conclusions are drawn as follows:

1. The low–strength concrete part plays a major role in the fracture behavior of CRCS under uniaxial compression loading. When the CRCS is failed, a sudden stress drop happens in CRCS and the rock part experiences a rapid strain recovery and plays an axial loading role on the concrete part, which intensifies the failure of the concrete part.

2. Compared with the pure concrete specimen, the strength and the strain at peak stress of the concrete part increase with the increment of rock part strength, which illustrates the load–bearing and deformation capacities of the concrete part are both enhanced. However, the load–bearing and deformation capacity of the rock part is weakened due to the influence of the concrete part, and the weakening effect becomes more obvious with the increase of rock strength.

3. Through the AE monitoring and signal analysis, it is found that the damage of CRCS mainly happens in the concrete part, and the damage extent of the concrete part has a positive correlation with the strength of the rock part.

4. A damage constitutive model of the concrete part is established by using a cascade system, including a damage body and a non–ideal elastic body and validated against the experimental data. This statistical damage constitutive model can be used to accurately

describe the effect of lithology on the mechanical response of the concrete part under uniaxial compression loading.

Author Contributions: Conceptualization, K.L.; methodology, S.J.; software, K.L. and S.J.; validation, K.L., S.J., Y.R. and J.H.; formal analysis, Z.Z.; investigation, Y.R., J.H. and Z.Z.; resources, S.J.; data curation, S.J.; writing—original draft preparation, K.L; writing—review and editing, Y.R.; visualization, J.H.; supervision, Z.Z.; project administration, K.L.; funding acquisition, S.J. and Y.R. All authors have read and agreed to the published version of the manuscript.

Funding: The research was supported by the National Natural Science Foundation of China, project No. 51974360.

Institutional Review Board Statement: Not applicable.

Informed Consent Statement: Not applicable.

Data Availability Statement: The experimental data cannot be shared at this time, because the data is also used in the ongoing study.

Acknowledgments: We thank anonymous reviewers for their constructive comments.

Conflicts of Interest: The authors declare no conflict of interest.

References

1. Armaghani, D.J.; Bayat, V.; Koopialipoor, M.; Pham, B.T. Investigating the effect of jointed environment on the cracked concrete arch dam in 3D conditions using FEM. *Bull. Eng. Geol. Environ.* **2021**, *80*, 55–70. [CrossRef]
2. de Granrut, M.; Simon, A.; Dias, D. Artificial neural networks for the interpretation of piezometric levels at the rock-concrete interface of arch dams. *Eng. Struct.* **2019**, *178*, 616–634. [CrossRef]
3. Fernandes, I. Role of granitic aggregates in the deterioration of a concrete dam. *Bull. Eng. Geol. Environ.* **2015**, *74*, 195–206. [CrossRef]
4. Fishman, Y.A. Features of shear failure of brittle materials and concrete structures on rock foundations. *Int. J. Rock. Mech. Min. Sci.* **2008**, *45*, 976–992. [CrossRef]
5. Fishman, Y.A. Stability of concrete retaining structures and their interface with rock foundations. *Int. J. Rock. Mech. Min. Sci.* **2009**, *46*, 957–966. [CrossRef]
6. Kebab, H.; Boumezbeur, A.; Rivard, P. Rock mass properties and their suitability as a foundation for a rolled compacted concrete gravity dam: Case study of Beni Haroun dam (Mila, NE Algeria). *Bull. Eng. Geol. Environ.* **2021**, *80*, 1729–1743. [CrossRef]
7. Rui, Y.C.; Zhou, Z.L.; Lu, J.Y.; Barkat, U.; Cai, X. A Novel AE Source Localization Method Using Clustering Detection to Eliminate Abnormal Arrivals. *Int. J. Min. Sci. Technol.* **2022**, *32*, 51–62. [CrossRef]
8. Cheng, R.; Cheng, R.; Chen, W.; Hao, H.; Li, J. Dynamic Response of Road Tunnel Subjected to Internal Boiling Liquid Expansion Vapour Explosion (BLEVE). *Tunn. Undergr. Space Technol.* **2022**, *123*, 104363. [CrossRef]
9. Wang, S.F.; Tang, Y.; Wang, S.Y. Influence of brittleness and confining stress on rock cuttability based on rock indentation tests. *J. Cent. South. Univ.* **2021**, *28*, 2786–2800. [CrossRef]
10. Rui, Y.C.; Zhou, Z.L.; Cai, X.; Dong, L.J. A Novel Robust Method for Acoustic Emission Source Location Using DBSCAN Principle. *Measurement* **2022**, *191*, 110812. [CrossRef]
11. Yang, J.C.; Liu, K.W.; Li, X.D.; Liu, Z.X. Stress initialization methods for dynamic numerical simulation of rock mass with high in–situ stress. *J. Cent. South. Univ.* **2020**, *27*, 3149–3162. [CrossRef]
12. Selçuk, L.; Aşma, D. Experimental investigation of the rock–concrete bi materials influence of inclined interface on strength and failure behavior. *Int. J. Rock. Mech. Min. Sci.* **2019**, *123*, 104119. [CrossRef]
13. Song, Z.Y.; Konietzky, H.; Herbst, M. Three–dimensional particle model based numerical simulation on multi–level compressive cyclic loading of concrete. *Constr. Build. Mater.* **2019**, *225*, 661–677. [CrossRef]
14. Zhao, B.Y.; Liu, Y.; Huang, T.Z.; Wang, X.P. Experimental study on strength and deformation characteristics of rock–concrete composite specimens under compressive condition. *Geotech. Geol. Eng.* **2019**, *37*, 2693–2706. [CrossRef]
15. Greco, O.; Ferrero, A.M.; Oggeri, C. Experimental and analytical interpretation of the behaviour of laboratory tests on composite specimens. *Int. J. Rock. Mech. Min. Sci.* **1993**, *30*, 1539–1543. [CrossRef]
16. Li, J.G.; Yu, Z.Q.; Zhou, Z.Y.; Wang, Y.C.; Li, J.W. Mechanical analysis and failure modes prediction of composite rock under uniaxial compression. *Sci. Rep.* **2021**, *11*, 22826. [CrossRef]
17. Bista, D.; Sas, G.; Johansson, F.; Lia, L. Influence of location of large–scale asperity on shear strength of concrete–rock interface under eccentric load. *J. Rock Mech. Geotech.* **2020**, *12*, 449–460. [CrossRef]
18. Saiang, D.; Malmgren, L.; Nordlund, E. Laboratory tests on shotcrete–rock joints in direct shear, tension and compression. *Rock Mech. Rock Eng.* **2005**, *38*, 275–297. [CrossRef]
19. Shen, P.W.; Tang, H.M.; Ning, Y.B.; Xia, D. A damage mechanics based on the constitutive model for strain–softening rocks. *Eng. Fract. Mech.* **2019**, *216*, 106521. [CrossRef]

20. Zhao, W.S.; Chen, W.Z.; Zhao, K. Laboratory test on foamed concrete–rock joints in direct shear. *Constr. Build. Mater.* **2018**, *173*, 69–80. [CrossRef]
21. Chang, X.; Lu, J.Y.; Wang, S.Y.; Wang, S.R. Mechanical performances of rock–concrete bi–material disks under diametrical compression. *Int. J. Rock. Mech. Min. Sci.* **2018**, *104*, 71–77. [CrossRef]
22. Chang, X.; Guo, T.F.; Lu, J.Y.; Wang, H. Experimental study on rock–concrete joints under cyclically diametrical compression. *Geomech. Eng.* **2019**, *17*, 553–564.
23. Dong, W.; Song, S.Z.; Zhang, B.S.; Yang, D. SIF–based fracture criterion of rock–concrete interface and its application to the prediction of cracking paths in gravity dam. *Eng. Fract. Mech.* **2019**, *221*, 106686. [CrossRef]
24. Qiu, H.; Zhu, Z.M.; Wang, M.; Wang, F.; Luo, C.S.; Wan, D.Y. Study of the failure properties and tensile strength of rock–mortar interface transition zone using bi–material Brazilian discs. *Constr. Build. Mater.* **2020**, *236*, 117551. [CrossRef]
25. Zhou, Z.L.; Lu, J.Y.; Cai, X. Static and dynamic tensile behavior of rock–concrete bi–material disc with different interface inclinations. *Constr. Build. Mater.* **2020**, *256*, 119424. [CrossRef]
26. Dong, W.; Wu, Z.M.; Zhou, X.M.; Wang, N.; Kastiukas, G. An experimental study on crack propagation at rock–concrete interface using digital image correlation technique. *Eng. Fract. Mech.* **2017**, *171*, 50–63. [CrossRef]
27. Javanmardi, F.; Léger, P.; Tinawi, R. Seismic structural stability of concrete gravity dams considering transient uplift pressures in cracks. *Eng. Struct.* **2005**, *27*, 616–628. [CrossRef]
28. Andjelkovic, V.; Pavlovic, N.; Lazarevic, Z.; Nedovic, V. Modelling of shear characteristics at the concrete–rock mass interface. *Int. J. Rock. Mech. Min. Sci.* **2015**, *76*, 222–236. [CrossRef]
29. Tian, H.M.; Chen, W.Z.; Yang, D.S.; Yang, J.P. Experimental and numerical analysis of the shear behaviour of cemented concrete–rock joints. *Rock Mech. Rock Eng.* **2015**, *48*, 213–222. [CrossRef]
30. Chavez, J.W.; Fenves, G.L. Earthquake analysis of concrete gravity dams including base sliding. *Earthq. Eng. Struct. Dyn.* **1995**, *24*, 678–686. [CrossRef]
31. Liu, X.S.; Tan, Y.L.; Ning, J.G.; Lu, Y.W.; Gu, Q.H. Mechanical properties and damage constitutive model of coal in coal–rock combined body. *Int. J. Rock. Mech. Min. Sci.* **2018**, *110*, 140–150. [CrossRef]
32. Lemaitre, J. Evaluation of dissipation and damage in metals submitted to dynamic loading. In Proceedings of the 1st International Conference on Mechanical Behavior of Materials, Kyoto, Japan, 15–20 August 1971.
33. Abu Al-Rub, R.K.; Kim, S.-M. Computational applications of a coupled plasticity–damage constitutive model for simulating plain concrete fracture. *Eng. Fract. Mech.* **2010**, *77*, 1577–1603. [CrossRef]
34. Ren, X.; Li, J. A unified dynamic model for concrete considering viscoplasticity and rate–dependent damage. *Int. J. Damage Mech.* **2013**, *22*, 530–555. [CrossRef]
35. Lemaitre, J. A continuous damage mechanics model for ductile fracture. *J. Eng. Mater. Technol.* **1985**, *107*, 83–89. [CrossRef]

mathematics

Article

Modeling Uranium Transport in Rough-Walled Fractures with Stress-Dependent Non-Darcy Fluid Flow

Tong Zhang [1,2,*], Xiaodong Nie [3,*], Shuaibing Song [1], Xianjie Hao [2,3] and Xin Yang [1]

1 State Key Laboratory of Mining Response and Disaster Prevention and Control in Deep Coal Mines, Anhui University of Science and Technology, Huainan 232001, China; songshuaibing@cumt.edu.cn (S.S.); yxaff727@163.com (X.Y.)
2 Institute of Energy, Hefei Comprehensive National Science Center, Hefei 230031, China; 2018013@aust.edu.cn
3 School of Mechanics and Civil Engineering, China University of Mining and Technology (Beijing), Beijing 100083, China
* Correspondence: zhangt_1990@sina.com (T.Z.); niexdcumtb@163.com (X.N.)

Abstract: The reactive-transportation of radioactive elements in fractured rock mass is critical to the storage of radioactive elements. To describe the reactive-transportation and distribution morphology of a uranium-containing solution, a stress-dependent reactive transport model was developed, and the simulator of FLAC3D-CFD was employed. The uranium transport experiment subjected to the variation of confining stress of 5–19 MPa and hydraulic pressure of 0.5–3.5 MPa was conducted in fractured rock mass. The results show that the uranium-containing solution transport and distribution is significantly dependent on the evolution of the connected channel in rough-walled fracture, which is significantly influenced by the confining stress and hydraulic pressure. In more detail, the increase of confining stress resulted in the anisotropic of seepage channel in aperture, and corresponding turbulence flow and uranium retention were presented at the fracture aperture of 2–5 μm. As the increase of hydraulic pressure, flow regime evolved from the inertial flow to vortex flow, and the transformation region is 16 MPa confining stress and 1.5 MPa hydraulic pressure. The evolution of loading paths also dominates the flow and solute transport, and high seepage speed and strong solute transport were presented at the $k = 1$ (ratio of vertical stress loading to horizontal stress unloading), and a laminar flow and weak solute transport were presented at $k = 0$.

Keywords: fractured rock mass; uranium-containing solution; multifield coupling; reactive transport; rough-walled fracture

MSC: 74L20; 74F10

Citation: Zhang, T.; Nie, X.; Song, S.; Hao, X.; Yang, X. Modeling Uranium Transport in Rough-Walled Fractures with Stress-Dependent Non-Darcy Fluid Flow. *Mathematics* **2022**, *10*, 702. https://doi.org/10.3390/math10050702

Academic Editors: Shaofeng Wang, Jian Zhou, Xin Cai, Xiaofeng Li and Zhengyang Song

Received: 2 January 2022
Accepted: 21 February 2022
Published: 23 February 2022

1. Introduction

The nuclear energy provides 16% clear energy for human development, while abandoning radioactive resources threatens the environment safety and human health. Uranium as a fundamental nuclear energy material attracts the attention of researchers and engineers in the field of mining and contaminant disposal. During the uranium mining, the original rock stress and flow state of the reservoir are destroyed, which is manifested by the propagation of fractures and the migration of radioactive pollutants. Natural rock mass is a multiphase medium in which solid, liquid, and gaseous phases co-exist. The geologic structure influenced by the mining dominates the strength capability and uranium transport behavior. Efforts are being made to study the structure change, uranium transport, desorption, adsorption, and precipitation in fractured rock masses, which significantly contributes to the mining of uranium and disposal of a uranium-containing solute [1–8].

The hydrogeological structure and interaction between the solid, gas, and fluid in the resource deposits are crucial to the uranium exploration and pollution prevention [9–12]. When methods of in situ leaching are employed in the mining industry, the fracturing of the host rock dominates the behaviors of the orebody during the recovery of the mined uranium. Specifically, the effect of fracture geometric properties, including the density, orientation, aperture, and roughness, and the flow regime, mechanical degeneration, containment transport, and retention, was investigated considering the external reactive activities on a multi-scale [13–16]. By combining the latest technological achievements in equipment, such as hydraulic pumping, computed tomography, three-dimensional (3D) profilers, materials test systems, and rapid triaxial rock (RTR)-1000 [17–19], in the field of in situ measurement and experimental study, novel discoveries were made, and empirical formulas were derived [20–26]. Kim et al. studied the reactive transport of uranium with bacteria in fractured rock and conducted a sensitivity analysis, based on a developed model [27]. Boissezon et al. modelled the uranium and ^{226}Ra mobility during and after an acidic in situ recovery test [28]. Feng et al. developed a fractal discrete fracture network model to study the radon migration in fractured media [29]. Wang et al. revealed that the concentration of HCO_3^- increased after CO_2 injection, and CO_2 was able to migrate toward the shallow aquifer through existing leakage pathways, and availability of Fe^{3+} is the main factor that limits mineralized uranium release [30,31]. Baghbanan developed a nonlinear algorithm for predicting the normal stress–normal displacement behavior of fractures [32]. Rong conducted water flow tests through non-mated rough-walled fractures under normal stresses ranging from 1.0 MPa to 5.0 MPa [33]. Zhang experimentally investigated the fluid flow regimes through deformable rock fractures by conducting water flow tests through both mated and non-mated sandstone fractures in triaxial cells [34].

On the other hand, the algorithms and mathematical model contribute to visualization of the geochemical-physical coupling process [35–38]. Rutqvist presented a linked multicontinuum and crack tensor approach for modeling coupled geomechanics, fluid flow, and solute transport in fractured rocks [39]. Lei reported a stress-induced variable aperture model to characterize the effect of polyaxial stress conditions on the fluid flow in 3D persistent fracture networks and performed the geomechanical modeling of the fractured rock by the finite-discrete element method (FEMDEM) [40]. Zhao presented a closed-form solution for modeling the coupled stress–flow–transport processes along a single fracture embedded in a porous rock matrix [41]. Crandall related the macroscopic roughness parameters to the effective flow through fractures, examining the relationship between wall roughness and fluid flow in rock fractures [42]. Akhavan quantified the effects of crack width, tortuosity, and roughness on the water permeability of cracked mortars [43].

The in situ leaching uranium and contaminants control is a complex geomechanical and geochemical process, and the transport and retention of the uranium-containing solution through a fractured rock mass significantly affects the underground environment safety. The characterization of transport and retention of the in situ leaching uranium in a fractured rock mass is important. In this study, a stress-dependent reactive transport model, describing the coupling evolution of geofractures and uranium-containing solute transport in fractured rock mass, was established. The transport and distribution of the uranium-containing solution in a fractured rock mass, located at an overlying resource reservoir of ordos basin, were studied, combined with the FLAC3D-CFD simulator. The factors, including fracture morphology, confining stress, and loading path, that influence the uranium-containing transport and distribution were analyzed. In the following section, the mathematical model was described, the application of the established model was showed in Section 3, the results and analysis were conducted in Section 4, and the conclusions were made in Section 5.

2. Mathematical Model for Uranium Transport in Rough-Walled Fractures

The reactive transport of the in situ leaching uranium is a complex process of diffusion, adsorption, desorption, and oxidation–reduction, which was deeply influenced by the variation of component, structure, hydraulic gradient and in situ stress. The geomechanical, geochemical, and hydro-mechanical effect on the transport of uranium should be taken into account, as the reactive transport of uranium is modelled.

2.1. Equivalent Hydromechanical Coupling Description

The flow regimes and morphology of the uranium-containing solute in fractured rock mass is dominated by fluid kinematical viscosity and the fracture geometric property. Changes in the fractured rock structure induced by normal stress and shear stress are characterized by the normal displacement and shear dilation, as shown in Figure 1. The fracture aperture, as a function of normal closure, dilatational strain and shear expansion, is governed by the following equation:

$$b = h_i - \delta + \Delta b_{dil} + \varepsilon + b_0 \quad (1)$$

where δ is the normal closure; h_i, the maximum closure; ε, the dilation strain; Δb_{dil}, the shear expansion; and b_0, is initial aperture.

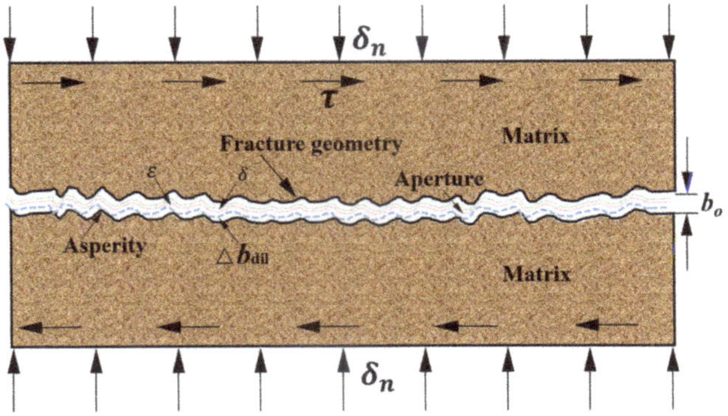

Figure 1. Schematic diagram of the fracture aperture model.

Considering the fracture geometry of contact area and connecting void, the equivalent mechanic aperture can be described as follows:

$$e = b(1 - 1.1w)^4(1 + \frac{2}{D})^{3/5} \quad (2)$$

$$w = w_1 + ne^{-\frac{\sigma_n}{k_n}} \quad (3)$$

where b is the mechanical aperture; e, the equivalent fracture aperture; w, the contact area; w_1, the initial contact area; n, the normal direction; D, the fractal dimension of the connecting void. In this work, the ultra-thin square plate (UTSP) covering method was used to calculate the fractal dimension, as shown in (Figure 2) [44].

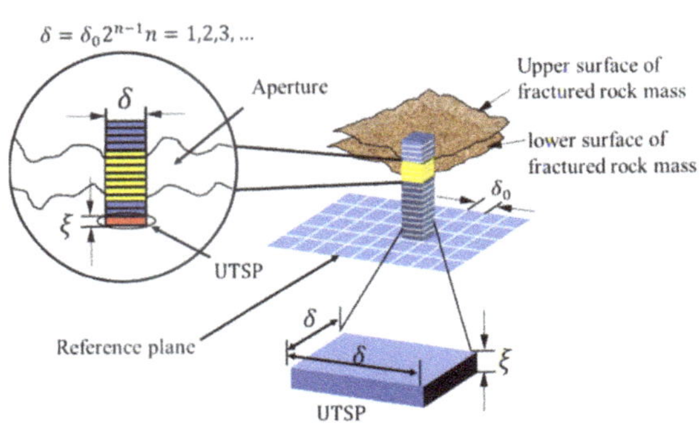

Figure 2. Schematic diagram of the UTSP covering method.

The description of aperture closure in the normal direction was studied by Bandis [45] in 1983 and progressively modified by Baghbanan and Jing [32], as shown below:

$$\delta = \frac{9\sigma_n h_i}{\sigma_{nc} + 10\sigma_n} \tag{4}$$

where σ_n is the normal stress; σ_{nc}, the critical normal stress to end-closed fracture; and h_i, the maximum closure. Furthermore, the normal stiffness is expressed as follows:

$$k_n = \frac{(10\sigma_n + \sigma_{nc})^2}{6\sigma_{nc} b_0} \tag{5}$$

where K_n is the normal stiffness. In addition, the effect of the shear stress on the fracture dilation was studied by Chen [16], and the following specific expression was obtained:

$$\Delta b_{dil} = \frac{1}{b_0}\left\{\frac{\psi_{\text{peak}}}{r}\left[1 - e^{-r(\delta-\delta_0)}\right] + \frac{\psi_{\text{peak}}^3}{9r}\left[1 - e^{-3r(\delta-\delta_0)}\right]\right\} \tag{6}$$

$$\psi_f^{\text{peak}} = \text{JRC}\log_{10}\frac{\text{JCS}}{-\sigma'_{zf}} \tag{7}$$

where Δb_{dil} is the fracture dilation caused by shear stress; JRC, the roughness coefficient of the fracture; JCS, the wall compressive strength of fracture; ψ_f^{peak}, the peak dilatancy angle of the fracture.

Furthermore, the dilation strain caused by the geochemical reaction is shown:

$$\varepsilon = \begin{cases} at^b & 0 < t < t_0 \\ at_0 & t > t_0 \end{cases} \tag{8}$$

where a and b are the fitting parameters related to the attribute of the rock mass; t_0, the threshold of dilatational strain; t, the experiment time.

On the other hand, structural changes in the fractured rock mass contribute to its self-mechanical attribute. To express the anisotropic and heterogeneous rock mass strength, the dynamic equivalent bulk modulus and shear modulus related to the fracture and matrix strength are governed by the following equation in the work of Gan [46]:

$$K = \frac{1}{\frac{1}{K_{int\ act}} + \sum^{fracnum} \frac{2V_{ratio}}{b}\left[\left(\frac{1}{K_{nf}} - \frac{1}{K_{sf}}\right)\left(1 - n_2^4\right) + \frac{1}{K_{sf}} n_1^2\right]} \tag{9}$$

$$G = \frac{1}{\frac{1}{G_{int\ act}} + \sum^{fracnum} \frac{2V_{ratio}}{b}\left[\left(\frac{1}{K_{nf}} - \frac{1}{K_{sf}}\right)\left(n_1^4 - n_1^2 n_2^2\right) + \frac{1}{K_{sf}} n_1^2\right]} \tag{10}$$

where K_{intact} is the bulk modulus of the intact rock; G_{intact}, the shear modulus of the intact rock; and V_{ratio}, the volumetric ratio of the truncated fracture over the element volume.

2.2. *Flow Regime in Fractured Structures*

In order to determine the flow state in a rough-walled fracture, the Darcy flow, non-Darcy flow, and turbulent flow subjected to changes in hydraulic gradient and fracture geometry were considered. The Forchheimer equation, given in the form shown in Equation (11), has been widely used as the empiric and theoretic formula to describe the flow behavior in porous media and rock fractures [15].

Forchheimer equation is described as:

$$-J = Av + Bv^2, \tag{11}$$

where J is the pressure gradient in the flow direction, and v, the seepage velocity. Furthermore, A and B are the coefficients describing energy losses due to viscous and inertial dissipation mechanisms, respectively. A is calculated by $\frac{\mu}{k}$, and B is expressed as $\beta\rho$, where μ is the dynamic viscosity of fluid; k, the intrinsic permeability; β, the non-Darcy coefficient; and ρ, the fluid density. Taking the equivalent mechanic aperture into account, the intrinsic permeability vector k_{ij} and non-Darcy coefficient vector β_{ij} are expressed as:

$$k_{ij} = \sum^{fracnum} \frac{1}{12}(P_{kk}\delta_{ij} - P_{ij}) = \sum^{fracnum} \frac{1}{12}\left(\frac{V_{ratio}}{b_{ini}} e^3 n_k^2 \delta_{ij} - \frac{V_{ratio}}{b_{ini}} e^3 n_i n_j\right) \quad (12)$$

$$\beta_{ij} = \sum^{fracnum} A_{ij} p_{ij} / \left(2^{b_{ij}} e^{b_{ij}+1}\right) \quad (13)$$

where A_{ij} and b_{ij} are dimensionless regression coefficients; P_{ij} represents the influence of asperity height; b_{ini} is the initial aperture of fracture.

2.3. Reactive Transport in Fractured Rock Mass

For in situ leaching of uranium, chemical reactions between the alkaline leaching solute and solid uranium can be described using the following expression:

$$\underset{I}{\underline{UO_2(S)}} + \underset{II}{\underline{\tfrac{1}{2}O_2(aq)}} + \underset{III}{\underline{CO_3{}^{2-}(aq)}} + \underset{IV}{\underline{2HCO_3{}^{-}(aq)}} \rightarrow \underset{V}{\underline{UO_2(CO)_3{}^{4-}(aq)}} + \underset{VI}{\underline{H_2O(l)}} \quad (14)$$

Variations in the components of the reactive transport are expressed as follows:

$$\partial_t c_{II} + \nabla(c_{II} U) = -\frac{\sigma(c_s)}{\rho\varphi} R_{II} \quad (15)$$

$$\partial_t c_{III} + \nabla(c_{III} U) = -\frac{\sigma(c_s)}{\rho\varphi} R_{III} \quad (16)$$

$$\partial_t c_{IV} + \nabla(c_{IV} U) = -\frac{\sigma(c_s)}{\rho\varphi} R_{IV} \quad (17)$$

$$\partial_t c_{V} + \nabla(c_{V} U) = -\frac{\sigma(c_s)}{\rho\varphi} R_{V} \quad (18)$$

$$\partial_t c_I = -\frac{\sigma(c_s)}{\rho_s(1-\varphi)} R_I \quad (19)$$

where C_I is the uranium oxide grade; C_{II}, C_{III}, C_{IV}, and C_V are the mass fractions of the solute; U is the transport velocity; $\sigma(c_S)$ is the effective reaction area of the uranium ore; φ is the porosity; R_I is the source of uranium oxide; and R_{II}, R_{III}, R_{IV}, and R_V are the solute source terms.

The migration of multispecies in a fluid flow is governed by the following reactive advection–dispersion equation:

$$\frac{\partial \alpha_l \rho_l \varphi_l^k}{\partial_t} + \nabla \cdot \left(\alpha_l \rho_l \vec{u_l} \varphi_l^k - \alpha_l \Gamma_l^k \cdot \nabla \cdot \varphi_l^k\right) = S_l^k k = 1, \ldots, N \quad (20)$$

where ${\varphi_l}^k$ is the component of the scalar k, defined as the ratio of k component mass to phase-l; α_l is the volume fraction, defined as the ratio of phase-l volume to the solution volume; ρ_l, and u_l are the density, and velocity of the species-l, respectively; and ${\Gamma_l}^k$ and ${S_l}^k$ are the diffusion coefficient and source item, respectively.

2.4. Transport Kinetic Equation

The governing equation of fluid flow in fractured rock followed Navier–Stokes (N-S) equations.

Continuity equation:

$$\frac{\partial \rho_f}{\partial t} + \nabla \cdot \left(\rho_f \mathbf{u}\right) = 0 \tag{21}$$

Momentum conservation equation:

$$\frac{\partial\left(\rho_f \mathbf{u}\right)}{\partial t} + \nabla \cdot \left(\rho_f \mathbf{u}\right)\mathbf{u} = -\nabla p + \nabla \cdot \left[\mu_f\left(\nabla \mathbf{u} + \nabla \mathbf{u}^T\right) - \frac{2}{3}\mu_f(\nabla \cdot \mathbf{u})I\right] \tag{22}$$

where ρ_f is the fluid density, $\mathbf{u}$ is the fluid velocity, p is the fluid pressure, μ_f is the hydrodynamic viscosity, I is the identity matrix.

2.5. Integrated Reactive Transport Model

Considering the structural change caused by physicomechanical effects, the transport of the uranium-containing solute through a rough-walled fracture is given as:

$$\begin{cases} UO_2(S) + \frac{1}{2}O_2(aq) + {CO_3}^{2-}(aq) + 2{HCO_3}^{-}(aq) \rightarrow UO_2{(CO)_3}^{4-}(aq) + H_2O(l) \\ \frac{\partial \rho_f}{\partial t} + \nabla \cdot \left(\rho_f \mathbf{u}\right) = 0 \\ \frac{\partial(\rho_f \mathbf{u})}{\partial t} + \nabla \cdot \left(\rho_f \mathbf{u}\right)\mathbf{u} = -\nabla p + \nabla \cdot \left[\mu_f(\nabla \mathbf{u} + \nabla \mathbf{u}^T) - \frac{2}{3}\mu_f(\nabla \cdot \mathbf{u})I\right] \\ \frac{\partial \alpha_l \rho_l \varphi_l^k}{\partial_t} + \nabla \cdot \left(\alpha_l \rho_l \vec{u_l} \varphi_l^k - \alpha_l \Gamma_l^k \cdot \nabla \cdot \varphi_l^k\right) = S_l^k, k = 1, \ldots, N \\ K = \dfrac{1}{\frac{1}{K_{int\ act}} + \sum^{fracnum} \frac{2V_{ratio}}{b}\left[\left(\frac{1}{K_{nf}} - \frac{1}{K_{sf}}\right)(1 - n_2^4) + \frac{1}{K_{sf}} n_1^2\right]} \\ G = \dfrac{1}{\frac{1}{G_{int\ act}} + \sum^{fracnum} \frac{2V_{ratio}}{b}\left[\left(\frac{1}{K_{nf}} - \frac{1}{K_{sf}}\right)(n_1^4 - n_1^2 n_2^2) + \frac{1}{K_{sf}} n_1^2\right]} \\ -J = \frac{\mu}{k}v + \beta \rho v^2, \\ k_{ij} = \sum^{fracnum} \frac{1}{12}\left(P_{kk}\delta_{ij} - P_{ij}\right) = \sum^{fracnum} \frac{1}{12}\left(\frac{V_{ratio}}{b_{ini}} e^3 n_k^2 \delta_{ij} - \frac{V_{ratio}}{b_{ini}} e^3 n_i n_j\right), \quad (14) \\ \beta_{ij} = \sum^{fracnum} A_{ij} p_{ij} / \left(2^{b_{ij}} e^{b_{ij}+1}\right). \end{cases}$$

2.6. Simulation Scheme

During the performance the developed model, the transformation script was edited using the fish and C language to call the FLAC3D and CFD. The stress tensor, hydraulic pressure, and fracture geometry were transferred between FLAC3D and CFD. The specific procedure is as follows, and scheme is shown in Figure 3.

The mesh model was established in the FLAC3D and CFD firstly, according to the simulation object. Then, the mechanical calculation was conducted in the FLAC3D, and a corresponding strain tensor in each block was obtained based on the Mohr–Coulomb criterion, and the convergence of the calculation is set as 10^{-5}; The strain tensor was transferred into the corresponding block in the CFD model based on the transformation script, and the permeability was defined based on Equations (12) and (13), and the pressure differences and fluid velocity were changed based on Equations (21) and (22). Simultaneously, the diffusion of the uranium-containing solution was re-calculated, combining the source of chemical solution. The calculation was conducted until the setting time. Finally, the fluid pressure was transferred into the corresponding block in FLAC3D through the UDS and fish script, and the effective stress was modified considering the fluid pressure. Then, the strain tensor was modified based on the Mohr–Coulomb criterion, and the convergence of the calculation was setup as 10^{-5}. Then, the next cycle was conducted, based on the above scheme.

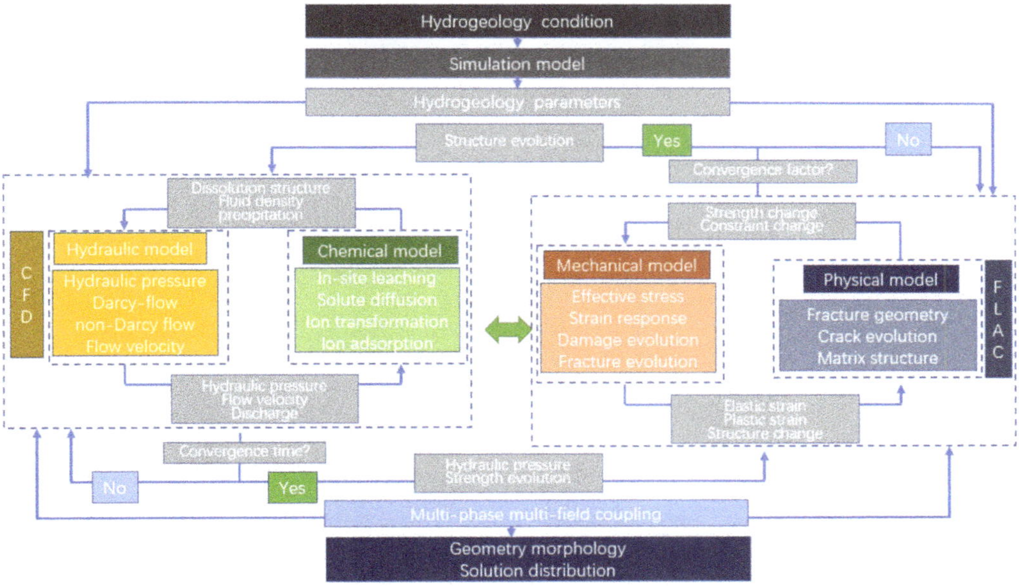

Figure 3. The simulation schematic.

2.7. Model Verification

To validate the application of the established model in the fractured rock mass, a results comparison between the model calculation and experiment work of Chen et al. [1] was conducted. In more detail, the sample is granite collected from a potential site for China's high-level radioactive waste disposal repository, with a density of 2.64–2.68 g/cm^3, porosity of 0.44–0.62%, uniaxial compressive strength of 97–161 MPa, and permeability of 10^{-20} and 10^{-19} m^2. The model with a diameter of 50 mm and length of 100 mm was established, and the fractured rock samples with mean asperity height of 0.62 mm, 0.95 mm, 1.3 mm, 1.5 mm, and JRC of 6.0, 7.1, 8.8, 10.1 were collected. The seepage experiment subjected to a series of hydraulic pressure ranging from 0.25 to 6.4 MPa was conducted at a certain boundary stress, and the corresponding boundary stress increases from 5 to 25 MPa. The geometry parameter, including asperity, aperture, and JCR, was transferred into the FLAC3D-CFD simulation. Detailed information about the relationship between the seepage velocity and hydraulic pressure difference is presented in Figure 4. The curve is the calculated result, and the point is the experimental result. Based on the results, it illustrates that the stress-dependent fracture geometry and hydraulic pressure-dependent flow regime can be well expressed. As boundary stress increased, the seepage channel decreased and corresponding hydraulic pressure difference increased, while the increase of hydraulic pressure promoted the flow transferred from Darcy flow to non-Darcy flow. As shown in Figure 4, the theoretical prediction is almost consistent with the experiment data with the R^2 (The coefficient of determination R^2 evaluates the accuracy of the match and represents the degree to which the regression line fits the data. The agreement is best at R^2 = 1.0) of 0.97–0.99. The seepage behavior and spatiotemporal coupling of stress–fracture–seepage in fractured rock masses can be well described using the proposed model.

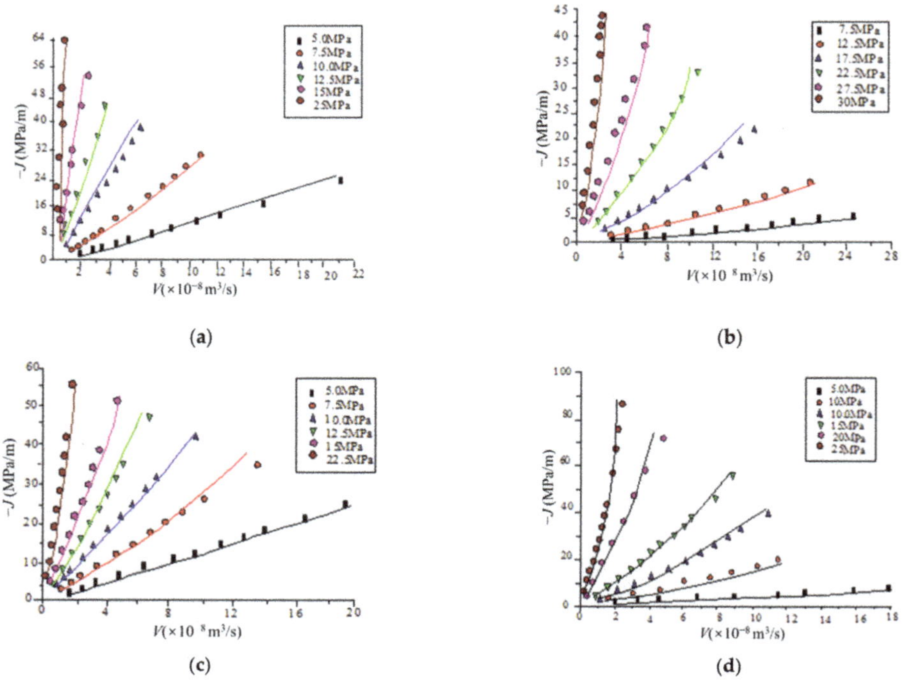

Figure 4. Comparison of the fracture model calculating and the experimental results: (**a**) sample 1, (**b**) sample 2, (**c**) sample 3, (**d**) sample 4.

3. Case Study

The Nalinggou uranium and Tarangaole coal mine is an overlay resources reservoir of Ordos Basin, in which the uranium is located at 410 m depth with the grade of 0.5%, and coal is located at a 600 m depth. The mathematical model was applied to simulate the migration of the uranium from a uranium layer to coal seam. The specimens were collected from the overlay resources reservoir, and artificial fractures were created. The fracture geometry, including the roughness, asperity, and aperture, were captured by a 3D laser scanner, as shown in Figure 5, and the geometry parameters of the experiment fracture are listed in Table 1. The fluid and mechanical parameter was confirmed based on the in situ leaching technology and uranium-containing layer. The model with dimensions of 100 mm length and 50 mm width was established in a FLAC3D-CFD simulator. Then, the hydromechanical parameters were imported, and radioactive transport and morphology were calculated. In more detail, the boundary was constrained by the confining stress of 16 MPa, and hydraulic pressure is set as 0.5–3.5 MPa, and the mass fraction of uranium-containing solution $UO_2(CO_3)_3{}^{4-}$ is set as 5×10^{-4} at the fracture inlet. Additionally, the preparation process is shown in Figure 6, and the mechanical and chemical parameters used in the numerical calculation are listed in Table 2.

Table 1. The geometry parameters of the experiment fracture.

No.	Fracture Length L (mm)	Fracture Width w (mm)	Contact Ratio (5 MPa)	Fracture Aperture (μm) (5 MPa)
1	100	49	0.20	1.64×10^{-5}
2	100	49	0.25	3.88×10^{-5}
3	100	49	0.25	2.47×10^{-5}
4	100	49	0.30	8.38×10^{-6}
5	100	49	0.26	8.86×10^{-6}
6	100	49	0.25	3.00×10^{-6}

Contact ratio: the area ratio of contact zone to rough-walled fracture.

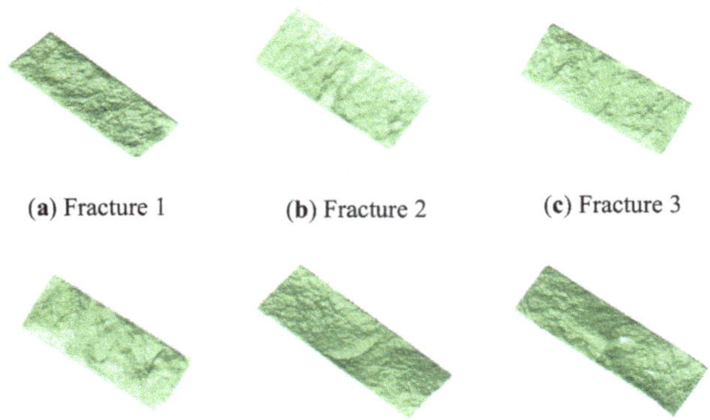

(**a**) Fracture 1 (**b**) Fracture 2 (**c**) Fracture 3

(**d**) Fracture 4 (**e**) Fracture 5 (**f**) Fracture 6

Figure 5. Surface scanning image of the sandy mudstone fracture.

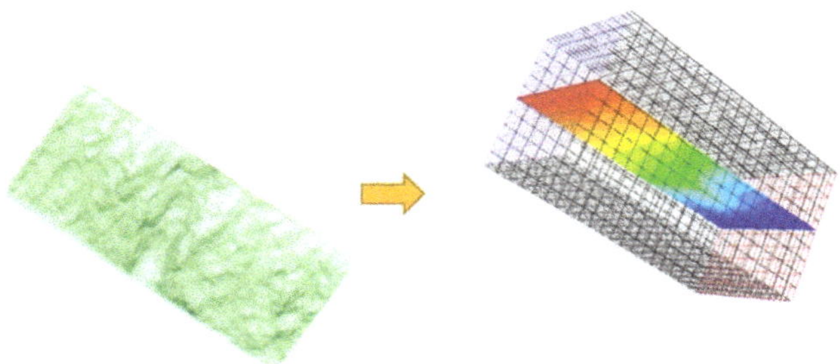

Figure 6. Transformation process of the fractured rock mass.

Table 2. The hydraulic and mechanic parameters of the fractured rock mass.

Stress Field	Density d (kg/m^3)	Bulk Modulus B (GPa)	Shear Modulus S (GPa)	Cohesion C (MPa)	Tensile Strength t (MPa)	Internal Friction Angle ψ (°)	Initial Permeability K (m^2)
	2660	33.94	22.4	4.0	22.52	35	
Chemical field	$UO_2(CO_3)_3{}^{4-}$		reaction rate (kg/m^{3}s)		Dispersion coefficient (m^2/s)		
	5.0×10^{-4}		10.5		0		
Fracture field	Initial mean fracture width (μm)		Initial permeability (m^2)		Initial non-darcy flow factor β (m^{-1})		
	30		7.0×10^{-12}		1.0×10^8		

Transport and Distribution

The fractured rock mass with an asperity of 0.62 mm, JRC of 6.0, and dimensions of 100 mm length × 500 mm diameter were collected. To analyze the effect of hydraulic pressure on uranium-containing solute transport and distribution, the boundary stress was fixed at 16 MPa, and the hydraulic pressure was set as 0.5, 1.5, 2.5, and 3.5 MPa, respectively. The fracture morphology was formed by the sealed and connected voids and contact areas, and variation of hydraulic pressure and uranium-containing solute occurred in the connected channel, as shown in Figure 7. In more detail, the transport path of uranium-containing solute was heterogenetically distributed, characterized by the regional concentration and dissipation of the uranium-containing solute in a fracture area.

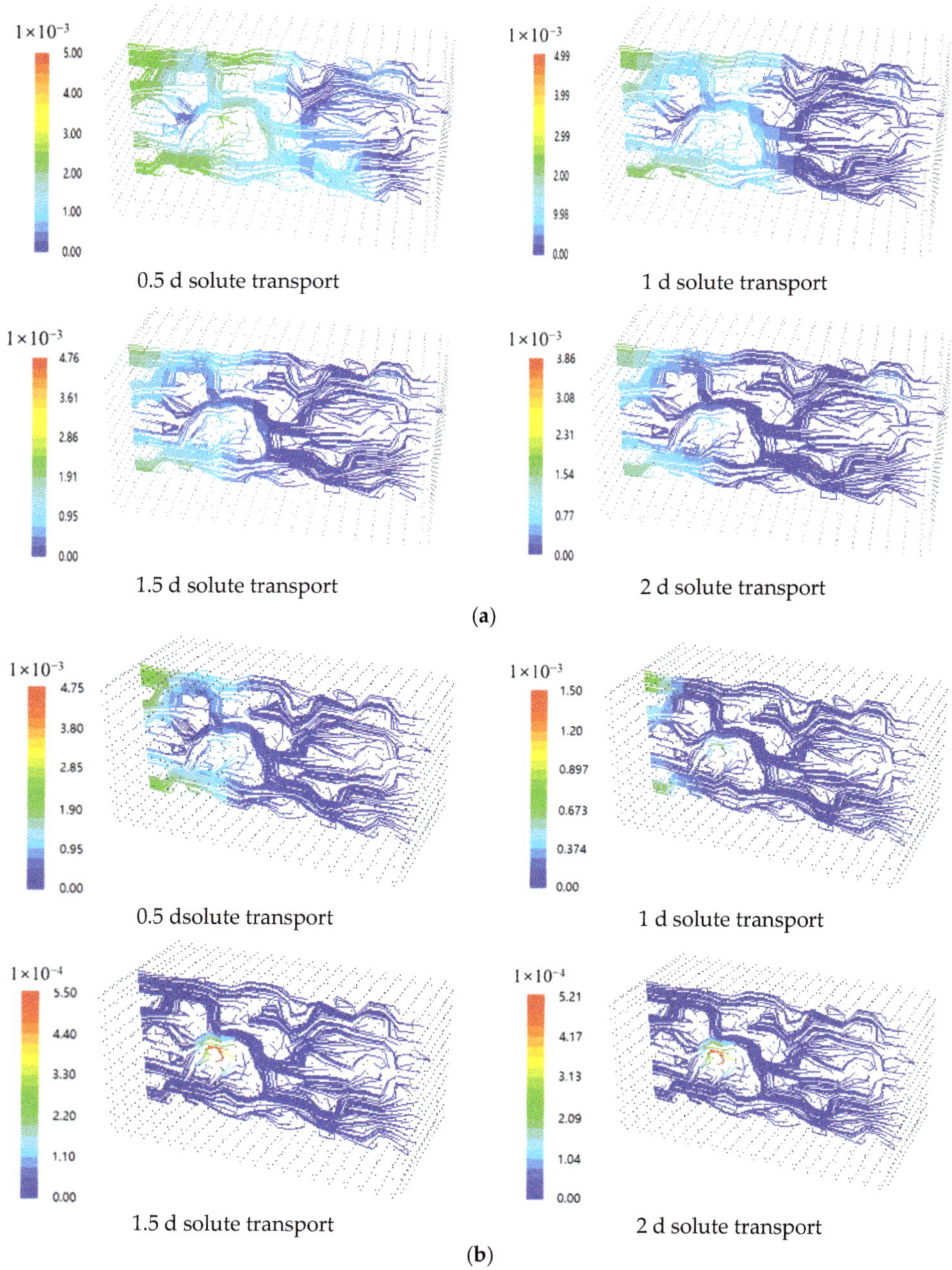

Figure 7. *Cont.*

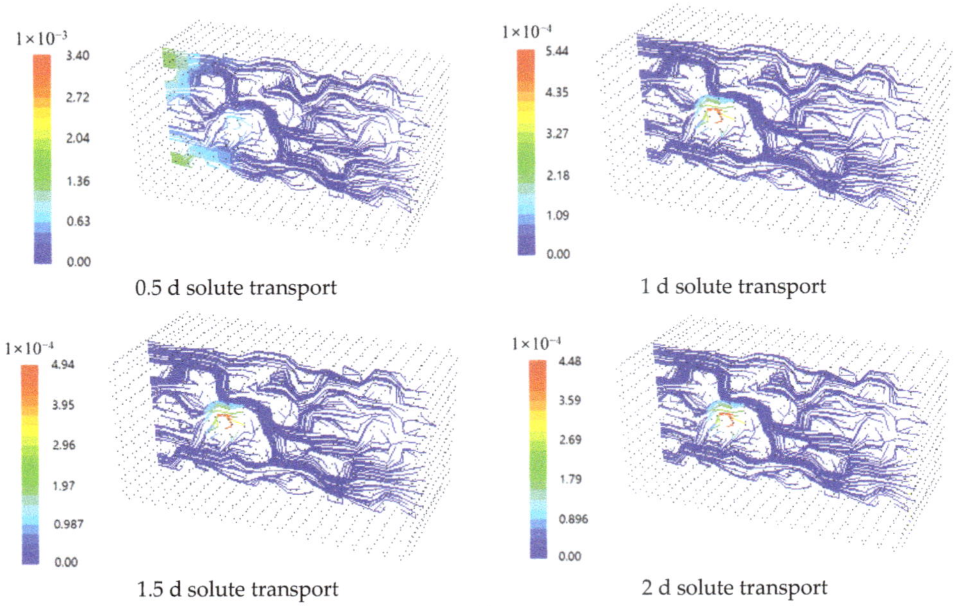

(c)

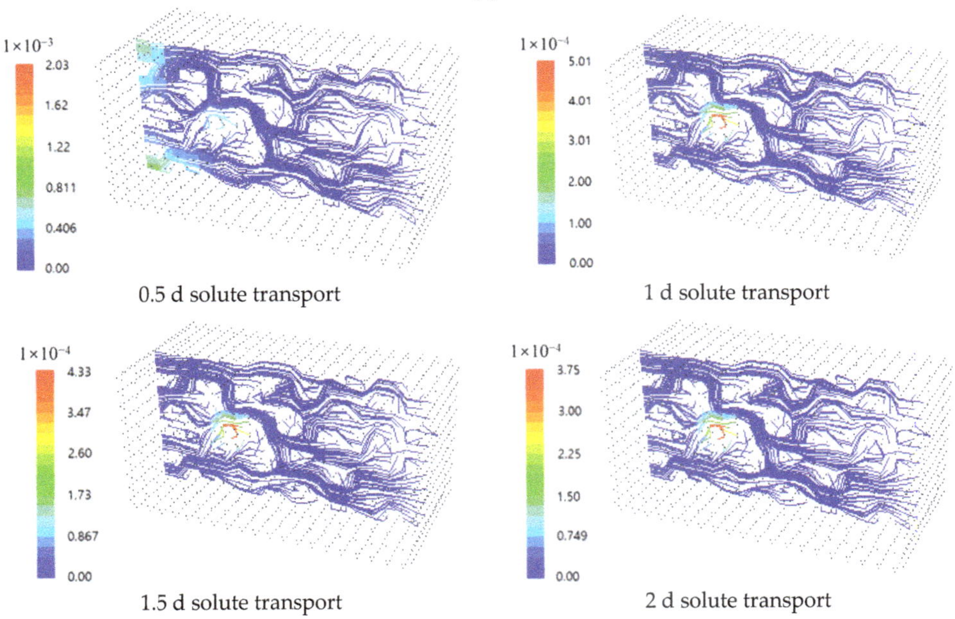

(d)

Figure 7. Fracture solute transport condition under different hydraulic pressures: (**a**) hydraulic pressure of 0.5 MPa, (**b**) hydraulic pressure of 1.5 MPa, (**c**) hydraulic pressure of 2.5 MPa, (**d**) hydraulic pressure of 3.5 MPa.

The rapid and following slow decrease in the uranium-containing solute responds to the continuing fluid flow in fracture. As the hydraulic pressure increased, the uranium-containing solute transport was increased, and 2 days, 1.5 days, 1 day, and 0.5 days were required for the uranium-containing solute migration in the hydraulic pressure of 0.5, 1.5, 2.5, and 3.5 MPa, respectively. In addition, the local retention of the uranium-containing solute presented in low-connected voids, combined with a fluid vortex phenomenon. The hydraulic pressure promoted the improvement of the void connectivity in the fracture, and corresponding seepage velocity and solute migration through the connected voids were increased.

4. Sensitivity Analysis

4.1. Confining Stress

The effect of confining stress on the solution migration and distribution was conducted, subjected to the boundary stress of 5–19 MPa and hydraulic pressure of 3.5 MPa. As shown in Figure 8, as the confining stress increases, the fracture aperture decreases in a negative exponential form, and the contact area ratio (the ratio of contact area in fracture to over total areas of fracture) increases in a logarithmic form.

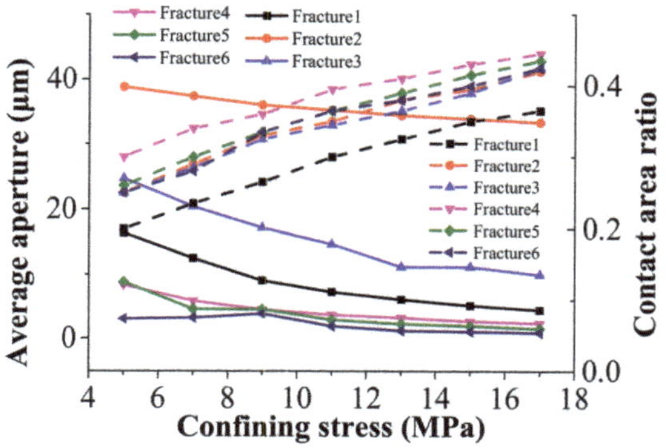

Figure 8. Correlation between confining stress, average aperture, and contact area ratio.

As the confining stress increases, the main seepage channels gradually decrease, whereas small seepage channels propagated along the fracture. The increase of confining stress triggered the decrease of the maximum average aperture from 5–40 μm to 1–35 μm, and dynamic and slight decrease corresponds to the confining stress range of 5–13 MPa and 13–17 MPa. In addition, a void connectivity decrease and contact area increase, responding to the increase of confining stress, were observed.

The different magnitude changes in contact area and fracture aperture occurred. In more detail, the larger contact area is associated with a smaller aperture, and dynamic fracture closure is presented in a larger initial aperture fracture.

4.2. Fracture Aperture

Figure 9a shows that the seepage velocity increases with the increase of fracture aperture. For the same average aperture, the seepage velocity increases slightly corresponding to an average aperture of 0–10 μm, and differences were observed in the seepage velocity for average apertures of 5 to 40 μm.

In Figure 9a, high seepage was observed in fracture4 with an 8 μm aperture. Meanwhile, fracture2 has a larger average aperture for the same seepage velocity. Figure 9b shows that the seepage velocity decreases with an increase in the fracture contact area ratio, unlike the dynamic decrease of the seepage velocity subjected to the increase of the contact area ratio from 0.2 to 0.35. In this case, a slight change in the seepage velocity corresponding to an increase in the contact area ratio from 0.35 to 0.45 was observed. The seepage velocity for the initial contact area ratio was 5–40 times that of the final state for a contact area ratio of 0.43; this observation indicates that the influence of the contact area ratio on the seepage velocity differs depending on the stages.

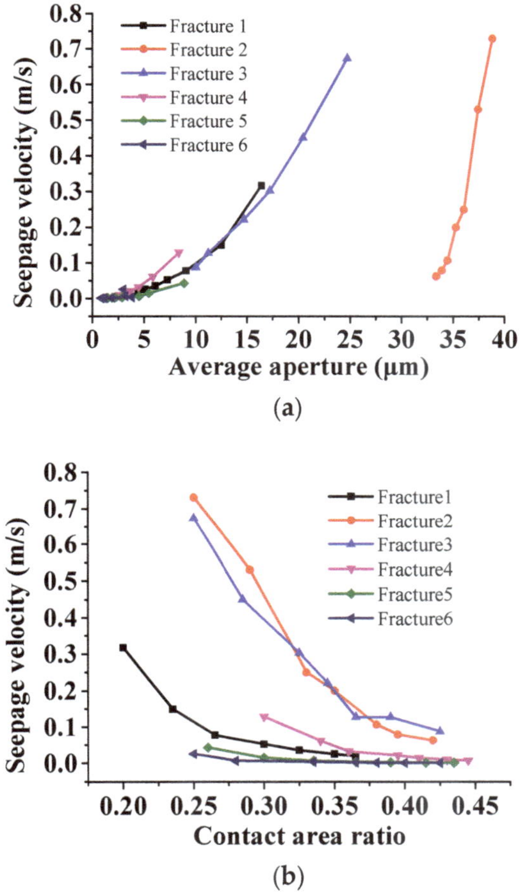

Figure 9. Correlation of seepage velocity with average aperture and contact area ratio. (**a**) correlation between average aperture and seepage velocity, (**b**) correlation between contact area ratio and seepage velocity.

In addition, for a certain contact area ratio, the seepage velocities of fracture2 and fracture3 were 5–8 times those of fracture1, fracture4, fracture5, and fracture6. On the other hand, the contact area ratios were different for all fractures, and fracture1 had small initial contact area ratio, and fracture4 had a large initial contact area ratio. The contact area ratio and fracture tortuosity also influenced the distribution of the connected voids, leading to differences between the seepage velocities of fracture1 and fracture2, thereby indicating that the average aperture is not the only factor dominating fracture seepage.

Figure 10 shows that the solute concentration changes with an increase in the average aperture, and an increment of 1×10^{-4} to 5×10^{-4} was observed for an increment of 2–5 μm in the average aperture; then, the solute concentration decreased by 1×10^{-4} to 10×10^{-4} in the following increment of 5–10 μm in the average aperture was observed. These observations imply that there exists a threshold in the fracture for the transportation between the lower and higher solute concentrations. The dynamic migration of the uranium-containing solute was presented in the early 12 h, and the migration seepage generally slowed in the later 12 h.

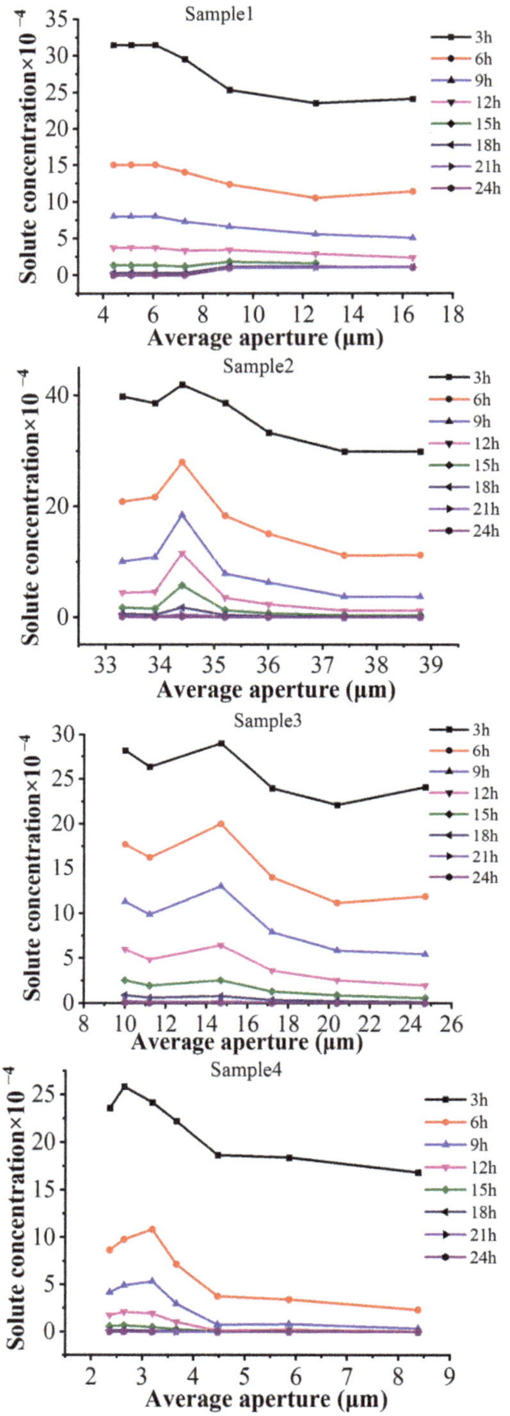

Figure 10. Relation between the solute concentration and time.

Considering fracture2 as an example, the characteristic of the solute migration is shown in Figure 11. The residual solute concentration varies spatiotemporally with changes in the average aperture and time. The residual solute is anisotropically distributed and the concentration decreased along the fracture. High and low solute concentrations were distributed at the inlet and outlet of the fracture, respectively. Furthermore, solute concentration decreased dynamically in the first 6 h, and slightly decreased during the following 6 h, and the morphology of the solute differed for different average apertures. This observation indicates that the transport of time-dependent uranium-containing solute is highly sensitive to the changes in the average aperture.

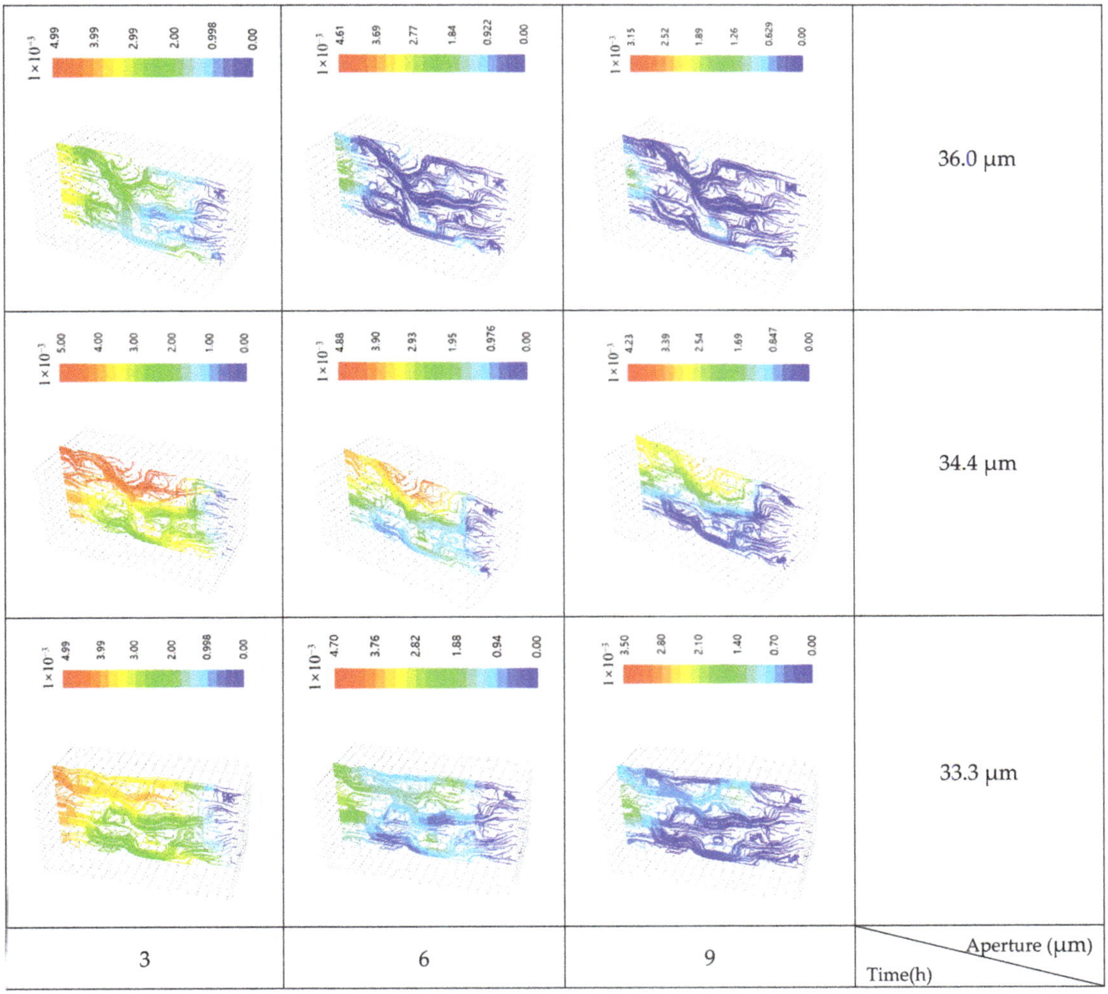

Figure 11. Characteristics of solute transport under different stresses.

4.3. Cross Channel of the Fracture

The experiment of fluid seepage and uranium-containing solute transport in cross channel of the fracture was conducted, based on the double-fracture model. Compared with the single-fracture model, the dual-fracture model adds a fracture perpendicular to the fracture mentioned above. The characteristics of seepage flow and uranium solute transport were studied in the scenarios of $k = 0$, $k = 0.5$, and $k = 1$ (the ratio of vertical stress loading and horizontal stress unloading, the horizontal stress direction is the direction perpendicular to the crack surface).

For the scenario of $k = 0$, the 18.3 MPa axial stress and 18.3 MPa, 14.3 MPa, 10.3 MPa, 7.3 MPa, and 6.3 MPa confining stresses were successively applied. As shown in Figure 12, the maximum seepage velocity is negatively related to the confining stress in multi-fractured rock masses, and that the maximum seepage velocities are 2.29×10^{-2}, 4.42×10^{-2}, 6.09×10^{-2}, 9.66×10^{-2}, and 1.11×10^{-1} corresponding to 18.3 MPa, 14.3 MPa, 10.3 MPa, 7.3 MPa, and 6.3 MPa confining stress, respectively. As the confining stress increases, the number and width of the flow paths for solute migration anisotropically decreased, and corresponding fracture permeability was decreased. The block and ribbon-shaped uranium-containing solute areas gradually decreased in inclined and vertical fractures. The local retention of the uranium-containing solute induced by the vortex and inertial flow caused by high-speed percolation and heterogeneous fracture morphology occurred in a short period; furthermore, slow solute release was observed under weak convection in a low permeability region.

As the confining stress decreases, the width of the double-fracture increases, while the solute concentration in the fracture decreases rapidly under the convection of higher-velocity fluids, and nonuniformly distributed block and ribbon-shaped solute concentration areas appeared in the fracture area. As the aperture increases, the influence of the asperity and roughness on the fluid flow state gradually decreases, and the vortex and inertial flow phenomena gradually evolved into laminar flow. Compared with the solute morphology under high confining stress, the time factor dominated the solute distribution under lower confining stress. With the increase of time, the number and area of solute-intensive areas decreased rapidly.

In the fracture intersections region, low-concentration solutes present a concentrated distribution, and the distribution area gradually increases as the confining stress decreases. Compared with the rough fracture surfaces, low-concentration cross fracture areas are more sensitive in smooth fractures.

4.4. Loading Path

Figure 13a–c show that the apertures of inclined fractures and vertical fractures exhibit different responses to the increase in confining stress at the same k. When $k = 0$, the confining pressure increases from 6.3 MPa to 18.3 MPa, the apertures of vertical fractures decrease from 31.4 to 24.2 μm, and the apertures of inclined fractures remain unchanged. Similar phenomena are also observed at $k = 0.5$ and $k = 1$, the confining pressure increases from 9.3 MPa to 18.3 MPa, the apertures of the vertical fractures decreased by 5.1 μm and 5.2 μm, and the apertures of inclined fractures decreased by 0.51 μm and 0.54 μm, respectively. As the confining stress increases, apertures of inclined fractures experienced an initial increase and then a dynamic decrease, while apertures of vertical fractures maintained a uniform decrease. This observation indicates that the k value affects different morphological fractures differently. As k increases, the apertures of inclined fractures decrease, while those of vertical fractures remain more or less unchanged. When the confining stress is 10.3 MPa and k increases from 0.5 to 1, the apertures of inclined fractures decrease by 7.3 μm, while those of vertical fractures only decrease by 0.7 μm. In addition, when $k = 0$, the apertures fluctuation amplitude of inclined fractures is 2 μm, while the apertures fluctuation amplitude of inclined fractures is 9.1 μm and 4.3 μm for $k = 0.5$ and $k = 1$, respectively. While apertures fluctuation amplitude vertical fractures remain more or less unchanged with the increases of k. This observation indicates that the inclined fractures are more sensitive than vertical fractures to changes of k. Thus, changes in axial stress have a more significant effect on inclined fractures than on vertical fractures.

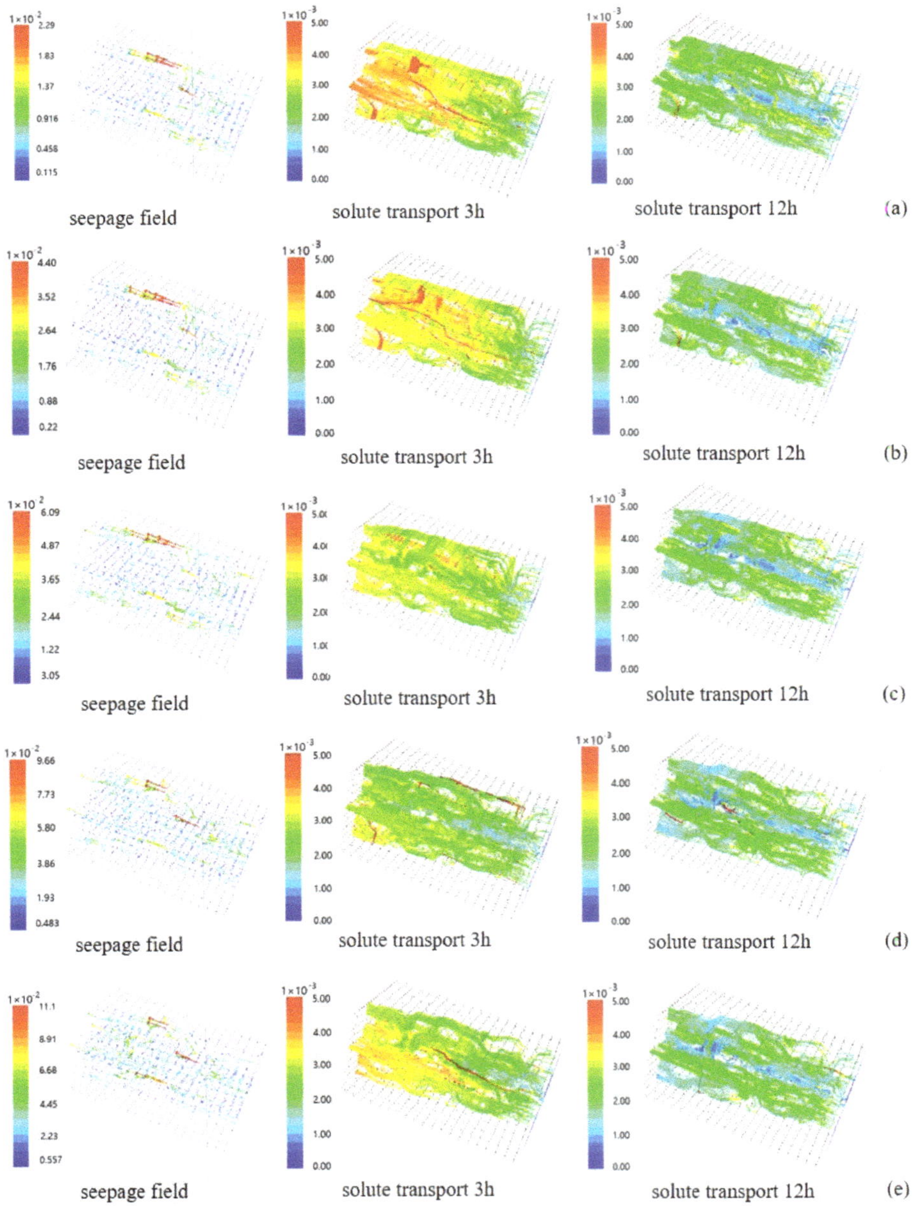

Figure 12. Seepage solute transfer characteristics under different horizontal vertical stress at $k = 0$ (**a**) confining stress 18.3 MPa, axial stress 18.3 MPa, (**b**) confining stress 14.3 MPa, axial stress 18.3 MPa, (**c**) confining stress 10.3 MPa, axial stress 18.3 MPa, (**d**) confining stress 7.3 MPa, axial stress 18.3 MPa, (**e**) confining stress 6.3 MPa, axial stress 18.3 MPa.

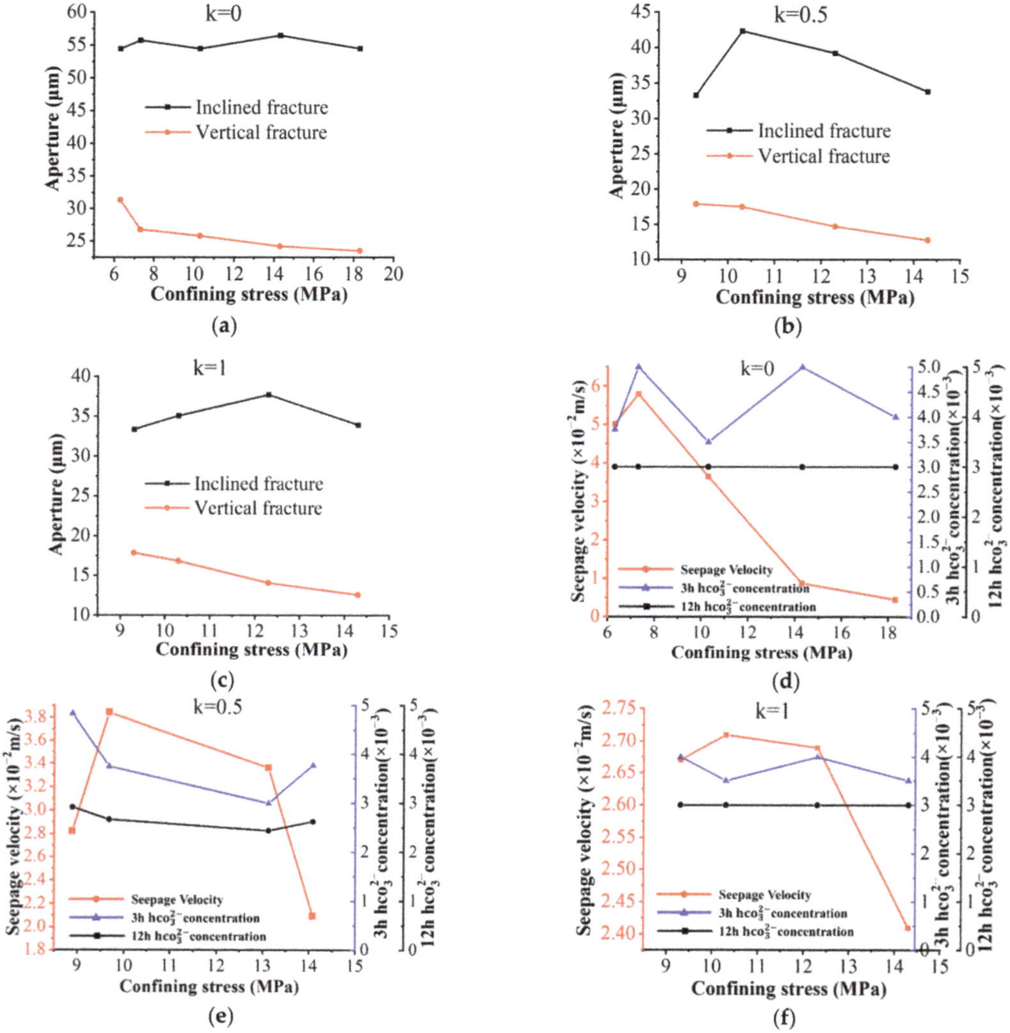

Figure 13. Relationship of the confining stress, aperture, seepage velocity, and concentration in different loading conditions. (**a**) confining stress for a k = 0, (**b**) confining stress for a k = 0.5, (**c**) confining stress for a k = 1, (**d**) the seepage velocity for a k = 0, (**e**) the seepage velocity for a k = 0.5, (**f**) the seepage velocity for a k = 1.

Figure 13d–f show that the seepage velocity of the double-fracture increases with the decrease in confining stress for a certain k. A fluctuation of the solute concentration induced by the increase of confining stress in solute concentration was observed at 3 h. This indicates that local retention and accelerated regional migration occur as solute is transported through nonuniform interconnected voids, resulting in the fluctuation of the solute concentration subjected to changes of confining stress. As k increases, the seepage velocity generally decreases, and the confining stress corresponding to the peak value of the seepage velocity increases from 8 MPa for k = 0 to 12.5 MPa for k = 1. In addition, as k increases, the fluctuation of the solute concentration responding to the change of confining stress gradually decreases.

5. Conclusions

Combined with the developed mathematical model and FLAC3D-CFD simulator, the migration of uranium-containing solute in fractured rock masses was studied, and the influence of the sensitive factor on the evolution of fracture and uranium-containing solution was analyzed, and the main achievements are as follows:

(1) A mathematical model describing the stress-dependent fracture structure and uranium-containing solute transport was established. The evolution of a connected channel in fracture aperture is influenced by the increase of confining stress, and a dynamic and slight decrease zone was confirmed for confining stress of 5–13 MPa and 13–17 MPa. The concentration of uranium-containing solution is directly influenced by the fracture aperture and hydraulic pressure, and a 2–5 μm fracture aperture was identified as a width threshold from a lower to higher uranium-containing solute concentration.

(2) The number and size of a connected channel decreased with the increase of confining stress in double-fractures. The turbulent flow was presented in a high fluid velocity and confining stress condition, and retention of uranium-containing solution characterized by block and ribbon-shaped solute concentration areas was observed in both inclined and vertical fractures, and dynamic decrease of uranium-containing solution was presented at the fracture intersections region. The dynamic decrease presented in the initial 12 h, and a slight decrease presented in the following period.

(3) The loading and unloading direction and rate significantly influence the fracture geometry and uranium-containing solute transportation. As the ration of vertical stress loading and horizontal stress loading k increases, the decrease in fracture aperture, seepage velocity, and uranium-containing solute concentration was observed. In addition, high seepage velocity and strong solute transport capacity for $k = 1$, and the laminar flow for $k = 0$ were observed.

Author Contributions: Conceptualization, Writing—review and editing, Project administration, Funding acquisition, T.Z.; Writing—original draft, X.N.; Methodology, Investigation, Formal analysis, T.Z., X.N. and S.S.; Data curation, X.H.; Supervision, T.Z. and X.H.; Software, Validation, T.Z., X.N. and X.Y. All authors have read and agreed to the published version of the manuscript.

Funding: This work was supported by the National Youth Science Foundation (Grant No. 51904011), the Institute of Energy, Hefei Comprehensive National Science Center (Grant No. 21KZS216), the Anhui Provincial Natural Science Foundation (Grant No. 1908085QE183), the University Synergy Innovation Program of Anhui Province (Grant No. GXXT-2021-019), and the Open Fund of State Key Laboratory of Water Resource Protection and Utilization in Coal Mining (Grant No. GJNY-18-73.7), the Open Fund of State Key Laboratory of Mining Response and Disaster Prevention and Control in Deep Coal Mines (Grant No.SKLMRDPC19ZZ05).

Institutional Review Board Statement: Not applicable.

Informed Consent Statement: Not applicable.

Data Availability Statement: Not applicable.

Conflicts of Interest: The authors declare no conflict of interest.

References

1. Chen, Y.; Zhou, J.; Hu, S.; Hu, R.; Zhou, C.-B. Evaluation of Forchheimer equation coefficients for non-Darcy flow in deformable rough-walled fractures. *J. Hydrol.* **2015**, *529*, 993–1006. [CrossRef]
2. Dangelmayr, M.A.; Reimus, P.W.; Wasserman, N.L.; Punsal, J.J.; Johnson, R.H.; Clay, J.T.; Stone, J.J. Laboratory column experiments and transport modeling to evaluate retardation of uranium in an aquifer downgradient of a uranium in-situ recovery site. *Appl. Geochem.* **2017**, *80*, 1–13. [CrossRef]
3. Zhang, T.; Gan, Q.; Zhao, Y.; Zhu, G.; Nie, X.; Yang, K.; Li, J. Investigations into Mining-Induced Stress–Fracture–Seepage Field Coupling Effect Considering the Response of Key Stratum and Composite Aquifer. *Rock Mech. Rock Eng.* **2019**, *52*, 4017–4031. [CrossRef]
4. Sims, R.; Lawless, T.A.; Alexander, J.L.; Bennett, D.G.; Read, D. Uranium migration through intact sandstone: Effect of pollutant concentration and the reversibility of uptake. *J. Contam. Hydrol.* **1996**, *21*, 215–228. [CrossRef]

5. Dittrich, T.M.; Reimus, P.W. Uranium transport in a crushed granodiorite: Experiments and reactive transport modeling. *J. Contam. Hydrol.* **2015**, *175–176*, 44–59. [CrossRef]
6. Dittrich, T.M.; Reimus, P.W. Reactive transport of uranium in fractured crystalline rock: Upscaling in time and distance. *J. Environ. Manag.* **2016**, *165*, 124. [CrossRef]
7. Panfilov, M.; Uralbekov, B.; Burkitbayev, M. Reactive transport in the underground leaching of uranium: Asymptotic analytical solution for multi-reaction model. *Hydrometallurgy* **2016**, *160*, 60–72. [CrossRef]
8. Read, D.; Ross, D.; Sims, R.J. The migration of uranium through Clashach Sandstone: The role of low molecular weight organics in enhancing radionuclide transport. *J. Contam. Hydrol.* **1998**, *35*, 235–248. [CrossRef]
9. Zhang, P.; Yang, T.; Yu, Q.; Xu, T.; Zhu, W.; Liu, H.; Zhou, J.; Zhao, Y. Microseismicity Induced by Fault Activation during the Fracture Process of a Crown Pillar. *Rock Mech. Rock Eng.* **2015**, *48*, 1673–1682. [CrossRef]
10. Cui, G.; Wei, J.; Feng, X.T.; Liu, J.; Elsworth, D.; Chen, T.; Xiong, W. Preliminary study on the feasibility of co-exploitation of coal and uranium-ScienceDirect. *Int. J. Rock Mech. Min. Sci.* **2019**, *123*, 104098. [CrossRef]
11. Du, L.; Li, S.; Li, X.; Wang, P.; Huang, Z.; Tan, Z.; Liu, C.; Liao, J.; Liu, N. Effect of humic acid on uranium(VI) retention and transport through quartz columns with varying pH and anion type. *J. Environ. Radioact.* **2017**, *177*, 142–150. [CrossRef]
12. Wang, G.; Wu, M.; Wang, R.; Xu, H.; Song, X. Height of the mining-induced fractured zone above a coal face. *Eng. Geol.* **2017**, *216*, 140–152. [CrossRef]
13. Ahn, H.; Oh, Y.; Ryu, J.H.; Jo, H.Y. Uranium sequestration in fracture filling materials from fractured granite aquifers. *J. Environ. Radioact.* **2020**, *225*, 106440. [CrossRef]
14. Tan, K.; Li, C.; Liu, J.; Qu, H.; Xia, L.; Hu, Y.; Li, Y. A novel method using a complex surfactant for in-situ leaching of low permeable sandstone uranium deposits. *Hydrometallurgy* **2014**, *150*, 99–106. [CrossRef]
15. Zeng, F.; Zhao, G. The optimal hydraulic fracture geometry under non-Darcy flow effects. *J. Pet. Sci. Eng.* **2010**, *72*, 143–157. [CrossRef]
16. Chen, Y.; Zhou, C.; Sheng, Y. Formulation of strain-dependent hydraulic conductivity for a fractured rock mass. *Int. J. Rock Mech. Min. Sci.* **2007**, *44*, 981–996. [CrossRef]
17. Espinoza, D.N.; Shovkun, I.; Makni, O.; Lenoir, N. Natural and induced fractures in coal cores imaged through X-ray computed microtomograph—Impact on desorption time. *Int. J. Coal Geol.* **2016**, *154*, 165–175. [CrossRef]
18. Lai, P.; Samson, C.; Bose, P. Visual enhancement of 3D images of rock faces for fracture mapping. *Int. J. Rock Mech. Min. Sci.* **2014**, *72*, 325–335. [CrossRef]
19. He, R.; Yang, Z.; Li, X.; Li, Z.; Liu, Z.; Chen, F. A comprehensive approach for fracability evaluation in naturally fractured sandstone reservoirs based on analytical hierarchy process method. *Energy Sci. Eng.* **2019**, *7*, 529–545. [CrossRef]
20. Zhou, Y.; Li, G.; Xu, L.; Liu, J.; Sun, Z.; Shi, W. Uranium recovery from sandstone-type uranium deposit by acid in-situ leaching—An example from the Kujieertai. *Hydrometallurgy* **2019**, *191*, 105209. [CrossRef]
21. Devaraj, N.; Panda, B.; Chidambaram, S.; Prasanna, M.V.; Singh, D.K.; Ramanathan, A.L.; Sahoo, S.K. Spatio-temporal variations of Uranium in groundwater: Implication to the environment and human health. *Sci. Total Environ.* **2021**, *775*, 145787.
22. Liu, H.; Wei, M.; Rutqvist, J. Normal-stress dependence of fracture hydraulic properties including two-phase flow properties. *Hydrogeol. J.* **2013**, *21*, 371–382. [CrossRef]
23. Li, Y.; Chen, Y.; Zhou, C. Hydraulic properties of partially saturated rock fractures subjected to mechanical loading. *Eng. Geol.* **2014**, *179*, 24–31. [CrossRef]
24. Song, Z.; Wang, Y.; Konietzky, H.; Cai, X. Mechanical behavior of marble exposed to freeze-thaw-fatigue loading. *Int. J. Rock Mech. Min. Sci.* **2021**, *138*, 104648. [CrossRef]
25. Wang, S.; Li, X.; Yao, J.; Gong, F.; Li, X.; Du, K.; Tao, M.; Huang, L.; Du, S. Experimental investigation of rock breakage by a conical pick and its application to non-explosive mechanized mining in deep hard rock. *Int. J. Rock Mech. Min. Sci.* **2019**, *122*, 104063. [CrossRef]
26. Wang, S.; Sun, L.; Li, X.; Wang, S.; Du, K.; Li, X.; Feng, F. Experimental investigation of cuttability improvement for hard rock fragmentation using conical cutter. *Int. J. Geomech.* **2021**, *21*, 06020039. [CrossRef]
27. Kim, J.W.; Baik, M.H.; Jung, H.; Jeong, J.T. Reactive transport of uranium with bacteria in fractured rock: Model development and sensitivity analysis. *J. Contam. Hydrol.* **2013**, *152*, 82–96. [CrossRef]
28. De Boissezon, H.; Levy, L.; Jakymiw, C.; Distinguin, M.; Guerin, F.; Descostes, M. Modeling uranium and 226Ra mobility during and after an acidic in situ recovery test (Dulaan Uul, Mongolia). *J. Contam. Hydrol.* **2020**, *235*, 103711. [CrossRef]
29. Feng, S.; Wang, H.; Cui, Y.; Ye, Y.; Liu, Y.; Li, X.; Wang, H.; Yang, R. Fractal discrete fracture network model for the analysis of radon migration in fractured media. *Comput. Geotech.* **2020**, *128*, 103810. [CrossRef]
30. Wang, H.; Zhang, L.; Lei, H.; Wang, Y.; Liu, H.; Li, X.; Su, X. Potential for uranium release under geologic CO2 storage conditions: The impact of Fe (III). *Int. J. Greenh. Gas Control* **2021**, *107*, 103266. [CrossRef]
31. Wang, Q.; Miao, X.; Wang, Y.; Gan, M.; Aftab, S.M.; Li, X.; Zhang, L.; Wang, Z. Simulation of uranium mobilization potential in a deep aquifer under geological carbon storage conditions. *Appl. Geochem.* **2020**, *118*, 104620. [CrossRef]
32. Baghbanan, A.; Jing, L. Stress effects on permeability in a fractured rock mass with correlated fracture length and aperture. *Int. J. Rock Mech. Min. Sci.* **2008**, *45*, 1320–1334. [CrossRef]
33. Rong, G.; Hou, D.; Yang, J.; Cheng, L.; Zhou, C. Experimental study of flow characteristics in non-mated rock fractures considering 3D definition of fracture surfaces. *Eng. Geol.* **2017**, *220*, 152–163. [CrossRef]

34. Zhang, Z.; Nemcik, J. Fluid flow regimes and nonlinear flow characteristics in deformable rock fractures. *J. Hydrol.* **2013**, *477*, 139–151. [CrossRef]
35. Yan, C.; Zheng, H. A two-dimensional coupled hydro-mechanical finite-discrete model considering porous media flow for simulating hydraulic fracturing. *Int. J. Rock Mech. Min. Sci.* **2016**, *88*, 115–128. [CrossRef]
36. Koyama, T.; Fardin, N.; Jing, L.; Stephansson, O. Numerical simulation of shear-induced flow anisotropy and scale-dependent aperture and transmissivity evolution of rock fracture replicas. *Int. J. Rock Mech. Min. Sci.* **2006**, *43*, 89–106. [CrossRef]
37. Zhao, Z.; Jing, L.; Neretnieks, I.; Moreno, L. Numerical modeling of stress effects on solute transport in fractured rocks. *Comput. Geotech.* **2011**, *38*, 113–126. [CrossRef]
38. Song, Z.; Konietzky, H.; Herbst, M. Bonded-particle model-based simulation of artificial rock subjected to cyclic loading. *Acta Geotech.* **2019**, *14*, 955–971. [CrossRef]
39. Rutqvist, J.; Leung, C.; Hoch, A.; Wang, Y.; Wang, Z. Linked multicontinuum and crack tensor approach for modeling of coupled geomechanics, fluid flow and transport in fractured rock. *Rock Mech. Geotech. Eng.* **2013**, *5*, 18–31. [CrossRef]
40. Lei, Q.; Latham, J.-P.; Xiang, J.; Tsang, C.-F. Polyaxial stress-induced variable aperture model for persistent 3D fracture networks. *Geomech. Energy Environ.* **2015**, *1*, 34–47. [CrossRef]
41. Zhao, Z.; Jing, L.; Neretnieks, I.; Moreno, L. Analytical solution of coupled stress-flow-transport processes in a single rock fracture. *Comput. Geosci.* **2011**, *37*, 1437–1449. [CrossRef]
42. Crandall, D.; Bromhal, G.; Karpyn, Z.T. Numerical simulations examining the relationship between wall-roughness and fluid flow in rock fractures. *Int. J. Rock Mech. Min. Sci.* **2010**, *47*, 784–796. [CrossRef]
43. Akhavan, A.; Shafaatian, S.; Rajabipour, F. Quantifying the effects of crack width, tortuosity, and roughness on water permeability of cracked mortars. *Cem. Concr. Res.* **2012**, *42*, 313–320. [CrossRef]
44. Chen, Y.; Liang, W.; Lian, H.; Yang, J.; Nguyen, V.P. Experimental study on the effect of fracture geometric characteristics on the permeability in deformable rough-walled fractures. *Int. J. Rock Mech. Min. Sci.* **2017**, *98*, 121–140. [CrossRef]
45. Bandis, S.; Lumsden, A.; Barton, N. Fundamentals of rock joints deformation. *Int. J. Rock Mech. Min. Sci. Geomech. Abstr.* **1983**, *20*, 249–268. [CrossRef]
46. Gan, Q.; Elsworth, D. A continuum model for coupled stress and fluid flow in discrete fracture networks. *Geomech. Geophys. Geo-Energy Geo-Resour.* **2016**, *2*, 43–61. [CrossRef]

mathematics

MDPI

Article

Regressive and Big-Data-Based Analyses of Rock Drillability Based on Drilling Process Monitoring (*DPM*) Parameters

Shaofeng Wang [1], Yu Tang [1,*], Ruilang Cao [2,*], Zilong Zhou [1,*] and Xin Cai [1]

1 School of Resources and Safety Engineering, Central South University, Changsha 410083, China; sf.wang@csu.edu.cn (S.W.); xincai@csu.edu.cn (X.C.)
2 State Key Laboratory of Simulation and Regulation of Water Cycle in River Basin, China Institute of Water Resources and Hydropower Research, Beijing 100048, China
* Correspondence: 195512136@csu.edu.cn (Y.T.); caorl@iwhr.com (R.C.); zlzhou@csu.edu.cn (Z.Z.)

Abstract: Accurate, rapid and effective analysis of rock drillability is very important for mining, civil and petroleum engineering. In this study, a method of rock drillability evaluation based on drilling process monitoring (*DPM*) parameters is proposed by using the field drilling test data. The revolutions per minute (*N*), thrust, torque and rate of penetration (*ROP*) were recorded in real time. Then, the two-dimensional regression analysis was utilized to investigate the relationships between the drilling parameters, and the three-dimensional regression analysis was used to establish models of *ROP* and specific energy (*SE*), in which the *N-F-ROP*, *N-T-ROP* and the improved *SE* model were obtained. In addition, the random forest (*RF*) and support vector machine combined with genetic algorithm (*GA-SVM*) were applied to predict rock drillability. Finally, a prediction model of uniaxial compressive strength (*UCS*) was established based on the *SE* and drillability index, I_d. The results show that both regression models and prediction models have good performance, which can provide important guidance and a data source for field drilling and excavation processes.

Keywords: rock drillability; *DPM* parameters; regression analysis; *RF*; *GA-SVM*; *UCS* prediction model

Citation: Wang, S.; Tang, Y.; Cao, R.; Zhou, Z.; Cai, X. Regressive and Big-Data-Based Analyses of Rock Drillability Based on Drilling Process Monitoring (*DPM*) Parameters. *Mathematics* **2022**, *10*, 628. https://doi.org/10.3390/math10040628

Academic Editor: Mario Versaci

Received: 17 January 2022
Accepted: 14 February 2022
Published: 17 February 2022

1. Introduction

Rock drillability plays an important role in mining, civil and petroleum engineering. The traditional rock drillability analysis requires the rock mechanics parameter. This means that the procedures of site drilling, coring, sample processing and laboratory test are needed, which will be time-consuming and costly and greatly affect the engineering efficiency. Moreover, sometimes, the rock core is impossible to be obtained owing to the fracture and fragmentation of the rock mass, which results in the rock mass properties being unable to be analyzed. However, the drilling process monitoring (*DPM*), as a new promising technology, can compensate for the above shortcomings, and it has been widely applied in recent years. Rock drilling process refers to the rock drilling relying on the bit's thrust and rotary cutting force. The thrust is used to push the drill pipe and make it in close contact with the rock mass so as to generate enough friction. The rotary cutting force is mainly used to break the rock mass. Meanwhile, the parameters fed back in the drilling process are closely related to the rock mass properties. Therefore, how to quickly and accurately obtain the rock mass properties is very important for the parameter design of rock breaking equipment.

The scholars in United States, Japan and France have been trying to find the relationships between the rock mass quality of engineering and drilling parameters by instrument measurement since 1970 [1–3]. However, it measures the speed based on the distance, which indicates that the achievement of drilling speed has great randomness when the rock formation is not uniform or the drilling rig is vibrating [4,5], Therefore, it is impossible to accurately obtain the rock physical properties and the related rock formation division. On this basis, Yue [6–9] developed the drilling process monitoring system (*DPM*), which effectively improved the accuracy of velocity measurement and was successfully applied to stratum identification and rock strength classification. However, the *DPM* system cannot measure torque, so He and Li et al. [10] improved it. The new drilling process monitoring apparatus can achieve the measurement of torque. Up to now, the development of a while-drilling system has been relatively mature, which can realize the real-time measurement of thrust, torque, rate of penetration (*ROP*) and revolutions per minute (*N*), and has been successfully applied to rock property analysis and formation analysis. Rodgers et al. [11–14] used while-drilling parameters to assess field rock strength and optimize core recovery. Karasawa [15,16] et al. conducted laboratory drilling tests, and they proposed the method to estimate the unconfined compressive strength of rock. He [17] et al. performed a series of drilling tests on sandstone, limestone, marble and granite in the field to predict the cohesion, internal friction angle and uniaxial compressive strength of rock. Li [18] et al. deduced the relationships between the drilling efficiency and drilling parameters based on the method of force limit of equilibrium and energy equilibrium. Li [19] et al. analyzed the influence of working parameters, such as impact power, propulsion force, rotating speed and drill bit type, on drilling efficiency and obtained the relationships between the drilling velocity and drilling parameters. Feng [20] et al. performed a field-drilling test to obtain the optimal drilling efficiency, and they found the optimal drilling conditions and rock drillability. Moreover, the combined thermo-mechanical drilling technology and acoustic emission method were utilized to investigate the rock drillability [21–23]. However, the fractured zones were not considered in the above studies. Therefore, Kalantari et al. [24,25] established a stress limit equilibrium analysis model for a T-shaped drag bit, which considered influence factors such as bit geometry parameters, fractured zone and contact friction during drilling. On this basis, the actual drilling data were used to estimate the rock strength parameters, such as cohesion, internal friction angle and uniaxial compressive strength. The results show that the borehole test results based on this model are in good agreement with the standard test results.

Meanwhile, a large number of models have been established to evaluate rock drillability. Hughes [26] and Mellor [27] et al. proposed the theoretical models of *SE* based on the uniaxial compressive strength and secant modulus of rocks. An empirical formula was put forward by Poane [28] et al. to evaluate the relationships between the drilling parameters and specific energy (*SE*). Feng [20] et al. improved the *SE* model using the controllable parameters (thrust and *ROP*). In addition, Zhang [29] et al. proposed a new rock drillability index, I_d, to evaluate rock classification, and it was applied to classify rock types successfully as an in situ test. Yu [30] et al. compared the effect of *SE* and I_d on rock strength assessment. The results present that the assessment performance of I_d is superior to *SE* due to the smaller overall fluctuation during the modeling process. However, the most widely used model by far is the *SE* model proposed by Teale [31]. The *ROP* is also an important index to evaluate rock drilling efficiency. Kahraman [32,33] conducted rotary and percussive drilling tests and obtained the prediction equation of *ROP* through regression analysis. Ataei [34] et al. established an empirical formula of *ROP* combined with the rock mass drillability index, and the results show that the model has a better prediction effect compared with the previous model. In addition, artificial intelligence (AI) techniques have been utilized to predict rock properties. Ocak [35] et al. used the multilayer perceptron neural network (MLPNN) to predict the elastic module of intact rocks, and the prediction results show that the MLPNN has good prediction capacity. Yesiloglu-Gultekin [36] et al. employed the artificial neural network and adaptive neuro fuzzy inference system to predict the uniaxial compressive strength of granite rocks, and the study indicates that the developed models have a high prediction performance. Sarkar [37] et al. adopted the feed-forward back-propagation neural network to estimate the strength parameters of rock, and the results show that the performance of the AI techniques is better than regression analysis. He [38] et al. utilized the deep convolutional neural network to predict the cohesion, internal friction angle and uniaxial compressive strength and obtained good prediction results. Therefore, the AI technique is a promising method to analyze rock properties. However, some explanations about the cutting actions of the drilling bits and TBM disc-cutters explained in many papers are not referred to, e.g., the explanations given by Roxborough and Phillips [39,40]. Some other researchers also used the concepts of thrust and rolling forces and the specific energy to study the mechanism of rock fragmentation by the cutters [41,42]. They used a higher order displacement discontinuity method for their analyses. The finite element method and discrete element method have also been used to analyze the drillability of the rock cutting heads.

Although many achievements have been obtained, the analysis of the relationships between the rock drillability and *DPM* parameters is still rare, and the AI algorithm is rarely used to analyze the relationships between them. Meanwhile, the relationships between the *DPM* parameters and rock properties have not been fully established. Therefore, the drilling test was conducted in this study, and the thrust (*F*), torque (*T*), rotating speed (*N*) and rate of penetration (*ROP*) were recorded in real time. Then, the two-dimensional and three-dimensional regression analyses were utilized to investigate the relationships between the rock drillability and *DPM* parameters, in which two *ROP* models are obtained and the *SE* model is improved. In addition, the random forest (*RF*) and support vector machine combined with genetic algorithm (*GA-SVM*) were used to predict the rock drillability. Finally, a model using *SE* and I_d for estimating the uniaxial compressive strength based on three drilling tools is established. The above efforts can achieve the accurate, rapid and effective analysis of rock drillability, and the model has very important guiding significance for field drilling and rock fragmentation.

2. Methodology

2.1. Drilling Process Monitoring (DPM) Parameters

In order to analyze the relationships between *DPM* parameters and rock drillability, a new drilling-monitoring system was utilized to perform drilling test. The system, as shown in Figure 1, can record the thrust (*F*), torque (*T*), rotating speed (*N*) and rate of penetration (*ROP*) in real time at 1-s interval through the corresponding sensor. These sensors mainly include pressure sensor, rotating speed sensor, torque sensor and laser displacement sensor. Finally, the recorded data will be transformed to the data processing system, where the data can be stored and processed. Meanwhile, the system can adjust the thrust and rotating speed artificially during the process of drilling. Moreover, all sensors are easily to be mounted and have high measurement precision, so it is convenient and reliable to conduct drilling test in field.

Figure 1. Digital drilling system and in situ testing.

Among the *DPM* parameters, the thrust refers to the force required to make the drill bit in close contact with the rock mass; the rotating speed refers to the number of rotating turns of drill bit per minute and the torque refers to the moment required to rotate the drill bit, which is used to cut rock mass; the rate of penetration refers to the length of drilling per unit time.

2.2. Drilling Conditions

Different thrust and rotating speed were applied to plain concrete with a 90 mm diameter drill bit. The range of thrust is from 8.5 to 70 kN, and that of rotating speed from 40 to 400 r/min. Through uniaxial compressive test and Brazilian tensile test, the uniaxial compressive strength and tensile strength of the concrete are 14.70 ± 0.50 MPa and 1.57 ± 0.32 MPa, respectively.

2.3. DPM Data

After drilling test, the corresponding torque and rotating speed values under different thrust and rotating speed drilling conditions were obtained, and the *SE* (energy required to break unit rock) while drilling was calculated following the *SE* model proposed by Teale [31], as shown in Equation (1). The average values of *DPM* parameters and *SE* are presented in Table 1.

$$SE = \frac{F}{A} + \frac{2\pi NT}{AV}, \tag{1}$$

where *SE* is specific energy of drilling, *F* is thrust, *N* is rotating speed, *T* is torque, *V* is rate of penetration and *A* is drilling area.

Table 1. The average value of *DPM* parameters and *SE*.

N/r·min^{-1}	F/kN	ROP/cm·min^{-1}	T/N·m	SE/MJ·m^{-3}
40	10	3.02	68.195	81.48
	11.5	3.31	72.011	78.76
	19	3.74	91.091	89.04
	28	3.69	113.987	113.48
	35	3.95	131.795	123.25
	40	4.03	144.515	132.79
	45	4.27	157.235	136.92
	50	3.61	169.955	173.99
	60	3.85	195.395	188.42
	65	4.27	208.115	181.99
	70	4.22	220.835	195.43
115	8.5	3.44	64.379	191.98
	19	4.44	91.091	211.82
	23.5	4.52	102.539	234.57
	29	4.81	116.531	251.06
	35.75	5.11	133.703	271.77
	42	6.38	149.603	244.96
	46.5	7.55	161.051	224.01
	53	7.01	177.587	265.73
	58.25	7.29	190.943	275.23
	60.5	5.87	196.667	350..07
	65	6.00	208.115	362.76

Table 1. *Cont.*

N/r·min^{-1}	*F*/kN	*ROP*/cm·min^{-1}	*T*/N·m	*SE*/MJ·m^{-3}
220	8.7	3.09	64.888	410.74
	18.25	4.784	89.183	366.12
	29	6.46	116.531	355.88
	34.85	7.71	131.4134	337.31
	40	6.68	144.515	427.54
	46.7	7.60	161.560	421.15
	51.5	7.83	173.771	440.06
	57.5	7.53	189.035	497.68
	60	7.21	195.395	536.97
400	8.5	4.36	64.399	525.06
	11.5	5.24	72.031	490.54
	18.8	6.55	90.602	492.93
	25	7.23	106.375	530.47
	30	8.09	119.095	551.47
	35.5	8.91	133.087	534.99
	42.5	13.31	150.895	435.13
	48.75	11.72	166.795	530.57
	53	10.61	177.607	584.46
	58	12.86	190.327	584.19

3. Regression Analysis

Two-dimensional and three-dimensional regression analysis were adopted to investigate the relationships between the *DPM* parameters and the influence of *DPM* parameters on rock drillability. It is worth noting that the data used for the regression analysis are the average values (Table 1) under their corresponding conditions.

3.1. Two-Dimensional Regression Analysis

The relationships between thrust, torque, *ROP* and *SE* were analyzed under different rotating speeds. There is a good positive linear relationship between thrust and torque, as shown in Figure 2, and the fitting functions are presented in Table 2. The relationship between thrust and *ROP* is powerful, and their regression curves and fitting functions are shown in Figure 3 and Table 2, respectively. Meanwhile, the relationship between torque and rotating speed was obtained, as shown in Figure 4 and Table 2, respectively.

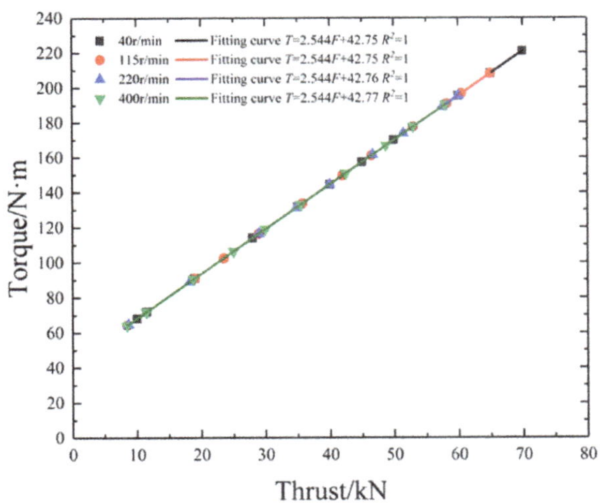

Figure 2. The relationship between thrust and torque under different rotating speed.

Table 2. Relationships between parameters of *DPM*.

Model Type			*T-F*		*V-F*		*V-T*		*SE-F*		*SE-T*	
Fitting model			$T = aF + b$	R^2	$V = aF^b$	R^2	$V = aT^b$	R^2	$SE = aF + b$	R^2	$SE = aT + b$	R^2
Rotating speed /r·min^{-1}	40	a b	2.544 42.75	1	2.386 0.1331	0.7057	1.364 0.2096	0.6666	2.049 55.2	0.967	0.8056 20.75	0.967
	115	a b	2.544 42.75	1	1.584 0.3536	0.7132	0.3744 0.5493	0.6941	2.363 167.4	0.6803	0.9288 127.7	0.6803
	220	a b	2.544 42.76	1	1.606 0.3934	0.8447	0.3172 0.6147	0.7830	2.659 319.1	0.5109	1.045 274.4	0.5109
	400	a b	2.544 42.77	1	1.214 0.5788	0.8803	0.09685 0.9337	0.8749	1.108 489.3	0.1826	0.4354 470.6	0.1826

where a and b are fitting parameters.

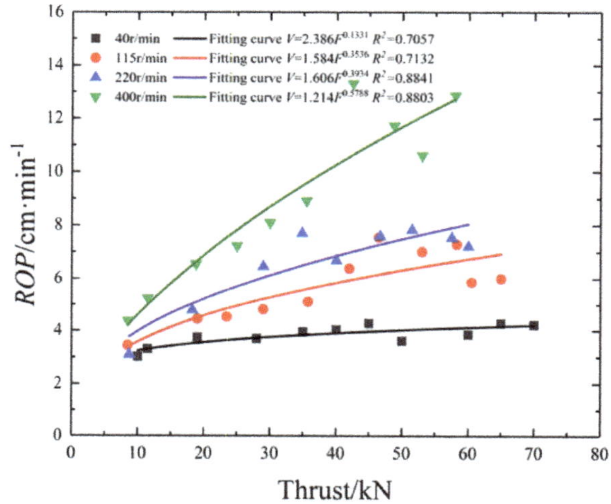

Figure 3. The relationship between thrust and *ROP* under different rotating speed.

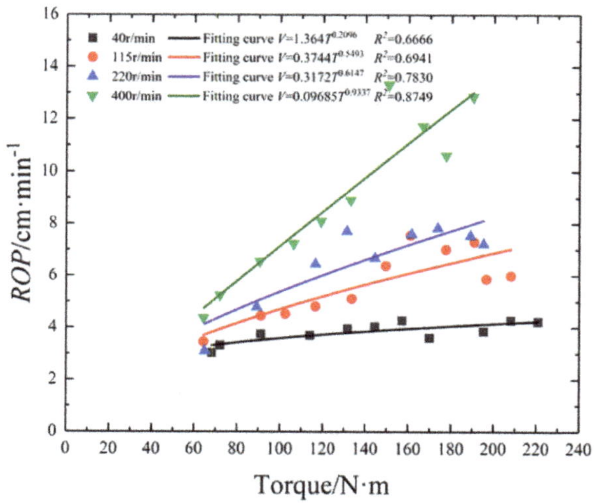

Figure 4. The relationship between torque and *ROP* under different rotating speed.

Similarly, the values reflecting the relationship between *SE* and thrust (or torque) under different rotating speeds were obtained, as shown in Figures 5 and 6 and Table 2. There is a good linear relationship when the rotating speed is 40 r/min. However, there is no significant statistical relationship under other rotating speed conditions. This means that the single variable is unable to reflect the *SE* of drilling effectively.

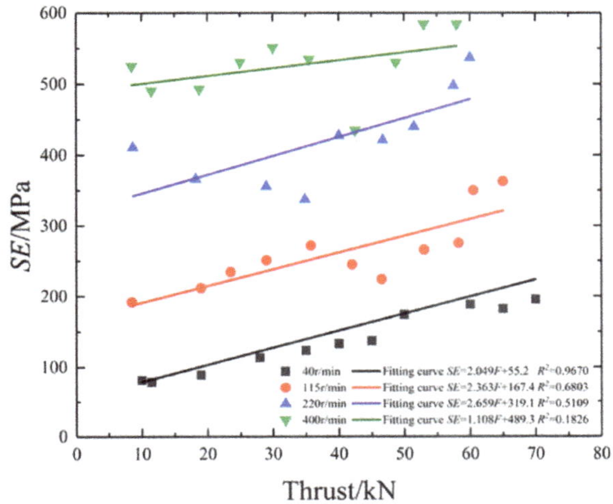

Figure 5. The relationship between thrust and *SE* under different rotating speed.

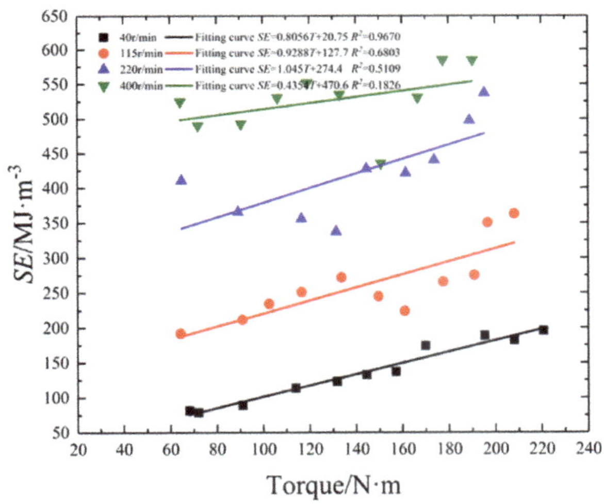

Figure 6. The relationship between torque and *SE* under different rotating speed.

3.2. Three-Dimensional Regression Analysis

The two-dimensional regression analysis results indicate that there is a significant statistical relationship between thrust and torque. However, for the regression analyses of *ROP* and *SE*, the fitting effects are incapable to meet the requirement of practical engineering. Therefore, the three-dimensional regression analyses were utilized to establish models based on *DPM* data.

The rotating speed, thrust and torque were selected to establish regression models of *ROP*. We defined them as *N-F-ROP* and *N-T-ROP* models. The three parameters correspond to the *X*-axis, *Y*-axis and *Z*-axis variables, respectively. Through regression analyses, the *N-F-*

ROP and *N-T-ROP* models were obtained, and their corresponding three-dimensional fitting curves and formulas are shown in Figures 7 and 8, and Equations (2) and (3), respectively.

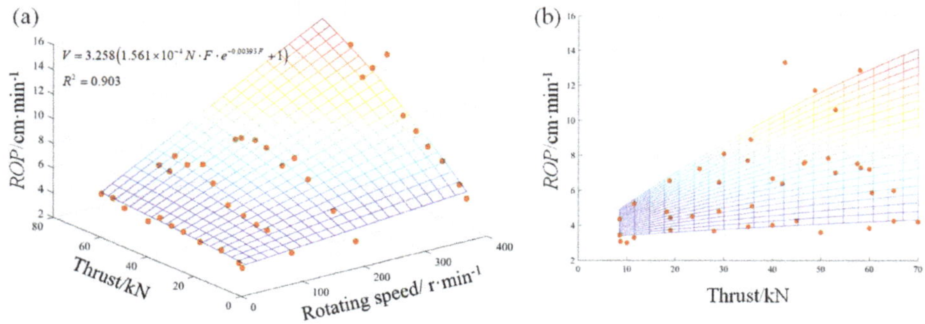

Figure 7. Regressed model of *N-F-ROP*, where (**a**) is three-dimensional surface view and (**b**) is the view of *Y-Z* direction.

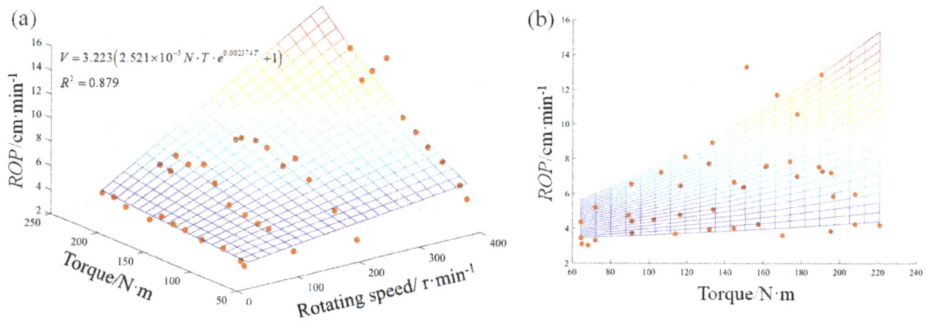

Figure 8. Regressed model of *N-T-ROP*, where (**a**) is three-dimensional surface view and (**b**) is the view of *Y-Z* direction.

Moreover, because there is a good linear relationship between thrust and torque, we improved the *SE* model proposed by Teale. The thrust, rotating speed, *ROP* and drilling area (A) were taken as independent variables to establish the regression model of *SE*. The results show that the above parameters have a significant statistical relationship, as presented in Figure 9 and Equation (4).

$$V = 3.258(1.561 \times 10^{-4} N \cdot F \cdot e^{-0.00393F} + 1) \tag{2}$$

$$V = 3.223(2.521 \times 10^{-5} N \cdot T \cdot e^{0.002374T} + 1) \tag{3}$$

$$SE = 0.2973\frac{F}{A} + 236.4\frac{N}{V \cdot A} + 94.87\frac{F}{A} \cdot \frac{N}{V \cdot A} \tag{4}$$

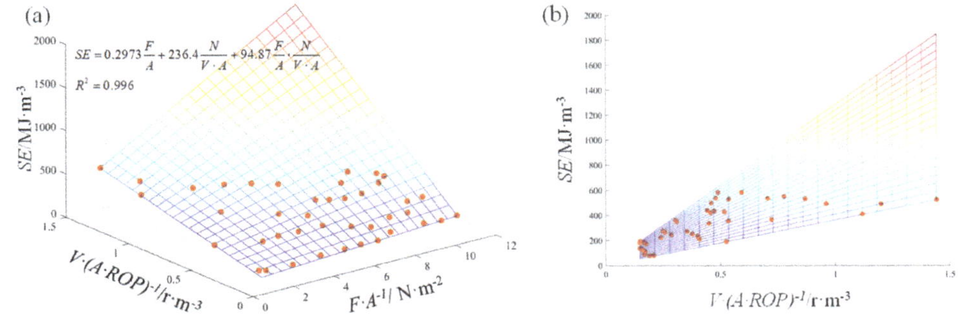

Figure 9. Regressed model of *SE*, where (**a**) is three-dimensional surface view and (**b**) is the view of Y-Z direction.

3.3. Regression Model Evalution

The determination coefficient (R^2) and root mean square error (*RMSE*) were utilized to evaluate regression performance of two-dimensional and three-dimensional regression models, and the corresponding calculation formulas are shown in Equations (5) and (6), respectively. The ultimately calculation results are shown in Table 3.

$$R^2 = \sum_{i=1}^{n} (y_{p(i)} - \overline{y_t})^2 / \sum_{i=1}^{n} (y_{t(i)} - \overline{y_t})^2 \tag{5}$$

$$RMSE = \sqrt{\frac{1}{n}\sum_{i=1}^{n} (y_{p(i)} - y_{t(i)})^2} \tag{6}$$

where $y_{t(i)}$, $y_{p(i)}$, $\overline{y_t}$ and n are the test value, predicted value, mean of all test values and the total number of values, respectively.

Table 3. The R^2 and *RMSE* values of regression models.

Regression Model	Rotating Speed /r·min^{-1}	R^2					*RMSE*				
		F-T	*F-V*	*T-V*	*F-SE*	*T-SE*	*F-T*	*F-V* (kN)	*T-V* (N·m)	*F-SE* (MJ·m^{-3})	*T-SE* (MJ·m^{-3})
Two-dimensional	40	1	0.7057	0.6666	0.9670	0.9670	0.005	0.207	0.2198	7.547	40.841
	115	1	0.7132	0.6941	0.6803	0.6803	0.005	0.675	0.6970	28.640	98.618
	220	1	0.8447	0.7830	0.5109	0.5109	0.00496	0.591	0.7047	43.047	43.983
	400	1	0.8803	0.8749	0.1826	0.1826	0.00498	1.032	1.0551	38.552	18.222
Three-dimensional			*N-F-V*	*N-T-V*	*SE=f* (*F*, *N*, *V*, *A*)		*N-F-V* (cm·min^{-1})		*N-T-V* (cm·min^{-1})	*SE=f* (*F*, *N*, *V*, *A*) (MJ·m^{-3})	
			0.903	0.879	0.996		0.794		0.890	10.11	

4. Analyses Using Machine Learning Methods

The *RF* and *GA-SVM* were used to analyze the relationships between the *SE* and *DPM* parameters. These two methods belong to the category of machine learning, and it has been successfully applied in many fields [43–46].

4.1. Random Forest

RF is an ensemble algorithm based on a decision tree, and it was developed by Breiman [47]. The *RF* model takes the bootstrap method to select a training set in the way of sampling with the replacement method. Then, the selected training sets are utilized to establish the classification or regression model based on a pre-designed number of decision trees (*ntree*) and node value (*mtry*). Finally, the test sets are taken to evaluate the model. In the prediction process, the classification model uses the method of classification voting to get the final result, while the prediction model adopts the method of regression mean.

A total of 11041 sets of real-time *DPM* data were used for training and testing in the *RF* model. Because there is a good linear relationship between torque and thrust, and the drilling area (*A*) is a constant, the rotating speed, thrust and *ROP* were taken as the input variables, and the output variable was *SE*. It should be noted that the average speed was used for the training and testing process in the *RF* model instead of the real-time speed under different thrust conditions due to the real-time speed possibly being a negative value caused by the slight rebound of the drill pipe during drilling. Of the *DPM* data, 70% were were selected randomly to be the training set, and the remaining 30% of the data were taken as the test set. The *ntree* and *mtry* were set to 150 and 2, respectively. The optimized prediction model was obtained by training, and the ultimate result was obtained through prediction. The architecture of *RF* is shown in Figure 10.

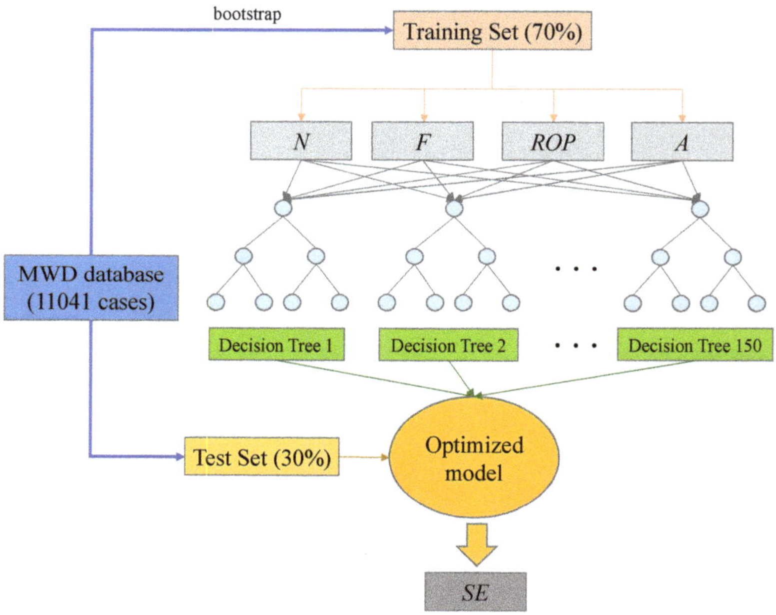

Figure 10. The architecture of random forest.

In order to evaluate the prediction accuracy of the established *RF* model, the R^2 and *RMSE* were calculated, as shown in Table 4. Meanwhile, the comparison of the original value and predicted value is shown in Figure 11. From the predicted results, we can find the R^2 and *RMSE* of the training set and test set are 0.9455 and 0.9591, 42.1341 and 48.7496, respectively, which indicates that the established *RF* model has a good prediction performance.

Table 4. The R^2 and *RMSE* of prediction models.

Prediction Models	Trian Set		Test Set	
	R^2	*RMSE* ($MJ \cdot m^{-3}$)	R^2	*RMSE* ($MJ \cdot m^{-3}$)
GA-SVM	99.99%	0.2578	99.98%	2.5208
RF	94.55%	42.1341	95.91%	48.7496

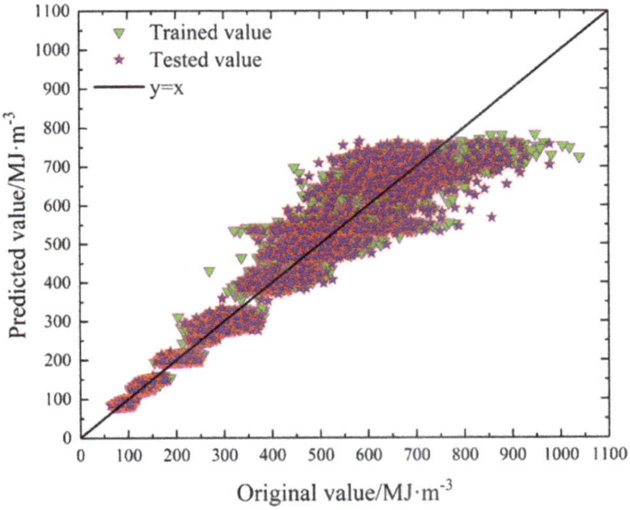

Figure 11. The comparison of original value and predicted value of *RF* model.

4.2. GA-SVM

The support vector machine (*SVM*) algorithm was invented by Cortes and Vapnik [48]. It is one of the machine learning methods based on statistical learning theory. The *SVM* model mainly includes the input layer, kernel function layer and output layer. Among them, the key function layer mainly includes radial basis function (*RBF*), polynomial and linear functions, which are the key factors to affect the prediction performance. In this paper, the genetic algorithm (*GA*) is combined with *SVM* to optimize the parameters. The *GA* [49] is a new method to find the optimal solution through the natural evolution process simulation, and the steps of *GA* mainly include population initialization, individual evaluation, selection operation, crossover operation, mutation operation and termination condition judgment. The *SVM* combined with *GA* can better optimize the parameter selection and improve the prediction accuracy.

Similar to the analysis of *RF*, the rotating speed, thrust and *ROP* are taken as the input variables, and *SE* as the output variable. Of the *DPM* data, 70% were randomly selected for model training, and the remaining 30% of the data were used for the model verification, as shown in Figure 12. In *GA*, the maximum evolutionary generation (*ga_option.maxgen*), maximum population (*ga_option.sizepop*), crossover validation (*ga_option.v*) and crossover probability (*ga_option.ggap*) were set as 100, 20, 10 and 0.9, respectively. By optimizing the parameters, the optimal penalty coefficient (c) and kernel radius (g) were 6.5961 and 27.2206, respectively. Then, the above parameters were used to establish the prediction model. Finally, the test sets were used to verify the model. The prediction results show the established model has good performance.

The R^2 and *RMSE* were used to evaluate the prediction accuracy of *GA-SVM*. By calculation, the corresponding evaluation results and comparison picture are shown in Table 4 and Figure 13, respectively. The results show that the prediction performance of the *GA-SVM* model is superior to the *RF* model. It also shows that it is feasible to use the thrust, rotating speed and *ROP* to evaluate the *SE* when the drilling area is a constant.

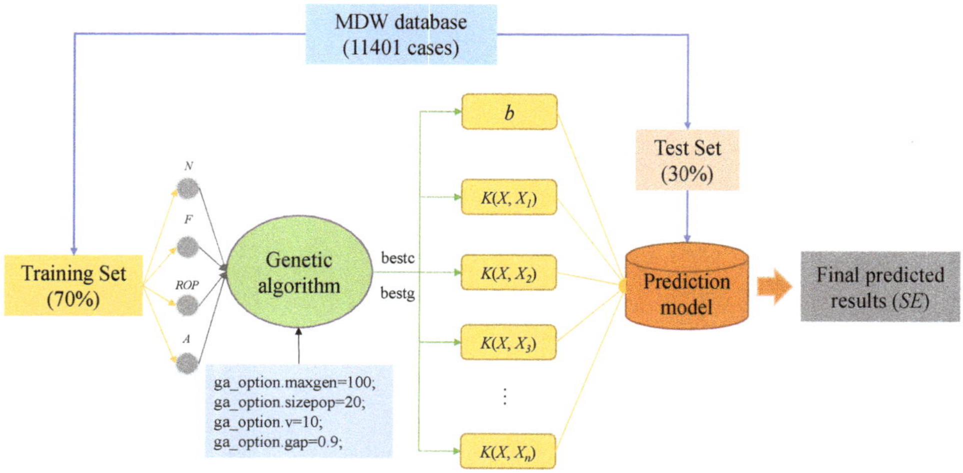

Figure 12. The architecture of *GA-SVM*.

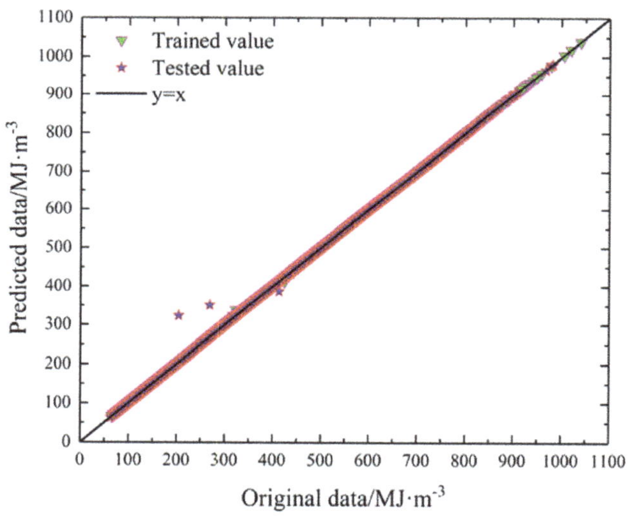

Figure 13. The comparison of original value and predicted value of *GA-SVM* model.

5. Relationships between Rock Properties and *DPM* Parameters

The rock properties and in situ stress conditions obviously affect rock cutting and drilling [50–53]. The fast and accurate prediction of the rock properties is the key to evaluating rock drillability and cuttability. Therefore, in order to find the relationships between the rock properties with *DPM* parameters, the *SE* and drillability index, I_d, were utilized to establish the rock strength prediction model. The *SE* can be obtained according to the improved model based on Teale's model. The drillability index, I_d, was proposed by Zhang [29], and it can be calculated based on Equation (7). Both *SE* and I_d can be calculated by *DPM* parameters.

The data used to establish the prediction models are present in Table 5. According to these data, the *SE* and I_d were obtained. Then, the prediction models of uniaxial compression strength of rock based on SE and I_d were established, which include the univariate model and two-variable model.

$$I_d = \gamma \pi_1^{\alpha} \pi_2^{\beta} \tag{7}$$

$$\pi_1 = \frac{DF}{T}, \pi_2 = \frac{V}{D\omega}, \omega = 2\pi N \tag{8}$$

where D is drill bit diameter, F is thrust force, T is torque, V is rate of penetration, ω is angular velocity, N is rotating speed and α, β, γ are fitting parameters. In this study, α, β, γ are determined to be 0.5, 0.6 and 1, respectively.

Table 5. Data source, quantity and bit types of *UCS* prediction model.

References	Bit Types	Bit Geometry Parameter	Data Size
Wang [54]	Standard diamond solid bit 	Radius is 7.5 mm	34
He [55]	Impregnated diamond bit 	External diameter is 70mm, and external diameter is 60 mm	17
Wang [56,57]	PDC bit 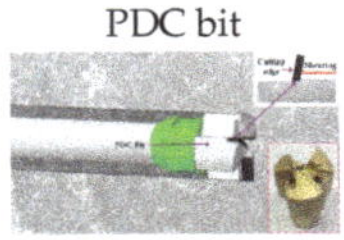	Cutting edge L_1, L_2 and L_3 are 18, 18 and 27 mm, respectively, and radius is 30 mm	42

5.1. Univariate Model of UCS

According to the calculated SE and I_d, the relationship between them and uniaxial compressive strength of rock under different cutting tools is established, respectively. When the standard diamond solid bit was used, the relationship between SE (or I_d) and UCS is present in Figure 14a, and the corresponding regression model is shown in Table 6. Similarly, the relationship between SE (or I_d) and UCS is present in Figure 14b,c, in which the impregnated diamond bit and PDC bit were performed to drilling. Their regression models are also shown in Table 6. Finally, all the data under the three types of bits were used to establish the relationship between the SE (or I_d) and UCS of rock. The regression model and fitting curve are shown in Table 6 and Figure 14d, respectively.

Table 6. Established *UCS* model based on *SE* and I_d separately under different bits.

Bit Types	*UCS* Prediction Model	
	SE-UCS	**I_d-*UCS***
Standard diamond solid bit	$SE = 0.5129UCS$	$I_d = 0.01519UCS^{-0.305}$
Impregnated diamond bit	$SE = 0.08552UCS$	$I_d = 0.0122UCS^{0.2208}$
PDC bit	$SE = 0.002172UCS$	$I_d = 0.03025UCS^{0.1704}$
Combine the above three bits	$SE = 0.4227UCS$	$I_d = 0.03247UCS^{-0.05269}$

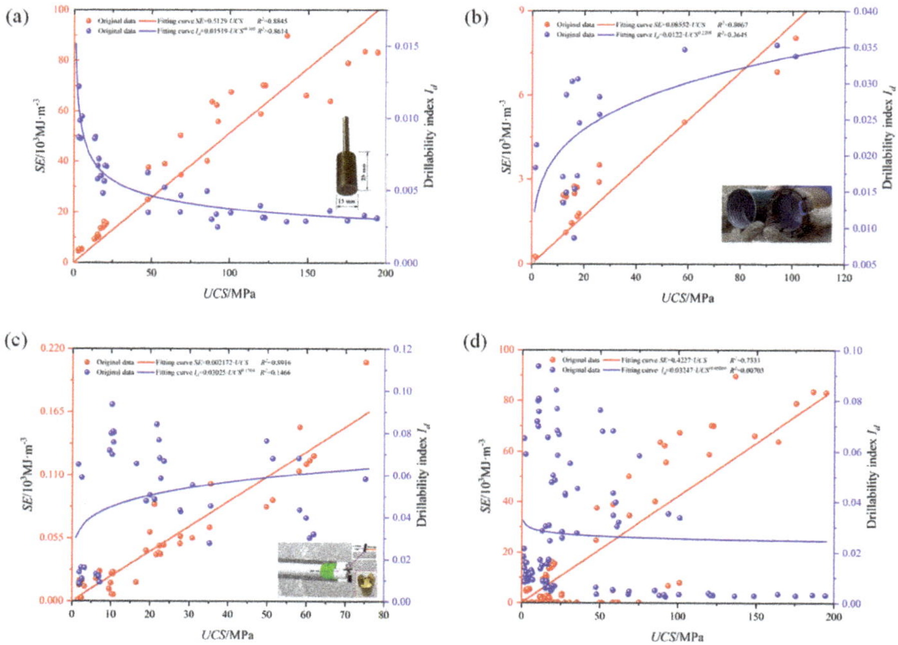

Figure 14. The relationship between *SE* (or I_d) and *UCS* for different bits of (**a**) standard diamond solid bit, (**b**) impregnated diamond bit, (**c**) PDC bit and (**d**) combine the above three bits.

From these univariate prediction models, we can find that there is a good relationship between *SE* (or I_d) and *UCS* for the standard diamond solid bit. The result indicates *SE* is positively correlated with the *UCS*, while the I_d is negatively correlated with *UCS*, which is consistent with Yu's [30] study. However, compared to the regression model for the standard diamond solid bit, the relationship between I_d and *UCS* presents an opposite law for the impregnated diamond bit and PDC bit, and the relationship between *UCS* and *SE* is obviously better than that between *UCS* and I_d.

5.2. Two-Variate Model of UCS

The univariate models of *UCS* indicate that the performances of the *UCS* models based on *SE* (or I_d) under different bits are not very good, especially for the model based on I_d. Therefore, the *UCS* prediction model was established by combining the *SE* and I_d. Similar to the univariate modeling, the *UCS* prediction models under different bits were established, respectively. The *UCS* prediction models under standard diamond solid bit, impregnated diamond bit and PDC bit are shown in Figure 15a–c, respectively, and the corresponding fitting equations are shown in Equations (9)–(11). Finally, all the data were combined to produce an *UCS* prediction model, and the regression curve and fitting equation are shown in Figure 15d and Equation (12). The results show that the performances of the established models combining *SE* and I_d are significantly superior to that of the univariate model.

$$UCS = 0.02973\left(19.96 \cdot SE \cdot e^{0.6472 I_d^{-0.1}} + 1\right) \tag{9}$$

$$UCS = 0.006402\left(1.081 \times 10^4 \cdot SE \cdot e^{-1.285 I_d^{-0.1}} + 1\right) \tag{10}$$

$$UCS = 1.844\left(967.5 \cdot SE \cdot e^{-1.07 I_d^{-0.1}} + 1\right) \tag{11}$$

$$UCS = 16.93\left(0.08349 \cdot SE \cdot e^{0.06913 I_d^{-0.1}} + 1\right) \tag{12}$$

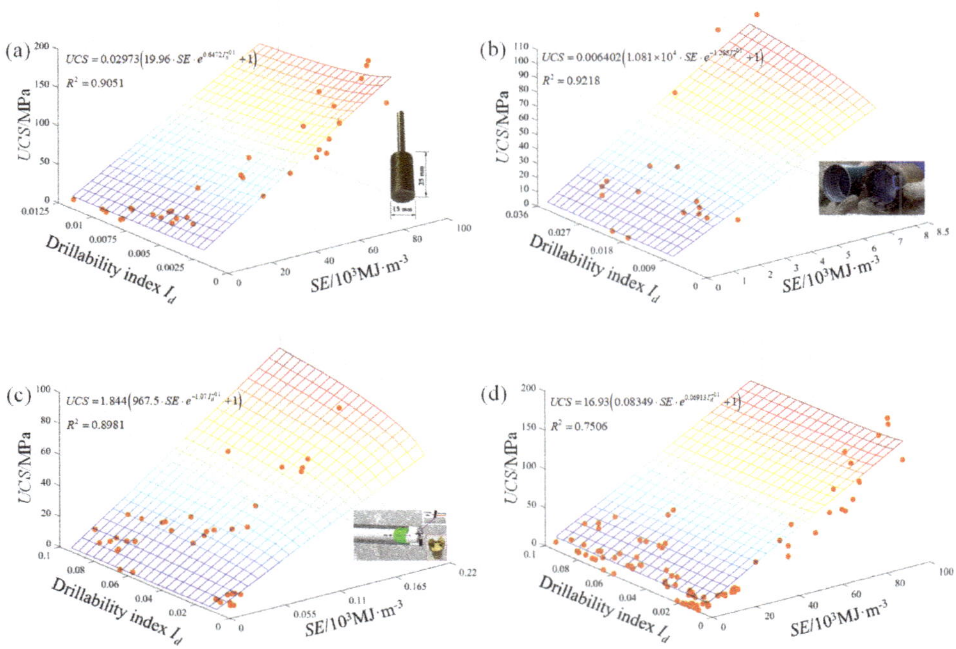

Figure 15. Established *UCS* models combine *SE* and I_d for different bits of (**a**) standard diamond solid bit, (**b**) impregnated diamond bit, (**c**) PDC bit and (**d**) combine the above three bits.

5.3. Evaluation of UCS Prediction Models

The determination coefficient (R^2) and root mean square error (*RMSE*) were used to evaluate the performance of univariate and two-variate models of *UCS*, and it can be calculated by Equations (5) and (6), respectively. The corresponding evaluation indexes obtained through calculation are shown in Table 7. The results show that the performance of the established *UCS* prediction model based on *SE* and I_d is better than that based on a single variable.

Table 7. The R^2 and *RMSE* of *UCS* models.

UCS Models		**Standard Diamond Solid Bit**		**Impregnated Diamond Bit**		**PDC Bit**		**Combine the Above Three Bits**	
		R^2	*RMSE*	R^2	*RMSE*	R^2	*RMSE*	R^2	*RMSE*
Univariate model	*SE-UCS*	0.8845	9.6273 (10^3 MJ·m³)	0.8067	0.8927 (10^3 MJ·m³)	0.8916	0.0151 (10^3 MJ·m³)	0.7333	12.7979 (10^3 MJ·m³)
	I_d-*UCS*	0.8614	0.0041	0.3645	0.0064	0.1466	0.0240	0.0070	0.0260
Two-variate model	*SE*-I_d-*UCS*	0.9051	18.3554 (MPa)	0.9218	7.9442 (MPa)	0.8981	6.5318 (MPa)	0.7506	22.8337 (MPa)

By substituting the drilling experiment data into the established *UCS* prediction model, it can be obtained that the prediction *UCS* of concrete material is 17.49 MPa. Compared to the tested *UCS* of concrete (14.7 MPa), the prediction accuracy is 81.02%.

6. Discussion

In order to investigate the relationships between the *DPM* parameters and rock drillability, the essential parameters, such as thrust, rotating speed, etc., needed to be collected during drilling. Therefore, the feasibility for evaluating rock drillability with *DPM* parameters is discussed in this chapter.

6.1. The Key DPM Parameters

According to the formula of *SE* and I_d, the thrust, rotating speed, torque and *ROP* are essential during drilling. They can be divided into two categories that are controllable and uncontrollable, respectively. Among these parameters, the thrust and rotating speed belong to the former, while the torque and *ROP* belong to the latter. In the process of drilling, the drilling performance can be improved by reasonably controlling the thrust and rotating speed. However, the torque and *ROP*, as the characterization parameters while drilling, change with the thrust and rotating speed, and they are also very sensitive to the mechanical properties of the rock mass. Therefore, it is very important to accurately measure the thrust, rotating speed, torque and *ROP* for the real-time analysis of rock drillability.

6.2. UCS Prediction Model

The established *UCS* prediction model by combining *SE* and I_d under three kinds of different drilling tools (standard diamond solid bit, impregnated diamond bit and PDC bit) has better prediction performance, and its performance is better than the univariate model established by *SE* or I_d separately. It can well reflect the relationships between the *UCS* of rock and *DPM* parameters. By substituting the drilling experimental data into the model, the prediction accuracy is 81.02%, which can meet the requirement of field application. However, the amount of data to establish the model is not very large, and the types of bits are not very comprehensive, so it is impossible to accurately predict the rocks with low *UCS*. Therefore, more *DPM* data in different rock layers and more kinds of drill bits should be considered in future study.

7. Conclusions

In order to investigate the relationships between the *DPM* parameters and rock drillability, drilling tests were performed. The rotating speed, thrust, torque and *ROP* were recorded in real time. Then, these *DPM* parameters were utilized to carry out regression analyses, and the *RF* and *GA-SVM* algorithms were applied to predict the *SE* for reflecting the rock drillability. Finally, the *UCS* prediction model using three types of drilling bits was established by combining the *SE* and I_d. Based on the above analyses, the following conclusions can be drawn:

(a) There is a good linear relationship between thrust and torque at a constant rotating speed condition, and the relationships between thrust and *ROP*, and torque and *ROP* are powerful, but their statistical relationship is not very significant. The relationships of the *SE* with thrust and torque are linear at a lower rotating speed, while there is no significant statistical relationship at medium and high rotating speeds.

(b) The established *N-F-ROP* and *N-T-ROP* regression models have better performance than the two-dimensional regression models. Meanwhile, the *SE* model of Teale has been improved based on the good relationship between thrust and torque. The new model has good performance to predict the *SE*. However, the improved *SE* model is not applicable to all the fields completely, and the general formula needs to be further studied.

(c) The prediction performance of the *GA-SVM* model is superior to the *RF* model, and it can reach more than 99%. Therefore, machine learning based on *GA-SVM* is a promising method to analyze the relationships between the *DPM* parameters and rock drillability. Moreover, it is feasible to evaluate the *SE* of drilling based on thrust, rotating speed, *ROP* and drilling area (*A*).

(d) The *UCS* prediction model was established by combining *SE* and I_d. It can reflect the relationships between the *UCS* and *DPM* parameters well for three types of bits (standard diamond solid bit, impregnated diamond bit and PDC bit).

Author Contributions: Conceptualization, S.W. and Y.T.; methodology, R.C., Y.T., S.W., Z.Z. and X.C.; validation, S.W. and Y.T.; writing—original draft preparation, Y.T. and S.W.; writing and review, S.W., Y.T., R.C., Z.Z. and X.C. All authors have read and agreed to the published version of the manuscript.

Funding: The work presented in this paper was financially supported by the National Natural Science Foundation of China (Grant No. 52174099, 51904333 and 52079150), the Natural Science Foundation of Hunan Province (No. 2021JJ30842), Young Talent Support Project of China Association for Science and Technology (No. YSXH 2020-QT002) and Changsha Outstanding Innovative Youth Program (No. kq2107003).

Institutional Review Board Statement: Not applicable.

Informed Consent Statement: Not applicable.

Data Availability Statement: Please email to request from Cao Ruilang at: caorl@iwhr.com.

Conflicts of Interest: The authors declare no conflict of interest.

References

1. Jan, Y.M.; Campbell, R.L.J. Borehole correction of MWD gamma ray and resistivity logs. In *SPWLA 25th Annual Logging Symposium*; Society of Professional Well Log Analysts Inc.: Houston, TX, USA, 1984; pp. 189–197.
2. Howarth, D.F.; Adamson, W.R.; Berndt, J.R. Correlation of model tunnel boring and drilling machine performances with rock properties. *Int. J. Rock Mech. Min. Sci. Geomech. Abstr.* **1986**, *23*, 171–175. [CrossRef]
3. Zacas, M.; Kavadas, M.; Mihalis, I. The use of recording drilling parameters in rock mass classification. In Proceedings of the 8th ISRM Congress, Tokyo, Japan, 25–29 September 1995; International Society for Rock Mechanics: Salzburg, Austria, 1995; Volume 3, pp. 1121–1124.
4. Gui, M.-W.; Soga, K.; Bolton, M.D.; Hamelin, J.P. Instrumented borehole drilling for subsurface investigation. *J. Geotech. Geoenviron.* **2002**, *128*, 283–291. [CrossRef]
5. Schunnesson, H. RQD predictions based on drill performance parameter. *Tunn. Undergr. Space Tech.* **1996**, *11*, 345–351. [CrossRef]
6. Yue, Z.Q.; Chen, J.; Gao, W. Automatic drilling process monitoring (DPM) for in-situ characterization of weak rock mass strength with depth. In *1st Canada-US Rock Mechanics Symposium (Rock Mechanics: Meeting Society's Challenges and Demands)*; ARMA: Vancouver, BC, Canada, 2007; pp. 199–206.
7. Tan, Z.Y.; Cai, M.F.; Yue, Z.Q.; Tham, L.G.; Lee, C.F. Application and reliability analysis of DPM system in site investigation of HK weathered granite. *J. Univ. Sci. Technol. B* **2005**, *12*, 481–488.
8. Yue, Z.Q.; Lee, C.F.; Law, K.T.; Tham, L.G. Automatic monitoring of rotary percussive drilling for ground characterization illustrated by a case example in Hong Kong. *Int. J. Rock Mech. Min. Sci.* **2004**, *41*, 573–612. [CrossRef]
9. Yue, Z.Q. Drilling process monitoring for refining and upgrading rock mass quality classification methods. *Chin. J. Rock Mech. Eng.* **2014**, *33*, 1977–1996.
10. He, M.M.; Li, N.; Zhang, Z.Q.; Yao, X.C.; Chen, Y.S.; Zhu, C.H. An empirical method for determining the mechanical properties of jointed rock mass using drilling energy. *Int. J. Rock Mech. Min. Sci.* **2019**, *116*, 64–74. [CrossRef]
11. Rodgers, M.; Mcvay, M.; Horhota, D.; Hernando, J. Assessment of Rock Strength from Measuring While Drilling Shafts in Florida Limestone. *Can. Geotech. J.* **2018**, *55*, 1154–1167. [CrossRef]
12. Rodgers, M.; Mcvay, M.; Ferraro, C.; Horhota, D.; Tibbetts, C.; Crawford, S. Measuring Rock Strength While Drilling Shafts Socketed into Florida Limestone. *J. Geotech. Geoenviron.* **2018**, *144*, 04017121. [CrossRef]
13. Rodgers, M.; Mcvay, M.; Horhota, D. Monitoring While Drilling Shafts in Florida Limestone. In Proceedings of the IFCEE 2018 Installation, Testing and Analysis of Deep Foundations (GSP 294), Orlando, FL, USA, 5–10 March 2018; American Society of Civil Engineers: Reston, VA, USA, 2018; pp. 613–621. [CrossRef]
14. Rodgers, M.; Mcvay, M.; Horhota, D.; Sinnreich, J.; Hernando, J. Assessment of Shear Strength from Measuring While Drilling Shafts in Florida Limestone. *Can. Geotech. J.* **2018**, *56*, 662–674. [CrossRef]
15. Karasawa, H.; Ohno, T.; Kosugi, M.; Rowley, J.C. Methods to estimate the rock strength and tooth wear while drilling with roller-bits-part 1: Milled-Tooth Tooth Bits. *Energy Resour. Technol.* **2002**, *124*, 125–132. [CrossRef]
16. Karasawa, H.; Ohno, T.; Kosugi, M.; Rowley, J.C. Methods to estimate the rock strength and tooth wear while drilling with roller-bits-part 2: Insert bits. *Energy Resour. Technol.* **2002**, *124*, 133–140. [CrossRef]
17. He, M.; Li, N.; Zhu, J.; Chen, Y. Advanced prediction for field strength parameters of rock using drilling operational data from impregnated diamond bit. *J. Pet. Sci. Eng.* **2019**, *187*, 106847. [CrossRef]
18. Li, Y.; She, L.; Wen, L.; Zhang, Q. Sensitivity analysis of drilling parameters in rock rotary drilling process based on orthogonal test method. *Eng. Geol.* **2020**, *270*, 105576. [CrossRef]
19. Li, H.; Liu, S.; Chang, H. Experimental research on the influence of working parameters on the drilling efficiency. *Tunn. Undergr. Space Technol.* **2020**, *95*, 103174. [CrossRef]
20. Shang, F.; Yu, W.; Guo, Z.; Yu, Z.; Shan, W.; Rui, C.; Shang, X. Estimation of optimal drilling efficiency and rock strength by using controllable drilling parameters in rotary non-percussive drilling. *J. Pet. Sci. Eng.* **2020**, *193*, 107376. [CrossRef]
21. Rossi, E.; Jamali, S.; Saar, M.O.; von Rohr, P.R. Field test of a Combined Thermo-Mechanical Drilling technology. Mode I: Thermal spallation drilling. *J. Pet. Sci. Eng.* **2020**, *190*, 107005. [CrossRef]
22. Rossi, E.; Jamali, S.; Schwarz, D.; Saar, M.O.; von Rohr, P.R. Field test of a Combined Thermo-Mechanical Drilling technology. Mode II: Flame-assisted rotary drilling. *J. Pet. Sci. Eng.* **2020**, *190*, 106880. [CrossRef]

23. Lakshminarayana, C.R.; Tripathi, A.K.; Pal, S.K. Prediction of Mechanical Properties of Sedimentary Type Rocks Using Rotary Drilling Parameters. *Geotech. Geol. Eng.* **2020**, *38*, 4863–4876. [CrossRef]
24. Kalantari, S.; Hashemolhosseini, H.; Baghbanan, A. Estimating rock strength parameters using drilling data. *Int. J. Rock Mech. Min. Sci.* **2018**, *104*, 45–52. [CrossRef]
25. Kalantari, S.; Baghbanan, A.; Hashemolhosseini, H. An analytical model for estimating rock strength parameters from small-scale drilling data. *Rock Mech. Geotech. Eng.* **2019**, *104*, 135–145. [CrossRef]
26. Hughes, H.M. Some aspects of rock machining. *Int. J. Rock Mech. Min. Sci. Geomech. Abstr.* **1972**, *9*, 205–211. [CrossRef]
27. Mellor, M. Normalization of specific energy values. *Int. J. Rock Mech. Min. Sci. Geomech. Abstr.* **1972**, *9*, 661–663. [CrossRef]
28. Poane, J.; Madson, D.; Bruce, W.E. *Drillability Studies: Laboratory Percussive Drilling*; US Department of the Interior, Bureau of Mines: Washington, DC, USA, 1969.
29. Zhang, K.; Hou, R.B.; Zhang, G.H.; Zhang, G.M.; Zhang, H.Q. Rock Drillability Assessment and Lithology Classification Based on the Operating Parameters of a Drifter: Case Study in a Coal Mine in China. *Rock Mech. Rock Eng.* **2016**, *49*, 329–334. [CrossRef]
30. Yu, B.S.; Zhang, K.; Niu, G.G.; Xue, X.R. Real-time rock strength determination based on rock drillability index and drilling specific energy: An experimental study. *B. Eng. Geol. Environ.* **2021**, *80*, 3589–3603. [CrossRef]
31. Teale, R. The concept of specific energy in rock drilling. *Int. J. Rock Mech. Min. Sci. Geomech. Abstr.* **1965**, *2*, 57–73. [CrossRef]
32. Kahraman, S. Rotary and percussive drilling prediction using regression analysis. *Int. J. Rock Mech. Min. Sci.* **1999**, *36*, 981–989. [CrossRef]
33. Kahraman, S.; Balci, C.; Yazici, S.; Bilgin, N. Prediction of the penetration rate of rotary blast hole drilling using a new drillability index. *Int. J. Rock Mech. Min. Sci.* **2000**, *37*, 729–743. [CrossRef]
34. Ataei, M.; KaKaie, R.; Ghavidel, M.; Saeidi, O. Drilling rate prediction of an open pit mine using the rock mass drillability index. *Int. J. Rock Mech. Min. Sci.* **2015**, *73*, 130–138. [CrossRef]
35. Ocak, I.; Seker, S.E. Estimation of elastic modulus of intact rocks by artificial neural network. *Rock Mech. Rock Eng.* **2012**, *45*, 1047–1054. [CrossRef]
36. Yesiloglu-Gultekin, N.; Gokceoglu, C.; Sezer, E.A. Prediction of uniaxial compressive strength of granitic rocks by various nonlinear tools and comparison of their performances. *Int. J. Rock Mech. Min. Sci.* **2013**, *62*, 113–122. [CrossRef]
37. Sarkar, K.; Tiwary, A.; Singh, T.N. Estimation of strength parameters of rock using artificial neural networks. *Bull. Eng. Geol. Environ.* **2010**, *69*, 599–606. [CrossRef]
38. He, M.M.; Zhang, Z.Q.; Ren, J.; Huan, J.Y.; Li, G.F.; Chen, Y.S.; Li, N. Deep convolutional neural network for fast determination of the rock strength parameters using drilling data. *Int. J. Rock Mech. Min. Sci.* **2019**, *123*, 104084. [CrossRef]
39. Roxborough, F.F.; Phillips, H.R. Rock excavation by disc cutter: Authors reply to the Discussion on F. F. Roxborough and H. R. Phillip's paper. *Int. J. Rock Mech. Min. Sci. Geomech. Abstr.* **1975**, *12*, 361–366. [CrossRef]
40. Jalali, S.E.; Zare, M. An analytical pre-feasibility study to generate rotary-percussive concept in hard rock TBMs. *J. Geol. Min. Res.* **2009**, *1*, 140–148.
41. Marji, M.F. Simulation of crack coalescence mechanism underneath single and double disc cutters by higher order displacement discontinuity method. *J. Cent. South Univ.* **2015**, *22*, 1045–1054. [CrossRef]
42. Marji, M.F. *Rock Fracture Mechanics with Displacement Discontinuity Method*; LAP Lambert Academic Publishing, Yazd Univ, Dept Min Engn: Yazd, Iran, 2014.
43. Goetz, J.N.; Brenning, A.; Petschko, H.; Leopolda, P. Evaluating machine learning and statistical prediction techniques for landslide susceptibility modeling. *Comput. Geosci-UK* **2015**, *81*, 1–11. [CrossRef]
44. Cutler, D.R.; Edwards, T.C., Jr.; Beard, K.H.; Cutler, A.; Hess, K.T.; Gibson, J.; Lawler, J.J. Random forests for classification in ecology. *Ecology* **2007**, *88*, 2783–2792. [CrossRef]
45. Chang, C.C.; Lin, C.J. *LIBSVM: A library for Support Vector Machines*; ACM Transactions on Intelligent Systems and Technology: New York, NY, USA, 2011.
46. Cutler, D.R.; Edwards, T.C., Jr.; Beard, K.H.; Cutler, A.; Hess, K.T.; Gibson, J.; Lawler, J.J. Landslide spatial modeling: Introducing new ensembles of ANN, MaxEnt, and SVM machine learning techniques. *Geofis. Internacional.* **2017**, *305*, 314–327.
47. Breiman, L. Random forests. *Mach. Learn.* **2001**, *45*, 5–32. [CrossRef]
48. Cortes, C.; Vapnik, V. Support-vector networks. *Mach. Learn.* **1995**, *20*, 273–297. [CrossRef]
49. Tang, K.S.; Man, K.F.; Kwong, S.; He, Q. Genetic Algorithms and Their Applications. *IEEE Signal Process.* **1996**, *13*, 21–37. [CrossRef]
50. Wang, S.F.; Li, X.B.; Yao, J.R.; Gong, F.Q.; Li, X.; Du, K.; Tao, M.; Huang, L.Q.; Du, S.L. Experimental investigation of rock breakage by a conical pick and its application to non-explosive mechanized mining in deep hard rock. *Int. J. Rock Mech. Min. Sci.* **2019**, *122*, 104063. [CrossRef]
51. Wang, S.F.; Sun, L.C.; Li, X.B.; Wang, S.Y.; Du, K.; Li, X.; Feng, F. Experimental investigation of cuttability improvement for hard rock fragmentation using conical cutter. *Int. J. Geomech.* **2021**, *21*, 06020039. [CrossRef]
52. Wang, S.F.; Tang, Y.; Li, X.B.; Du, K. Analyses and predictions of rock cuttabilities under different confining stresses and rock properties based on rock indentation tests by conical pick. *Trans. Nonferr Metal. Soc.* **2021**, *31*, 1766–1783. [CrossRef]
53. Wang, S.F.; Tang, Y.; Wang, S.Y. Influence of brittleness and confining stress on rock cuttability based on rock indentation tests. *J. Cent. South Univ.* **2021**, *28*, 2786–2800. [CrossRef]

54. Wang, Y.J.; She, L.; Zhao, Y.F.; Cao, R.L. Experimental study on measurement of rock strength parameters based on digital drilling technology. *Chin. J. Geotech. Eng.* **2020**, *42*, 1659–1668.
55. He, M.M. Research on the prediction of rock mass mechanics characteristics based on the rotary penetration technology. *Xi'an: Xi'an Univ. Technol.* **2021**, *28*, 2786–2800.
56. Wang, Y.J.; She, L.; Zhao, Y.F.; Cao, R.L. Method for Measuring Rock Mass Characteristics and Evaluating the Grouting-Reinforced Effect Based on Digital Drilling. *Rock Mech. Rock Eng.* **2018**, *52*, 841–851. [CrossRef]
57. Wang, Q.; Qin, Q.; Gao, S.; Li, S.C.; Gao, H.K.; He, M.C.; Jiang, B.; Zhang, C. Relationship between rock drilling parameters and rock uniaxial compressive strength based on energy analysis. *J. China Coal Soc.* **2018**, *43*, 1289–1295.

mathematics

Article

Acoustic Emission *b* Value Characteristics of Granite under True Triaxial Stress

Longjun Dong, Lingyun Zhang, Huini Liu, Kun Du and Xiling Liu *

School of Resources and Safety Engineering, Central South University, Changsha 410083, China; lj.dong@csu.edu.cn (L.D.); 165511003@csu.edu.cn (L.Z.); lhn1722421759@163.com (H.L.); dukuncsu@csu.edu.cn (K.D.)

* Correspondence: lxlenglish@163.com

Abstract: The acoustic emission *b* value is an important and widely used parameter for the early prediction of rock fractures. In this study, five groups of true triaxial compression tests were conducted on granite specimens to analyze changes in *b* value during the process of rock failure, and to investigate the *b* value characteristics of acoustic emission events. First, the acoustic emission events that simultaneously triggered at least four sensors were located using P-wave arrivals and sensor coordinates. Then, considering various intervals of acoustic emission event counts, stress magnitude, and stress proportion, *b* values were calculated using the values of the maximum amplitude, average amplitude, maximum absolute energy, and average absolute energy of the acoustic emission events. In addition, the goodness of the fitting curves was used to evaluate the fitting reliability of the *b* values. The results indicated higher accuracy of *b* value when calculated using the average amplitude setting for intervals of acoustic emission event counts of 200 or greater, stress magnitude of 20 MPa or greater, and stress proportion of 10% or greater. Moreover, the interval of event counts of 200 is suggested as a window parameter for *b* value calculations, and the *b* values are observed to exhibit a decreasing trend before fracture for more than 80% of the specimens. Furthermore, the *b* value tends to decrease with an increase in confining pressure. Thus, the *b* value can be used as an indicator for validating the stress concentration area, including magnitudes and accumulative probability density distribution of events, which is a beneficial complement to clarifying precursor information of rock mass instability.

Keywords: true triaxial compression test; acoustic emission; *b* value

Citation: Dong, L.; Zhang, L.; Liu, H.; Du, K.; Liu, X. Acoustic Emission *b* Value Characteristics of Granite under True Triaxial Stress. *Mathematics* **2022**, *10*, 451. https://doi.org/10.3390/math10030451

Academic Editor: Pastor Manuel

Received: 22 December 2021
Accepted: 27 January 2022
Published: 30 January 2022

1. Introduction

The concept of *b* value originated in the study of seismology. In 1944, Gutenberg and Richter first proposed that *b* value reflects the relationship between the frequency and magnitude of earthquakes, i.e., the number of earthquakes exponentially decreasing with the increase of the magnitude [1], and the *b* value is an important scale parameter for operational earthquake forecasting during earthquake sequences [2–8]. Several engineering disasters and great earthquakes have shown that *b* value decreased before the time of occurrence of higher magnitude events (e.g., main shocks, rock bursts and roof falls) [9–14]. The failure mechanism of brittle rocks is similar to that of earthquakes, so *b* value analysis for the microseism and acoustic emission (AE) signals obtained from the failure process of brittle rocks has become an important research issue [14–16].

In the past few decades, the *b* value has been widely used in rock failure prediction and rock damage analysis [17–21]. In an attempt to reveal the onset precursors of rock damage and failure, the uniaxial and conventional triaxial stress tests have been conducted by several researchers on small-scale rock specimens, and the *b* value characteristics of the process of rock failure have been investigated. For instance, Mogi first proposed that the *b* value reflects the internal heterogeneity of rocks, and that more heterogeneous rocks have a higher *b* value [22]. Scholz first observed that the *b* value is related to the stress state of the rocks, and decreases as the stress increases [23]. Weeks et al. reported that the b value is negatively correlated with stress [12]. As the stress level increases beyond 93% and 100%, the b value begins to decrease [7]. Moreover, the b value can capture the inception of severe cracking in concrete [24]. However, studies on the b value characteristics for rocks under different true triaxial stresses have been rarely reported thus far.

Because the b value is used as a warning index for earthquakes and rock damage, it is important to capture the trend in its variations with the change of time and stress. While the methods used by scholars for calculating the b value are different, the amplitude and energy of AE hits are most commonly used to estimate b value [25]. At least four methods for obtaining the magnitude of completeness (Mc) have been used to improve the calculation precision of the b value [26–29]. Meanwhile, when a single AE sensor is used to detect AE signals, it cannot use the location method to determine the AE events and the noise signals from the testing machine and environment, which are used to analyze the b value and other properties of the AE; this may result in a high AE hit rate (HR) and an inaccurate b value.

In this study, five groups of true triaxial compressive tests were conducted on granite specimens. The minimum principal stress (σ_3) values used for the test were 10, 20, 30, 50, and 100 MPa, and the intermediate principal stress (σ_2) was varied from test to test at the same σ_3 value. Six sensors were attached to the loading mode to detect the AE signals during the loading. AE localization without a premeasured velocity was applied [30]. Based on the location result of the AE signals, the AE events located inside the specimen were selected, and the values of A_{max}, A_{avg}, E_{max}, and E_{avg} of the AE events were chosen for analysis. A new b value estimation method which employs apparent frequency-amplitude distribution was used to calculate various b values in this loading test, and the b value characteristics of granite specimens under different true triaxial stresses were discussed. The results of this paper are of great help for deeper understanding the relationship between the b value and the differential stress, as well as the damage accumulation assessment through temporal variation of b value in the true triaxial stress state.

2. Experimental Details

The true triaxial testing system (Model TRW-3000) used in this study was developed by the Advanced Research Center at Central South University, China, and the detailed parameters of the testing system have been described by several literatures [31–35]. The test system can apply quasi-static loading along the direction of the three principal stresses independently through hydraulically driven pistons, with a maximum load capacity of 3000, 2000, and 2000 kN along the vertical (z), and horizontal (x and y) directions, respectively. The loading system is shown in Figure 1a,b. The plate has extremely high stiffness and rigidity in order to transfer the load into the rock specimens. Five groups of tests for σ_3 values of 10, 20, 30, 50 and 100 MPa were conducted. In the test settings, σ_1 was applied along the z direction, σ_2 was applied along the y direction, and σ_3 was applied along the x direction. The loading control mode was set to a loading speed of 0.3 MPa/s. The loading paths of the tests, as shown in Figure 1c, are as follows:

a. σ_1, σ_2, and σ_3 were loaded with 1 MPa to keep the rock specimens attached to the plate; and subsequently the three loads (i.e., σ_1, σ_2, and σ_3) were increased until σ_3 reached its predefined value.
b. σ_1 and σ_2 were loaded until σ_2 reached its predefined value, while σ_3 was kept constant.
c. σ_1 was loaded until the failure of the rock specimen, while σ_2 and σ_3 were kept constant.

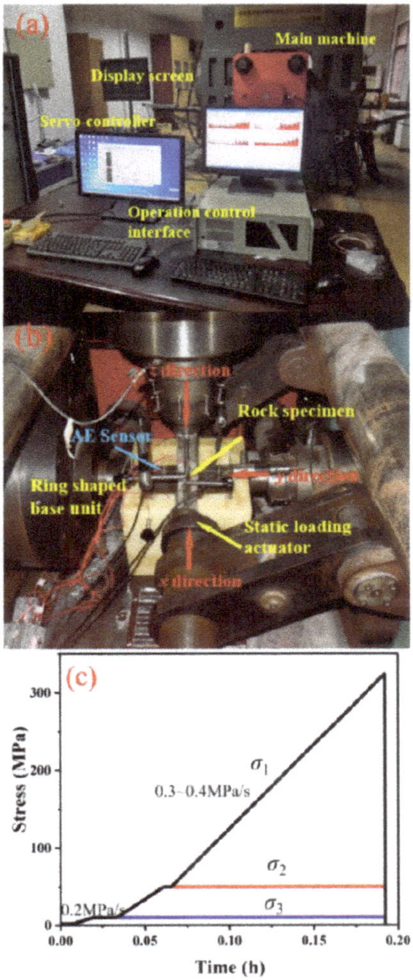

Figure 1. Illustration of the true triaxial test system: (**a**) an overview of the testing system; (**b**) image of the true triaxial machine and the designed loading apparatus; and (**c**) representative loading path. AE in the figure represents acoustic emission.

The granite specimens used in these experiments were obtained from a granite quarry in the Hunan Province, China, the average uniaxial compressive strength was 113.7 MPa. The main mineral composition of the granite includes feldspar, quartz, and black mica. The rock blocks were cut into cubic specimens with a side length of 50 ± 0.02 mm.

A PCI-2 AE test system with the AE-win software was used to monitor the AE signals of the granite specimens. The NANO-type resonant narrowband sensors with a center response frequency of 125 (750) kHz and 2/4/6 preamplifiers were chosen for the tests. Six AE sensors were uniformly fixed on six platens in six directions around the specimen. During the tests, both the gain of the preamplifier and the threshold were set to 40 dB, the sampling rate was set to 10 MHz, and the threshold was 40 dB.

3. Analysis of *b* Value Characteristics

3.1. AE Localization

During the experiments, the triaxial stress applied to the rock specimens increased gradually; accordingly, the velocity field inside the specimens changed with the change in triaxial stress [36]. Many factors, such as sonic speed [37], temperature [38], abnormal arrivals [39], have negative effect on the localization of AE events. As a result, the localization results with large errors will significantly affect the event type discrimination [40], empty region identification [41], and abnormal region tomography [42]. To eliminate the influence of wave velocity error on localization results, a localization method without premeasured wave velocity was proposed [30,43–45]. The advantage of analytical algorithm that it could obtain the precious solution was properly applied to remove the influence of abnormal arrivals in collaboration with the Iterative Solutions [39]. Furthermore, Dong et al. [46] optimized the A* path search algorithm and proposed a velocity-free localization method (VFH) for complex three-dimensional structures. The localization accuracy was greatly improved using the proposed VFH localization method. In this study, six sensors were used and their coordinates were assumed to be S_i (x_i, y_i, z_i) (i = 1, 2, 3, 4, 5, 6). The localization method without premeasuring the wave velocity was used to obtain the location of AE events during the experiments. For an AE event, assuming that the focal coordinate is (x, y, z), the origin time is t_0, and the P-wave arrival at sensor S_i is t_i (i = 1, 2, 3, 4, 5, 6), the following equation is satisfied:

$$(x_i - x)^2 + (y_i - y)^2 + (z_i - z)^2 = v^2(t_i - t_0)^2 \tag{1}$$

where the symbol v represents the unknown velocity of the P-wave in the travel path. When at least four sensors are triggered, the simultaneous Equation (1) can be used to determine the solution; and the coordinates of the focus, i.e., (x, y, z) inside the specimen space is filtered as a valid AE event.

3.2. AE Signals

Rock fracture is the initiation and propagation of microcracks until they form macrocracks, inducing rock fraction [44]. When a microcrack is formed, the AE sensor is triggered multiple times to the vibration wave whose voltage exceeds the threshold. An AE hit is a vibration wave detected by an AE sensor, as shown in Figure 2, in which the analog signal is converted into a digital signal by the collector. A complete wave is an AE hit. Theoretically, the number of AE hits is positively correlated with the degree of internal damage of the rock.

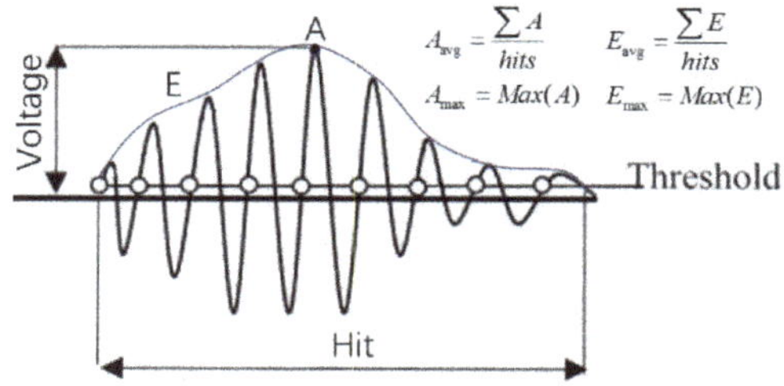

Figure 2. Waveform of AE electrical acoustic emission signals. A represents the amplitude of the signal. E represents the energy of the signal, which is the area enclosed by the blue envelope curve.

Figure 3a,b shows the relationship between the AE hit rate (HR), AE event rate (ER), cumulative AE hit rate (CHR), cumulative AE event rate (CER), and the true triaxial stress of the granite specimen G-10-10 (the specimen No. was designed as "G-digit1-digit2", where "G" represents Granite, "digit1" represents the applied stress value of σ_3, and "digit2" represents the applied stress value of σ_2; here, "G-10-10" means that the applied stress value of σ_3 and σ_2 on this granite specimen are 10 MPa and 10 MPa, respectively). As the same trends were found in all specimens, the testing results of specimen G-10-10 were mostly used for variation analysis of AE parameters hereafter. HR is the total number of AE hits of all sensors per second divided by the total number of sensors, and ER is the number of located AE events per second. Clearly, HR and ER have roughly the same variation trend—an initial increase, followed by oscillation, and finally a decrease. Initially, the value of ER ranged between 40 and 80, while that of HR ranged between 100 and 200. When σ_1 was 164.7 MPa at 263 s, the HR and ER values gradually decreased; and when σ_1 is 76% of its peak value (217.7 MPa), the ER and HR values decreased to less than 10 and less than 50, respectively. Furthermore, the slopes of CHR and CER are observed to decrease with the increase in σ_1, indicating a change in the trend of the HR and ER values; the ER value remained at less than 10 until 356 s, when the specimen was fractured. The other specimens were also observed to exhibit a trend similar to that of the granite specimen G-10-10.

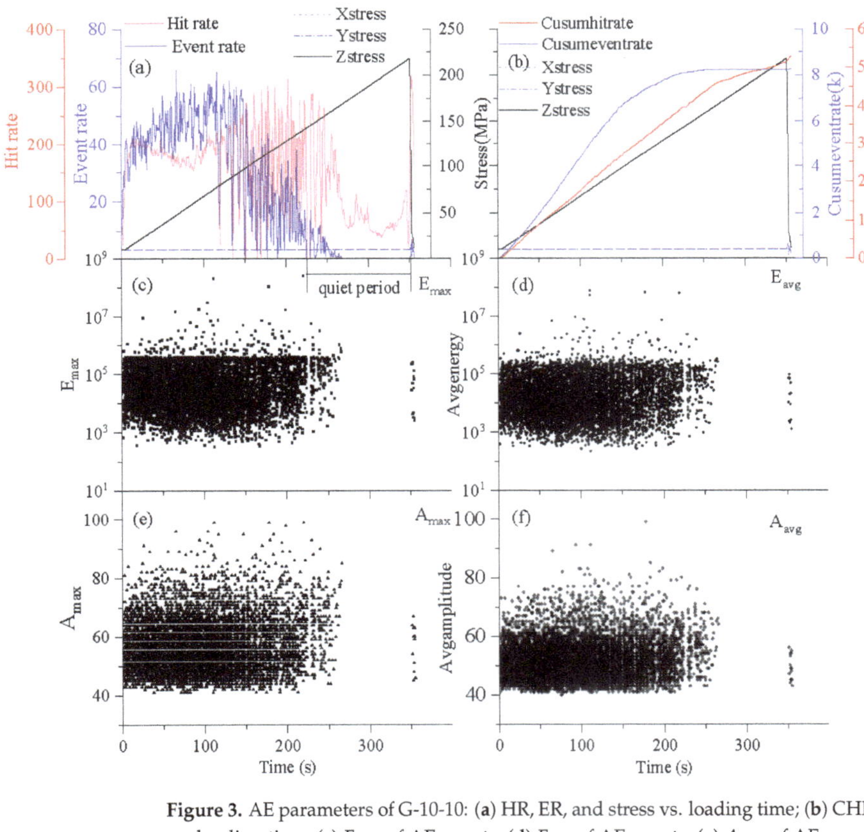

Figure 3. AE parameters of G-10-10: (**a**) HR, ER, and stress vs. loading time; (**b**) CHR, CER, and stress vs. loading time; (**c**) E_{max} of AE events; (**d**) E_{avg} of AE events; (**e**) A_{max} of AE events; and (**f**) A_{avg} of AE events.

3.3. Quiet Period

The change trends of different parameters, i.e., the absolute energy (E_{max}, E_{avg}) and amplitude (A_{max}, A_{avg}) values of AE events in specimen G10-10, are shown in Figure 3c–f. E_{max}, E_{avg}, A_{max} and A_{avg} are only meaningful for one AE event. E_{max} and A_{max} are the maximum value and selected among the AE sensors triggered by the same AE events, and E_{avg} and A_{avg} are the average value of the AE sensors triggered by the same AE events. In this study, E_{max} and E_{avg} have a similar change trend; A_{max} and A_{avg} also have a similar change trend. The total AE process can be divided into "active period" and "quiet period" according to the distribution of ER and the E and A values. As σ_1 approached its peak value, and the values of HR and ER decreased to approximately 50 and less than 10, respectively (which was the quiet period). The HR curve was observed to fluctuate slightly, while the ER remained stable. Therefore, it is more scientific and reasonable to use ER to determine the quiet period.

Figure 4 shows that the value of ER did not increase linearly with increasing σ_1, and some blanks were observed before the specimen fracture. However, the ER value increased significantly after the stress state of the rock specimen changed. Furthermore, when σ_3 reached its predefined value and σ_1 and σ_2 continued to increase, the value of ER was observed to increase and then decrease gradually. Similarly, when σ_2 reached its predefined value and σ_1 continued to increase, the value of ER increased and then decreased gradually. Thus, the stress state can cause changes in ER. Therefore, ER is expected to reduce gradually to finally disappear at the same stress station. To some extent, ER has a great relationship with the stress state of the rock specimen.

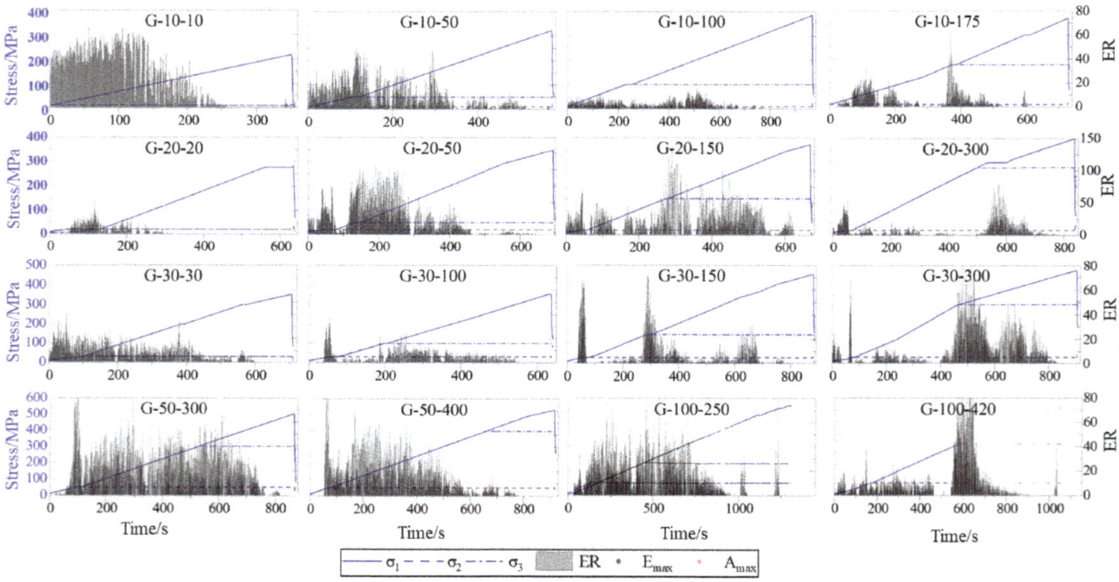

Figure 4. ER, stress vs. time of all specimens in this study. The specimen No. was designed as "G-digit1-digit2", where "G" represents granite, "digit1" represents the applied stress value of σ_3, and "digit2" represents the applied stress value of σ_2; for example, "G-10-50" means that the applied stress value of σ_3 and σ_2 on this granite specimen are 10 MPa and 50 MPa, respectively.

The length of the quiet period is shown in Table 1. When σ_1 reached between 50% and 93% of its peak value, the ER and HR values significantly declined, and AE entered the quiet period. Figure 5 shows the length of the quiet period, and a large dispersion in the distribution of the quiet period can be observed; the duration was from 93–452 s, the mean was 239.79 s, and the variance was 114.81. The stress at the beginning of the quiescent period was 27–91% of the maximum stress (mean is 69%). Because there is no obvious rule between the length and start time of the quiet period under different stress states, it is difficult to predict rock failure.

Table 1. Relationship between AE and σ_1 in the quiet period.

Specimen No.	ST	ET	ST/ET	DT	σ_{1qp}/MPa	σ_{1max}/MPa	$\sigma_{1qp}/\sigma_{1max}$
G-10-10	226	356	0.37	130	142.35	217.90	0.65
G-10-50	334	575	0.42	241	186.21	323.00	0.58
G-10-100	583	966	0.40	383	231.51	351.00	0.66
G-10-175	450	727	0.38	277	213.91	370.00	0.58
G-20-20	224	633	0.65	409	75.12	293.20	0.26
G-20-50	470	692	0.32	222	242.90	348.40	0.70
G-20-150	571	669	0.15	98	325.16	390.00	0.83
G-20-300	699	837	0.16	138	344.18	399.03	0.86
G-30-30	455	705	0.35	250	236.66	353.39	0.67
G-30-50	387	669	0.42	282	197.17	340.69	0.58
G-30-100	510	636	0.20	126	281.73	357.33	0.79
G-30-150	427	879	0.51	452	224.36	455.17	0.49
G-30-200	586	952	0.38	366	307.85	500.21	0.62
G-30-300	808	901	0.10	93	437.32	474.57	0.92
G-50-200	773	982	0.21	209	409.69	535.07	0.77
G-50-300	767	873	0.12	106	439.68	503.08	0.87
G-50-400	764	921	0.17	157	458.77	532.66	0.86
G-100-260	919	1303	0.29	384	548.55	750.00	0.73
G-100-420	821	1054	0.22	233	538.00	815.00	0.66

Note: "ST" denotes the start time of the quiet period, "ET" means the end time of the quiet period, and "DT" denotes the duration of the quiet period.

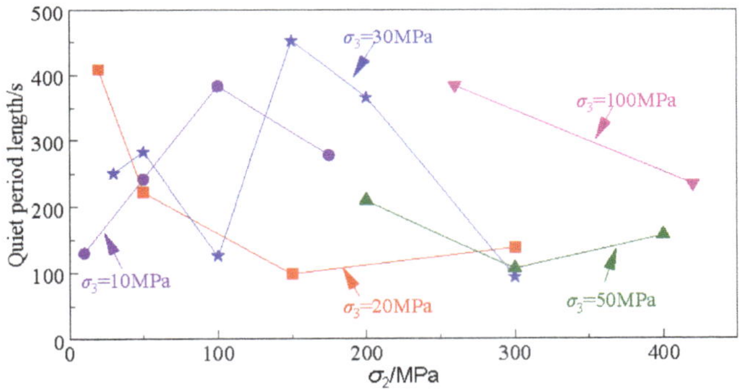

Figure 5. Quiet period under different true triaxial stresses.

3.4. *b* Value

The *b* value is defined as the log-linear slope of the cumulative frequency magnitude distribution of the AE, and the mathematical equation for determining the *b* value is as follows:

$$\lg N = a - b\,M \tag{2}$$

where *M* is the magnitude of the AE events, and *N* is the cumulative frequency of the AE events with magnitude of *M* or greater; *a* is an empirical constant; and *b* is the gradient. This equation is always recognized as the G–R relationship [1]. In the case of the AE technique, the G–R relationship is modified as follows [3,25]:

$$\lg N(A/20) = a - b * (A/20) \tag{3}$$

$$\lg N(\lg(E)) = a - b * \lg(E) \tag{4}$$

where *A* is the peak amplitude of the AE hit, *E* is the absolute energy of the AE hit, and *N* is the cumulative number of the AE hit with an amplitude greater than *A* or absolute energy greater than *E*. The magnitude of the event is a logarithmic scale of the instrumentally measured amplitude, while the value of *A* recorded in dB is divided by 20, and *E* is used in its logarithmic form; $\Delta(A/20) = \Delta(\lg E) = 0.1$ is set for the calculation to produce the same magnitude as that of an earthquake; a is an empirical constant; and b is the b value of the AE. Because the amplitudes used for b value estimation in rock AE test are the apparent amplitudes measured on the sample boundary by acoustic sensor which is attenuated from the source, the corresponding apparent frequency-amplitude distribution does not necessarily represent the underlying size distribution of the sources because attenuation may modify the b value. Recently, attenuation effect on apparent amplitude-frequency distribution and b value was investigated from a statistical point of view by Liu et al. [4] and Chen et al. [47]; the authors theoretically proved that the b value is unchanged within a specific interval before and after attenuation. This indicates that if the apparent frequency-amplitude distribution after attenuation follows the G–R relationship in a certain interval of amplitude, the b value inferred in that interval is the same as that of the true source amplitude distribution, and the authors also proposed a new method named FGS for b value estimation through apparent amplitude-frequency distribution in a rock AE test. The FGS is a new b value estimation procedure which specify the minimum data volume, data counting procedure, bin width and linear regression method, and employ the Fisher optimal split and the global search algorithm to determine the logarithmic linear segment in the apparent frequency-amplitude distribution. The use of the FGS method for b value estimation actually means that once we obtain the located AE events, we can directly use the FGS method for log-linear segment determination in the apparent frequency-amplitude distribution, and further for b value estimation, without doing the work of attenuation compensation; moreover, the b value estimated by FGS can represent the real source distribution characteristics. The FGS is also used in this loading test for b value estimation.

While both the energy and amplitude of the AE event can indicate the magnitude of an AE event, the b value estimated using the energy data should be taken into consideration [48]. Figure 6 shows the b value estimated from different intervals of AE event counts of the granite specimen G-10-10. The intervals were set at 60, 80, 100, 200, 300, 400, 500, 600, 700, 800, 900, and 1000 events, such that the start value of the next interval is the mean value of the previous interval, as (0 60), (30 90), (90 150) et al. Furthermore, Figure 7 shows that most b value curves exhibit the same trend: first decreasing, and then increasing or continuing to decrease, which was distinctly observed for count intervals at 1000 events. Similarly, the values of b_{Emax} and b_{Eavg} of all specimens exhibited the same trend as b_{Amax} and b_{Aavg}. However, the values of R_{Emax} and R_{Eavg} were significantly smaller than the values of R_{Amax} and R_{Aavg}, which exhibited a very similar behavior, and in most cases, the latter values are greater than 0.95. In general, the b value ranged between 0.3 and 3, and the value of R^2 was greater than 0.8. When the intervals were 60, 80, and 100, the b value was greater than 3 and R^2 was less than 0.8, which indicates that the b value estimated using a small number of AE events has poor applicability. With an increase in interval (i.e., more than 200 events), almost all R^2 values were in the range 0.8–1 and the b values were in the range 0.5–3. The distribution of the b value was observed to be sparser with the increase in the interval, and 200 is the lower limit of this interval. Therefore, to analyze the damage development condition of the rock specimen during the loading process, it is better to choose an interval with a small value. The fluctuating trend of the b value indicates the damage growth in the rock specimen. Table 2 summarizes the mean of R^2 at different AE event count intervals of the specimen G-10-10. Apparently, the R^2 value of b_E (b_{Emax} and b_{Eavg}) is smaller than that of b_A (b_{Amax} and b_{Aavg}), which indicates that the b value calculated using amplitude is better than that using energy. Moreover, the values of b_{Amax} and b_{Aavg} were nearly equal, and interestingly, the mean R^2 value of b_{Emax} increased as the interval of the AE event count increased, while that of b_{Eavg} increased first and then slightly decreased. When the interval of the AE event count was set at 200, the mean of the R^2 of b_{Eavg} a b_{Emax} and b_{Eavg} were 0.968 and 0.971, respectively, and their standard deviations were all 0.020. Thus, we conclude that the b value calculated using E_{avg} at an interval of an AE event count of 200 takes both real-time and reliability into account.

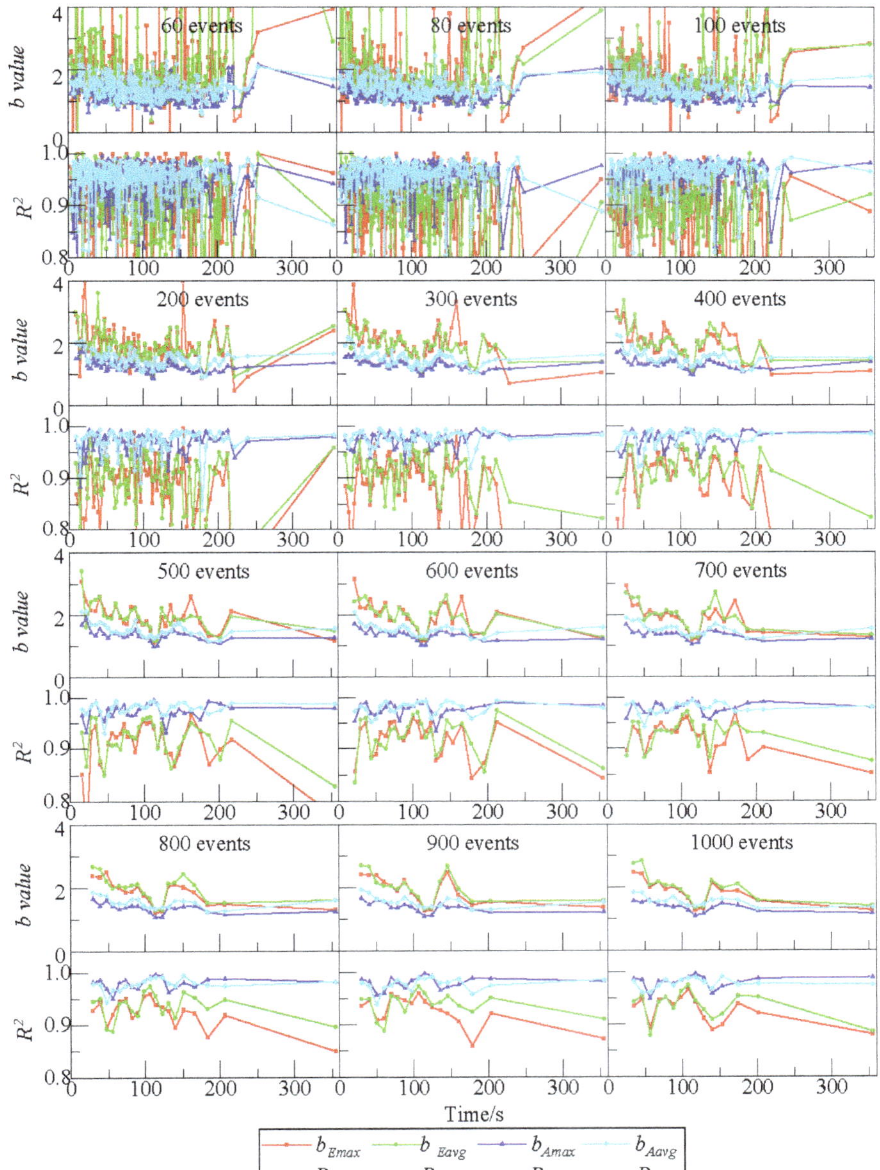

Figure 6. b value and its fitting goodness at different AE event count intervals for specimen G-10-10. The range of b_{Amax} values is between 0.63 to 2.17.

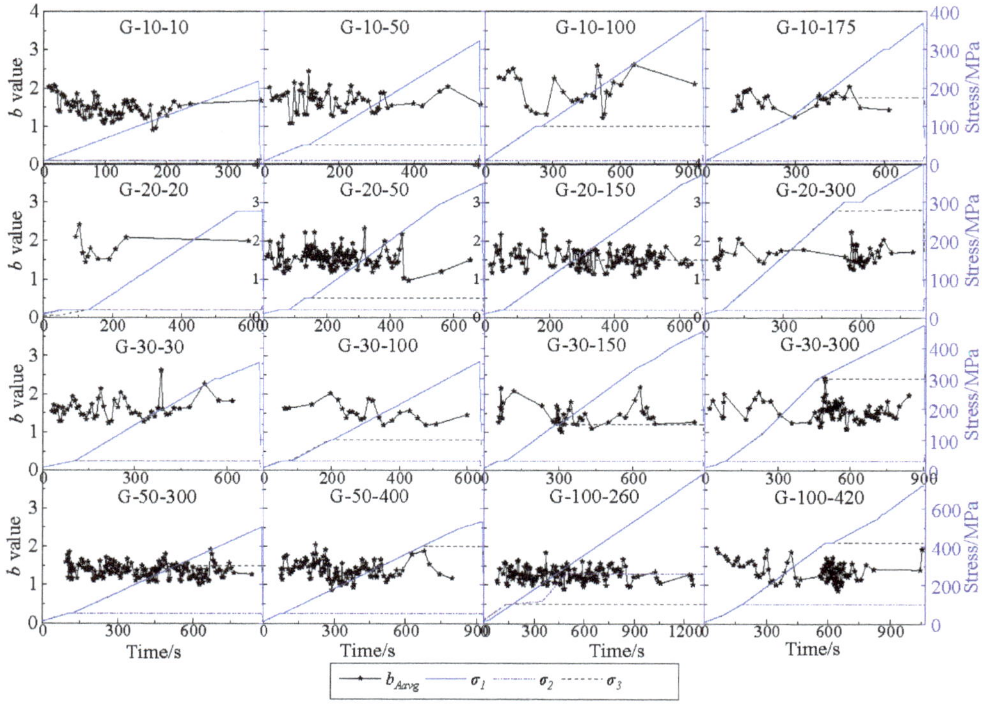

Figure 7. b value and stress vs. time under different true triaxial stresses. The range of b values is between 0.85 to 2.62.

Table 2. The mean of R^2 at different AE event count intervals for specimen G-10-10.

AE Event Interval	Mean of R_{Emax}	Mean of R_{Eavg}	Mean of R_{Amax}	Mean of R_{Aavg}
60	0.890	0.902	0.945	0.945
80	0.883	0.900	0.955	0.954
100	0.879	0.892	0.959	0.960
200	0.881	0.895	0.968	0.971
300	0.888	0.910	0.973	0.977
400	0.897	0.917	0.974	0.978
500	0.905	0.917	0.974	0.980
600	0.916	0.923	0.976	0.980
700	0.920	0.929	0.979	0.979
800	0.923	0.935	0.979	0.978
900	0.926	0.939	0.980	0.976
1000	0.927	0.935	0.980	0.977

In order to distinguish the b value at each loading stage of the rock specimen, the AE events contained in different intervals of stress proportion were used to estimate the b value. Figure 8 shows the variation of b value at the chosen stress proportion intervals of 2%, 4%, 6%, 8%, 10% and 12% of the peak value of σ_1 and a threshold at 200 for a minimum number of AE events in a stress proportion interval was set to ensure the accuracy of the b value. Figure 10 shows that there is a blank space on the right side of the b value curve, and the trend of the b value is obvious when the stress proportion interval was less than 10%—it first decreased and then increased, which is, however, not consistent with Figure 8. The portion of the b value curves falling in the final stage was missing, mainly because the AE event count intervals were less than 200 in the quiet period. The mean of the R^2 at different AE stress proportion intervals of G-10-10 is listed in Table 3; the value of b_E (b_{Emax} and b_{Eavg}) was observed to be smaller than that of b_A (b_{Amax} and b_{Aavg}), and the mean R^2 of b_{Emax} increased as the stress proportion interval increased, while that of b_{Eavg} first increased and then slightly decreased. According to Table 3, calculating the b value using A_{max} and setting the stress proportion interval at 10%, σ_1 is better for distinguishing the b value at each loading stage.

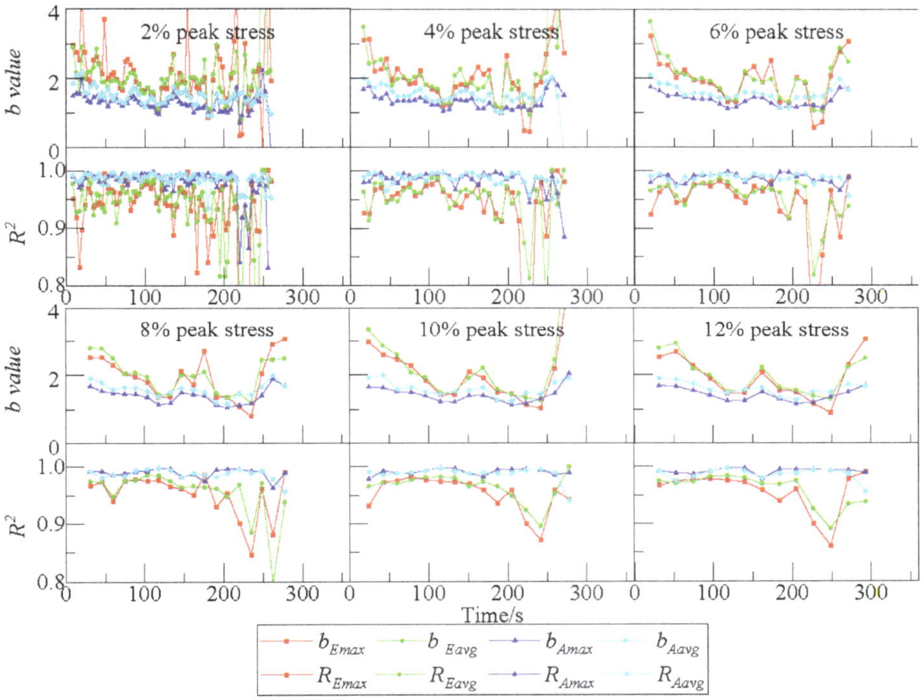

Figure 8. b value and its fitting goodness at different stress proportion intervals for the specimen G-10-10. The range of b_{Amax} values is between 0 to 2.21.

Table 3. The mean of R^2 at different AE stress proportion intervals for the specimen G-10-10.

Stress Proportion Interval	Mean of R_{Emax}	Mean of R_{Eavg}	Mean of R_{Amax}	Mean of R_{Aavg}
2%	0.867	0.875	0.955	0.971
4%	0.883	0.896	0.970	0.973
6%	0.887	0.907	0.973	0.974
8%	0.904	0.909	0.977	0.973
10%	0.906	0.928	0.980	0.974
12%	0.914	0.919	0.982	0.971

In order to distinguish the changes in the b value with time when true triaxial stress is loaded, the b value was estimated using the AE data during a certain stress proportion interval. Because the stress had a positive relationship with time, the loading speed was fixed at 0.3 KN/s. Figure 9 shows the result of the b value estimated according to the AE data during a certain stress interval. The stress intervals of σ_1 were 5, 10, 15, 20, 25, and 30 MPa, and a threshold for a minimum number of AE events was set at 200. The trend of the b value is obvious: it first dropped and then increased, which is consistent with Figure 8. Table 4 lists the mean of R^2 at different AE stress proportion intervals for the specimen

G-10-10, and the mean of R^2 for both R_{Amax} and R_{Aavg} perform well, such that both values are greater than 0.95, and hence, difficult to choose. Combined with Figure 9, setting stress intervals at 20 MPa takes both data size and reliability into account.

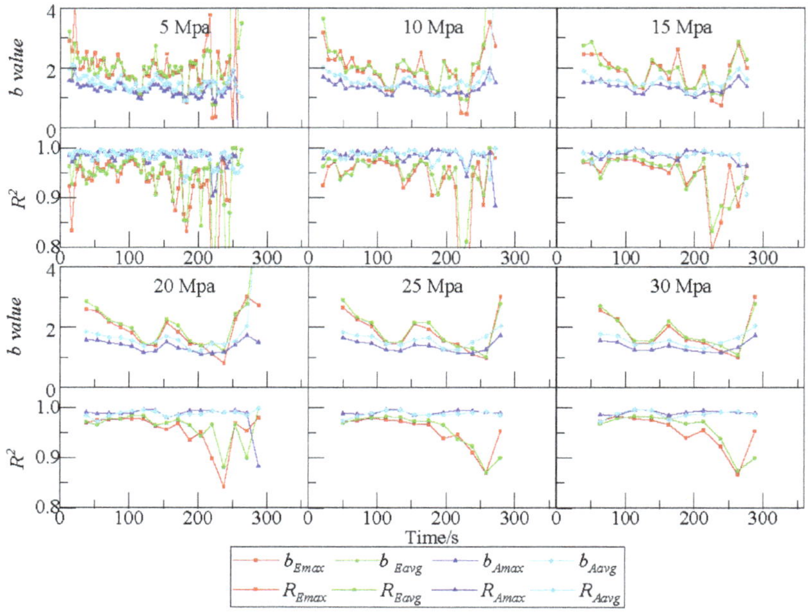

Figure 9. b value and its fitting goodness at different stress intervals for specimen G-10-10. The range of b_{Amax} values is between 1.01 to 1.99.

Table 4. The mean of R^2 at different AE stress proportion intervals for specimen G-10-10.

Stress Interval/MPa	Mean of R_{Emax}	Mean of R_{Eavg}	Mean of R_{Amax}	Mean of R_{Aavg}
5	0.859	0.882	0.967	0.971
10	0.878	0.904	0.967	0.973
15	0.897	0.898	0.972	0.970
20	0.912	0.927	0.970	0.978
25	0.908	0.909	0.981	0.976
30	0.907	0.913	0.982	0.975

A comparison of the interval of the AE event count at 200, the stress proportion at 10% σ_1, and stress magnitude at 20 MPa, showed that all their R^2 values performed well; while the interval of stress proportion at 10% σ_1 required the acquisition of the maximum value of σ_1, the interval of stress magnitude at 20 MPa may be affected by the load speed, and the values of σ_2 and σ_3. Therefore, the AE event count at 200 is considered to be better for real-time monitoring of rock damage; hence, the AE event interval method is more reliable than the stress interval and the stress proportion interval methods. Therefore, the b value calculated using A_{avg} and setting the interval of the AE event count at 200 was chosen for plotting all b value curves for the granite specimens, as shown in Figure 9. The b value was observed to decrease first and then increase or oscillate, and finally decrease. Moreover, the b value exhibited a decreasing trend before fracture in approximately 80% of the specimens.

Figure 10 shows the b value for the whole process of five groups of granite specimens under different true triaxial stresses, calculated using the average amplitude of the AE event. Apparently, the b value ranged between 1 and 2, and generally decreased with increasing confining pressure, as observed in the aqua rectangle area shown in Figure 10. Furthermore, in the specimen group σ_3 at 10 MPa, the b value was observed to decrease after an initial increase with increasing σ_2. In the specimen group σ_3 at 20, 30, 50, and 100 MPa, the b value tended to decrease with increasing σ_2. In addition, when σ_3 was 50 and 100 MPa, the b value was approximately 1.1, which is smaller than that of the other groups (~1.4); this means that a small b value may associate with high confining pressures. Since a large number of AE events with large amplitude will occur with the increase of differential stress, especially in the final loading stage in rock AE tests, this will in turn result in a decrease of b value. Therefore, temporal variation of the b value can be used for damage accumulation assessment in rock AE tests, while spatial variation of the b value can be used to determine the stress concentration area, and the temporal-spatial variation of the b value is an important reference for precursory analysis of rock instability.

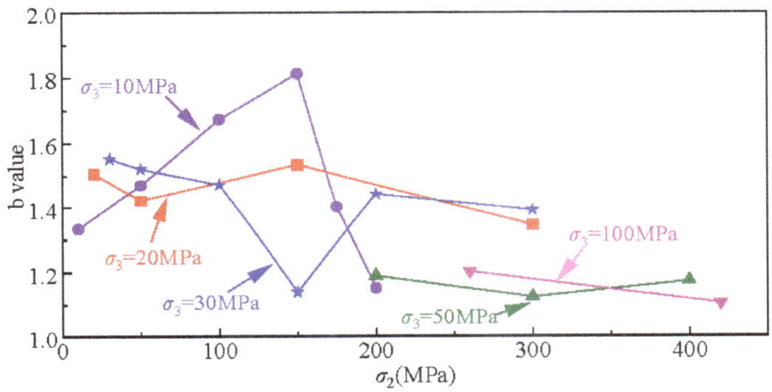

Figure 10. b value in different granite specimens under true triaxial stress.

4. Conclusions

This study conducted five groups of true triaxial compressive tests on granite specimens, and detected and analyzed their AE behavior from loading to rock failure. A large number of AE signals were observed to be generated at the moment of stress state change, and the AE events were fewer at constant stress conditions, namely no loading stress, steady loading stress, or steady increase in loading stress. Therefore, it is believed that a change in the stress conditions can be judged by the sudden occurrence of a large number of AE events. Meanwhile, a random disturbance that appears during machine loading may affect the length of the quiet period; and the length of the quiet period may be affected by the loading time because the specimen is directly broken when the loading stress reaches the peak value. Consequently, future research should investigate the variations in the length of the quiet period under different loading speed conditions, and to determine the length of the quiet period in the critical state before rock failure at reduced loading speed. In addition, when σ_2 stress gradually increases, the rock transforms from brittle deformation to ductile deformation, and the peak stress first increases and then decreases, causing a significant change in AE parameters.

The b value is a critical parameter to describe spatial and temporal size distribution characteristics in laboratory rock AE tests. Since the b value is negatively correlated with the differential stress, therefore, temporal variation of the b value is an important indicator for damage accumulation assessment, and the decrease trend of b value can be used for precursory analysis of instability in rock AE tests. At the same time, rock samples always have complex fault systems which display a scale-invariant hierarchy or a fractal geometry, and the faults will form discontinuous branching structures distributed in a three-dimensional volume of the rock sample. This geometrical distribution of faults is in concert with stress heterogeneities and interactions, and influences the extent of the b value. Therefore, a combination analysis of fault distribution and b value can better describe the damage evolution, which in turn helps to understand the fracture mechanism of rock material. This requires a detailed detection of internal structure of rock samples before loading, as well as the detection of damage evolution during the loading process through various techniques, for example, non-destructive testing techniques, and such investigation will be the focus of future rock AE research.

Author Contributions: L.D. is responsible for the experimental part and the paper writing; L.Z. and H.L. are responsible for the data processing and part of the paper writing; X.L. is responsible for experimental setup and part of data processing; K.D. is responsible for experimental results analysis. All authors have read and agreed to the published version of the manuscript.

Funding: The work is supported in part by the National Science Foundation for Excellent Young Scholars of China (51822407), in part by the National Natural Science Foundation of China (51774326, 51504288), in part by the Natural Science Foundation for Distinguished Young Scholars of Hunan Province, China (2018JJ1037), and in part by the Huxiang Young Talents Plan of Hunan Province, China (2018RS3001). Kun Du was supported by the open fund of Hebei Mining Development and Safety Technology Laboratory (grant number 2016kykf03) and the open fund of Mining Disaster Prevention and Control Ministry Key Laboratory at Shandong University of Science and Technology (grant number MDPC201917).

Informed Consent Statement: Not applicable.

Data Availability Statement: Not applicable.

Conflicts of Interest: The authors declare no conflict of interest.

Abbreviations

σ_1	maximum principal stress (MPa)
σ_2	intermediate principal stress (MPa)
σ_3	minimum principal stress (MPa)
A_{max}	maximum amplitude of an acoustic emission event (dB)
A_{avg}	average amplitude of an acoustic emission event (dB)
E_{max}	maximum absolute energy of an acoustic emission event (aJ)
E_{avg}	average absolute energy of an acoustic emission event (aJ)
b_A	*b* value based on the amplitude
b_E	*b* value based on the energy
b_{Amax}	*b* value based on the maximum amplitude of multiple hits
b_{Aavg}	*b* value based on the average amplitude of multiple hits
b_{Emax}	*b* value based on the maximum energy of multiple hits
b_{Eavg}	*b* value based on the average energy of multiple hits
R^2	goodness of fitting
R_{Amax}	fitting goodness of *b* value based on the maximum amplitude of multiple hits
R_{Aavg}	fitting goodness of *b* value based on the average amplitude of multiple hits
R_{Emax}	fitting goodness of *b* value based on the maximum energy of multiple hits
R_{Eavg}	fitting goodness of *b* value based on the average energy of multiple hits

References

1. Gutenberg, B.; Richter, C.F. Frequency of earthquakes in California. *Bull. Seismol. Soc. Am.* **1994**, *34*, 185–188. [CrossRef]
2. Smith, W.D. The b-value as an earthquake precursor. *Nature* **1981**, *289*, 136–139. [CrossRef]
3. Rao, M.V.M.S.; Lakshmi, K.J.P. Analysis of b-value and improved b-value of acoustic emissions accompanying rock fracture. *Curr. Sci.* **2005**, *89*, 1577–1582.
4. Liu, X.L.; Han, M.S.; He, W.; Li, X.B.; Chen, D.L. A new b-value estimation method in rock acoustic emission testing. *J. Geophys. Res. Solid Earth* **2020**, *125*, e2020JB019658. [CrossRef]
5. Jordan, T.H.; Chen, Y.T.; Gasparini, P.; Madariaga, R.; Main, I.G.; Marzocchi, W.; Papadopoulos, G.; Sobolev, G.; Yamaoka, K.; Zschau, J. Operational earthquake forecasting: State of Knowledge and Guidelines for Utilization. *Ann. Geophys. Italy* **2011**, *54*, 361–391.
6. Sammonds, P.R.; Meredith, P.G.; Main, I.G. Role of pore fluids in the generation of seismic precursors to shear fracture. *Nature* **1992**, *359*, 228–230. [CrossRef]
7. Kwiatek, G.; Goebel, T.H.W.; Dresen, G. Seismic moment tensor and *b* value variations over successive seismic cycles in laboratory stick-slip experiments. *Geophys. Res. Lett.* **2014**, *41*, 5838–5846. [CrossRef]

8. Goh, A.T.C.; Zhang, W. Reliability assessment of stability of underground rock caverns. *Int. J. Rock Mech. Min. Sci.* **2012**, *55*, 157–163. [CrossRef]
9. Jafari, M.A. The distribution of b-value in different seismic provinces of Iran. In Proceedings of the 14th World Conference on Earthquake Engineering, Beijing, China, 12–17 October 2008.
10. Okal, E.A.; Romanowicz, B.A. On the variation of b -values with earthquake size. *Phys. Earth Planet. Inter.* **1994**, *87*, 55–76. [CrossRef]
11. Scholz, C.H. Microfractures, aftershocks, and seismicity. *Bull. Seismol. Soc. Am.* **1968**, *58*, 1117–1130.
12. Weeks, J.; Lockner, D.; Byerlee, J. Change in b-values during movement on cut surfaces in granite. *Bull. Seismol. Soc. Am.* **1978**, *68*, 333–341. [CrossRef]
13. Lockner, D.A.; Byerlee, J.D.; Kuksenko, V.; Ponomarev, A.; Sidorin, A. Quasi-static fault growth and shear fracture energy in granite. *Nature* **1991**, *350*, 39. [CrossRef]
14. Mondal, D.; Roy, P. Fractal and seismic b-value study during dynamic roof displacements (roof fall and surface blasting) for enhancing safety in the longwall coal mines. *Eng. Geol.* **2019**, *253*, 184–204. [CrossRef]
15. Liu, X.; Liu, Z.; Li, X.; Gong, F.; Du, K. Experimental study on the effect of strain rate on rock acoustic emission characteristics. *Int. J. Rock Mech. Min. Sci.* **2020**, *133*, 104420. [CrossRef]
16. Wang, S.; Tang, Y.; Wang, S. Influence of brittleness and confining stress on rock cuttability based on rock indentation tests. *J. Cent. South Univ.* **2021**, *28*, 2786–2800. [CrossRef]
17. Lei, X.L. How do asperities fracture? An experimental study of unbroken asperities. *Earth Planet. Sci. Lett.* **2003**, *213*, 347–359. [CrossRef]
18. Main, I.G.; Meredith, P.G.; Jones, C. A reinterpretation of the precursory seismic b-value anomaly from fracture mechanics. *Geophys. J. Int.* **1989**, *96*, 131–138. [CrossRef]
19. Main, I.G.; Sammonds, P.R.; Meredith, P.G. Application of a modified Griffith criterion to the evolution of fractal damage during compressional rock failure. *Geophys. J. Int.* **1993**, *115*, 367–380. [CrossRef]
20. Schroeder, M. Fractals, Chaos, Power Laws: Minutes From an Infinite Paradise. *Phys. Today* **1991**, *44*, 91. [CrossRef]
21. Clauset, A.; Shalizi, C.R.; Newman, M.E.J. Power-Law Distributions in Empirical Data. *Siam Rev.* **2009**, *51*, 661–703. [CrossRef]
22. Mogi, K. Study of elastic shocks caused by the fracture of heterogeneous materials and its relation to earthquake phenomena. *Bull. Earthq. Res. Inst. Univ. Tokyo* **1962**, *40*, 125–173.
23. Scholz, C. The frequency-magnitude relation of microfracturing in rock and its relation to earthquakes. *Bull. Seismol. Soc. Am.* **1968**, *58*, 399–415. [CrossRef]
24. Zitto, M.E.; Piotrkowski, R.; Gallego, A.; Sagasta, F.; Benavent-Climent, A. Damage assessed by wavelet scale bands and b-value in dynamical tests of a reinforced concrete slab monitored with acoustic emission. *Mech. Syst. Signal Process.* **2015**, *60–61*, 75–89. [CrossRef]
25. Sagasta, F.; Zitto, M.E.; Piotrkowski, R.; Benavent-Climent, A.; Suarez, E.; Gallego, A. Acoustic emission energy b-value for local damage evaluation in reinforced concrete structures subjected to seismic loadings. *Mech. Syst. Signal Process.* **2018**, *102*, 262–277. [CrossRef]
26. Wiemer, S.; Katsumata, K. Spatial variability of seismicity parameters in aftershock zones. *J. Geophys. Res.* **1999**, *104*, 13135–13151. [CrossRef]
27. Wiemer, S.; Wyss, M. Minimum magnitude of completeness in earthquake catalogs: Examples from Alaska, the western United States, and Japan. *Bull. Seismol. Soc. Am.* **2000**, *90*, 859–869. [CrossRef]
28. Cao, A.; Gao, S.S. Temporal variation of seismic b-values beneath northeastern Japan island arc. *Geophys. Res. Lett.* **2002**, *29*, 48-1–48-3. [CrossRef]
29. Woessner, J.; Wiemer, S. Assessing the quality of earthquake catalogues: Estimating the magnitude of completeness and its uncertainty. *Bull. Seismol. Soc. Am.* **2005**, *95*, 684–698. [CrossRef]
30. Dong, L.; Li, X.; Tang, L.; Gong, F. Mathematical Functions and Parameters for Microseismic Source Location without Pre-Measuring Speed. *Chin. J. Rock Mech. Eng.* **2011**, *30*, 2057–2067.
31. Kda, B.; Cy, A.; Rui, S.A.; Ming, T.A.; Sw, A. Failure properties of cubic granite, marble, and sandstone specimens under true triaxial stress. *Int. J. Rock Mech. Min. Sci.* **2020**, *130*, 104309.
32. Du, K.; Tao, M.; Li, X.B.; Zhou, J. Experimental Study of Slabbing and Rockburst Induced by True-Triaxial Unloading and Local Dynamic Disturbance. *Rock Mech. Rock Eng.* **2016**, *49*, 3437–3453. [CrossRef]
33. Luo, Y.; Gong, F.; Liu, D.; Wang, S.; Si, X. Experimental simulation analysis of the process and failure characteristics of spalling in d-shaped tunnels under true-triaxial loading conditions. *Tunn. Undergr. Space Technol.* **2019**, *90*, 42–61. [CrossRef]
34. Gong, F.Q.; Si, X.F.; Li, X.B.; Wang, S.Y. Experimental investigation of strain rockburst in circular caverns under deep three-dimensional high-stress conditions. *Rock Mech. Rock Eng.* **2019**, *52*, 1459–1474. [CrossRef]
35. Wang, S.; Sun, L.C.; Li, X.; Wang, S.; Feng, F. Experimental investigation of cuttability improvement for hard rock fragmentation using conical cutter. *Int. J. Geomech.* **2020**, *21*, 06020039. [CrossRef]
36. Nur, A.; Simmons, G. Stress-induced velocity anisotropy in rock: An experimental study. *J. Geophys. Res.* **1969**, *74*, 6667–6674. [CrossRef]
37. Qi-Yue, L.I.; Dong, L.J.; Xi-Bing, L.I.; Yin, Z.Q.; Liu, X.L. Effects of Sonic Speed on Location Accuracy of Acoustic Emission Source in Rocks. *Trans. Nonferrous Met. Soc. China* **2011**, *21*, 2719–2726.

38. Dong, L.J.; Tao, Q.; Qing-Chun, H.U. Influence of temperature on acoustic emission source location accuracy in underground structure. *Trans. Nonferrous Met. Soc. China* **2021**, *31*, 2468–2478. [CrossRef]
39. Dong, L.; Zou, W.; Li, X.; Shu, W.; Wang, Z. Collaborative Localization Method Using Analytical and Iterative Solutions for Microseismic/Acoustic Emission Sources in the Rockmass Structure for Underground Mining. *Eng. Fract. Mech.* **2019**, *210*, 95–112. [CrossRef]
40. Dong, L.J.; Tang, Z.; Li, X.B.; Chen, Y.C.; Xue, J.C. Discrimination of mining microseismic events and blasts using convolutional neural networks and original waveform. *J. Cent. South Univ.* **2020**, *27*, 3078–3089. [CrossRef]
41. Dong, L.; Tong, X.; Hu, Q.; Tao, Q. Empty region identification method and experimental verification for the two-dimensional complex structure. *Int. J. Rock Mech. Min. Sci.* **2021**, *147*, 104885. [CrossRef]
42. Dong, L.; Tong, X.; Ma, J. Quantitative investigation of tomographic effects in abnormal regions of complex structures. *Engineering* **2021**, *7*, 1011–1022. [CrossRef]
43. Dong, L.J.; Li, X.B.; Zhou, Z.L.; Chen, G.H.; Ma, J. Three-dimensional analytical solution of acoustic emission source location for cuboid monitoring network without pre-measured wave velocity. *Trans. Nonferrous Met. Soc. China* **2015**, *25*, 293–302. [CrossRef]
44. Dong, L.; Shu, W.; Li, X.; Han, G.; Wei, Z. Three dimensional comprehensive analytical solutions for locating sources of sensor networks in unknown velocity mining system. *IEEE Access* **2017**, *5*, 11337–11351. [CrossRef]
45. Dong, L.; Sun, D.; Li, X.; Du, K. Theoretical and experimental studies of localization methodology for AE and microseismic sources without pre-measured wave velocity in mines. *IEEE Access* **2017**, *5*, 16818–16828. [CrossRef]
46. Dong, L.; Hu, Q.; Tong, X.; Liu, Y. Velocity-free ms/ae source location method for three-dimensional hole-containing structures. *Engineering* **2020**, *6*, 827–834. [CrossRef]
47. Chen, D.; Liu, X.; He, W.; Xia, C.; Gong, F.; Li, X.; Cao, X. Effect of attenuation on amplitude distribution and b value in rock acoustic emission tests. *Geophys. J. Int.* **2021**. [CrossRef]
48. Du, K.; Su, R.; Tao, M.; Yang, C.; Momeni, A.; Wang, S. Specimen shape and cross-section effects on the mechanical properties of rocks under uniaxial compressive stress. *Bull. Eng. Geol. Environ.* **2019**, *78*, 6061–6074. [CrossRef]

mathematics

MDPI

Article

Predictive Modeling of Short-Term Rockburst for the Stability of Subsurface Structures Using Machine Learning Approaches: t-SNE, K-Means Clustering and XGBoost

Barkat Ullah [1], Muhammad Kamran [2] and Yichao Rui [1,*]

[1] School of Resources and Safety Engineering, Central South University, Changsha 410083, China; barkat_ullah@csu.edu.cn
[2] Department of Mining Engineering, Institute Technology of Bandung, Bandung 40132, Indonesia; muhdkamran62@gmail.com
* Correspondence: ruiyichao@csu.edu.cn

Abstract: Accurate prediction of short-term rockburst has a significant role in improving the safety of workers in mining and geotechnical projects. The rockburst occurrence is nonlinearly correlated with its influencing factors that guarantee imprecise predicting results by employing the traditional methods. In this study, three approaches including including t-distributed stochastic neighbor embedding (t-SNE), K-means clustering, and extreme gradient boosting (XGBoost) were employed to predict the short-term rockburst risk. A total of 93 rockburst patterns with six influential features from micro seismic monitoring events of the Jinping-II hydropower project in China were used to create the database. The original data were randomly split into training and testing sets with a 70/30 splitting ratio. The prediction practice was followed in three steps. Firstly, a state-of-the-art data reduction mechanism t-SNE was employed to reduce the exaggeration of the rockburst database. Secondly, an unsupervised machine learning, i.e., K-means clustering, was adopted to categorize the t-SNE dataset into various clusters. Thirdly, a supervised gradient boosting machine learning method i.e., XGBoost was utilized to predict various levels of short-term rockburst database. The classification accuracy of XGBoost was checked using several performance indices. The results of the proposed model serve as a great benchmark for future short-term rockburst levels prediction with high accuracy.

Keywords: rock burst; t-SNE; unsupervised learning; supervised learning; XGBoost

Citation: Ullah, B.; Kamran, M.; Rui, Y. Predictive Modeling of Short-Term Rockburst for the Stability of Subsurface Structures Using Machine Learning Approaches: t-SNE, K-Means Clustering and XGBoost. *Mathematics* **2022**, *10*, 449. https://doi.org/10.3390/math10030449

Academic Editor: Ezequiel López-Rubio

Received: 15 December 2021
Accepted: 24 January 2022
Published: 30 January 2022

1. Introduction

Rockburst is an abrupt and violent failure of the rock mass that results in personnel injury and economic loss in underground rock excavations [1,2]. It is generally believed that because of the sudden release of stored elastic energy, rockburst causes an adverse phenomenon of ejecting, spalling, slabbing, and bursting at a high speed in a very short time, which greatly endangers worker safety and also damages field equipment and established structures [3,4]. Rockburst has been a serious threat to many engineering projects (i.e., mining and geotechnical) around the globe. In China, with the extensive depth of underground coal mines and underground rock excavations [5], the rockburst hazard is becoming more severe and frequent for rock engineering [3,4]. Rockburst has been widely reported in several countries around the globe. Likewise, in Canada, rockburst cases are reported in more than 15 mines [6]. From 1936 to 1993, the United States documented more than 172 rockburst cases in which more than 78 fatalities and 158 injuries occurred [6,7]. Despite reducing the mining activities, Germany still documented rockbursts from 1983 to 2007, and some serious injuries and deaths were delineated in more than 40 cases [8]. China, as the current world's largest coal producer, is facing a linear increase in rockburst cases with the increase of coal production from underground mining. According to Zhang et al. [9], over 100 Chinese coal mines have recorded rockburst disasters. Despite the fact that many prevention and control exertions have been undertaken, the rockburst disaster still remains an unsolved universal issue for underground rock excavations.

A large amount of experimental research is now being undertaken with the goal of better understanding the mechanical behavior of rock mass under various engineering situations [10–12] The rockburst mechanism, types, and some useful control measures are also proposed following theoretical analysis, field studies, and laboratory tests [13]. In addition, some updated monitoring methods including microgravity, microseismic and geological radar are implemented for monitoring and forecasting the rockburst danger [14]. These methods can monitor and forecast the rockburst danger before it occurs. Nevertheless, the accurate determination of rockburst prediction is still a strenuous challenge because it has several influencing factors including rock properties, geological conditions, stress levels, and energy accumulation [9]. Rockburst prediction is classified into two categories: short-term rockburst prediction and long-term prediction [8]. Short-term rockburst prediction is usually followed by installing on-site monitoring systems, i.e., electromagnetic radiation, microseismic, infrared radiations, and microgravity methods [6]. By analyzing and monitoring the microseismic wave released during rock fracturing, some precursory features of rockbursts were discovered that were helpful for the prediction of rockburst. The microseismic indicators that are commonly used for rockburst prediction are the energy indicator [15], the events number [16], the *b* value which is defined as the slope of the commutative hit with respect to the amplitude [17], and apparent volume [18]. Conversely, the long-term rockburst prediction can be estimated by following rockburst potential and field conditions. Various predictive indicators are recommended by the researchers for the prediction of rockburst potential, e.g., strain energy storage index (W_{et}) proposed by [19], defined as the ratio of stored strain energy (W_{sp}) to dissipated strain energy (W_{st}). Wattimena et al. [20] considered an elastic strain energy density as a measuring indicator of rockburst potential. Altindag [21] introduced the rock brittleness coefficient as a burst liability index that is defined as the ratio of uniaxial compressive stress (UCS) to tensile stress (σ_t). According to Wang and Park. [22], the tangential stress criterion defined as the ratio between tangential stress (σ_θ), and UCS of rock mass (σ_c) is another useful index to quantify the risk of rockburst. The rockburst occurrence is generally influenced by many factors that may include rock properties, stress domination, groundwater conditions, excavation methods, etc. The rockburst intensity is nonlinearly correlated with the influencing factors [23] that guarantee imprecise predicting results by employing the traditional methods [24]. Hence, soft computing methods have been recently implemented in monitoring and predicting the dynamic disaster of rockburst.

With the growth in the use of computers in applied sciences over the past few years, machine learning methods are adopted for predicting the rockburst risk more effectively. Researchers have recommended several machine learning methods. For example, Wojtecki et al. [25] applied a variety of algorithms, i.e., decision tree (DT), random forest (RF), gradient boosting (GB), and artificial neural network (ANN), to evaluate the rockburst in the upper Silesian coal basin, Poland. A convolutional neural network (CNN) based data-driven model was built by Zhao et al. [26] and the performance of the model was then compared with the traditional neural network. Zhao et al. [1] recommended a model for rockburst prediction by implementing a DT model on microseismic monitoring data. Various classification models were adopted to predict the occurrence and intensity of rockburst in the form of distinct data-driven classification problems [27]. Zhou et al. [28] classified a long-term rockburst by adopting support vector machine (SVM) model and their results were recommended for underground rocks excavation. A study was conducted on predicting the rockburst intensity by applying an extreme learning machine (ELM). Furthermore, a particle swarm optimization (PSO) model was implemented to optimize the hidden layer bias and input weight matrix [29]. Li et al. [30] established a hybrid model (KPCA-APSO-SVM), that was based on three different models including kernel principal component analysis (KPCA), the adaptive-PSO, and SVM. Several influencing parameters, i.e., the ratio of tangential stress (σ_θ) to UCS (σ_c), the ratio of UCS (σ_c) to the tensile stress (σ_t) and strain energy storage index (W_{et}) were taken as input parameters and the results depicted that the KPCA-APSO-SVM model has strong reliability in rock burst prediction. In order to predict and categorize the sensitivity of rockburst, multivariate adaptive regression splines (MARS) and deep forest algorithms were applied [31]. Additionally, the dimensional reduction and visualization of input features were carried out by t-SNE. Zhou et al. [32] studied and compared the forecasting outcomes of 12 different machine learning algorithms in long-term rockburst prediction. A C5.0 DT algorithm has been used as the main classifier for rockburst classification and evaluation [33]. A locally weighted C4.5 DT algorithm has also been introduced for predicting the risk of rockburst in coal mines [34]. Ahmad et al. [35] investigated the potential of J48 and random tree algorithms to predict the rockburst classification levels. Wang et al. [36] developed a bagging and boosting tree-based ensemble technique to predict rockburst disasters in hard rock mines. Pu et al. [37] adopted SVM to evaluate the rockburst liability in Kimberlite diamond mine. Pu et al. [24] studied the long-term rockburst predictivity using an unsupervised learning method and SVM at Kimberlite diamond mine. Sun et al. [3] has proposed a RF and firefly algorithm (FA) based ensemble classifier to attain an optimal rockburst prediction model.

So far, the above-mentioned literature revealed that rockburst risk is investigated using different supervised and DT approaches. Almost all studies have been conducted on long-term rockburst prediction and classification, whereas few among them have focused on investigating short-term rockburst. Liang et al. [38] evaluated the predictability of short-term rockburst using microseismic data obtained from the tunnels of Jinping-II hydropower project in China. Several ensemble learning algorithms including RF, adaptive boosting (AdaBoost), gradient boosting decision tree (GBDT), XGBoost, and light gradient boosting machine (LightGBM) have been evaluated and, among them, the RF and GBDT have shown good performance. Zhou et al. [39] considered the predictive performance of the stochastic gradient boosting (SGB) approach in the prediction of rockburst. Feng et al. [40] employed an optimized probabilistic neural network (PNN) on microseismic monitoring data to forecast the rockburst risk. The model was modified by combining the mean impact value algorithm (MIVA), the modified firefly algorithm (MFA), and PNN (MIVA-MFA-PNN model). Ji et al. [41] developed a genetic algorithm (GA) and SVM based model (GA-SVM) to analyze microseismic data to predict rockburst occurrence. Table 1 depicts the traditional supervised machine learning approaches proposed by the researchers for predicting rockburst. The traditional supervised classification algorithms have major limitations in complex phenomena such as rockburst potential due to the difficulty of obtaining a large number of good quality labeled samples. One interesting contender for overcoming this issue is a combination with an unsupervised technique to enhance the results of a classification algorithm.

Table 1. Traditional supervised machine learning approaches proposed by the researchers for predicting rockburst.

S.No	References	Machine Learning Models	Dataset Size	Year
1	Zhou et al. [32]	KNN	246	2016
2	Li et al. [42]	LR	135	2017
3	Afraei et al. [43]	LR	188	2018
4	Faradonbeh et al. [44]	DT	134	2019
5	Pu et al. [45]	DT	132	2018
6	Ghasemi et al. [33]	DT	174	2020
7	Faradonbeh et al. [44]	ANN	134	2019
8	Adoko et al. [46]	ANFIS	174	2013
9	Zhou et al. [32]	SVM	246	2016
10	Guo et al. [31]	MARS	344	2021

Note: KNN, k-nearest neighbors; LR, Logistic regression; DT, Decision tree; ANFIS, adaptive neuron fuzzy inference system; ANN, Artificial neural network; SVM, Support vector machines; multivariate adaptive regression splines.

2. Significance of the Study

In reality, the predictive characteristics of rockburst levels are not constant throughout many geotechnical and geomechanical engineering domains. Despite the fact that numerous diverse results are attained in the broad anatomies of rockburst prediction, the underlying influence of each uncertainty level remains unknown. There is currently no accurate method for anticipating the complex phenomena, i.e., short-term rockburst intensity levels. This paper provides a three-step mechanism for predicting the intensity level of short-term rockburst as follows:

(1) To begin, a cutting-edge data depletion process called t-distributed stochastic neighbor embedding (t-SNE) was developed to lessen the magnification of original rockburst database;
(2) Second, an unsupervised machine learning, namely K-means clustering, was used to classify the t-SNE dataset in order to reduce the inconsequential spectral dissimilarity effect in homogeneous localities;
(3) Finally, XGBoost, a supervised gradient boosting machine learning algorithm, has been developed to forecast various levels of short-term rockburst database. Figure 1 depicts a flowchart of this work.

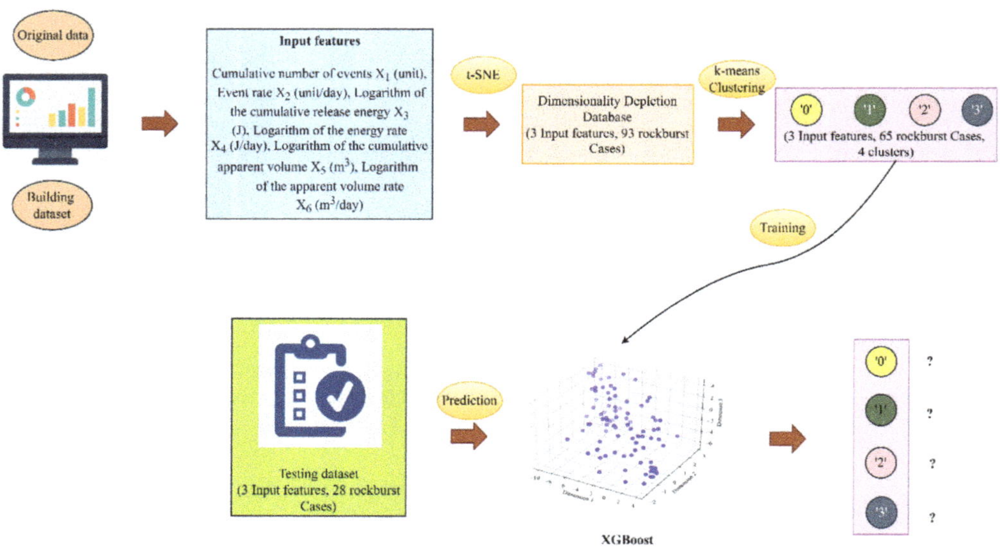

Figure 1. Flowchart of the study.

3. Material and Methods

3.1. Data Acquisition

In order to build the database of this work, a total of 93 short-term rockburst patterns with six influential features were collected from genuine microseismic monitoring events of the Jinping-II hydropower project in China [47]. The dataset used in this paper has been taken from the work of Liang et al. [38] based on the dataset provided by Feng et al. [47]. The rockburst intensity has been classified into four levels, i.e., no rockburst level (0) depicts that the rock specimens has no significant fracture on the free face, slight rockburst level (1) elucidates small specimen with minor fragment displacement and kinetic energy release, moderate rockburst level (2) shows the block spalling of the rock mass in the diverticulum and roadway wall whereas violent rockburst level (3) represent massive rock mass spalling, promptly distorting the surrounding rock mass. Figure 2 shows the distribution of various rockburst levels in this study.

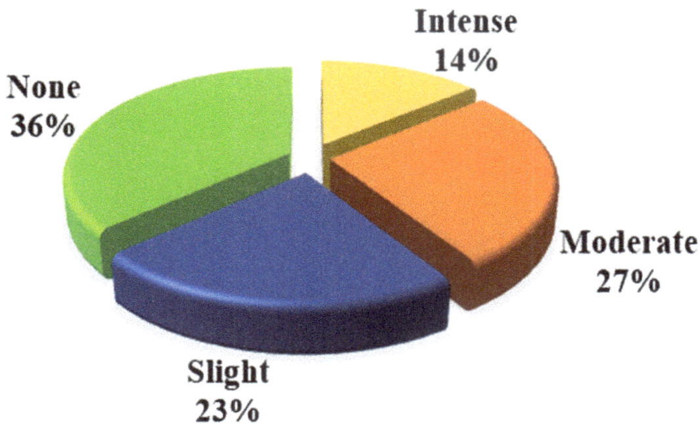

Figure 2. Distribution of each rockburst level.

From Table 2, it is clear that six influential features are designated in this study. In order to make the execution more appropriate, the values of X_3, X_4, X_5 and X_6 are selected in logarithmic scale. The main aim of the log function is to respond to the skewness toward large values in rockburst database.

Table 2. Statistical description of rockburst database.

Descriptive Statistics	Cumulative Number of Events X_1 (Unit)	Event Rate X_2 (Unit/Day)	Logarithm of the Cumulative Release Energy X_3 (J)	Logarithm of the Energy Rate X_4 (J/Day)	Logarithm of the Cumulative Apparent Volume X_5 (m^3)	Logarithm of the Apparent Volume Rate X_6 (m^3/Day)
Mean	13.011	1.735	4.389	3.562	4.150	3.334
Standard deviation	13.690	1.738	1.441	1.332	0.660	0.558
Minimum	1	0.111	0.780	0.178	2.511	1.666
Maximum	70	12.250	7.094	5.890	5.168	4.393

The box plot of each feature for the four rockburst levels is shown in Figure 3. From Figure 3, it is depicted that the rockburst is positively correlated with each feature. The larger values of features indicate the higher level of rockburst. Moreover, some outliers are present in the entire features of short-term rockburst dataset under each corresponding rockburst level, which shows the complexity of rockburst phenomenon. Hence, the effect of all the features is incorporated in this study to enhance the overall accuracy of rockburst database.

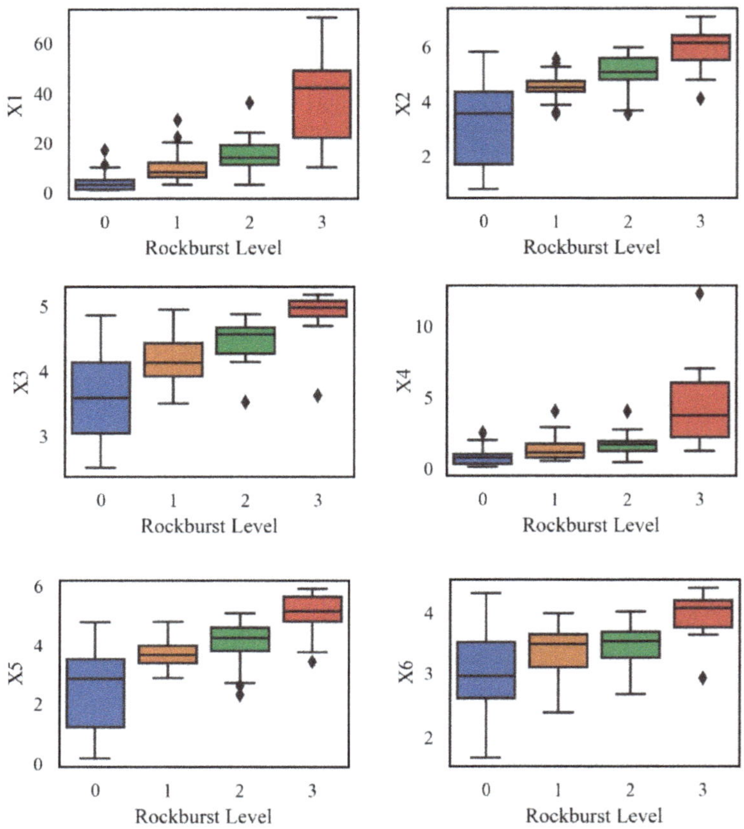

Figure 3. Boxplot of each influencing feature to corresponding rockburst level.

3.2. SNE Based t-SNE Algorithm

Hinton and Roweis [48] developed an enhanced stochastic neighbor embedding (SNE) based t-SNE algorithm.

The SNE operates in the following two steps: (1) Firstly, the SNE permutes the distance between points (data points) to a conditional probability in high-dimensional space attributing their resemblance. (2) Lastly, the SNE matches that conditional probability (probability of points in high-dimensional space) to the conditional probability of other points (map points) in low-dimensional space [49].

3.3. K-Means Clustering

Clustering analysis has been the best choice to avoid artificial division and supervision. In clustering, a dataset is generally grouped by a similar number and keeps the higher similarity in each group. The division of the dataset has happened according to the distance between the data points. Furthermore, the similarity and dissimilarity criteria also have an important role in the data division process. An unsupervised machine learning approach called K-means clustering [50,51] has wide and significant applications in dividing n observations into K clusters. Each observation in K-means clustering is related to the cluster with the nearby mean. The working principle of the algorithm consists of two dispersed phases. The first phase selects the K centers randomly with an already selected value of K, while the second phase collects each data object in the vicinity of the nearest center [52]. The most widely employed clustering criterion is known as the sum of the squared Euclidean distances. The main focus of this criterion is to measure the distance between each data point and cluster center [53].

3.4. Extreme Gradient Boosting (XGBoost)

XGBoost is abbreviated as extreme gradient boosting, which is an ensemble learning algorithm of machine learning techniques [54]. It includes simple classification and regression trees (CARTs) by integrating statistical boosting methods. Boosting improves the estimation precision of the model by constructing multiple trees as an alternative to constructing a single tree, and then combining them to build a consensus prediction framework [55]. XGBoost generates the tree by consecutively employing the residuals of past trees as contributions to the resultant tree. As such, the resulted tree develops the overall prediction by showing the errors of the past trees. At the point when the loss function is minimal, this consecutive model structure interaction can be articulated as a kind of gradient descent that advances the prediction by emerging another tree at each stage to ultimately decrease the fall [56]. The expansion of the new tree halts when the pre-determined most extreme number of trees is reached, or when the training error cannot be raised to a pre-indicated number of consecutive trees. Both the estimation precision and execution promptness of gradient boosting can be greatly enhanced by including random sampling; this comprehensive approach is designated probabilistic boosting [57]. In particular, for each tree in alignment, an irregular subsample of the training data is taken from the complete set of training data, excluding substitution. This irregularly specified subsample is then applied instead of the complete sample to appropriate the tree and determine the update of the model. XGBoost is an upgraded decentralized gradient boosting that can accomplish state-of-the-art prediction exhibitions [54]. XGBoost employs second-order estimation of the loss function, which is faster to combine than conventional GBMs. XGBoost has been effectively applied to mine gene articulation data [58]. The general architecture of XGBoost is depicted in Figure 4.

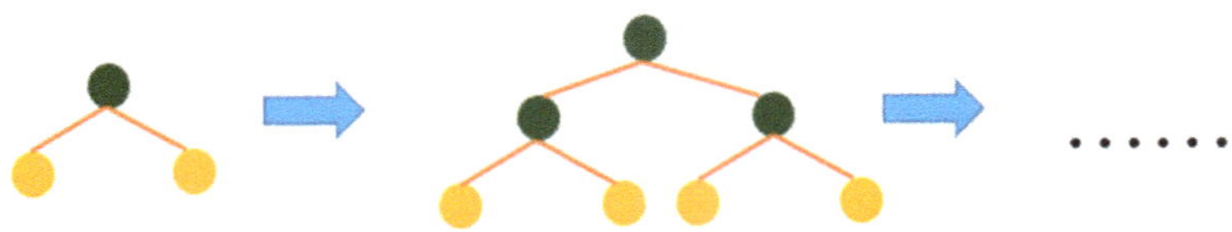

Figure 4. Level-wise tree model in XGBoost algorithm.

3.5. Hyperparameter Tunning

The hyperparameter in the machine learning algorithms need to be optimized. These hyperparameters should be calibrated contingent on the data in reference to defining it manually. As the short-term rock burst dataset is limited, we employed the cross-validation method based on normalizing data. Several cross-validation methods are applied by the researchers to optimize the hyperparameter.

Choubineh et al. [59] proposed the splitting of data into training, validation, and testing datasets to authenticate the machine learning algorithm. The validation dataset is employed to optimize the hyperparameters, whereas training on test datasets and training datasets are applied to evaluate the final performance of the model [59]. Nevertheless, a single contingent splitting of the data on various subsets is inadequate for ideal model evaluation because of the non-linearity of the datasets. If other contingent splitting is employed, it will compute the other values for performance indicators. The single splitting of data is only logical in large data set circumstances.

Among the hyperparameter tunning methods, the other most common method is the k-fold technique. In the k-fold method, the whole data is divided into k segments, then the first segment is employed for testing the execution of machine learning algorithms following training the data on the supplementary k-1 segment. Afterward, the second segment is taken for testing and the remaining data is employed as a training dataset. In the last different values of performance metrics are computed for all the k-fold. Hence cross-validation assists in attaining the average and standard deviation values of the metrics.

The random permutation method is also employed as hyperparameter optimization. This method involves irregular splitting of the data into training and testing datasets, after which the data is reorganized, and a new splitting of training and testing datasets is attained. This technique is repeated for n number of times and at every turn metrics are computed. Correspondingly, in the last, the average and standard deviation values of the metrics are calculated. Hence cross-validation not only computes the performance criteria for the testing dataset but accomplishes it multiple times by employing autonomous data to divide it into training and testing datasets. As in our case, the data is limited, so cross-validation was employed multiple times. The algorithm of 5-folds cross validation is shown in Algorithm 1. The grid search CV has been used to build the model, evaluate its performance, and make the short-term rockburst prediction level.

Algorithm 1: 5-folds XGBoost cross validation

Input	I(t), $I(t_1) \in I(t)$: Initial Dataset Extreme Gradient Boosting (XGBoost): Decision Algorithm L: Loss Function 5: Fold Number
Step 1	$U_1 \oplus U_2 \oplus \ldots\ldots T$ $U_1 + U_2 + \ldots\ldots T$ $\Leftrightarrow U_i \cap \ldots\ldots$
Step 2	**for** I from 1 to 10 **do**
Step 3	$F_i = XGBoost(T/U_i)$
Step 4	**for** $S(A_i)$ in U_i **do**
Step 5	$e_j = L(F_i, S(A_i)$
Step 6	End for
Step 7	End for
Step 8	Returne

3.6. Grid Search CV

A comprehensive grid search was followed for hyperparameter tunning [60]. This method authorizes search within specified hyperparameters range and describes the best value which results in the optimum value of evaluation criterion. GridSearchCV() has been implemented in scikit-learn python programing language in order to compute this method. This technique purely computes the cross validation (CV) score for all hyperparameter combinations in a specific range. The flowchart of algorithm's parameters optimization using grid search is shown in Figure 5. GridSearchCV() not only permits calculation of the optimal hyperparameter but also estimates the metric to its best value. In our case, all the other parameters of the python programing language were used as a default in order to implement Grid Search CV.

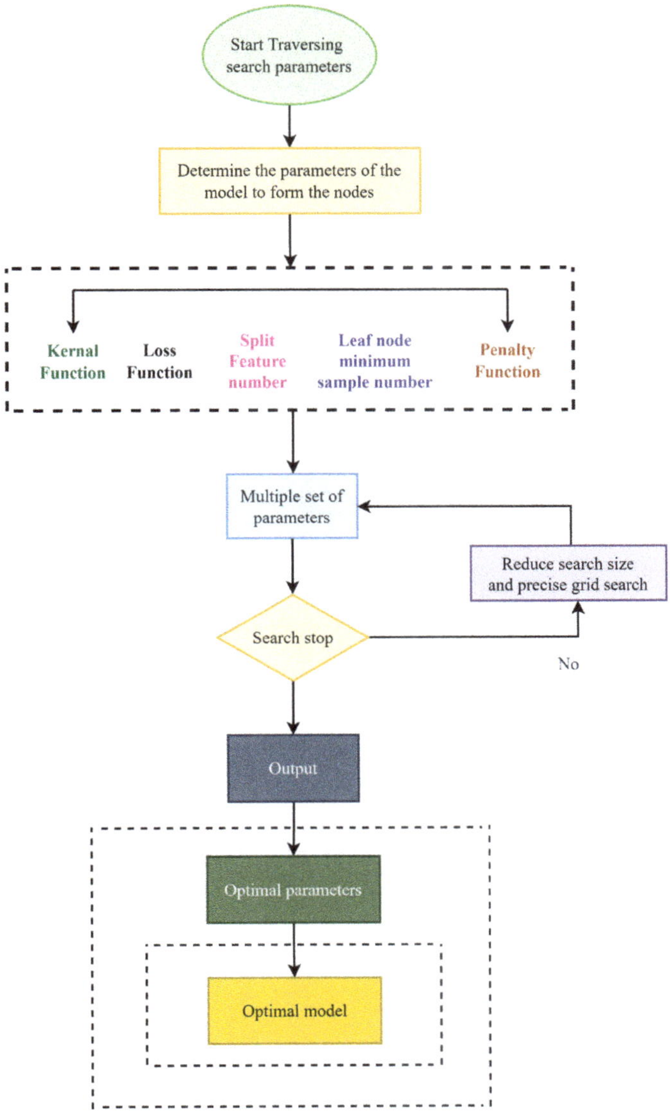

Figure 5. The flowchart of parameters optimization using grid search.

4. Result and Discussion

4.1. Rockburst Database Reduction Using t-SNE

Consider that the data points r_p and r_q in rockburst dataset select their corresponding neighbors based on conditional probability, shown as $S_{q|p}$ in Equation (1) [49,61]. The Gaussian kernel is used to define conditional probability.

$$S_{q|p} = \begin{cases} \dfrac{\exp(-||r_p - r_q||_2/2\sigma_p{}^2)}{\sum_{k\neq p}\exp - ||r_p - r_k||_2/(2\sigma_p)^2} & \text{whereas } p \neq q \\ 0 \; p = q \end{cases} \tag{1}$$

whereas $||r_q - r_p||$ represents the Euclidean distance between data points r_p and r_q while σ_p shows the Gaussian distribution variance choosing r_p as the center position, which is established by binary search by employing the mechanism of perplexity. The perplexity is given in Equation (2).

$$\text{Perp}(S_p) = 2^{E(S_p)} \tag{2}$$

where $E(S_p)$ is the Shannon entropy of S_p computed in bits and S_p induces a probability distribution for any value of σ_p. The $E(S_p)$ is given in Equation (3).

$$E(S_p) = -\sum S_{q|p} \log_2 S_{q|p} \tag{3}$$

Assume that b_p and b_q are allocated in a low dimension that are resembled to r_p and r_q in the high dimension. It is possible to compute a similar conditional probability ($T_{q|p}$) for the map points b_p and b_q in low-dimensional (corresponding to the datapoints r_p and r_q in high-dimensional space). In this case, the Gaussian distribution is stated as $\frac{1}{\sqrt{2}}$. Succeeding the resemblance of $S_{q|p}$ of r_q to r_p is given in Equation (4).

$$T_{q|p} = \begin{cases} \frac{\exp(-||b_p - b_q||^2)}{\sum_{k \neq p} \exp(-||b_p - b_k||^2)} & p \neq q \\ 0\ p = q \end{cases} \tag{4}$$

If dimensionality depletion outcome is satisfactory, then the resemblance in high dimensionality space is assumed to be identical to that in low dimensionality in $S_{q|p} = T_{q|p}$. When the conditional uncertainty between r_p and all other points are examined, the conditional uncertainty distribution S_q can be established. Correspondingly, the identical uncertainty distribution T_q is established as S_q low dimensionality space. To measure the resemblance between two points, the Kullback–Leibler divergence is employed. Hence, a cost function J is established as shown in Equation (5).

$$J = \sum_p KL(S_p||T_p) = \sum_p \sum_q S_{q|p} \log \frac{S_{q|p}}{T_{q|p}} \tag{5}$$

In Equation (5), the distribution of conditional probabilities of data point r_p and map point b_p over other data points, and map points are represented as S_p and T_p, respectively. The SNE is amended to t-SNE with the addition of two major improvements [62]. Firstly, for pairwise estimation of likenesses in both low and high-dimensional spaces, the symmetric version of SNE is introduced. The improved t-SNE for data points r_p and r_q is depicted in Equation (6).

$$S_{pq} = \frac{S_{q|p} + S_{p|q}}{2n} \tag{6}$$

By employing the symmetric property ($S_{pq} = S_{qp}$), the data point r_p will have the probability to pick the data point r_q as its neighbor, where n shows total data points. Secondly, the Gaussian kernel is replaced by the t-distribution to evaluate the likeliness between the map points. More precisely, the t-SNE uses a heavy-tailed t-distribution for b_p and b_q (map points) in low-dimensional space. This process takes place with 1 degree of freedom, then the T_{pq} can be obtained by using Equation (7):

$$T_{pq} = \frac{1 + (||r_p - r_q||_2)^{-1}}{\sum_{k \neq l} (1 + ||r_k - r_l||_2)^{-1}} \tag{7}$$

To make it more precise, the comprehensive mechanism of t-SNE is given as:

Stage 1: Get data $S = S_1, S_2, S_3, \ldots, S_n$ in high dimension region, and give the dimensionality reduction consequences as $B^{(T)} = T_1, T_2, T_3, \ldots, T_n$;

Stage 2: Compute perplexity, and assign iteration times T, momentum of $\alpha(t)$ and learning rate η;
Stage 3: Calculate $S_{p|q}$ as given in Equation (1);
Stage 4: Estimate S_{pq} as depicted in Equation (7);
Stage 5: Arbitrarily choose Y with N;
Stage 6: Compute T_{pq} as stated in Equation (7), estimate the gradient as stated in Equation (9);
Stage 7: Finally repeat the stage 6 so that the iteration number is remarkable than T.

The Jupyter notebook has been utilized using Scikit-learn module in order to accomplish the t-SNE. In the first stage, the rockburst database is visualized from high-resolution amplitude to low-resolution amplitude. The initial rockburst dataset is tabulated into four clusters. In this study, the event related features, i.e., the cumulative number of events X_1 (unit) and event rate X_2 (unit/day) are considered in the first group (Dimension 1). The energy associated features including the logarithm of the cumulative release energy X_3 (J) and the logarithm of the energy rate X_4 (J/day) are categorized in the second group (Dimension 2). The apparent volume related features, i.e., the logarithm of the cumulative apparent volume X_5 (m^3) and the logarithm of the apparent volume rate X_6 (m^3/day) are collected in the third group (Dimension 3). In order to reflect the initial rockburst dataset, the learning rate = 100 is executed with the Matplotlib in the Python programming language (all the other parameters are kept as a default). Following the rockburst data dimensionality reduction technique, the feature established amplitude was formed in such a way that the initial rockburst database may keep the originality to high scalability. The rockburst dataset after the dimensionality reduction is depicted in Figure 6. After the adoption of the t-SNE mechanism, the actual rockburst dataset (93 × 6 matrix) is renovated to a (93 × 3) matrix, as revealed in Table 3. Figure 6 demonstrates a low-resolution amplitude visualization of the rockburst dataset following the t-SNE data reduction mechanism.

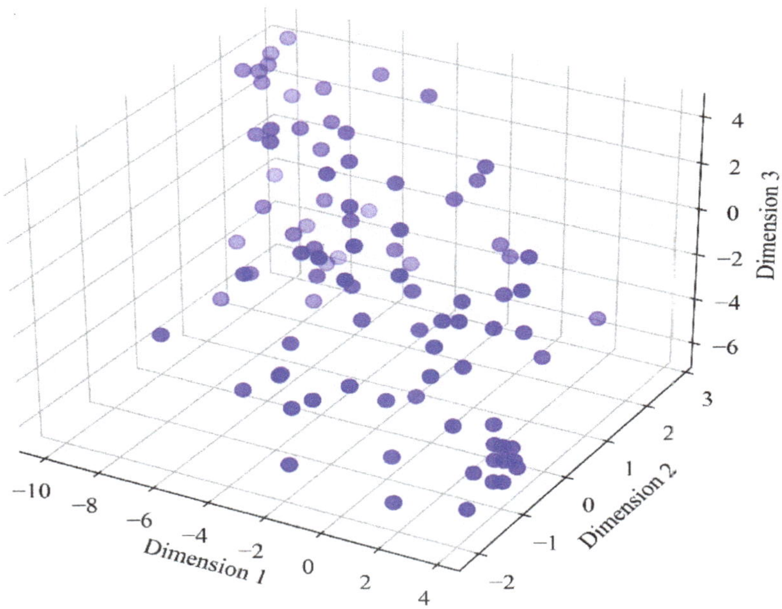

Figure 6. 3D low-resolution amplitude of rockburst database.

Table 3. Rockburst database after low-resolution amplitude with t-SNE.

Samples	Dimension 1	Dimension 2	Dimension 3
1	−9.1895	1.876923	3.533078
2	−5.25797	1.386265	2.998773
3	−6.33402	0.83398	−0.95647
4	−6.6661	1.667999	1.523691
5	−3.36939	0.296317	1.838995
	…..	…..	…..
88	−8.27044	1.192174	2.389334
89	−8.87826	1.073105	−2.3535
91	−2.44182	−0.94443	1.698488
92	−5.97327	1.043975	−4.14844
93	−0.7725	−1.40264	1.910676

4.2. K-Means Clustering on t-SNE Based Rockburst Database

In K-means clustering, the completion of early rockburst level grouping occurs when all the data objects are appended in some clusters and the average of the primitive clusters is then recalculated. This iteration happens many times until the criterion function is reduced to its minimum. Based on the target object r and average of cluster J_i that is r_i, the criterion function can be obtained using an Equation (8) [63]:

$$C = \sum_{i=1}^{k} \sum_{r \in J_i}^{k} |r - r_i|^2 \tag{8}$$

where C indicates the sum of squared error of all objects in the database. In this study, to compute the adjacent distance between data points and cluster center, the Euclidean distance is considered as a criterion function. The Euclidean distance between one vector $r = (r_1, r_2, r_n)$ and another vector $s = (s_1, s_2, \ldots s_n)$, the Euclidean distance $D(r_i, s_i)$ can be obtained by the following Equation (9):

$$D(r, s) = \left[\sum_{i=1}^{n} (r_i - s_i) \right]^{1/2} \tag{9}$$

The Jupyter notebook has been utilized using Scikit-learn module in order to accomplish the K-means clustering. Rousseeuw [64] have established the generalization of the cluster monitoring. Silhouette mechanism is contingent on balancing the objects tightness and separation. The silhouette coefficient can show that the t-SNE data is grouped in a good manner reflecting that the objects are organized into the groups that they match. This is an index to evaluate that the authentication of the clustering to be used for selecting the optimal k in the cluster. Based on the four different rockburst levels, we assume the number of clusters = 4 for K-means clustering. Several iterations stages were computed in this study as shown in Figure 7. Various studies have shown that a silhouette coefficient of more than 0.5 is an acceptable model for K-means clustering [65–68]. The silhouette coefficient of 0.53 shows that the clusters was reliable following 10th iteration in the t-SNE obtained short-term rockburst dataset.

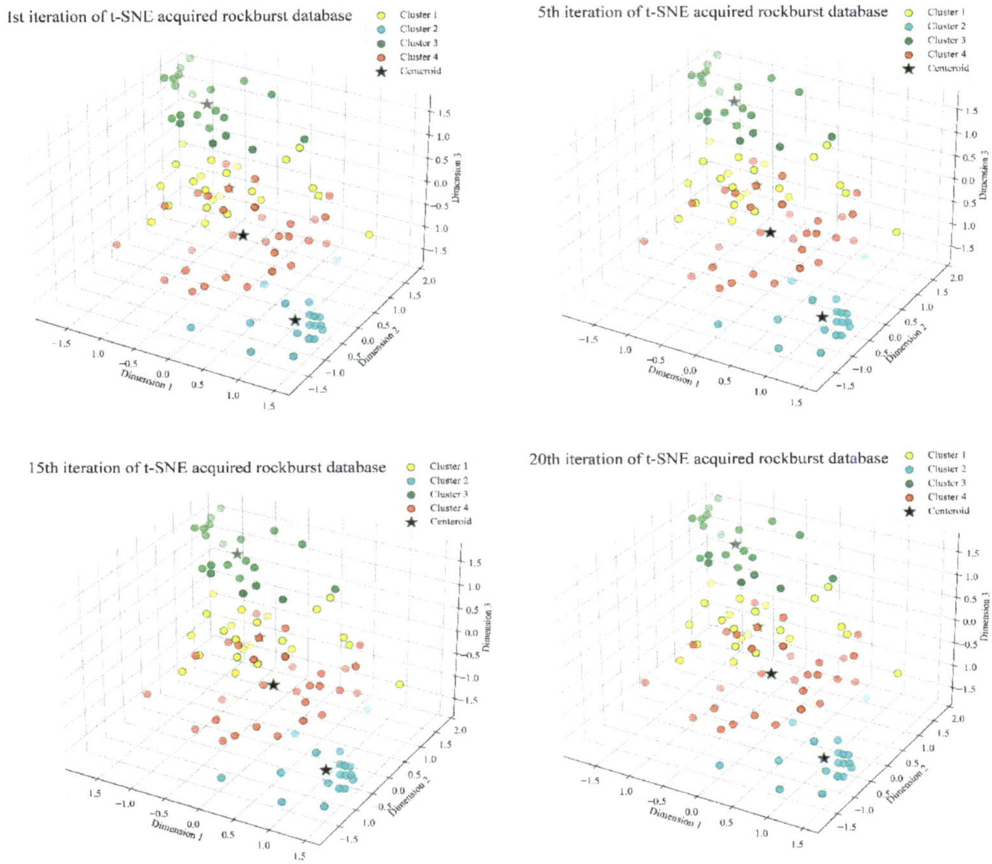

Figure 7. K-means clustering mechanism of low-resolution amplitude.

4.3. Extreme Gradient Boosting (XGBoost) Prediction Model

Consider $\overline{v_m}$ as the forecasted rockburst prediction level result of the nth number of data for which the characteristics vector is U_n; P denotes the number of estimators, with q_s (s ranging from 1 to P) corresponding to individual tree anatomy; and v_n^0 denotes the preliminary assumption that is the average of the measured characteristics in the learning information. To forecast the results, Equation (10) uses a variety of expansion functions.

$$\overline{v_m} = v_n^0 + \gamma \sum_{s=1}^{P} q_s(U_n) \tag{10}$$

whereas γ is the learning rate, which is included to better model implementation, execute rhythmically while connecting the most recent tree, and avoid overfitting.

In Equation (9), a character S_{th} is linked to the model at the S_{th} state, and the S_{th} forecasted value v_n^{-s} is implemented from the preceding state forecasted value $v_n^{-(s-1)}$, and the augmented q_s of the character of the attached S_{th} character is illustrated in Equation (11).

$$v_n^{-s} = v_n^{-(s-1)} + \gamma\, q_s \tag{11}$$

whereas q_s represents the weight of leaves created by decreasing the objective function of the S_{th} tree

$$obj = \eta K + \sum_{\alpha=1}^{K} [T_\alpha \beta_\alpha + \frac{1}{2}(L_\alpha + \mu)\beta_\alpha^2] \quad (12)$$

wherein K indicates the leaves of the S_{th} tree and β_α represents the weight of the leaves from 1 to K, η and μ are the uniformity characteristics that are used to apply the coherence to the anatomy in order to avoid the model overfitting. The parameters L_α and T_α represent the sum of all data associated with a leaf of the previous and subsequent loss function gradients, respectively.

A single leaf is divided into distinct numeration leaves in order to form the S_{th} tree. The anatomy of using the gain settings is seen in Equation (13). Consider the interdependent right leaf R_C and B_C and the interdependent left leaf R_W and B_W achieving the divergence. The diverging benchmark is generally assumed when the gain parameter is close to zero. The uniformity characteristics and are periphrastically susceptible on the gain attribute, i.e., a greater regularization parameter will result in a lower gain parameter, which will prevent the slope of the leaf from converging. However, it will reduce the framework's capacity to adapt to the rockburst training dataset.

$$gain = \frac{1}{2}\left[\frac{R_W^2}{B_W + \mu} + \frac{R_C^2}{B_C + \mu} + \frac{(R_W + R_C)^2}{B_W + B_C + \mu}\right] \quad (13)$$

In order to forecast the rockburst intensity level, a gradient boosting machine learning algorithm has been applied on the k-means clustering dataset. It was noted that employing an entire dataset to train the XGBoost model may arise the over-fitting issues. More specifically, the framework may adjust magnificently in addition to the dataset that employed for the training stage, but it is unable to predict new data. For the avoidance of doubt, the rockburst dataset is split into training and testing sets with the relative size of 7:3, meaning that 70% of the entire data is chosen for training and 30% of the entire data is selected for testing the trained framework. The samples order in the dataset must be randomly adjusted before the splitting to overcome the localization of the training set.

The XGBoost model was employed to predict the rockburst intensity level. For the XGBoost model, the online Jupyter platform was executed in python. The python program language 3.6.6 that was accessible on the Jupyter program was executed to accomplish the XGBoost. A standard XGBoost model with default attributes that are developed in XGBoost module: M = 100 estimators, the regularization attribute of $\gamma = 0$, $\lambda = 1$, a learning rate of $\eta = 0.3$ was implemented in this study. We assumed a repeated 5-fold cross-validation setup and ensured that the argument from the same essay is not distributed over the training and testing datasets as shown in Figure 8. The cross-validation was repeated 3 times on standard scalar normalized data, which yielded a total of 15 folds. For other parameters, the default values of the XGBoost model are implemented in this study.

The classification accuracy of XGBoost was checked using precision, recall, and f_1-score measures. Precision can properly predict the datasets; recall interpret the capability of accurately predicting the actual features to the maximum level, and f_1-score demonstrates a universal metric that implements the performance of both recall and precision. Therefore, the aforementioned performance indicators are implemented in this study to estimate the performance of the model. Assume the confusion matrix is defined by Equation (14). A confusion matrix is usually implemented as a standard to demonstrate the performance of a classification model on a testing dataset for which the true values are already defined.

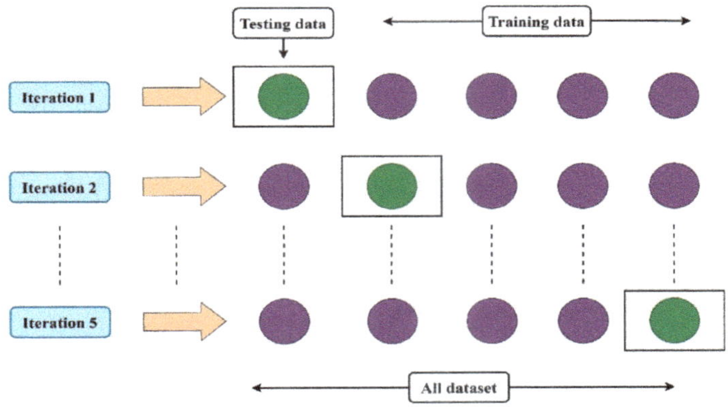

Figure 8. Five-fold cross-validation employed in the study.

$$S = \begin{bmatrix} s_{11} & s_{12} & \cdots & s_{1t} \\ s_{21} & s_{22} & \cdots & s_{2t} \\ \vdots & \vdots & \ddots & \cdots \\ s_{t1} & s_{t2} & \cdots & s_{tt} \end{bmatrix} \quad (14)$$

where t represents the number of rockburst levels, s_{11} is the number of features accurately predicted for the class m, and S_{mn} denotes the number of features of class that is categorized to class n.

On the basis of the confusion matrix, the precision, recall, and f_1-score measure for each rockburst level are determined by Equations (15)–(17), respectively.

$$Pr = \frac{s_{mm}}{\sum_{m=1}^{t} s_{mn}} \quad (15)$$

$$Re = \frac{s_{mm}}{\sum_{n=1}^{t} s_{mn}} \quad (16)$$

$$f_1 - score = \frac{2{*}Pr * Re}{Pr + Re} \quad (17)$$

To further analyses the accuracy of XGboost, the accuracy is given by Equation (18)

$$Accuracy = \frac{1}{\sum_{m=1}^{k} \sum_{n=1}^{k} S_{tt}} \sum_{m=1}^{k} S_{mm} \quad (18)$$

$$macro - Pr = (\sum_{n=1}^{t} \frac{S_{mm}}{\sum_{n=1}^{E} S_{mn}})/E \quad (19)$$

$$macro - Re = (\sum_{n=1}^{t} \frac{s_{mm}}{\sum_{n=1}^{E} s_{mn}})/E \quad (20)$$

$$macro - f_1 = \frac{2{*}macro - Pr{*}macro - Re}{macro - Pr + macro - Re} \quad (21)$$

The prediction results of XGBoost algorithms were acquired on the testing dataset. In order to forecast the results of the proposed XGBoost algorithm combined with t-SNE and K-means clustering, three different performance indices have been employed in this study. The classification report for the testing dataset was computed using python programing language. The classification report gives a perspective of the proposed framework performance on the rockburst dataset as shown in Table 4. The precision values were calculated using Equation (15). The precision value for no rockburst level achieved better outcomes as compared to slight rockburst level, moderate rockburst level and violent rockburst level. The precision value for no rockburst, slight rockburst, moderate rockburst and violent rockburst were 100%, 60%, 100% and 88%, respectively. Equation (16) was employed to measure the recall value for each rockburst level. The recall value of slight rockburst performed better as compared to no rockburst level, moderate rockburst level and violent rockburst level. No rockburst, modest rockburst, moderate rockburst, and strong rockburst have recall values of 86 percent, 100%, 83%, and 88%, respectively. To measure f_1-score for each corresponding rockburst level, Equation (17) was employed in this study. The f_1-score for no rockburst level outperformed slight rockburst level, moderate rockburst level and violent rockburst level. The f_1-score for no rockburst, slight rockburst, moderate rockburst and violent rockburst were 92%, 75%, 91% and 88%, respectively. In order to measure the overall accuracy of the framework on the testing dataset, Equation (18) was utilized in this study. The accuracy for the overall testing dataset was 88 percent, indicating that the XGBoost combined with t-SNE and K-means clustering performed well in this study.

Table 4. Classification report of XGBoost algorithm.

Class	XGBoost Model		
	Precision %	Recall %	f_1-Score %
No rockburst	100	86	92
Slight rockburst	60	100	75
Moderate rockburst	100	83	91
Violent rockburst	88	88	88
Accuracy			88
macro avg	87	89	66
Weighted avg	91	88	88

The model's accuracy is measured as a whole, while recall and precision are calculated for each class separately. For the rockburst phenomenon, we employ macro average of precision, recall, f_1-score for our model as shown by Equations (19)–(21). The macro-average scores are the simple mean of scores of all rockburst levels. Hence, macro- average precision is the mean of the precision of four different levels of rockburst. The macro- average recall depicts the mean of the recall of four different levels of rockburst. Whereas macro- average f_1-score represents the mean of the f_1-score of four different levels of rockburst. So, the mean of precision, recall and f_1-score were 87, 89 and 66, respectively. The weighted average scores are the sum of the scores of all levels after multiplying their respective levels proportions. Hence, the weighted average of precision, recall and f_1-score were 91, 88 and 88, respectively.

In addition, a confusion matrix of the XGBoost algorithm was established, as shown in Figure 9. The values on the main diagonal show the samples number correctly predicted by the XGBoost. It can be seen that most rockburst samples were accurately classified using the XGBoost. Based on the confusion matrix (see Figure 9) only two rockburst levels have been mis-predicted in the entire short-term rockburst dataset. More precisely, one moderate rockburst (2) level is misclassified as violent rockburt (3) level, whereas one violent rockburst (3) level is misclassified as slight rockburst (2) level. According to the results, the XGBoost algorithm showed good performances in predicting the rockburst intensity level.

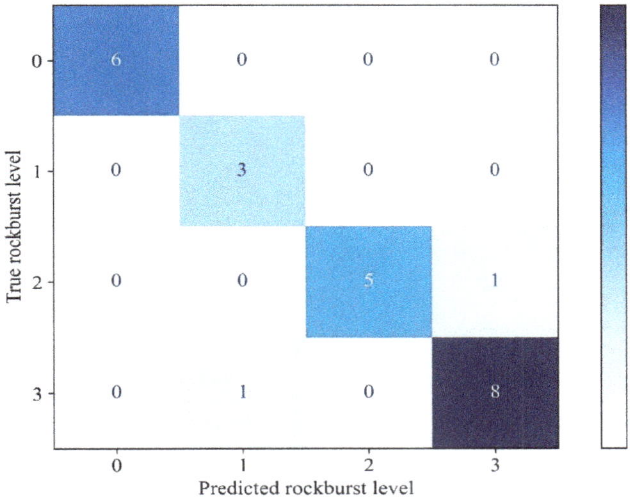

Figure 9. Confusion matrix of testing dataset.

5. Conclusions

This research work developed t-SNE+K-means clustering+XGBoost to predict the predict rockburst levels efficiently and accurately. The robustness of the obtained framework was authenticated by analyzing the outcomes for the proposed framework using different performance indices. As for predicting the rockburst level, three methods including t-SNE, K-means clustering, and XGBoost model, which are broadly employed in geotechnical engineering, were applied during the study. More precisely, the data employed in this research work were obtained from genuine microseismic events. The short-term rockburst level is evaluated by the statistical performance to approximate the robust framework for the best effective model in connection with data prediction. The results of t-SNE+K-means clustering+XGBoost model shows that it can estimate the return rockburst level with high accuracy.

Hence, the t-SNE+K-means clustering+XGBoost model acquired in this study is recommended as an accurate and efficient model for the prediction of rockburst intensity levels. It can be employed as a rockburst prevention and warning system, owing to the fact that the proposed model will have reliable prediction performance in different rock conditions. Therefore, the model can be generalized by maintaining some additional rock mechanics data and geological information. This model can be merged into the initiation of the rockburst level of the microseismical events that are continuously disseminated.

The range and number of trainings should be taken into consideration, which is has a consequential effect on the logical reasoning of the data-driven models. The current research will be further extended by establishing some cutting-edge machine learning algorithms and comparing the outcome of those models with the outcome of the model acquired in this research work. The state-of-the-art machine learning technique can comprise hybrid, metaheuristic, and ensemble machine learning models.

Author Contributions: Conceptualization, M.K. and B.U.; methodology, M.K. and B.U.; data curation, M.K.; writing—original draft preparation, M.K. and B.U.; writing—review and editing, Y.R. All authors have read and agreed to the published version of the manuscript.

Funding: This research received no external funding.

Institutional Review Board Statement: Not applicable.

Informed Consent Statement: Not applicable.

Data Availability Statement: All the data and models employed and/or generated during the study appear in the submitted article.

Conflicts of Interest: The authors declare no conflict of interest.

References

1. Zhao, H.; Chen, B.; Zhu, C. Decision Tree Model for Rockburst Prediction Based on Microseismic Monitoring. *Adv. Civ. Eng.* **2021**, *2021*, 8818052. [CrossRef]
2. Feng, X.-T.; Yashun, X.; Guangliang, F. Mechanism, warning and dynamic control of rockburst evolution process. In Proceedings of the ISRM Regional Symposium—7th Asian Rock Mechanics Symposium, Seoul, Korea, 15–19 October 2012.
3. Sun, Y.; Li, G.; Zhang, J.; Huang, J. Rockburst Intensity Evaluation by a Novel Systematic and Evolved Approach: Machine Learning Booster and Application. *Bull. Eng. Geol. Environ.* **2021**, *80*, 8385–8395. [CrossRef]
4. Cai, M. Principles of Rock Support in Burst-Prone Ground. *Tunn. Undergr. Space Technol.* **2013**, *36*, 46–56. [CrossRef]
5. Cai, X.; Cheng, C.; Zhou, Z.; Konietzky, H.; Song, Z.; Wang, S. Rock Mass Watering for Rock-Burst Prevention: Some Thoughts on the Mechanisms Deduced from Laboratory Results. *Bull. Eng. Geol. Environ.* **2021**, *80*, 8725–8743. [CrossRef]
6. Pu, Y.; Apel, D.B.; Liu, V.; Mitri, H. Machine Learning Methods for Rockburst Prediction-State-of-the-Art Review. *Int. J. Min. Sci. Technol.* **2019**, *29*, 565–570. [CrossRef]
7. Mark, C. Coal Bursts in the Deep Longwall Mines of the United States. *Int. J. Coal Sci. Technol.* **2016**, *3*, 1–9. [CrossRef]
8. Pu, Y.; Apel, D.B.; Wei, C. Applying Machine Learning Approaches to Evaluating Rockburst Liability: A Comparation of Generative and Discriminative Models. *Pure Appl. Geophys.* **2019**, *176*, 4503–4517. [CrossRef]
9. Zhang, J.; Jiang, F.; Yang, J.; Bai, W.; Zhang, L. Rockburst Mechanism in Soft Coal Seam within Deep Coal Mines. *Int. J. Min. Sci. Technol.* **2017**, *27*, 551–556. [CrossRef]
10. Zhou, Z.; Cai, X.; Li, X.; Cao, W.; Du, X. Dynamic Response and Energy Evolution of Sandstone Under Coupled Static–Dynamic Compression: Insights from Experimental Study into Deep Rock Engineering Applications. *Rock Mech. Rock Eng.* **2020**, *53*, 1305–1331. [CrossRef]
11. Wang, S.; Tang, Y.; Wang, S. yong Influence of Brittleness and Confining Stress on Rock Cuttability Based on Rock Indentation Tests. *J. Cent. South Univ.* **2021**, *28*, 2786–2800. [CrossRef]
12. Wang, S.; Tang, Y.; Li, X.; Du, K. Analyses and Predictions of Rock Cuttabilities under Different Confining Stresses and Rock Properties Based on Rock Indentation Tests by Conical Pick. *Trans. Nonferrous Met. Soc. China (Engl. Ed.)* **2021**, *31*, 1766–1783. [CrossRef]
13. Li, X.; Gong, F.; Tao, M.; Dong, L.; Du, K.; Ma, C.; Zhou, Z.; Yin, T. Failure Mechanism and Coupled Static-Dynamic Loading Theory in Deep Hard Rock Mining: A Review. *J. Rock Mech. Geotech. Eng.* **2017**, *9*, 767–782. [CrossRef]
14. Lu, C.P.; Dou, L.M.; Liu, B.; Xie, Y.S.; Liu, H.S. Microseismic Low-Frequency Precursor Effect of Bursting Failure of Coal and Rock. *J. Appl. Geophys.* **2012**, *79*, 55–63. [CrossRef]
15. Liu, J.P.; Feng, X.T.; Li, Y.H.; da Xu, S.; Sheng, Y. Studies on Temporal and Spatial Variation of Microseismic Activities in a Deep Metal Mine. *Int. J. Rock Mech. Min. Sci.* **2013**, *60*, 171–179. [CrossRef]
16. Srinivasan, C.; Aroras, S.K.; Yajq, R.K. Use of Mining and Seismological Parameters as Premonitors of Rockbursts. *Int. J. Rock Mech. Min. Sci.* **1997**, *34*, 1001–1008. [CrossRef]
17. Ma, X.; Westman, E.; Slaker, B.; Thibodeau, D.; Counter, D. The B-Value Evolution of Mining-Induced Seismicity and Mainshock Occurrences at Hard-Rock Mines. *Int. J. Rock Mech. Min. Sci.* **2018**, *104*, 64–70. [CrossRef]
18. Ma, T.H.; Tang, C.A.; Tang, S.; Kuang, L.; Yu, Q.; Kong, D.Q.; Zhu, X. Rockburst Mechanism and Prediction Based on Microseismic Monitoring. *Int. J. Rock Mech. Min. Sci.* **2018**, *110*, 177–188. [CrossRef]
19. Kidybiiqski, A. Bursting Liability Indices of Coal. *Int. J. Rock Mech. Min. Sci. Geomech. Abstr.* **1981**, *18*, 295–304. [CrossRef]
20. Wattimena, R.K.; Sirait, B.; Widodo, N.P.; Matsui, K. Evaluation of Rockburst Potential in a Cut-and-Fill Mine Using Energy Balance. *Int. J. JCRM* **2012**, *8*, 19–23.
21. Altindag, R. Correlation of Specific Energy with Rock Brittleness Concepts on Rock Cutting. *J. S. Afr. Inst. Min. Metall.* **2003**, *103*, 163–171.
22. Wang, J.-A.; Park, H.D. Comprehensive Prediction of Rockburst Based on Analysis of Strain Energy in Rocks. *Tunn. Undergr. Space Technol.* **2001**, *16*, 49–57. [CrossRef]
23. Cai, M. Prediction and Prevention of Rockburst in Metal Mines—A Case Study of Sanshandao Gold Mine. *J. Rock Mech. Geotech. Eng.* **2016**, *8*, 204–211. [CrossRef]
24. Pu, Y.; Apel, D.B.; Xu, H. Rockburst Prediction in Kimberlite with Unsupervised Learning Method and Support Vector Classifier. *Tunn. Undergr. Space Technol.* **2019**, *90*, 12–18. [CrossRef]
25. Wojtecki, Ł.; Iwaszenko, S.; Apel, D.B.; Cichy, T. An Attempt to Use Machine Learning Algorithms to Estimate the Rockburst Hazard in Underground Excavations of Hard Coal Mine. *Energies* **2021**, *14*, 6928. [CrossRef]
26. Zhao, H.; Chen, B.; Zhang, Q. Data-Driven Model for Rockburst Prediction. *Math. Probl. Eng.* **2020**, *2020*, 5735496. [CrossRef]
27. Afraei, S.; Shahriar, K.; Madani, S.H. Developing Intelligent Classification Models for Rock Burst Prediction after Recognizing Significant Predictor Variables, Section 2: Designing Classifiers. *Tunn. Undergr. Space Technol.* **2019**, *84*, 522–537. [CrossRef]

28. Zhou, J.; Li, X.; Shi, X. Long-Term Prediction Model of Rockburst in Underground Openings Using Heuristic Algorithms and Support Vector Machines. *Saf. Sci.* **2012**, *50*, 629–644. [CrossRef]
29. Xue, Y.; Bai, C.; Qiu, D.; Kong, F.; Li, Z. Predicting Rockburst with Database Using Particle Swarm Optimization and Extreme Learning Machine. *Tunn. Undergr. Space Technol.* **2020**, *98*, 103287. [CrossRef]
30. Li, Y.; Wang, C.; Xu, J.; Zhou, Z.; Xu, J.; Cheng, J. Rockburst Prediction Based on the KPCA-APSO-SVM Model and Its Engineering Application. *Shock Vib.* **2021**, *2021*, 7968730. [CrossRef]
31. Guo, D.; Chen, H.; Tang, L.; Chen, Z.; Samui, P. Assessment of Rockburst Risk Using Multivariate Adaptive Regression Splines and Deep Forest Model. *Acta Geotech.* **2021**, 1–23. [CrossRef]
32. Zhou, J.; Li, X.; Mitri, H.S. Classification of Rockburst in Underground Projects: Comparison of Ten Supervised Learning Methods. *J. Comput. Civ. Eng.* **2016**, *30*, 04016003. [CrossRef]
33. Ghasemi, E.; Gholizadeh, H.; Adoko, A.C. Evaluation of Rockburst Occurrence and Intensity in Underground Structures Using Decision Tree Approach. *Eng. Comput.* **2020**, *36*, 213–225. [CrossRef]
34. Wang, Y. Prediction of Rockburst Risk in Coal Mines Based on a Locally Weighted C4.5 Algorithm. *IEEE Access* **2021**, *9*, 15149–15155. [CrossRef]
35. Ahmad, M.; Hu, J.L.; Hadzima-Nyarko, M.; Ahmad, F.; Tang, X.W.; Rahman, Z.U.; Nawaz, A.; Abrar, M. Rockburst Hazard Prediction in Underground Projects Using Two Intelligent Classification Techniques: A Comparative Study. *Symmetry* **2021**, *13*, 632. [CrossRef]
36. Wang, S.; Zhou, J.; Li, C.; Armaghani, D.J.; Li, X.; Mitri, H.S. Rockburst Prediction in Hard Rock Mines Developing Bagging and Boosting Tree-Based Ensemble Techniques. *J. Cent. South Univ.* **2021**, *28*, 527–542. [CrossRef]
37. Pu, Y.; Apel, D.B.; Wang, C.; Wilson, B. Evaluation of Burst Liability in Kimberlite Using Support Vector Machine. *Acta Geophys.* **2018**, *66*, 973–982. [CrossRef]
38. Liang, W.; Sari, A.; Zhao, G.; McKinnon, S.D.; Wu, H. Short-Term Rockburst Risk Prediction Using Ensemble Learning Methods. *Nat. Hazards* **2020**, *104*, 1923–1946. [CrossRef]
39. Zhou, J.; Shi, X.Z.; Huang, R.D.; Qiu, X.Y.; Chen, C. Feasibility of Stochastic Gradient Boosting Approach for Predicting Rockburst Damage in Burst-Prone Mines. *Trans. Nonferrous Met. Soc. China (Engl. Ed.)* **2016**, *26*, 1938–1945. [CrossRef]
40. Feng, G.; Xia, G.; Chen, B.; Xiao, Y.; Zhou, R. A Method for Rockburst Prediction in the Deep Tunnels of Hydropower Stations Based on the Monitored Microseismicity and an Optimized Probabilistic Neural Network Model. *Sustainability* **2019**, *11*, 3212. [CrossRef]
41. Ji, B.; Xie, F.; Wang, X.; He, S.; Song, D. Investigate Contribution of Multi-Microseismic Data to Rockburst Risk Prediction Using Support Vector Machine with Genetic Algorithm. *IEEE Access* **2020**, *8*, 58817–58828. [CrossRef]
42. Li, N.; Jimenez, R. A Logistic Regression Classifier for Long-Term Probabilistic Prediction of Rock Burst Hazard. *Nat. Hazards* **2018**, *90*, 197–215. [CrossRef]
43. Afraei, S.; Shahriar, K.; Madani, S.H. Statistical Assessment of Rock Burst Potential and Contributions of Considered Predictor Variables in the Task. *Tunn. Undergr. Space Technol.* **2018**, *72*, 250–271. [CrossRef]
44. Faradonbeh, R.S.; Taheri, A. Long-Term Prediction of Rockburst Hazard in Deep Underground Openings Using Three Robust Data Mining Techniques. *Eng. Comput.* **2019**, *35*, 659–675. [CrossRef]
45. Pu, Y.; Apel, D.B.; Lingga, B. Rockburst Prediction in Kimberlite Using Decision Tree with Incomplete Data. *J. Sustain. Min.* **2018**, *17*, 158–165. [CrossRef]
46. Adoko, A.C.; Gokceoglu, C.; Wu, L.; Zuo, Q.J. Knowledge-Based and Data-Driven Fuzzy Modeling for Rockburst Prediction. *Int. J. Rock Mech. Min. Sci.* **2013**, *61*, 86–95. [CrossRef]
47. Feng, X.T.; Chen, B.R.; Zhang, C.Q.; Li, S.J.; Wu, S.Y. *Mechanism, Warning and Dynamic Control of Rockburst Development Processes*; Science Press: Beijing, China, 2013. (In Chinese)
48. Hinton, G.; Roweis, S. *Stochastic Neighbor Embedding*; Advances in Neural Information Processing Systems; The MIT Press: Cambridge, MA, USA, 2002; pp. 833–840.
49. Liu, H.; Yang, J.; Ye, M.; James, S.C.; Tang, Z.; Dong, J.; Xing, T. Using T-Distributed Stochastic Neighbor Embedding (t-SNE) for Cluster Analysis and Spatial Zone Delineation of Groundwater Geochemistry Data. *J. Hydrol.* **2021**, *597*, 126146. [CrossRef]
50. Hartigan, J.A.; Wong, M.A. Algorithm AS 136: A K-Means Clustering Algorithm. *J. R. Stat. Soc. Ser. C (Appl. Stat.)* **1979**, *28*, 100–108. [CrossRef]
51. Zhu, X.; Jin, X.; Jia, D.; Sun, N.; Wang, P. Application of Data Mining in an Intelligent Early Warning System for Rock Bursts. *Processes* **2019**, *7*, 55. [CrossRef]
52. Zhao, Y.; Song, J. GDILC: A Grid-Based Density-Isoline Clustering Algorithm. In Proceedings of the 2001 International Conferences on Info-Tech and Info-Net. Proceedings (Cat. No.01EX479), Beijing, China, 29 October–1 November 2001.
53. Likas, A.; Vlassis, N.; Verbeek, J.J. The Global K-Means Clustering Algorithm. *Pattern Recognit.* **2003**, *36*, 451–461. [CrossRef]
54. Chen, T.; Guestrin, C. XGBoost: A Scalable Tree Boosting System. In Proceedings of the ACM SIGKDD International Conference on Knowledge Discovery and Data Mining, San Francisco, CA, USA, 13–17 August 2016; Association for Computing Machinery: New York, NY, USA, 2016; pp. 785–794.
55. Schapire, R.E. The Boosting Approach to Machine Learning: An Overview. In *Nonlinear Estimation and Classification. Lecture Notes in Statistics*; Denison, D.D., Hansen, M.H., Holmes, C.C., Mallick, B., Yu, B., Eds.; Springer: New York, NY, USA, 2003; Volume 171. [CrossRef]

56. Elith, J.; Leathwick, J.R.; Hastie, T. A Working Guide to Boosted Regression Trees. *J. Anim. Ecol.* **2008**, *77*, 802–813. [CrossRef]
57. Friedman, J.H. Stochastic Gradient Boosting. *Comput. Stat. Data Anal.* **2002**, *38*, 367–378. Available online: http://www.elsevier.com/locate/csda (accessed on 25 November 2021). [CrossRef]
58. Wang, Z.; Monteiro, C.D.; Jagodnik, K.M.; Fernandez, N.F.; Gundersen, G.W.; Rouillard, A.D.; Jenkins, S.L.; Feldmann, A.S.; Hu, K.S.; McDermott, M.G.; et al. Extraction and Analysis of Signatures from the Gene Expression Omnibus by the Crowd. *Nat. Commun.* **2016**, *7*, 12846. [CrossRef] [PubMed]
59. Choubineh, A.; Helalizadeh, A.; Wood, D.A. Estimation of Minimum Miscibility Pressure of Varied Gas Compositions and Reservoir Crude Oil over a Wide Range of Conditions Using an Artificial Neural Network Model. *Adv. Geo-Energy Res.* **2019**, *3*, 52–66. [CrossRef]
60. Bergstra, J.; Ca, J.B.; Ca, Y.B. Random Search for Hyper-Parameter Optimization Yoshua Bengio. *J. Mach. Learn. Res.* **2012**, *13*, 281–305.
61. Kamran, M. A State of the Art Catboost-Based T-Distributed Stochastic Neighbor Embedding Technique to Predict Back-Break at Dewan Cement Limestone Quarry. *J. Min. Environ. JME* **2021**, *12*, 679–691. [CrossRef]
62. van Der Maaten, L.; Hinton, G. Visualizing Data Using T-SNE. *J. Mach. Learn. Res.* **2008**, *9*, 2579–2605.
63. Shi, N.; Liu, X.; Guan, Y. Research on K-Means Clustering Algorithm: An Improved k-Means Clustering Algorithm. In Proceedings of the 3rd International Symposium on Intelligent Information Technology and Security Informatics (IITSI 2010), Jian, China, 2–4 April 2010; pp. 63–67.
64. Rousseeuw, P.J. Silhouettes: A Graphical Aid to the Interpretation and Validation of Cluster Analysis. *J. Comput. Appl. Math.* **1987**, *20*, 53–65. [CrossRef]
65. Kim, S.W.; Gil, J.M. Research Paper Classification Systems Based on TF-IDF and LDA Schemes. *Hum.-Cent. Comput. Inf. Sci.* **2019**, *9*, 30. [CrossRef]
66. Sarno Riyanarto, G.H.; Pamungkas, E.W.; Sunaryono, D. Clustering of ERP Business Process Fragments. In Proceedings of the 2013 International Conference on Computer, Control, Informatics and Its Applications (IC3INA), Jakarta, Indonesia, 19–21 November 2013.
67. Rani Usha, S.S. Comparison of Clustering Techniques for Measuring Similarity in Articles. In Proceedings of the 3rd IEEE International Conference on Computational Intelligence and Communication Technology, Ghaziabad, India, 9–10 February 2017.
68. Ma, Y.; Peng, M.; Xue, W.; Ji, X. A Dynamic Affinity Propagation Clustering Algorithm for Cell Outage Detection in Self-Healing Networks. In Proceedings of the 2013 IEEE Wireless Communications and Networking Conference (WCNC), Shanghai, China, 7–10 April 2013; pp. 2266–2270. [CrossRef]

mathematics

Article

Physical and Mechanical Properties Evolution of Coal Subjected to Salty Solution and a Damage Constitutive Model under Uniaxial Compression

Min Wang [1], Qifeng Guo [1,2,3,*], Yakun Tian [4] and Bing Dai [4]

1 School of Civil and Resource Engineering, University of Science and Technology Beijing, Beijing 100083, China; b20170003@xs.ustb.edu.cn
2 State Key Laboratory of Coal Resources in Western China, Xi'an University of Science and Technology, Xi'an 710054, China
3 State Key Laboratory of Water Resource Protection and Utilization in Coal Mining, China Energy Investiment, Beijing 100000, China
4 School of Resource Environment and Safety Engineering, University of South China, Hengyang 421001, China; 2017000012@usc.edu.cn (Y.T.); daibingusc@usc.edu.cn (B.D.)
* Correspondence: guoqifeng@ustb.edu.cn

Citation: Wang, M.; Guo, Q.; Tian, Y.; Dai, B. Physical and Mechanical Properties Evolution of Coal Subjected to Salty Solution and a Damage Constitutive Model under Uniaxial Compression. *Mathematics* **2021**, 9, 3264. https://doi.org/10.3390/math9243264

Academic Editor: Tinh Quoc Bui

Received: 5 November 2021
Accepted: 8 December 2021
Published: 16 December 2021

Abstract: Many underground reservoirs for storing water have been constructed in China's western coal mines to protect water resources. Coal pillars which work as dams are subjected to a long-term soaking environment of concentrated salty water. Deterioration of the coal dam under the attack of the salty solution poses challenges for the long-term stability and serviceability of underground reservoirs. The evolution of the physical and mechanical properties of coal subjected to salty solutions are investigated in this paper. Coal from a western China mine is made to standard cylinder samples. The salty solution is prepared according to chemical tests of water in the mine. The coal samples soaked in the salty solution for different periods are tested by scanning electron microscope, nuclear magnetic resonance, and ultrasonic detector techniques. Further, uniaxial compression tests are carried out on the coal specimens. The evolutions of porosity, mass, microstructures of coal, solution pH values, and stress–strain curves are obtained for different soaking times. Moreover, a damage constitutive model for the coal samples is developed by introducing a chemical-stress coupling damage variable. The result shows that the corrosion effect of salty solution on coal samples becomes stronger with increasing immersion time. The degree of deterioration of the longitudinal wave velocity (v_p) is positively correlated with the immersion time. With the increase in soaking times, the porosity of coal gradually increases. The relative mass firstly displays an increasing trend and then decreases with time. The peak strength and elastic modulus of coal decreases exponentially with soaking times. The developed damage constitutive model can well describe the stress–strain behavior of coal subjected to salty solution under the uniaxial compression.

Keywords: coal; deterioration characteristics; chemical-stress coupling factor; damage constitutive model

1. Introduction

Coal mining is normally accompanied by discharge of a large amount of water. This drained mine water with a high level of salt, in some cases, may damage ecosystem balance. Some theories and techniques, including the adsorbent method, evaporation, bioremediation, and irrigation of croplands, have been proposed and applied to treat the sodic–salty associated water from coal seams [1]. The coal resources of China are mainly distributed in western mining areas where the ecological environment is fragile and water resources are scarce. Large-scale coal mining could result in environmental problems, such as surface water waste and environmental pollution [2,3]. To protect and utilize water resources, some underground mines [4] built underground reservoirs in the underground coal mines for water storage. However, mine water in some of these mines [4] has high salinity level with K^+, Na^+, Cl^-, and SO_4^{2-}, etc. As the dams of underground reservoirs, coal pillars are in a long-term soaking environment of concentrated saltwater that could significantly change the microstructure of coal and affect the performance of coal dams. The immersion of salty solutions poses a serious threat to the stability of the coal dams. Therefore, it is crucial to understand the mechanical behavior of coal subjected to saltwater for the design and management of underground coal dams.

In past decades, considerable research has been carried out on the water–rock interaction. The chemical solutions mainly affect the rocks through hydro-physical and hydro-chemical interactions. The physical effects of water on rocks include water lubrication, water wedge, etc. The presence of water promotes the dissolution of soluble salts, accelerates the hydrolysis of colloids, weakens the connection force between mineral particles, and finally reduces the strength of the rock. Scholars have conducted a lot of research on the influence of water on the physical and mechanical properties of rock materials. Dyke and Dobereiner [5] studied the influence of water content on the strength and deformation characteristics of sandstone and found that small changes in water content can significantly affect the mechanical response of the rock. The increase in water content promotes the propagation of microcracks and the occurrence of dilatancy, which results in a decrease in rock strength. Hawkins and McConnell [6] discussed the sensitivity of different types of sandstone to water. It showed that the water sensitivity of sandstone increases with the increase in the proportion of quartz and clay minerals, and the strength of sandstone is in an exponential relationship with water content. Vásárhelyi [7] carried out experiments on the physical and mechanical properties of dry and saturated limestone and found that the porosity, elastic modulus, uniaxial compressive strength, and tensile strength of limestone under saturated conditions are all 66% of the dry specimens. Erguler [8] conducted uniaxial compression tests and Brazilian splitting tests on clay-bearing rocks with different water content. With the increase in water content, the uniaxial compressive strength, elastic modulus, and tensile strength of the rock were reduced by 90%, 93%, and 90%, respectively, compared with the specimen in the dry state. Zhao [9] studied coal's dynamic tensile failure characteristics in dry and saturated states, and found that saturated coal samples have higher tensile strength compared with dry coal samples. Gu [10,11] tested the static and dynamic mechanical properties of coal samples under different water-bearing conditions. The results showed that with the increase in water content, the static mechanical parameters of coal samples deteriorated to a certain degree, while the dynamic mechanical parameters increased first and then decreased.

The chemical effects of water on rocks include ion exchange, dissolution, and hydrolysis. The water and rock chemical action changes the mineral composition in the rock mass and increases the pore structure, which leads to the change in macro-mechanical properties. M. G. Karfakis and M. Akram [12] discussed the influence of water–rock interactions on rock fracture toughness. Fencht et al. [13] carried out triaxial compression tests on fractured quartz sandstones with different pH levels of NaCl and $CaCl_2$, and studied the effects of different solutions on the friction strength of sandstone fracture surfaces. Feng and Chen et al. [14–16] carried out experiments on the mechanical properties of sandstone under different corrosion conditions, discussed the microscopic failure mechanism of the rock under the action of corrosion, and carried out a quantitative analysis of the damage evolution inside the rock. Han et al. [17] conducted mechanical tests on sandstone after soaking in different chemical solutions and analyzed the corrosion effect of different solutions on the rock. Hutchinson and Johnson [18] studied the corrosion and degradation of limestone under the action of acid. Qiao [19] carried out uniaxial compression tests of sandstone after soaking in aqueous solutions with different ion concentrations, and analyzed the influence of water chemistry on the microstructure and macro-mechanical properties of the rock. Lin [20] studied the evolution of mechanical damage of rocks under the coupled chemical-stress conditions, and found that chemical corrosion increased the porosity of the rocks, which led to the decline of mechanical properties. Xie [21] tested the influence of chemical degradation on the mechanical behavior of limestone and found that chemical degradation enhanced the deformation of rock and significantly increased the permeability. Li [22] employed the nuclear magnetic imaging technology and mechanical tests method to study the effect of chemical solutions on the degradation of the microstructure and mechanical properties of limestone.

However, most existing research on the effects of water and chemistry on rock mechanical behaviors were focused on rock, e.g., limestone, granite, sandstone, etc. Some studies on the influence of salty solution on coal have been carried out in surfactant absorption and mineral flotation. Ozdemir et al. [23] used a series of experimental measurement methods to study the influence of hypersaline on surface chemistry characteristics of coal flotation. They found that the hypersaline reduced the surface tension between coal particles and air bubbles. Zhang et al. [24] studied the effect of different ions on interfacial tension between water and kerosene and found that the surface tension decreased with the increasing salt concentration. Ni et al. [25] studied the effect of NaCl–SDS compound solution on the wetting performance of coal. It found that the addition of sodium salt can efficiently decrease the surface tension of coal and improve the wettability. To date, research on coal's physical and mechanical properties subjected to salty solutions is rarely reported. The physics and mechanisms behind coal damage affected by salty mine water are unclear, bringing difficulties to design, management, and stability control of coal dams in underground reservoirs.

This paper aims to study the evolution of the physical and mechanical properties of coal subjected to salty solution. Firstly, coal samples from Ningdong mining area in western China are prepared to standard cylinders. A salty solution is prepared based on chemical test results of mine water. The scanning electron microscope and nuclear magnetic resonance technology are employed to study the changes in porosity, mass, and microstructure of coal samples for different soaking times. Then, uniaxial compression tests are carried out to obtain the stress–strain curves of coal samples for different soaking times. The effects of salty solution on the damage of coal samples are discussed. Moreover, a damage constitutive model for the coal samples is developed by introducing a chemical-stress coupling damage variable. The developed model is verified with the stress–strain curves from uniaxial compression tests.

2. Experimental Materials and Methods

2.1. Preparation of Coal Samples

The coal used in the test is taken from Lingxin coal mine in Ningdong Mining Area, China. The coal samples are drilled from the same coal block to ensure uniformity. According to the ISRM rock preparation standard [26], the coal samples are made into a standard cylinder with a diameter of 50 mm and a height of 100 mm. The unevenness of the two end faces should not exceed ± 0.05 mm. The end face is perpendicular to the axis of the rock sample, and the allowable deviation is $\pm 0.25^{\circ}$.

2.2. Salt Solution

Groundwater is a complex chemical solution containing various ionic components. The cations in the mine water retrieved from the Lingxin mining area were determined by using inductively coupled plasma atomic emission spectroscopy (ICP-OES 730, Agilent, Santa Clara, CA, USA) and the anions were tested by using Chromatograph (LC-2010 PLUS, SHIMADZU, Kyoto, Japan). The chemical test results are presented in Table 1. The main ion components of the mine water are Na^{+}, SO_4^{2-}, and Cl^{-} and the calculated concentration ratio is about 5:1:3. In this study, the mixed anions of SO_4^{2-}-Cl^{-} and the cation Na^{+} are selected to study the degradation effect of the salty solution on coal samples. Since the water–coal reaction is a long-term process, we used a higher ion concentration in the configuration solution than that in mine water to shorten the experimental period. The ratio of the cation (Na^{+}) to mixed anions (SO_4^{2-}-Cl^{-}) is the same as the ratio in mine water. In the final salty solution, the concentration of sulfate ions is 0.1 mol/L, and the solution pH is 9. The chemical reagents used in this study are sodium chloride (≥99.5% purity), sodium sulfate (≥99% purity), and sodium hydroxide (≥96% purity) that are all made by Sinopharm Chemical Reagent Co. Ltd. (Shanghai, China). High-purity distilled water is used throughout the experiment.

Table 1. Chemical test results of mine water.

PH	TDS	K^+	Na^+	Ca^{2+}	Cl^-	$SO_4{}^{2-}$
	mg/L	mg/L	mg/L	mg/L	mg/L	mg/L
8.79	4720	10.7	1350	86	1230	1145

2.3. *Experimental Procedure*

After the coal samples were prepared, the longitudinal wave velocities of coal samples were measured to ensure uniformity of samples. The coal samples with close wave velocities were divided into seven groups (US1-US6 and P) according to the immersion time. Each group contained three samples, and a total of 21 samples were used for further experiments. The sample group US1-US6 were used for mechanical tests and the sample group P was used for physical tests. The coal samples were dried before immersion and the original mass were weighed. The solution was regularly agitated during the soaking process to make the coal and solution fully interact. During the soaking process, the pH value of the solution, the mass, the longitudinal wave velocity, and the porosity of the coal samples were measured, and the experimental temperature was 20 °C. The measurement time interval was determined according to the change rate of pH value of the salty solution. When the pH value of the solution becomes stable, it is considered that the water–coal interaction has reached a steady state. An acidity meter PHS-3E from China Shanghai INESA Scientific Instrument Co. Ltd. was used to measure the pH value of the solution. The accuracy of the acidity meter is 0.01. The longitudinal wave velocity was tested by a ZBL-U5200 non-metallic ultrasonic detector produced by China Beijing ZBL SCI & TECH Co. Ltd. An electronic scale with accuracy of 0.01 g was used for the measurement of coal mass. The sample group P was chosen for the mass and the longitudinal wave velocity test throughout the whole immersion time.

The coal sample P-2 was selected for the porosity test, and the evolution law of porosity throughout the immersion period was analyzed. The low-field nuclear magnetic resonance analyzer produced by China Suzhou Newmarket analytical instrument company was applied for porosity tests.

In order to understand the microscopic mechanism of the effect of salty solution on coal's pore feature, SEM technology was used to analyze the microstructure of coal. The coal was cut into a 1 cm × 1 cm × 1 cm square piece to meet the requirements of the SEM sample size. In this study, the SEM of ZEISS EVO 18 (Jena, Germany) was used for the microscopy experiment. The related parameters are: acceleration voltage, 200 V–30 Kv; magnification times, 5–10^6 times; focused working distance, 2–145 mm; and sample size, diameter $\leq$ 250 mm, height $\leq$ 145 mm.

For the uniaxial compression tests, the MTS815 rock mechanics testing machine was applied. The maximum axial load of the MTS815 testing machine is 2700 kN. The ranges of the axial and circular extensometers are respectively 5 mm and 8 mm. A combination of stress loading and hoop displacement loading was adopted in the test. The loading rate is 100 N/s. When the axial load increases to 30 KN, it converts to displacement loading with a 0.01 mm/min loading rate. Figure 1 shows the detailed flow of experiments.

Figure 1. Diagrammatic sketch showing the flow of experiments.

3. Results and Discussion

3.1. Changes in PH Values of Salt Solution

Figure 2 shows the relationship between the soaking times and pH value of the solution. It can be seen that as the soaking time increasing, the pH value of the solution gradually tends to be neutral, indicating that the PH value of the solution has the ability of self-balancing during the process of the water–rock interaction. The pH value of the solution first changed very obviously, and then the changing rate gradually decreased. This indicates that the water–rock interaction is time dependent, i.e., the water–rock reaction will gradually weaken with the increase in the soaking time, and eventually tend to be stable.

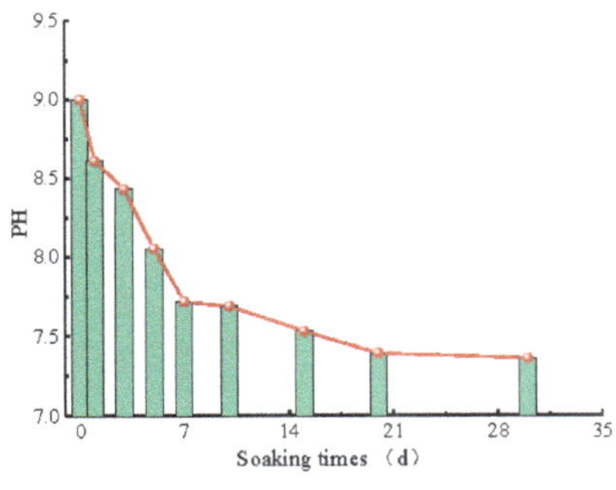

Figure 2. The relationship curve between soaking times and PH values of solution.

3.2. Variation Law of Longitudinal Wave Velocity of Coal Samples

Many studies have shown that the longitudinal wave velocity of rock samples is sensitive to the development of microstructures such as internal pores and defects [27–32]. In order to eliminate the errors caused by the differences in samples, the longitudinal wave velocity of the same sample was tested during the immersion period. Therefore, the changes in longitudinal wave velocity can be used to characterize the influence of salty solution on the damage to the internal microstructure of the coal samples.

The test results indicated that the longitudinal wave velocity of coal samples displays different degrees of deterioration with the increasing soaking times. In this paper, the change rate of longitudinal wave velocity (v_{cr}) was used to characterize the damage law of v_p caused by chemical corrosion. The larger the value of v_{cr}, the higher the degree of damage. The expression of the change rate of v_{cr} is shown as follows:

$$v_{cr} = \frac{\left|v_{pt} - v_{p0}\right|}{v_{p0}} \times 100\% \tag{1}$$

In the above, v_{p0} and v_{pt} are the longitudinal wave velocities of coal samples before and after the different soaking times.

Figure 3 displays the relationship between soaking times and v_{cr} of sample P-2. As the soaking time increases, the longitudinal wave velocities of the coal sample exhibit a deteriorating trend. At the beginning of the test, the longitudinal wave velocity of coal samples deteriorated at a high rate and the water–coal reaction was violent, which caused the increase in pores of the coal sample. Macroscopically, the longitudinal wave velocity decreased to varying degrees. Table 2 displays the mass and v_p of coal samples under different soaking times.

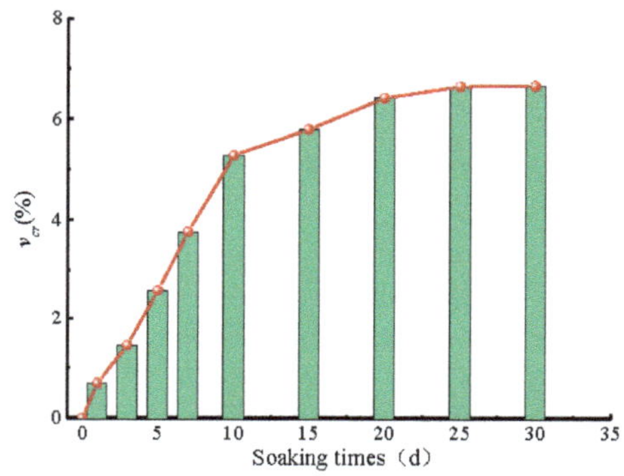

Figure 3. The relationship curve between the soaking times and the change rate of longitudinal wave velocities.

Table 2. The mass and longitudinal wave velocity (v_p) of coal samples under different soaking times.

Soaking Times/d	P-1		P-2		P-3	
	Mass/g	v_p/km·s^{-1}	Mass/g	v_p/km·s^{-1}	Mass/g	v_p/km·s^{-1}
0	257.79	1.83	255.32	1.76	253.07	1.80
0.25	261.15	-	258.85	-	256.12	-
0.5	265.33	-	263.13	-	261.18	-
1	270.91	1.82	268.24	1.75	265.86	1.79
3	269.92	1.80	267.06	1.73	264.99	1.77
5	269.54	1.79	267.30	1.71	264.44	1.76
7	269.04	1.76	266.50	1.69	264.49	1.73
10	268.40	1.73	265.71	1.67	263.32	1.71
15	268.02	1.73	265.73	1.66	262.73	1.70
20	266.85	1.72	264.47	1.65	262.04	1.69
25	266.80	1.71	264.28	1.64	262.00	1.68
30	266.77	1.71	264.26	1.64	262.02	1.68

3.3. *Mass Changes of Coal Samples*

During the immersion process, the water–coal reaction dissolves the mineral components in the coal samples and finally changes the mass of the coal sample. The mass differences between the original dried samples and saturated samples with different soaking times were analyzed, which can indirectly reflect the water–coal process's degradation law. The prepared coal samples were dried for 24 h with temperature of 50 °C and the original mass m_0 were weighed. After different soaking times, the coal samples were taken out from the salt water, the moisture on the surface of the samples was wiped off, and the mass of the saturated samples (m_t) were weighed. The relationship between the m_0 and m_t is shown as follows:

$$\begin{cases} m_0 + m_l - m_r = m_t \\ \Delta m = m_t - m_0 = m_l - m_r \end{cases} \quad (2)$$

where m_l represents the mass of pore water inside the coal sample; m_r represents the mass of the dissolved substance.

Figure 4 shows the relationship between soaking times and the relative mass difference of sample P-2. It can be seen that the relative mass difference of the coal sample increases rapidly at the beginning of the test, then gradually decreases. At the initial state of immersion, the solution diffuses through the pores accompanied by the water–coal reaction, but the mass of the immersed solution was obviously greater than that of the dissolved substance, which led to a rapid increase in the relative mass difference. As the soaking time increased, the coal sample reached a saturated state. The pore volume of the coal sample gradually increased with the dissolution, and the immersed solution in the pores continued to increase. However, the strong water–coal reaction made the mass of dissolved substance greater than that of the added immersion solution, which decreased the relative mass difference. After that, the interaction between solution and coal sample gradually weakened, and the variation range of relative mass difference gradually reduced.

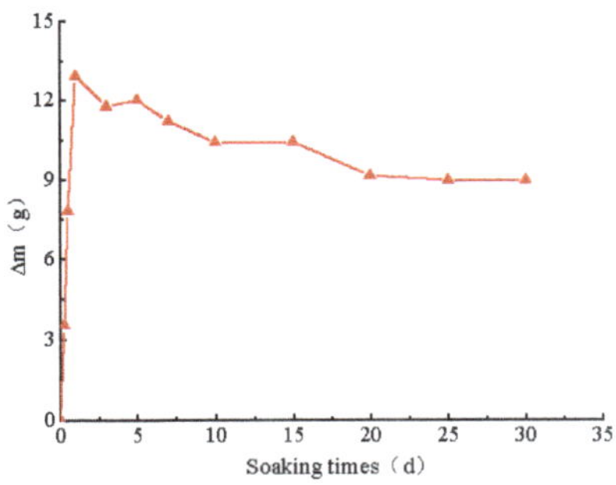

Figure 4. The relationship between Δm and soaking times.

3.4. Porosity of Coal Samples

The nuclear magnetic resonance (NMR) method was used to observe the change of porosity and pore size distribution of coal samples under different soaking times. The signal source in the NMR test is hydrogen ion. Based on the hydrogen ion signal detected in the pores, the analysis software obtains the relaxation time (T_2) and the initial magnetic field vectors of pores with different sizes through a special sequence (CPMG). Finally, the distribution of the relaxation time can be obtained through inversion. Different relaxation times represent different sizes of pores and the larger relaxation times represent larger pore sizes. The area of the relaxation peak in the T_2 curves reflects the number of pores [33]. Equation (3) represents the relationship between T_2 and parameters of pores [33].

$$1/T_2 = \rho(S/V) = F_s(\rho/r) \quad (3)$$

where T_2 represents the inversion relaxation time, ms; ρ represents lateral relaxation density, um/ms; S represents the surface area of pore, cm^2; V represents pore volume, cm^3; r represents pore size, nm; and F_s represents shape factor of pore.

Figure 5 exhibits the T_2 distribution of coal samples under different soaking times. The pore structure of coal is divided into three regions: the adsorption pores with a pore size of 1–100 nm (corresponding T_2 smaller than 2.5 ms), seepage pores with a pore size of 100–10,000 nm (corresponding T_2 larger than 2.5 ms and smaller than 50 ms), and fractures with a pore size larger than 10,000 nm (corresponding T_2 larger than 50 ms) [34–37]. Figure 5 shows that the T_2 distribution of coal under different soaking times has a bimodal characteristic. After the immersion in salty solution, the T_2 distribution curves of coal move towards the right, indicating that the immersion effect increases the connectivity of the micropores and results in the appearance of pores with larger sizes. The area of T_2 distribution curves corresponding to the relaxation times of 0–2.5 ms, 2.5–50 ms, and 50 ms–1000 ms were calculated and the values represent the number of pores. Table 3 displays the pore number of coal samples after different soaking times. It is obvious that the number of seepage pores and fractures increased with the increase in soaking times. While the number of adsorption pores exhibits a trend of rapid increase first and then a slow decline. It is mainly because the small pores gradually connect to form larger pores in the later stage of immersion.

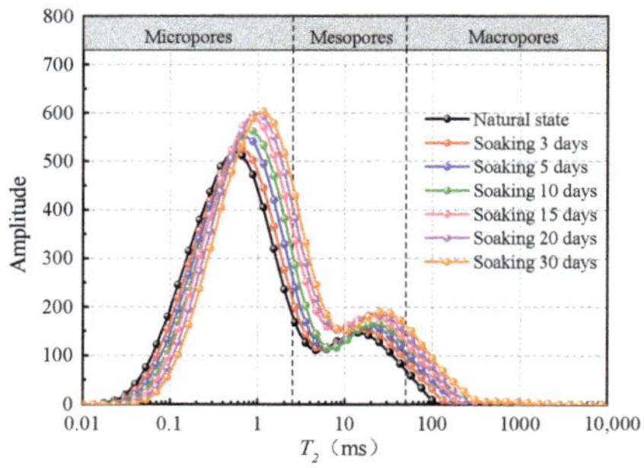

Figure 5. The T_2 distribution curves of coal samples.

Table 3. Number of pores in coal samples after different soaking times.

Soaking Times/d	Porosity/%	Adsorption Pores	Seepage Pores	Fractures
0	21.28	9809	2570	213
3	21.91	9819	2765	354
5	22.96	10,041	2983	517
10	23.66	10,062	3172	688
15	24.88	10,385	3749	710
20	25.25	10,208	3982	807
30	25.58	9915	4210	1026

Under the effect of the water–coal interaction, the minerals in the coal sample dissolved, leading to increased porosity. Figure 6 shows the relationship curve between soaking times and porosity of coal sample.

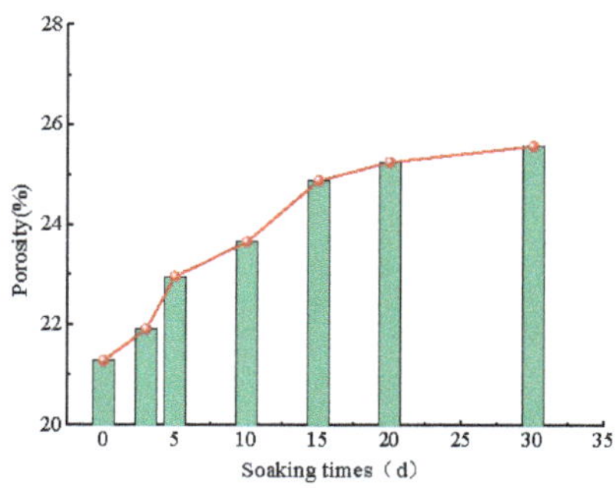

Figure 6. The change law of porosity as a function of soaking times.

3.5. Microscopic Morphological Characteristics of Coal Samples

The structure of coal is very complex due to the many mineral components in the coal. During the immersion process, the salty solution and the coal undergo a series of water–rock reactions, which lead to the dissolution of minerals, changes in the microscopic structure, and the increase in microcracks inside the samples.

It can be seen from Figure 7 that a large number of mineral particles are distributed on the surface of the coal sample under the natural state. At the same time, it can be observed that microfractures and micropores exist on the surface of coal. Most pores are blocked by mineral components, resulting in poor connectivity between pores. After being soaked for different times, the coal samples showed different degrees of corrosion, and the microscopic morphology changed to a certain extent. As the immersion time increases, the number of mineral particles on the surface of the coal samples decreases significantly. The mineral particles originally embedded in the coal body were corroded into holes of different sizes. Under the effect of the salty solution, microfractures gradually appeared and connected with the internal structure. The number of pores and fractures in the coal samples increased after being soaked with the salty solution. At the same time, the strength of the coal samples will gradually decrease with the increasing number of pores and cracks.

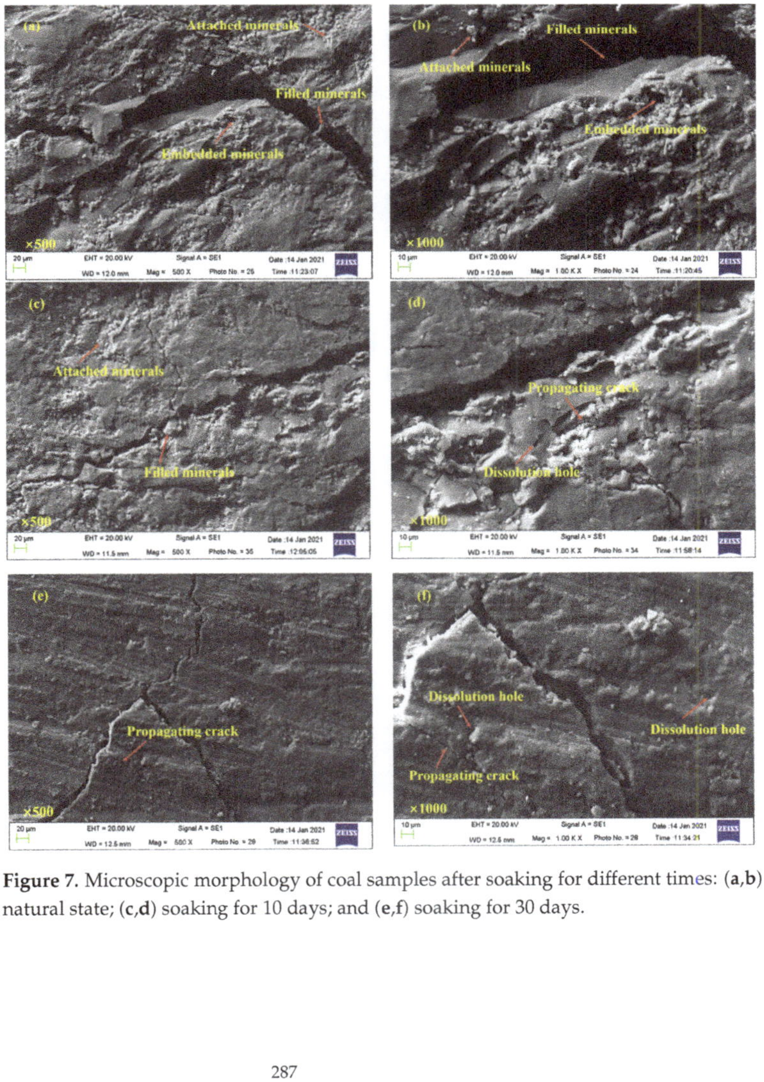

Figure 7. Microscopic morphology of coal samples after soaking for different times: (**a**,**b**) represent natural state; (**c**,**d**) soaking for 10 days; and (**e**,**f**) soaking for 30 days.

3.6. Mechanical Properties of Coal Samples

In the mechanical test, the stress and the strain of coal samples during the failure process were obtained. The stress–strain curves of coal samples for different soaking times (i.e., ST) are illustrated as Figure 8. The mechanical parameters calculated according to the stress–strain curve are shown in Table 4. It can be seen from Figure 8 that the compaction stage of the coal samples after immersion is longer than that of the natural state. As the soaking time increases, the compaction stage becomes longer. This is mainly because the internal defects such as pores and microfractures increased with the soaking time and more defects result in longer initial compaction stage of the stress–strain curve.

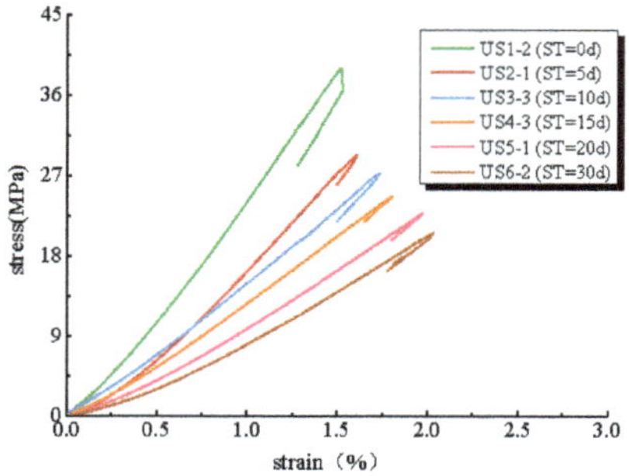

Figure 8. The stress–strain curve of coal samples.

Table 4. Mechanical parameters of coal samples.

Sample No.	Soaking Time (d)	Peak Strength (MPa)	Elastic Modulus (GPa)	Failure Strain (%)
US1-1	0	40.980	2.957	1.483
US1-2	0	38.974	2.865	1.519
US1-3	0	37.584	2.651	1.538
US2-1	5	29.269	2.172	1.609
US2-2	5	32.427	2.361	1.581
US2-3	5	27.985	1.886	1.664
US3-1	10	26.087	1.785	1.763
US3-2	10	27.953	1.603	1.709
US3-3	10	27.266	1.636	1.737
US4-1	15	23.758	1.508	1.874
US4-2	15	25.481	1.545	1.792
US4-3	15	24.663	1.475	1.804
US5-1	20	22.806	1.250	1.976
US5-2	20	23.469	1.347	1.895
US5-3	20	21.568	1.257	1.983
US6-1	30	21.357	1.166	2.113
US6-2	30	20.605	1.235	2.034
US6-3	30	18.624	1.013	2.164

Figure 9 displays the relationship between soaking times and several mechanical parameters of coal samples. It appears that the peak strength and elastic modulus of the coal samples gradually decrease with the soaking time, while the failure strain exhibits an increasing trend with soaking time. The reduction in the peak strength of coal under the salty solution immersion is related to the adsorption of ions or molecules in the solution to the coal surface. The adsorbed ions or molecules on the coal surface reduce the surface energy and its fracture strength. The result manifests that the immersion of salty solution significantly influences the mechanical properties at the initial state, and the declining amplitude becomes smaller with the increasing soaking time.

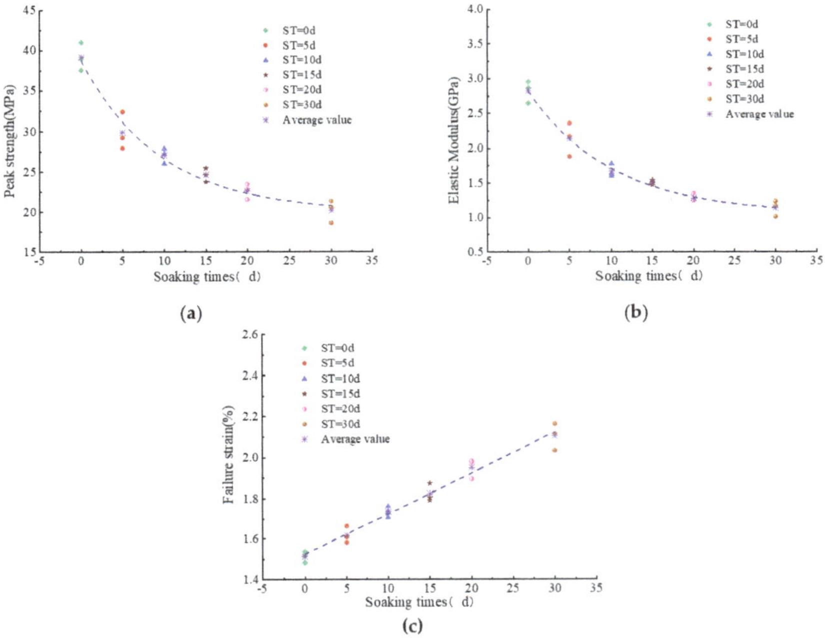

Figure 9. The relationship between soaking times and mechanical parameters of coal: (**a**) peak strength; (**b**) elastic modulus; and (**c**) failure strain.

4. Statistical Damage Constitutive Relationship of Coal Samples under the Coupling Action of Salty Solution and Uniaxial Compression

4.1. Damage Constitutive Relationship and Damage Evolution Equation under Uniaxial Compression

According to the principle of strain equivalence [38,39], the strain produced by the damaged material under the stress σ is equivalent to the strain produced by the undamaged material under the effective stress σ'.

$$\varepsilon = \frac{\sigma}{E'} = \frac{\sigma'}{E} = \frac{\sigma}{E_0(1 - D_s)} \tag{4}$$

In the above, σ and σ' are nominal stress and effective stress, respectively; ε is strain; E_0 is the elastic modulus of the sample in the initial state; E' is the elastic modulus of the sample in a damaged state; and D_s is the damage variable under loading.

According to Equation (4), the damage constitutive model of coal samples during uniaxial compression can be obtained as follows:

$$\sigma = E\varepsilon(1 - D_s) \tag{5}$$

The rock material contains various defects, and the microstructure is nonuniform. The random distribution of these defects in the rock causes the damage to distribute randomly inside the rock during the compression process, resulting in great differences in the mechanical properties of the rock [40,41]. It can be considered that the mechanical properties of the rock are a random variable. The rock can be divided into micro-element bodies containing several defects, and the relationship between the statistical distribution density of material failure and the damage variable is assumed as in [41].

$$dD_s = P(\varepsilon)d\varepsilon \tag{6}$$

There, $P(\varepsilon)$ is a measure of the damage rate of the micro-element body during the loading process and F is the strength parameters of rock micro-element body.

Assuming that the strength of the micro-element body obeys the Weibull distribution during the loading process, the probability density function can be expressed as in [41].

$$P(\varepsilon) = \frac{m}{\varepsilon_0}\left(\frac{\varepsilon}{\varepsilon_0}\right)^{m-1} \exp\left[-\left(\frac{\varepsilon}{\varepsilon_0}\right)^{m}\right] \tag{7}$$

In the above, ε_0 is the average value of the strength of the micro-element body, and m is the shape factor of the distribution function, representing the uniformity of the rock material.

Substituting Equation (7) into Equation (6), the internal damage evolution equation during compression can be obtained as:

$$dD_s = P(\varepsilon)d\varepsilon = \frac{m}{\varepsilon_0}\left(\frac{\varepsilon}{\varepsilon_0}\right)^{m-1} \exp\left[-\left(\frac{\varepsilon}{\varepsilon_0}\right)^{m}\right]d\varepsilon \tag{8}$$

$$\begin{aligned} D_s &= \int_0^{\varepsilon} \tfrac{m}{\varepsilon_0}\left(\tfrac{\varepsilon}{\varepsilon_0}\right)^{m-1} \exp\left[-\left(\tfrac{\varepsilon}{\varepsilon_0}\right)^{m}\right]d\varepsilon \\ &= 1 - \exp\left[-\left(\tfrac{\varepsilon}{\varepsilon_0}\right)^{m}\right] \end{aligned} \tag{9}$$

Substituting Equation (9) into Equation (5), the damage constitutive relation of rock under uniaxial compression is:

$$\sigma = E\varepsilon(1 - D_s) = E\varepsilon \exp\left[-\left(\frac{\varepsilon}{\varepsilon_0}\right)^{m}\right] \tag{10}$$

The stress–strain relationship curve of the rock should meet the following boundary conditions:

$$\begin{cases} \varepsilon = 0,\ \sigma = 0 \\ D_s = 0, d\sigma/d\varepsilon = 0 \\ \sigma = \sigma_p,\ \varepsilon = \varepsilon_p \\ \varepsilon = \varepsilon_p,\ d\sigma/d\varepsilon = 0 \end{cases} \tag{11}$$

In the above, σ_p and ε_p are the peak stress and peak strain of the rock, respectively. By combining Equations (10) and (11), the following Equation can be obtained:

$$\begin{cases} m = \dfrac{1}{\ln\left(\frac{E\varepsilon_p}{\sigma_p}\right)} \\ \varepsilon_0 = \dfrac{\varepsilon_p}{(1/m)^{1/m}} \end{cases} \tag{12}$$

Substituting Equation (12) into Equation (9), the damage evolution Equation of the rock is obtained as:

$$D_s = 1 - \exp\left[-\frac{1}{m}\left(\frac{\varepsilon}{\varepsilon_p}\right)^{m}\right] \tag{13}$$

Substituting Equation (13) into Equation (10), the damage constitutive Equation of the rock can be obtained:

$$\sigma = E\varepsilon(1 - D_s) = E\varepsilon \exp\left[-\frac{1}{m}\left(\frac{\varepsilon}{\varepsilon_p}\right)^{m}\right] \tag{14}$$

4.2. The Chemical Damage Variable under the Action of Salty Solution

The damage of the coal sample after immersion in salt solution includes not only the load damage, but also the chemical damage caused by the immersion. According to the principle of strain equivalence, the constitutive relationship of the internal damage of the coal sample after immersion in the salt solution is expressed as:

$$\sigma = (1 - D_c)E_0\varepsilon \quad (15)$$

There, E_0 is the elastic modulus of a sample in the initial state and D_c is defined as the chemical damage variable under salt solution immersion.

According to Equation (15), the expression of D_c can be obtained:

$$E_c = (1 - D_c)E_0 \quad (16)$$

$$D_c = 1 - \frac{E_c}{E_0} \quad (17)$$

In the above, E_c is the elastic modulus of a coal sample after immersion in salty solution.

Figure 10 shows the relationship between the average values of chemical damage variable and immersion time. With the increase in the immersion time, the chemical damage variable of the coal sample firstly grows rapidly, and then the growth rate gradually slows down. By fitting the test data of the damage variable, the relationship expression between the chemical damage variable and the immersion time is obtained as shown in Equation (18). Under the action of the salty solution, the chemical damage variable of the coal sample is in an exponential relationship with the immersion time and the correlation coefficient is 0.977. The increase in immersion time causes more accumulated damage inside the coal sample, resulting in the deterioration of the coal sample's compressive strength.

$$D_c = 0.582 - 0.712 \times 0.862^t \ \mathrm{R}^2 = 0.977 \quad (18)$$

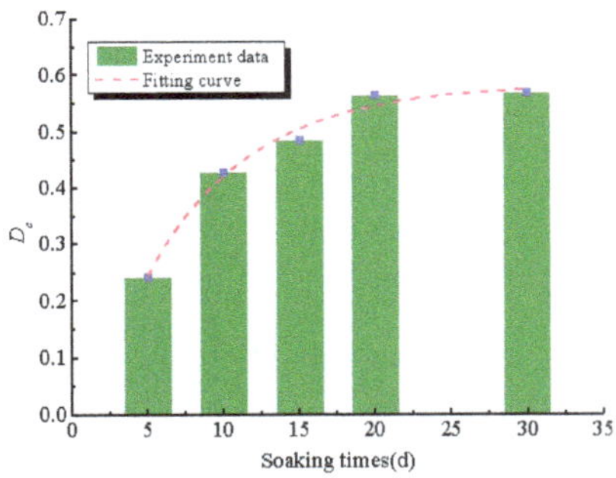

Figure 10. The relationship curve between average value of chemical damage variable and soaking times.

4.3. *The Damage Evolution Law under the Coupling Action of Uniaxial Compression and Salty Solution Immersion*

The failure of the salty solution immersed coal sample under the action of the uniaxial load is caused by the mutual coupling and mutual influence of chemical damage and load damage. For the coal sample after soaking in salty solution, the total damage consists of two parts. The chemical damage caused by the salty solution immersion is considered as the first type of damage, and the damage caused by the load is regarded as the second type of damage.

$$\sigma = (1 - D_s)E_c\varepsilon \quad (19)$$

Substituting Equation (16) into Equation (19), the damage constitutive relationship of the coal sample under the coupling effect of salty solution immersion and uniaxial load can be obtained.

$$\sigma = (1 - D_s)(1 - D_c)E_0\varepsilon \quad (20)$$

The total damage variable D under the coupling action of chemical and load is expressed as Equation (21):

$$D = D_s + D_c - D_s D_c \tag{21}$$

Combing Equations (13), (17), and (21), the final form of total damage variable D is shown as Equation (22):

$$D = 1 - \frac{E_c}{E_0} \exp\left[-\frac{1}{m} \left(\frac{\varepsilon}{\varepsilon_p} \right)^m \right] \tag{22}$$

The typical damage evolution curve under the coupling action of salty solution corrosion and uniaxial compression is shown in Figure 11. The damage degree of the specimen gradually increases with the increase in axial strain. The damage evolution of the coal sample under the coupling action of salty solution corrosion and load has obvious nonlinear characteristics. In the initial stage of compression, the internal microcracks and pores of the specimen are compressed. The degree of damage in the compaction stage is relatively small, resulting in a linear damage evolution curve. The continued action of the external load leads to the expansion of the microcracks, and increases the damage of the internal structure. Finally, crack penetration caused the specimen to be destroyed. Comparing the length of the linear section of the damage evolution curve, it is found that the length of the linear section of the curve gradually increases with the increase in the immersion time. It indicates that the porosity inside the coal sample increases with the increase in immersion time, resulting in the linear segment being more pronounced.

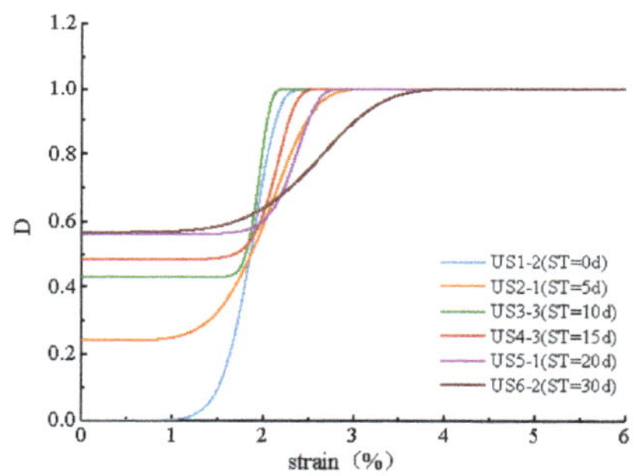

Figure 11. The typical damage evolution curve.

4.4. Damage Constitutive Model Considering Chemical-Stress Coupling Factor

Combining Equations (12) and (20)–(22), the damage constitutive equation of the coal sample under the coupling action of chemical and uniaxial load can be expressed as Equation (23):

$$\begin{cases} \sigma = E_c \varepsilon \exp\left[-\frac{1}{m} \left(\frac{\varepsilon}{\varepsilon_p} \right)^m \right] \\ m = \frac{1}{\ln \frac{E_c \varepsilon_p}{\sigma_p}} \\ \varepsilon_0 = \varepsilon_p m^{1/m} \end{cases} \tag{23}$$

In order to verify the damage constitutive model of the coal sample under the coupling effect of salty solution immersion and the uniaxial load, the stress–strain curve of the uniaxial test was compared with the curve obtained by the damage constitutive which is expressed as Equation (23). The damage constitutive curve is shown as the dotted line in Figure 12. It is found that the unmodified damage constitutive relationship curve is in poor agreement with the test curve. The characteristics of the stress–strain curve caused by the corrosion of the salty solution are not well described by using the unmodified damage constitutive (Equation (23)). Further, a chemical-stress coupling factor μ was proposed to modify the damage constitutive. Literature [42] has introduced a thermal-mechanical coupling factor in the damage constitutive model to display the damage process of rock under high temperatures. Like the rock treated with high temperature, the number of pores and cracks in the coal sample increased under the effect of the corrosion of salty solution. This leads to an increase in the length of the compaction stage. In this study, the expression of chemical-stress coupling factor μ refers to that in literature [42]. The expression of coupling factor μ and modified damage constitutive mode. are expressed as follows:

$$\mu = \mu_0 + A \exp\left[-2(\varepsilon - \varepsilon_p)^2/\omega^2\right]/\omega\sqrt{\pi/2} \tag{24}$$

$$\sigma = \mu(1 - D_s)(1 - D_c)E_0\varepsilon \tag{25}$$

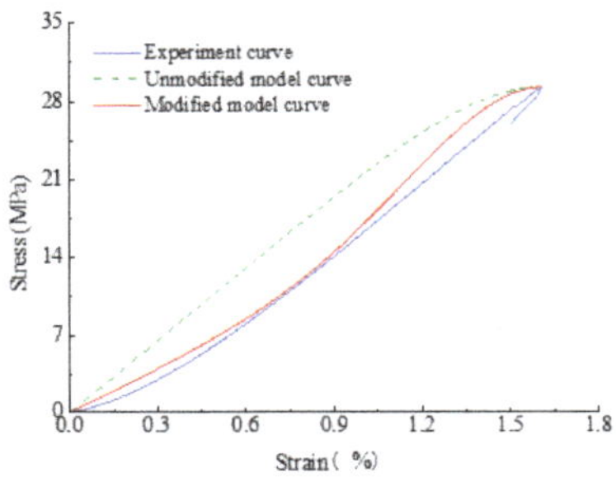

(**a**) US2-1

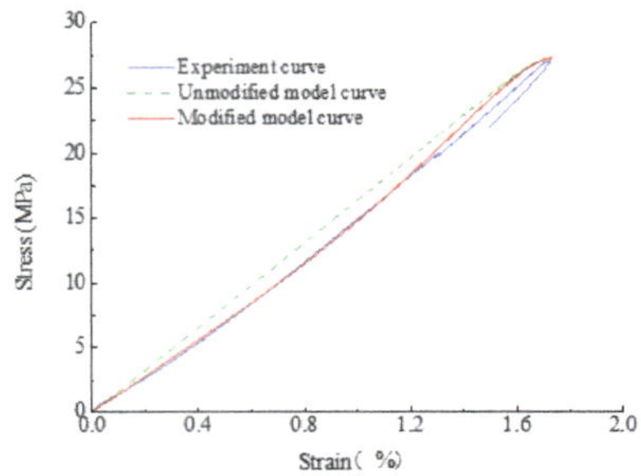

(**b**) US3-3

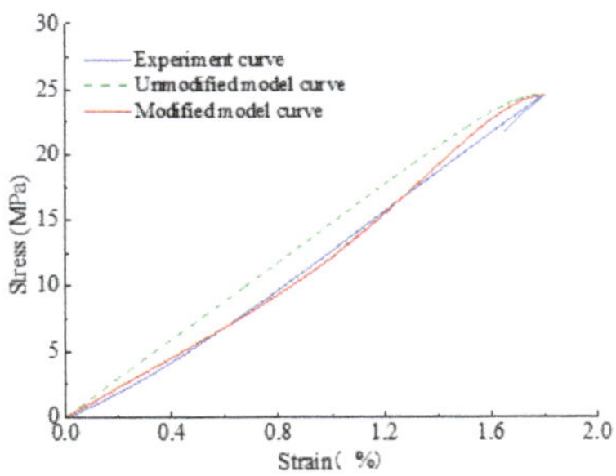

(**c**) US4-3

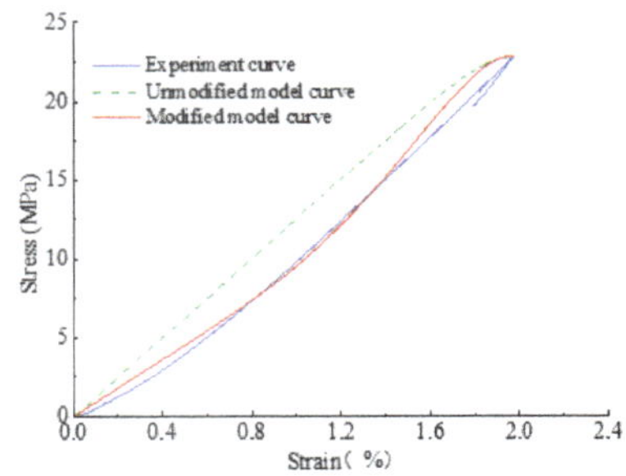

(**d**) US5-1

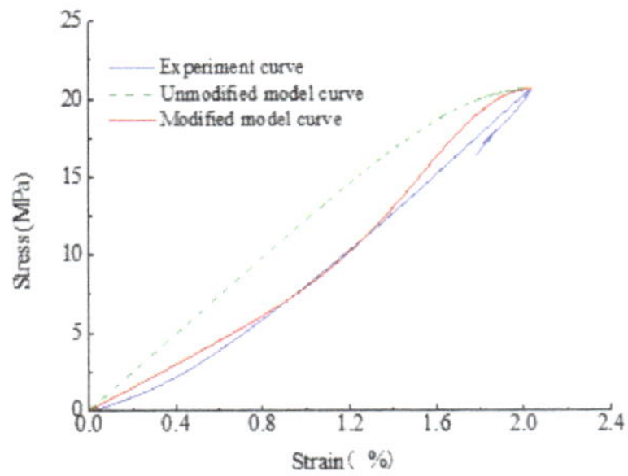

(**e**) US6-2

Figure 12. Comparison of stress–strain test curve and theoretical curve: (**a**) ST = 5 d, (**b**) ST = 10 d, (**c**) ST = 15 d, (**d**) ST = 20 d, and (**e**) ST = 30 d.

In the above, ε_p is the peak strain, μ_0 is the lower limit of the chemical-stress coupling factor, A is the integrated area of the function curve above the baseline, and ω is the standard deviation of the function, which indicate the concentration of the chemical-stress coupling effect. The related parameters (μ_0, A, w) in coupling factor μ need to be determined by fitting the experimental data. The fitting parameters of coupling factor μ are shown in Table 5.

Table 5. Fitting parameters in coupling factor μ.

Sample	$\varepsilon_c/10^{-2}$	$A/10^{-2}$	$w/10^{-2}$	μ_0	Correlation Coefficient
US2-1	1.609	0.28	0.56	0.60	0.93
US2-2	1.581	0.26	0.46	0.55	0.95
US2-3	1.664	0.28	0.55	0.59	0.96
US3-1	1.763	0.51	0.81	0.50	0.94
US3-2	1.709	0.42	0.82	0.60	0.98
US3-3	1.737	0.15	0.80	0.85	0.98
US4-1	1.874	0.32	0.78	0.68	0.97
US4-2	1.792	0.23	0.80	0.78	0.98
US4-3	1.804	0.25	0.85	0.76	0.98
US5-1	1.976	0.30	0.85	0.72	0.96
US5-2	1.895	0.35	0.83	0.68	0.94
US5-3	1.983	0.34	0.81	0.66	0.95
US6-1	2.113	0.41	0.81	0.60	0.93
US6-2	2.034	0.40	0.80	0.60	0.95
US6-3	2.164	0.21	0.85	0.82	0.99

As shown in Figure 12, the modified damage constitutive relationship (Equation (25)), considering the chemical-stress coupling factor μ, is in good agreement with the test results and the change of compaction stage can be well described. In addition, the experimental data of Gu [10] were applied to verify the effectiveness of the modified damage constitutive model. The mechanical response of coal under different soaking times was analyzed in Gu's study. Figure 13 displays the verification of the modified damage constitutive model with the experimental data of Gu. It is obvious that the modified model is highly consistent with the test curve. The modified model well described the stress–strain curves of coal under the different soaking times.

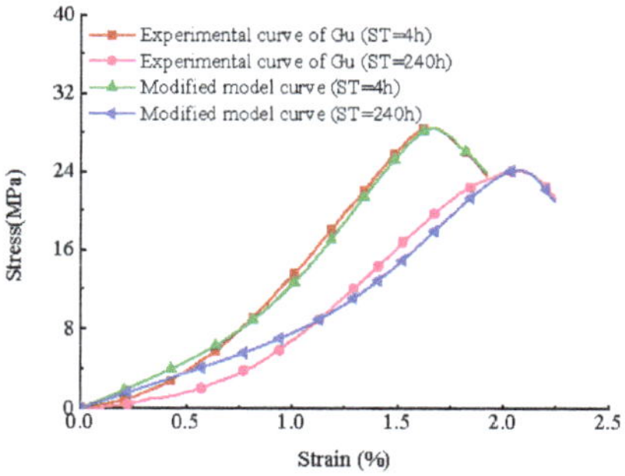

Figure 13. Verification of modified damage constitutive model with experimental data of Gu [10].

5. Conclusions

In this paper, physical and mechanical tests were carried out on coal samples subjected to a salty solution. The degradation mechanisms of the physical and mechanical properties of the coal samples were studied. A damage constitutive model for coal subjected to the salty solution under uniaxial compression was proposed by introducing the chemical-stress coupling factor. The relevant results are summarized as follows:

(1) The physical characteristics of coal under the action of salty solution show varying degrees of deterioration. Under the attack of salty solution, the relative mass difference of the coal samples displays an increasing trend at the beginning of the test, then gradually decreases. With the increase in immersion time, the pH value gradually tends to

be neutral. The mineral particles of coal samples are dissolved by the salty solution, resulting in aggravated chemical damage to the microstructure and an increase in the coal sample's porosity.

(2) The microstructure morphology of coal samples for different immersion times were analyzed. With the effect of the salty solution, the microscopic morphology has changed to a certain extent. As the immersion time increases, the number of mineral particles on the surface of the coal decreases significantly and the connectivity of the internal structure increases.

(3) The peak strength and elastic modulus of the coal samples changes exponentially as a function of soaking time. The failure strain exhibits an increasing trend with soaking time. As the soaking time increases, the initial damage of the coal sample increases exponentially. According to the theory of damage mechanics, the chemical-stress coupling damage variable was introduced, and the variation law of the coupled damage variable with time was obtained. Under the effect of salty solution, the changes of the stress–strain curve can be well described by modified damage constitutive model that considers the coupling factor μ.

In the present study, the deterioration mechanism of coal under the action of salty solution was studied and a damage model for coal subjected to the salty solution under uniaxial compression was proposed. The modified damage model can well describe the changes in the compaction stage of the stress–strain curve caused by the water–rock reaction. The pre-peak behavior can also be well described by the modified model. However, the post-peak behavior cannot be fully characterized. In order to fully describe the mechanical behavior during rock failure, it is necessary to characterize the post-peak behavior. We will focus on this issue in future work.

Author Contributions: Conceptualization, methodology, formal analysis, software, validation, writing—original draft preparation, M.W.; writing—review and editing, Q.G.; supervision, project administration, Y.T.; funding acquisition, Q.G. and B.D. All authors have read and agreed to the published version of the manuscript.

Funding: This work was supported by the Open Fund of the Key Laboratory of Western Mine Exploitation and Hazard Prevention, Ministry of Education (SKLCRKF1901), Open Fund of State Key Laboratory of Water Resource Protection and Utilization in Coal Mining (grant no. GJNY-18-73.11), the National Natural Science Foundation of China (grant no. 51804163).

Institutional Review Board Statement: Not applicable.

Informed Consent Statement: Not applicable.

Data Availability Statement: Not applicable.

Conflicts of Interest: The authors declare no conflict of interest.

References

1. Hamawand, I.; Yusaf, T.; Hamawand, S.G. Coal seam gas and associated water: A review paper. *Renew. Sustain. Energy Rev.* **2013**, *22*, 550–560. [CrossRef]
2. Gu, D.Z.; Li, J.F.; Cao, Z.G.; Wu, B.Y.; Jiang, B.B.; Yang, Y.; Yang, J.; Chen, Y.P. Technology and engineering development strategy of water protection and utilization of coal mine in China. *J. China Coal Soc.* **2021**, *46*, 3079–3089. (In Chinese)
3. Xu, H.; Lai, X.; Zhang, S.; Zhang, Y.; Shan, P.; Zhang, X.; Liu, B.; Wan, P. Multiscale Intelligent Inversion of Water-Conducting Fractured Zone in Coal Mine Based on Elastic Modulus Calibration Rate Response and Its Application: A Case Study of Ningdong Mining Area. *Lithosphere* **2021**, *2021*. [CrossRef]
4. Gu, D.Z. Theory framework and technological system of coal mine underground reservoir. *J. China Coal Soc.* **2015**, *40*, 239–246. (In Chinese)
5. Dyke, C.; Dobereiner, L. *Evaluating the Strength and Deformability of Sandstones*; Geological Society of London: London, UK, 1991.
6. Hawkins, A.B.; McConnell, B.J. Sensitivity of sandstone strength and deformability to changes in moisture content. *Q. J. Eng. Geol. Hydrogeol.* **1992**, *25*, 115–130. [CrossRef]
7. Vasarhelyi, B. Statistical Analysis of the Influence of Water Content on the Strength of the Miocene Limestone. *Rock Mech. Rock Eng.* **2004**, *38*, 69–76. [CrossRef]

8. Erguler, Z.; Ulusay, R. Water-induced variations in mechanical properties of clay-bearing rocks. *Int. J. Rock Mech. Min. Sci.* **2008**, *46*, 355–370. [CrossRef]
9. Zhao, Y.; Liu, S.; Jiang, Y.; Wang, K.; Huang, Y. Dynamic Tensile Strength of Coal under Dry and Saturated Conditions. *Rock Mech. Rock Eng.* **2015**, *49*, 1709–1720. [CrossRef]
10. Gu, H.; Tao, M.; Li, X.; Li, Q.; Cao, W.; Wang, F. Dynamic response and failure mechanism of fractured coal under different soaking times. *Theor. Appl. Fract. Mech.* **2018**, *98*, 112–122. [CrossRef]
11. Gu, H.; Tao, M.; Li, X.; Cao, W.; Li, Q. Dynamic tests and mechanical model for water-saturated soft coal with various particle gradations. *Int. J. Rock Mech. Min. Sci.* **2020**, *132*, 104386. [CrossRef]
12. Karfakis, M.G.; Akram, M. Effects of chemical solutions on rock fracturing. *Int. J. Rock Mech. Min. Sci. Geomech. Abstr.* **1993**, *30*, 1253–1259. [CrossRef]
13. Feucht, L.; Logan, J.M. Effects of chemically active solutions on shearing behavior of a sandstone. *Tectonophysics* **1990**, *175*, 159–176. [CrossRef]
14. Feng, X.; Ding, W.X. Coupled chemical stress processes in rock fracturing. *Mater. Res. Innov.* **2011**, *15*, s547–s550. [CrossRef]
15. Feng, X.-T.; Chen, S.; Zhou, H. Real-time computerized tomography (CT) experiments on sandstone damage evolution during triaxial compression with chemical corrosion. *Int. J. Rock Mech. Min. Sci.* **2004**, *41*, 181–192. [CrossRef]
16. Feng, X.-T.; Ding, W. Experimental study of limestone micro-fracturing under a coupled stress, fluid flow and changing chemical environment. *Int. J. Rock Mech. Min. Sci.* **2007**, *44*, 437–448. [CrossRef]
17. Han, T.; Wang, X.; Li, Z.; Li, D.; Xing, F.; Han, N. Laboratory investigation of the mode-I fracture of sandstone caused by a combination of freeze-thaw cycles and chemical solutions. *Bull. Int. Assoc. Eng. Geol.* **2020**, *79*, 3689–3706. [CrossRef]
18. Hutchinson, A.; Johnson, J.; Thompson, G.; Wood, G.; Sage, P.; Cooke, M. Stone degradation due to wet deposition of pollutants. *Corros. Sci.* **1993**, *34*, 1881–1898. [CrossRef]
19. Qiao, L.; Wang, Z.; Huang, A. Alteration of Mesoscopic Properties and Mechanical Behavior of Sandstone Due to Hydro-Physical and Hydro-Chemical Effects. *Rock Mech. Rock Eng.* **2016**, *50*, 255–267. [CrossRef]
20. Lin, Y.; Gao, F.; Zhou, K.; Gao, R.; Guo, H. Mechanical Properties and Statistical Damage Constitutive Model of Rock under a Coupled Chemical-Mechanical Condition. *Geofluids* **2019**, *2019*, 1–17. [CrossRef]
21. Xie, S.; Shao, J.; Xu, W. Influences of chemical degradation on mechanical behaviour of a limestone. *Int. J. Rock Mech. Min. Sci.* **2011**, *48*, 741–747. [CrossRef]
22. Li, H.; Zhong, Z.; Liu, X.; Sheng, Y.; Yang, D. Micro-damage evolution and macro-mechanical property degradation of limestone due to chemical effects. *Int. J. Rock Mech. Min. Sci.* **2018**, *110*, 257–265. [CrossRef]
23. Ozdemir, O.; Taran, E.; Hampton, M.; Karakashev, S.; Nguyen, A. Surface chemistry aspects of coal flotation in bore water. *Int. J. Miner. Process.* **2009**, *92*, 177–183. [CrossRef]
24. Zhang, Y.; Zhu, H.; Zhu, J.; Yang, F.; He, H.; Qin, Z.; Shi, Q.; Pan, G. Experimental and emulational study on the role of ion in coal adsorbing kerosene: Water–kerosene interface and catenoid characteristics. *Fuel* **2021**, *294*, 120540. [CrossRef]
25. Guanhua, N.; Qian, S.; Meng, X.; Hui, W.; Yuhang, X.; Weimin, C.; Gang, W. Effect of NaCl-SDS compound solution on the wettability and functional groups of coal. *Fuel* **2019**, *257*, 116077. [CrossRef]
26. ISRM. *SM for Determining the Uniaxial Compressive Strength and Deformability of Rock Materials*; ISRM Lisbon: Lisbon, Portugal, 1979; pp. 137–140.
27. Huang, Y.-H.; Yang, S.-Q.; Hall, M.R.; Zhang, Y.-C. The Effects of NaCl Concentration and Confining Pressure on Mechanical and Acoustic Behaviors of Brine-Saturated Sandstone. *Energies* **2018**, *11*, 385. [CrossRef]
28. Song, Z.; Frühwirt, T.; Konietzky, H. Inhomogeneous mechanical behaviour of concrete subjected to monotonic and cyclic loading. *Int. J. Fatigue* **2019**, *132*, 105383. [CrossRef]
29. Zheng, H.; Feng, X.-T.; Pan, P.-Z. Experimental investigation of sandstone properties under CO2–NaCl solution-rock interactions. *Int. J. Greenh. Gas Control.* **2015**, *37*, 451–470. [CrossRef]
30. Song, Z.; Wang, Y.; Konietzky, H.; Cai, X. Mechanical behavior of marble exposed to freeze-thaw-fatigue loading. *Int. J. Rock Mech. Min. Sci.* **2021**, *138*, 104648. [CrossRef]
31. Wang, S.; Tang, Y.; Wang, S. Influence of brittleness and confining stress on rock cuttability based on rock indentation tests. *J. Cent. South Univ.* **2021**, *2021. 28*, 2786–2800. [CrossRef]
32. Wang, S.; Sun, L.; Li, X.; Wang, S.; Du, K.; Li, X.; Feng, F. Experimental Investigation of Cuttability Improvement for Hard Rock Fragmentation Using Conical Cutter. *Int. J. Geomech.* **2021**, *21*, 06020039. [CrossRef]
33. Xu, J.; Zhai, C.; Liu, S.; Qin, L.; Wu, S. Pore variation of three different metamorphic coals by multiple freezing-thawing cycles of liquid CO_2 injection for coalbed methane recovery. *Fuel* **2017**, *208*, 41–51. [CrossRef]
34. Zhang, Z.; Weller, A. Fractal dimension of pore-space geometry of an Eocene sandstone formation. *Geophysics* **2014**, *79*, D377–D387. [CrossRef]
35. Zhou, S.; Liu, D.; Cai, Y.; Yao, Y. Fractal characterization of pore–fracture in low-rank coals using a low-field NMR relaxation method. *Fuel* **2016**, *181*, 218–226. [CrossRef]
36. Chen, S.; Tang, D.; Tao, S.; Ji, X.; Xu, H. Fractal analysis of the dynamic variation in pore-fracture systems under the action of stress using a low-field NMR relaxation method: An experimental study of coals from western Guizhou in China. *J. Pet. Sci. Eng.* **2018**, *173*, 617–629. [CrossRef]

37. Hongchao, X.; Guanhua, N.; Shang, L.; Qian, S.; Kai, D.; Jingna, X.; Gang, W.; Yixin, L. The influence of surfactant on pore fractal characteristics of composite acidized coal. *Fuel* **2019**, *253*, 741–753. [CrossRef]
38. Lemaître, J. A continuous damage mechanics model for ductile fracture. *J. Eng. Mater. Technol. Trans. ASME* **1985**, *107*, 83–89. [CrossRef]
39. Lemaitre, J. *A Course on Damage Mechanics*; Springer Science & Business Media: Cham, Switzerland, 2012.
40. Song, Z.; Konietzky, H.; Herbst, M. Bonded-particle model-based simulation of artificial rock subjected to cyclic loading. *Acta Geotech.* **2018**, *14*, 955–971. [CrossRef]
41. Tang, C.A. *Catastrophe in the Process of Rock Rupture*; China Coal Industry Publishing House: Beijing, China, 1993. (In Chinese)
42. Zhang, Z.Z.; Gao, F.; Xu, X.L. Experimental study of temperature effect of mechanical properties of granite. *Rock Soil Mech.* **2011**, *32*, 2346–2352. (In Chinese)

mathematics

MDPI

Article

Research on Deep-Hole Cutting Blasting Efficiency in Blind Shafting with High In-Situ Stress Environment Using the Method of SPH

Bo Sun [1], Zhiyu Zhang [1,2,†], Jiale Meng [1], Yonghui Huang [3,*], Hongchao Li [4,*,†] and Jun Wang [5]

1 Faculty of Land Resources Engineering, Kunming University of Science and Technology, Kunming 650093, China; sunbo595158771@163.com (B.S.); 11301052@kust.edu.cn (Z.Z.); Meng13190454849@163.com (J.M.)
2 Yunnan Key Laboratory of Sino-German Blue Mining and Utilization of Special Underground Space, Kunming University of Science and Technology, Kunming 650093, China
3 Faculty of Electric Power Engineering, Kunming University of Science and Technology, Kunming 650500, China
4 City Collage, Kunming University of Science and Technology, Kunming 650051, China
5 Yuxi Dahongshan Mining Co. Ltd., Yuxi 650302, China; wangjun17787718394@163.com
* Correspondence: 20130151@kust.edu.cn (Y.H.); 20160032@kust.edu.cn (H.L.)
† Disclosure: Zhiyu Zhang is the co-first author. Hongchao Li is co-corresponding authors.

Abstract: This article aiming at the lack of research on the influence of rock clamp production on cutting blasting under high in-situ stress conditions and the lack of rock damage criteria for RHT constitution in numerical simulation. Combined with the critical rock damage criterion and the embedded function of RHT constitution, the criterion for determining the critical damage of rock in RHT constitutive was studied, and the mechanical parameters of Metamorphic sodium lava were substituted to obtain the critical damage threshold of rock in numerical simulation. The smooth particle hydrodynamics (SPH) method was used to numerically simulate and analyze the influence of different rock clamping coefficients on the rock damage range and the cavity area in the cutting blasting. The stress state applied by the numerical simulation was inversely deduced by the field test scanning results to simulate the rock clamping coefficient K_r at the corresponding depth. The relationship between the cavity area S_c and the free surface distance D_f is analyzed and established. The results show that the rock clip production has an inhibitory effect on the development and propagation of blast-induced cracks. The stress applied in the numerical simulation affects the range and development degree of cracks, and the cracks generated by the explosion are mainly circumferential cracks. The larger coefficient of rock clip production, the more obvious the inhibitory effect on cut blasting, the less the blast-induced cracks and the smaller the rock damage circle. The fitting results show that the curve fitting degree is about 0.94, which proves the accuracy of S_c-D_f curve, and provides important reference value for the design of one-time completion blasting of upward blind shaft.

Keywords: numerical methods; deep rock mechanics; rock damage judgment criteria; SPH; blind shaft cutting blasting

Citation: Sun, B.; Zhang, Z.; Meng, J.; Huang, Y.; Li, H.; Wang, J. Research on Deep-Hole Cutting Blasting Efficiency in Blind Shafting with High In-Situ Stress Environment Using the Method of SPH. *Mathematics* **2021**, *9*, 3242. https://doi.org/10.3390/math9243242

Academic Editors: Shaofeng Wang, Jian Zhou, Xin Cai, Xiaofeng Li and Zhengyang Song

Received: 1 November 2021
Accepted: 8 December 2021
Published: 14 December 2021

1. Introduction

The demand for metal materials in various industries is gradually increasing as science and technology advance. After years of mining, the mineral endowment in the shallow surface is diminishing, and the mine construction is gradually shifting towards the deep part of the mine. Deep rock is frequently accompanied by high in-situ stress, and drilling and blasting are currently the main methods of deep mining excavation. Scholars through a large number of field tests concluded that deep rock blasting is the main source of mining.

As a result, it is critical to investigate the impact of high in-situ stress on the construction effects of rock crushing, blasting into shafts and roadway excavation.

Currently, many scholars have carried out a series of field tests and numerical simulations for rock blasting under high in-situ conditions [1–5]. Yan et al. [6] used wavelet packet analysis to analyze the vibration signals during tunnel blasting excavation under high in-situ stress conditions and obtained the energy distribution of the vibration signals in different frequency bands. Wei et al. [7] analyzed the effect of in-situ stress on the crack extension by simulating the tangential pack blasting under different in-situ stress conditions based on a damage mechanics model. Yang et al. [8,9] established a computational model for double-hole blasting of high in-situ stress rock masses based on the SPH-FEM algorithm, and concluded that the hole spacing should be appropriately shortened under high-stress conditions to facilitate crack expansion by analyzing the stress field distribution as well as the dynamic evolution process. Liu et al. [10] used the hollow core envelope method to measure the in-situ stress in the roadway and combined it with FLAC3D software for stability analysis to derive the influence of the tectonic stress field on the roadway. Dai [11] based on blasting funnel theory, the study of high-stress roadway avalanche blasting, and concluded that the choice of high explosive velocity explosives under high in-situ stress is more effective. Tang et al. [12] based on the decay of blasting frequency in perimeter rock excavation to define the damage variables and derive the decay law of blasting damage amount with the change of blast core distance. LI et al. [13–16] investigated the effect between dynamic unloading effects of in-situ stress and high in-situ stress conditions during blasting excavation using FLAC3D and PFC5.0 numerical software. Yang et al. [17] investigated the relationship between the lateral pressure coefficient and the effect of unloading rate and damage extent using the finite difference method, and verified its accuracy by field testing. Luo et al. [18] proposed a blast damage simulation method based on normal impact load to study the blast damage incubation mechanism under different stress conditions, and concluded that a certain degree of in-situ stress plays a suppressive role on blast damage. Lu et al. [19,20] used LS-DYNA finite element simulation software to simulate the confining effect of rock tunnel to be excavated by applying nodal reaction forces to analyze the damaging effect of blasting excavation on the surrounding rock, and concluded that the damaged area was mainly caused by the initial stress redistribution in the surrounding rock, and the blast load action increased its extent. Li et al. [21] used numerical simulation methods to introduce damage state indicators and proposed two indicators to reflect the effect of trenching blasting, and studied the effect of stress wave loading rate and in-situ stress on cutting blasting. In the above-mentioned analytical studies, for the study of rock mass blasting under high in-situ stress conditions, the rock mass damage is mainly reflected by instrument measurement and numerical simulation. However, there are different constitutive models such as TCK, HJC, Yang, RHT, etc. in numerical simulations, and their damage assessment criteria for specific intrinsic structures have not been studied much. In addition, there are few studies on the effect of rock entrapment on cavity formation and damage patterns in trenching and blasting.

In summary, this paper will employ the rock critical damage parametric theory and the RHT damage model to establish the RHT intrinsic critical damage assessment criteria based on the embedded plastic strain and ultimate strain equations in the RHT constitution. Furthermore, it is impossible to predict the results of the upward-oriented once-in-a-whole cutting blasting into the cavity state and crack expansion under high in-situ stress conditions. To study the characteristic law of rock damage and cutting cavity formation under high in-situ stress in upward blind cutting blasting using the smooth hydrodynamic method (SPH method), and the results were compared with the field cavity scanning results to verify the accuracy of the numerical simulation, and the inverse extrapolation of the rock clamping performance in the test area was performed based on the field test results to investigate the influence law of rock clamping production.

2. Establishment of RHT Constitutive Damage Criterion

2.1. Blast Damage Variable Definition

According to the research results of previous scholars [22], the damage of the rock body under the blast load is the combined effect of the blast pressure, the expansion of the blast-generated gas and the unloading stress, while taking into account the anisotropy of the rock body, the expansion of primary cracks and the development of new cracks. The commonly used rock damage variables are:

$$D = 1 - \frac{\overline{E}}{E} \tag{1}$$

where: D is the rock damage variable; $\overline{E}$ is the equivalent modulus of elasticity of the damaged rock after blasting; E is the modulus of elasticity of the intact rock before blasting.

According to the elastic stress wave theory, there is a certain relationship between the elastic modulus of rock and the longitudinal wave velocity before and after blasting excavation, as in Equations (2) and (3):

$$E = \rho c_p^2 \frac{(1-\mu)(1-2\mu)}{1-\mu} \tag{2}$$

$$\overline{E} = \overline{\rho}\overline{c_p}^2 \frac{(1+\overline{\mu})(1-2\overline{\mu})}{1-\overline{\mu}} \tag{3}$$

where: ρ, $\overline{\rho}$ represent the density of the rock before and after blasting excavation, kg/m^3; c_p, $\overline{c}_p$ are the longitudinal velocity of the rock before and after blasting, m/s; μ, $\overline{\mu}$ are the Poisson's ratio of the rock before and after blasting excavation, respectively.

In this paper, it is assumed that the density and Poisson's ratio of the rock before and after blasting do not change, i.e.,: $\rho = \overline{\rho}$, $\mu = \overline{\mu}$. Therefore, the finishing of Equations (2) and (3) into Equation (1) can be obtained as follows:

$$D = 1 - \left(\frac{\overline{c_p}}{c_p}\right)^2 \tag{4}$$

where the rate of change of the longitudinal wave velocity of the rock body before and after blasting η can be expressed by Equation (5):

$$\eta = \frac{c_p - \overline{c_p}}{c_p} = 1 - \frac{\overline{c_p}}{c_p} \tag{5}$$

Thus, the damage variable D can be integrated as:

$$D = 1 - (1-\eta)^2 \tag{6}$$

Wu [23] investigated the damage failure behavior of rock materials based on the existing research results for the rock damage variable D combined with the damage failure criterion of rock materials, and the damage value corresponding to the peak strength of the rock material is regarded as the rock critical damage parameter D_{cr}, while the material damage is mainly caused by the plastic deformation of the material. Therefore, the relationship between the plastic strain of the rock and the critical damage parameters of the rock can be expressed by Equation (7):

$$D_{cr} = \frac{\varepsilon_p}{\varepsilon_{max}} \tag{7}$$

where: ε_{max}, ε_p are the peak strain and plastic strain under uniaxial compression, respectively.

2.2. Blast Damage Action Zoning

The primary rock mass has initial damage D_0 due to the presence of joints, fissures and structural surfaces, etc. Explosive blast generates shock waves to make the blast hole wall around a certain range of rock crushing, the crushing area rock damage variable $D_2 = 1$, outside the crushing zone is considered as the rock fracture and damage zone (see Figure 1), and its different zoning damage variables are listed in Table 1:

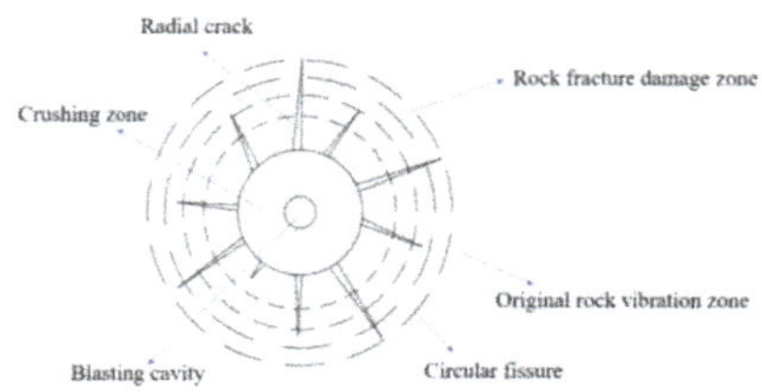

Figure 1. Damage Zoning of Rock Mass under Blasting Load.

Table 1. Damage Zone of Rock under Blasting Load.

Damage Zone	Rock Damage Volume D	Blasting Center Distance r/m
Crushing zone	$D = 1$	$r < r_0$
Partition threshold	1	r_0
fracture and damage zone	$D_{cr} < D < 1$	$r_0 < r < r_1$
Partition threshold	D_{cr}	r_1
Primary rock stress zone	$D \leq D_{cr}$	$r > r_1$

2.3. RHT Constitutive Damage Criterion

The RHT damage constitution was chosen for this numerical simulation, which was proposed by Riedel, Hiermaier, and Thoma in 1997 based on the HJC constitution, by introducing the elastic limit surface, failure strength surface, and residual strength surface to describe the relationship between the elastic limit strain, material failure strength, and residual strength and hydrostatic pressure [24].

The RHT constitutive model is defined for the damage parameter D as:

$$D = \sum\left(\Delta\varepsilon_p/\varepsilon_p^f\right) \tag{8}$$

Which:

$$\Delta\varepsilon_p = \varepsilon_p^f - \varepsilon_p \tag{9}$$

$$\varepsilon_p^f = \begin{cases} D_1(p^* - (1-D)p_t^*)^{D_2} & p^* \geq (1-D)p_t^* + \left(\varepsilon_p^m/D_1\right)^{1/D_2} \\ \varepsilon_m^p & (1-D)p_t^* + \left(\varepsilon_p^m/D_1\right)^{1/D_2} > p^* \end{cases} \tag{10}$$

In Equations (8)–(10): $\Delta\varepsilon_p$ is the equivalent plastic strain increment; $\varepsilon_p{}^f$ is the final failure equivalent plastic strain; $\varepsilon_p{}^m$ is the minimum equivalent plastic strain when the material is damaged; D_1, D_2 are the initial damage parameters of the material and the complete damage parameters of the material, respectively; p^* is rock shear strength and $p_t{}^*$ is failure cut-off pressure.

And for the material in the p^*, $p_t{}^*$ is usually expressed using the Formulas (11) and (12):

$$p^* = \frac{p}{f_c} \tag{11}$$

$$p_t^* = \frac{F_r Q_2 f_s^* f_t^*}{3(Q_1 f_t^* - Q_2 f_s^*)} \tag{12}$$

Which:

$$Q_1 = R_3\left(\frac{\pi}{6}, 0\right) \tag{13}$$

$$Q_2 = Q(p^*) = Q_0 + Bp^* \tag{14}$$

$$R_3(\theta, p^*) = \frac{2(1-Q_2^2)cos\theta + (2Q_2-1)\sqrt{4(1-Q_2^2)cos^2\theta + 5Q_2^2 - 4Q_2}}{4(1-Q_2^2)cos^2\theta + (1-2Q_2)^2} \tag{15}$$

In Equations (11)–(15): p is the pressure on the material, MPa; f_c is the uniaxial compressive strength of the rock, MPa; F_r is the dynamic strain rate increment factor; Q_0 is the initial tensile-compression radial ratio parameter; f_s^*, f_t^* are the tensile and shear strength of the concrete relative to the compressive strength; B represents the Rhodes angle-related parameter.

The dynamic strain rate increment factor is related to the material strain rate and shear strength, and its expression is:

$$F_r = \begin{cases} F_r^c & 3p^* \geq F_r^c \\ F_r^c - (3p^* - F_r^c)(F_r^t - F_r^c)/F_r^c + F_t^c f_t^* & F_r^c > 3p^* \geq -F_r^t f_t^* \\ F_r^t & -F_r^t f_t^* > 3p^* \end{cases} \tag{16}$$

$F_r{}^c$ and $F_r{}^t$ represent the dynamic compression strain rate enhancement factor and dynamic tensile strain rate enhancement factor, respectively.

Where $\dot{\varepsilon}_0^c$ and $\dot{\varepsilon}_0^t$ are the compressive reference strain rate and tensile reference strain rate, respectively, and the values are 3.0×10^{-5} s^{-1} and 3.0×10^{-6} s^{-1} for the given parameters of the present structure, while the expressions of the compressive strain rate index β_c and tensile strain rate index β_t are [19]:

$$\beta_c = \frac{4}{20 + 3f_c} \tag{17}$$

$$\beta_t = \frac{2}{20 + f_c} \tag{18}$$

The elastic limit surface in the RHT structure is set by the equation of the elastic limit surface and the "cap function" that constrains the overflow of the elastic limit stress in the material under high hydrostatic conditions. Where the elastic limit surface equation is:

$$\sigma_{el}^*\left(p, \theta, \dot{\varepsilon}\right) = \sigma_y^* \times F_e \times F_c(p^*) \tag{19}$$

The elastic strength parameter used above is given by:

$$F_e = \begin{cases} g_c^* & 3p^* \geq F_r^c g_c^* \\ g_c^* - (3p^* - F_r^c g_c^*)(g_t^* - g_c^*)/F_r^c g_c^* + F_r^t g_t^* f_t^* & F_r^c g_c^* > 3p^* \geq -F_r^t g_t^* f_t^* \\ g_t^* & -F_r^t g_t^* f_t^* \leq 3p^* \end{cases} \tag{20}$$

F_c is the "cap function", which is introduced to reduce the volume expansion caused by shear and to limit the elastic limit stress of the material at high hydrostatic pressure, while the cap of the yield surface is represented by:

$$F_c = \begin{cases} 1 & p^* \geq p_c^* \\ \sqrt{1 - \left(\frac{p^* - p_u^*}{p_c^* - p_u^*}\right)^2} & p_c^* > p^* > p_u^* \\ 0 & p_u^* \geq p^* \end{cases} \tag{21}$$

In Formulas (22) and (23): $g_c{}^*$ and $g_t{}^*$ are compression yield surface parameters and tensile yield surface parameters, respectively.

Which:

$$g_c^* = f_{c,el}/f_c \tag{22}$$

$$g_t^* = f_{t,el}/f_t \tag{23}$$

Form (24), (25): $f_{c,el}$ and $f_{t,el}$ are uniaxial compressive ultimate stress and uniaxial tensile elastic ultimate stress, respectively.

According to the above RHT embedded formulas, it can be seen that the description of the damage parameters in the present structure model only has the initial damage value D_1 and the complete damage value D_2, and there is no criterion for determining the critical damage of the rock. Therefore, to establish the mathematical relationship between the rock critical damage parameters in the RHT constitution and the parameters in the principal structure model, the RHT principal structure-function is organized according to Equation (7), and the state of the rock at the beginning of crushing is selected as the rock critical damage threshold judging criterion. The values of each of the above equations of state taken at the start of crushing conditions are collated as follows: $F_r = F_r{}^c$, $F_e = g_c{}^*$, $F_c = 1$, so the plastic strain and ultimate strain in Equation (7) can be collated as:

$$\varepsilon_p = \frac{f_c - f_c \times R_3\left(\frac{\pi}{6}\right) \times F_r \times g_c^*}{3G\zeta} \tag{24}$$

$$\varepsilon_{max} = D_1\left[p^* - (1 - D_{cr})\frac{F_r Q_2 f_s^* f_t^*}{3(Q_1 f_t^* - Q_2 f_s^*)}\right] \tag{25}$$

The lithology of the test area is mainly Metamorphic sodium lava, and the parameters of the RHT intrinsic model of metamorphic lava can be obtained through relevant mechanical tests and theoretical calculations, see Table 2.

Table 2. RHT constitutive parameters.

Constitutive Parameters	Parameter Values	Access Approaches
$\rho/\text{kg/m}^3$	2860	testing methods
f_c/MPa	41.5	uniaxial compression test
f_s^*	0.09	
f_t^*	0.02	
G/GPa	3.8	theoretical calculation
β_c	0.028	
β_t	0.033	
g_c^*	0.53	
g_t^*	0.70	
Q_0	0.681	Constitutive self-contained
B	0.011	
D_1	0.04	
D_2	1	

The determined RHT constitutive parameters of the variable metamorphic lava were substituted into Equations (7), (26) and (27) to obtain the critical parameter D_{cr} = 0.106 for the RHT intrinsic damage of the variable metamorphic lava in this numerical simulation, and this was used as the criterion for determining the rock damage in the subsequent numerical simulation.

3. Field Cutting Blasting Test

The test area is located in Yunnan Dahongshan iron ore mine 400 m platform 8# penetration. The roadway size is 4 m × 3 m, the site construction layout is shown in Figure 2a. Furthermore, the DL421 cart was used for the preliminary small-diameter precision drilling operation, with an empty hole depth of 9.5 m, a loaded hole depth of 10 m, a fill length of 0.5 m, and a hole diameter of 106 mm. The empty hole diameter was expanded from 105 mm to 205 mm by using a T150 drilling rig with reaming bit.

The explosive was injected into the charging hole through the charging device (Figure 2b) by wind pressure, and the ANN-2 viscous ammonium nitrate explosive was adhered to the charging hole due to its own characteristics. The charge method adopts the coupling charge method, and the explosive is detonated by detonating cord and a millisecond delay detonator. The field after charge is shown in Figure 2c. The blasting was carried out by bottom detonation, and the blasting effect is shown in Figure 2d. After blasting, the broken rock formed the blasting pile as shown in Figure 2e. Through observation, it can be seen that the rock fragmentation in the blasting pile is small, indicating that the dosage is too much. In further research, the dosage can be adjusted by realizing the interval charge, whereas the overall slot cavity blasting effect is shown in Figure 2f after scanning of the blasted slot cavity using the winged HM100 special UAV.

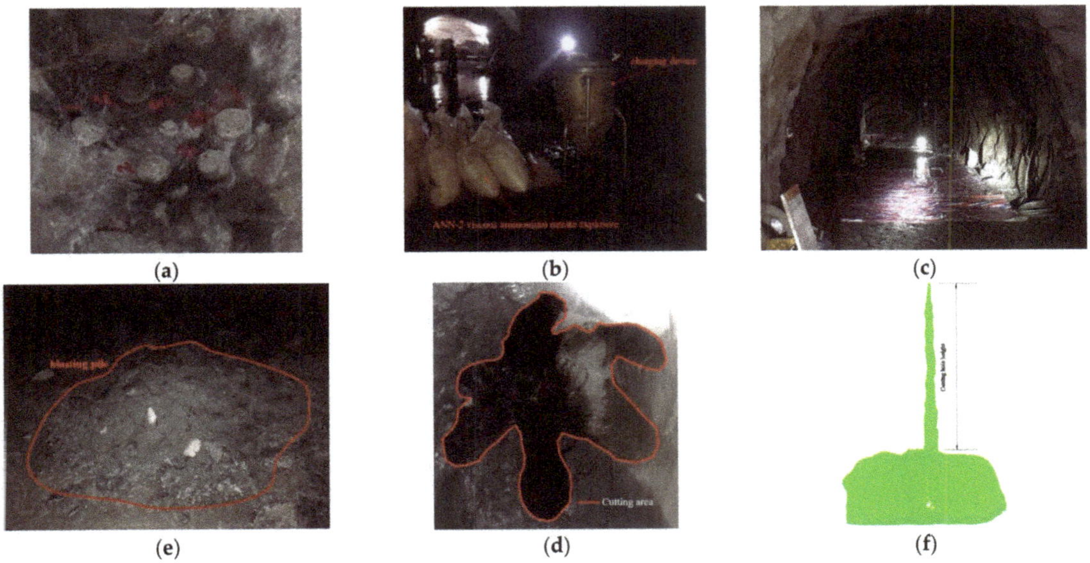

Figure 2. Blasting effect of trenching field test and UAV 3D scan. (**a**) Hole layout; (**b**) Charging device; (**c**) Field after charge; (**e**) Blasting pile; (**d**) Cutting blasting effect; (**f**) Scanning effect.

Since there is only one free surface in the cutting blast, the rock clip production cost has a great influence on the blasting effect, and the rock clip production cost increases linearly with the depth of the cutting hole. In addition, under medium to high in-situ stress conditions, due to the complexity of the coupling mechanism between blast load and rock clamping mechanism, there is no effective means to measure rock clamping performance at the test site. Therefore, the slot cavity area was scanned by using UAV for every 0.5 m section of the slot cavity. To analyze the effect of rock clamping on the effect of cutting blasting, numerical simulation was used to simulate rock clamping by changing the stress state of the numerical model to approximate the field scan results, and the field section scan results at different depths are shown in Figure 3.

It can be seen from Figure 3 that the rock clamping cost increases as the depth of the slot cavity increases, and the area of the slot cavity at the wellhead is larger at 0–0.5 m because it is close to the free surface. Then the area of the slot cavity decreases gradually at 1–8.5 m with the increase of the rock clamping cost and the coupling effect of the empty hole compensation and the rock clamping cost, but the decay of the slot cavity area is not large. However, because there are no effective and direct measurement methods for the complex mechanism of rock clamping production, this paper will use numerical simulation to invert the rock clip production by changing the stress state, to study the impact law of rock clip production on cutting blasting.

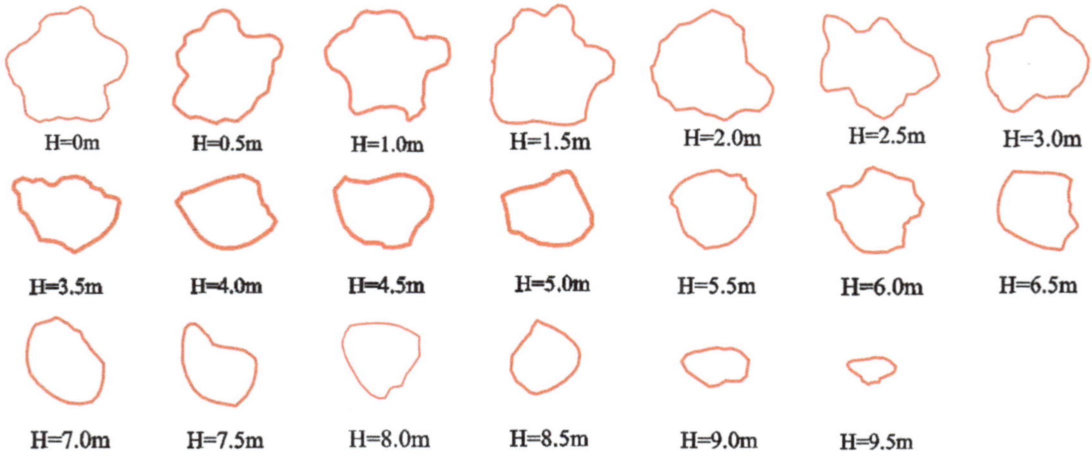

Figure 3. Scanned area of trenching cross-section at different depths.

4. Numerical Simulation Analysis for Rock Clip Production

Since rock clamping production cannot be measured directly, such as by indirect methods (e.g., rock wave velocity, stress peak, etc.), the sensors buried near the blasting area will be damaged and costly due to blasting impact and other factors. The measurement points set at the distant end of the blasting area will lead to inaccurate data due to distance attenuation, so this paper will use numerical simulation to model the rock clamping production at different depths and also, numerical simulation to model the hollowing section at different depths, and invert the rock clamping production at the corresponding hollowing depth by changing the stress state, to study its influence on the rock damage and cavity-forming law.

4.1. Numerical Model

The geometric model and numerical model are shown in Figure 4a,b. The numerical model is designed by LS-DYNA finite element analysis software, which mainly consists of rock and explosive. The numerical simulation mainly focuses on the analysis of the rock clip production, the formation of the slot cavity and the rock damage in the hollowing section at different depths. The numerical model is a quasi-3D model with the size of 5 m × 5 m × 0.01 m, the diameter of the charge hole Φ = 0.1 m, the diameter of the hollow hole Φ = 0.2 m, and the six hollow holes in the form of barrel type hollowing with the hole spacing d = 0.35 m.

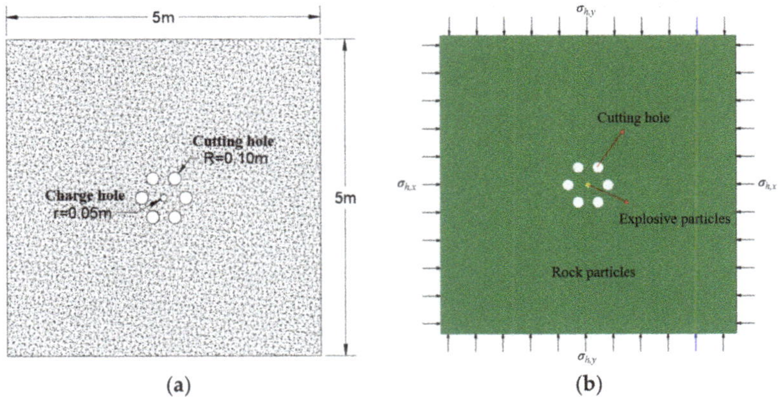

Figure 4. (**a**) Schematic diagram of geometric. (**b**) numerical models and stress application in numerical simulation.

4.2. Stress Application Method

The method of applying stress on the model boundary in this simulation is shown in Figure 5. In order to prevent the particles near the model boundary from being truncated by the kernel function in the calculation process, it is necessary to set multi-layer virtual particles outside the model boundary and generate five-layer virtual particles through the keyword *BOUNDARY_SPH_SYMMETRY_PLANE in the boundary mirror. The density, velocity and mass of virtual particles are the same as those of real particles. The stress is linearly increased to the confining pressure p_0 by using the relevant keywords, so that the model is in a quasi-static dynamic load state. The stress expression of the virtual particle in the stress loading process is:

$$\sigma(t) = \begin{cases} \frac{t}{t_0} p_0, & t \leq t_0 \\ p_0, & t > t_0 \end{cases} \tag{26}$$

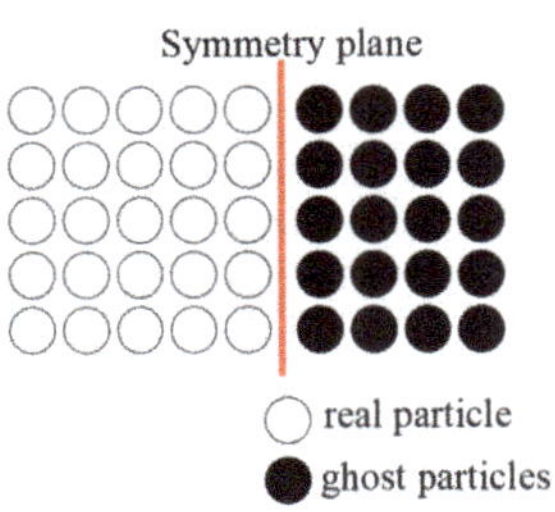

Figure 5. Diagram of stress application.

4.3. Stress Condition Analysis

Deep rock blasting is mainly achieved through the combined effect of explosive blast load and high in-situ stress, so horizontal in-situ stress σ_h and vertical in-situ stress σ_v need to be considered, and E.T. Brown et al. [25] summarized the fitting equation between vertical stress σ_v and depth H:

$$\sigma_v = 0.027H \tag{27}$$

While the lateral pressure coefficient K between horizontal in-situ stress and vertical in-situ stress reveals a certain relationship with depth, Zhao et al. compiled the relationship between the lateral pressure coefficient K and depth H for magmatic rocks in China based on a large amount of data as follows:

$$K = \frac{200}{H} + 0.7 \tag{28}$$

It is difficult to obtain accurate data due to high consumables, improper measurement and instrument error in deep in-situ stress measurement. This paper will impose stress based on previous scholars' summaries of the macroscopic in-situ stress field distribution law and invert the distribution law based on field data for rock clip production in the test area. Once again, from the geological data provided by the mine geology department, the test area is 400 m platform 8# penetration, the burial depth is about 640 m, and the vertical in-situ stress σ_v = 17.28 MPa and horizontal in-situ stress σ_h = 17.5 MPa can be calculated with Equations (29) and (30).

4.4. Explosive Constitution

Explosives provided by the mine ANN-2 viscous ammonium nitrate explosives, explosives parameters are shown in Table 3. The numerical simulation of explosives used through the keyword *MAT_HIGH_EXPLOSIVE_BURN to achieve, while introducing the JWL equation of state to describe the explosive process of volume, pressure and energy change characteristics. The expression equation is given by Formula (29) as:

$$P = A\left(1 - \frac{\omega}{R_1 V}\right)e^{-R_1} + B\left(1 - \frac{\omega}{R_2 V}\right)e^{-R_2 V} + \frac{\omega E_0}{V} \tag{29}$$

where: P is the explosive burst pressure, V represents the relative volume of explosives in the burst process, E_0 refers to the explosive initialized internal energy, A, B, R_1, R_2, ω for the material gauge constants.

Table 3. ANN-2 Viscous ammonium nitrate explosive parameters.

Density/kg/m^3	Detonation Velocity/m/s	C-J Pressure/GPa	Ferocity/mm	Blasting Force/mL	JWL State Equation Parameters				
					A/GPa	B/GPa	R_1	R_2	ω
950	2800	6.5	18	280	326	5.8	5.80	1.56	0.57

The explosive parameters and the parameters related to the JWL equation of state can be determined by experimental fitting, and the explosive and JWL equation of state parameters are shown in Table 3.

4.5. Clamping Factor Simulation Analysis

To investigate the effect of various rock clamping production methods on cavity formation and rock damage in cutting blasting. Different rock clips production was applied to the numerical model, and the rock damage circle after blasting was delineated according to the previously determined critical damage parameters. Figure 6a–f show the damage clouds of the cutting wellhead section under different applied stress conditions, respectively.

Figure 6a–f can reveal that, when there is no rock clamping effect, the rock damage range is larger, and the explosion-generated cracks are very visible. The explosion-generated fissures are primarily circular, whiles the rock damage in the region of microfracture development expansion phenomenon is visible and the distribution is more uniform. When the rock clamping is 0.5–2.0, the area of slot cavity and the range of rock damage relative to no rock clamping effect is reduced, then the burst crack length gradually reduced, and rock fracture development, expansion phenomenon is significantly weakened. When the rock clip production is up to 2.5, burst cavity around the surrounding rock damage to a small extent. Obvious explosive cracks are not observed, and mostly in the form of microfractures concentrated in the cutting hole near the rock damage range are significantly reduced. Therefore, it can be concluded that as the rock clamping production increases, the slot cavity area and rock damage range gradually decrease under the influence of rock clamping production inhibition. The rock damage range and slot cavity area are then measured and analyzed, and the changing trend is plotted as showed Figure 7.

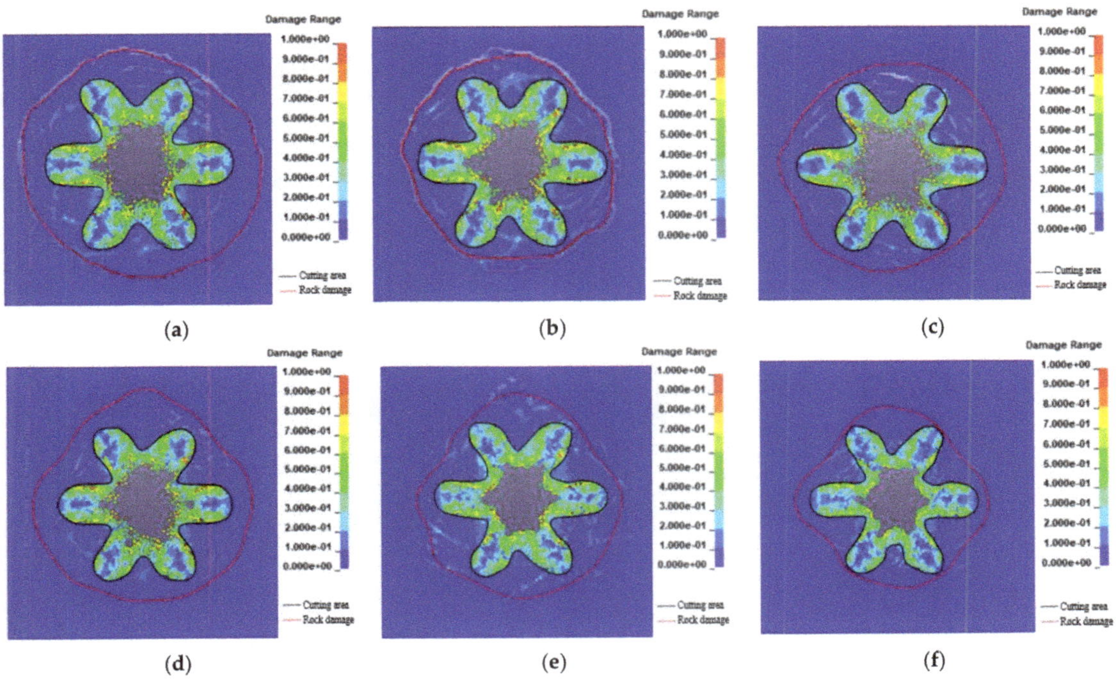

Figure 6. The area of the slot cavity under the application of different rock clamping factors, and the rock damage circle is determined by the critical damage parameters obtained above. (**a**) $K_r = 0$; (**b**) $K_r = 0.5$; (**c**) $K_r = 1.0$; (**d**) $K_r = 1.5$; (**e**) $K_r = 2.0$; (**f**) $K_r = 2.5$.

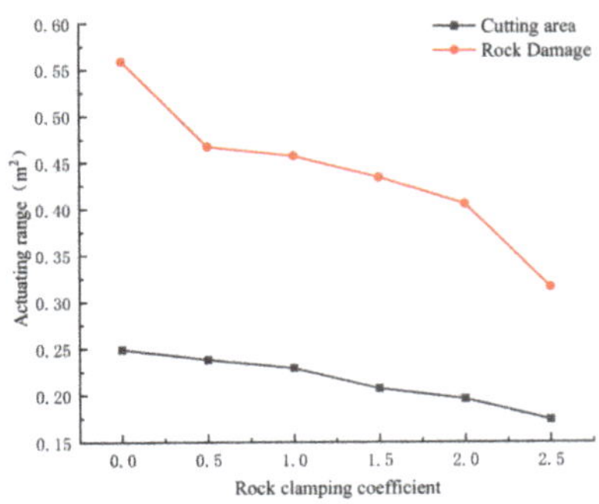

Figure 7. Trend of rock damage range and slot cavity area with different rock clamping factor applied.

From the curves, it can be seen that the rock damage extent was reduced by 9.3%, 16.5%, 22.4%, 27.5% and 43.5%, respectively, and the slot cavity area was reduced by 4.4%, 8.0%, 16.9%, 21.3% and 30.1%, respectively, after the rock clamping mechanism was applied. Due to the coupling effect of explosive load and rock clip production, the area of the slot cavity is relatively less affected, while the rock damage is more affected. In addition, the larger the rock clip production, the greater effect on the blasting effect, and the greater inhibition of rock fracture formation and expansion. The main reason is that rock clip production has a certain inhibitory effect on the propagation of explosive stress waves, so the impact of high rock clip production on rock damage is relatively small. The blasting effect on rock damage under the influence of high rock clip production is relatively small, while the rock clip production inhibits the expansion of primary cracks and the development of new cracks in the rock-damaged area. The excavation unloading under the action of blasting load coupling has consumed most of the energy in the process of breaking the surrounding rock between the charge hole and the empty hole. The blasting stress wave energy propagating to the outside of the cutting hole is not sufficient to achieve the effect of crack expansion on the rock outside the empty hole.

The field scan results were compared with the numerical simulation results under different stress conditions applied, and the rock clip production was inferred by varying the stress applied in the numerical simulation to correspond to the depth of the slot cavity. Figure 8 shows some of the numerical simulation comparison.

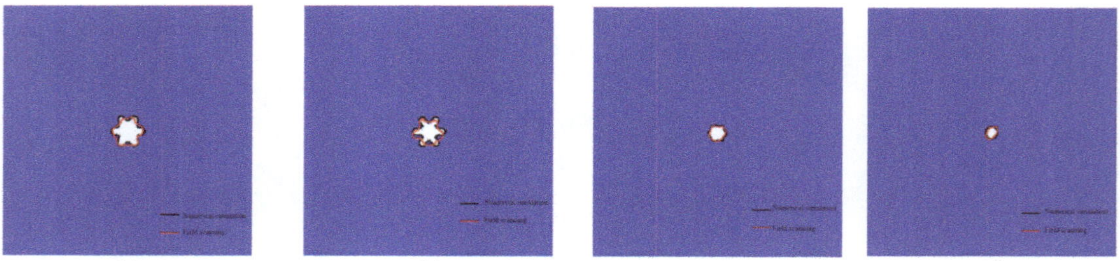

(a) (b) (c) (d)

Figure 8. The area of the slot cavity under different stress conditions applied and compared with the field effect. (**a**) $D_f = 0$ m; (**b**) $D_f = 0.5$ m; (**c**) $D_f = 7.0$ m; (**d**) $D_f = 10$ m.

It can be seen from Figure 8 that the numerical simulation effect is similar to the field test effect, and the simulation effect is more regular because the numerical simulation cannot simulate this property of rock anisotropy. The area of the simulated slot cavity and the area of the scanned slot cavity in the field are counted separately. The simulated area is 0.45, 0.39, 0.16, 0.09 m^2, and the scanned area is 0.44, 0.38, 0.15, 0.08 m^2, respectively. The numerical simulation fit was 94.5%, which verified the accuracy of the RHT constitutive parameters and the reliability of the numerical simulation.

5. Analysis of the Influence Law of Rock Clamping

To evaluate the effect of cutting blasting under high rock clip production, the effect of rock clip production on the formation of slot cavity was analyzed from the perspective of the law of rock clip production on the formation of slot cavity. Different stress states were applied to the cavity at different depths to invert the rock clip production at the corresponding depth, and the scanned area of slot cavity in the field at different depths was circled and recorded with the numerical simulation area to draw the cavity-depth curve (see Figure 9a). Here the blasting slot cavity changes are mainly divided into three stages, the depth of hollowing in the 0–2.5 m due to the compensation of the free surface, the slot cavity area is larger and the area decay rate is small; and then the depth of hollowing in 3–7.5 m, as the distance away from the free surface becomes larger, the free surface to provide a gradual reduction in the role of rock clamping becomes larger, so the area of the blasting cavity decay rate increases. When the depth of hollowing is 8–9.5, there is no empty hole around the compensation effect and far from the free surface so it cannot be through the compensation effect. Where the blasting effect can be regarded as an infinite media blasting, the rock clamping effect on the blasting effect is very large, so the slot cavity area decay phenomenon. The stresses applied by the numerical simulation at different depths were counted to find the rock clamping factor K_r at the corresponding depths, and the variation curve of slot area S_c of the hollowing section with the rock clamping factor K_r was plotted (see Figure 9b). The variation law between S_c, K_r and D_f was plotted based on the inverse rock clamping stress coefficient K_r at different slot depths, combined with the free surface distance D_f (as shown in Figure 9c).

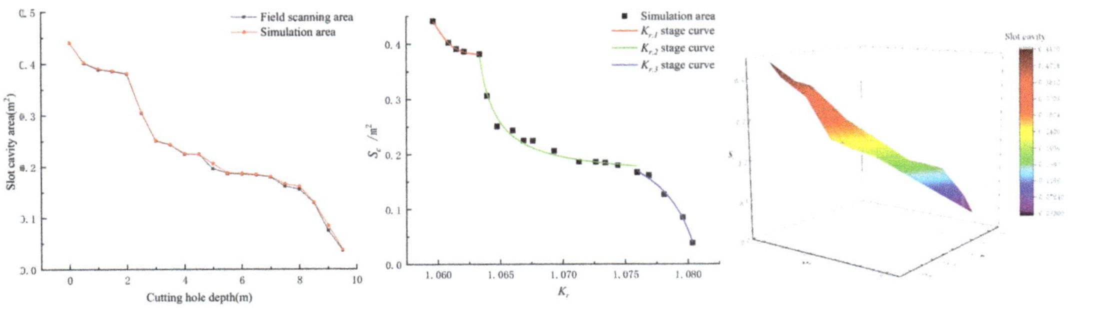

Figure 9. (**a**) Numerical simulation backpropagation effect compared with field effect; (**b**) 3 stages of variation of slot cavity area with rock clamping production; (**c**) The variation pattern of slot cavity area with rock clamping factor and free surface distance.

Regarding the Hawk-Brown world stress distribution law and the related analysis method of K_r. Fuchs world stress map WSM2000 [26], the regression analysis of the rock clamping production coefficient K_r with the variation law of the free surface distance D_f in this trenching and blasting test was further conducted. According to the change curve of slot cavity area, the change law of 3 different stages is analyzed, and the relationship between rock clamping production coefficient K_r and free surface distance D_f is derived as:

$$K_r = F\left(D_f\right) = \begin{cases} 1.06 + 0.002D_{f,1} & 0.5 \leq D_{f,1} < 2.5 \\ 1.052D_{f,2}^{0.01} & 2.5 \leq D_{f,2} < 7.5 \\ 1.036D_{f,3}^{0.019} & 7.5 \leq D_{f,3} < 10 \end{cases} \tag{30}$$

The regression analysis of the three influential stages of the change of S_c with K_r, and its change law under different stages are:

$$S_c = F(K_r) = \begin{cases} 0.381 + 0.108/\left[1 + (K_{r,1}/1.06)^{1411.5}\right] & 1.059 \leq K_{r,1} < 1.063 \\ (0.161 - 0.148K_{r,2})/1 - 0.942K_{r,2} & 1.063 \leq K_{r.2} < 1.076 \\ (0.264 - 0.245K_{r,3})/(1 - 0.923K_{r,3}) & 1.079 \leq K_{r,3} < 1.080 \end{cases} \tag{31}$$

Combining Equation (31) with Equation (32), the relationship between S_c and D_f can be obtained as

$$S_c = F\left(D_f\right) = \begin{cases} 0.392D_{f,1}^{-0.04} & 0.5 \leq D_{f,1} < 2.5 \\ 5.45 - 5.12\left(1 - e^{-D_{f,2}/0.525}\right) - 0.25\left(1 - e^{-D_{f,2}/6.96}\right) & 2.5 \leq D_{f,2} < 7.5 \\ \left[0.177/\left(1 + e^{D_{f,3} - 8.97/0.44}\right)\right] - 0.002 & 7.5 \leq D_{f,3} < 10 \end{cases} \tag{32}$$

Therefore, Equation (32) reflects the relationship between S_c and D_f. By substituting the free surface distance D_f into Equation (32), the regression area $S_c{}'$ of the cavity at different free surface distances is obtained and compared with the field test scan results. According to the curve results (Figure 10), it can be seen that the S_c-K_r curve established in this paper has high accuracy in predicting the blasting cavity area values, and its actual values are within 99% confidence interval of the theoretically predicted values, and the goodness of fit of the theoretical curve is about 0.94, which fully proves that the S_c-K_r curve constructed in this paper has high accuracy and provides an important judgment basis for further prediction of blasting effect in trenching and blasting tests.

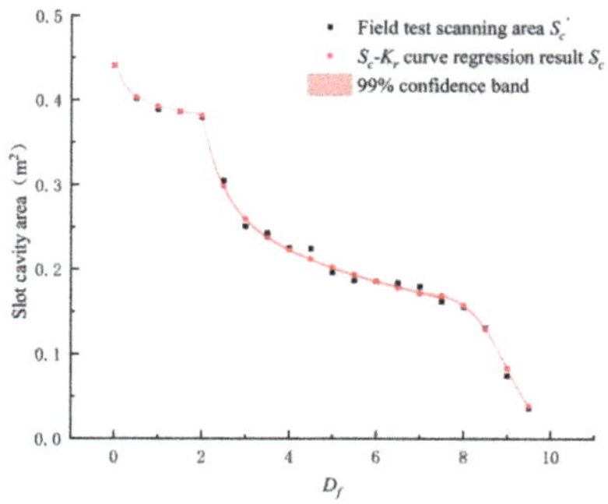

Figure 10. Comparison of actual and regression values.

6. Discussion

In this paper, aiming at the influence of high in-situ stress conditions on the cavity state of the blasting cavity of the one-time well completion cutting in the upward blind shaft, the rock clip production is analyzed. The cavity area measured by UAV scanning after the field test is corrected for the stress state applied in the numerical simulation to achieve the effect close to the field, so as to reverse the rock clip production under different sections. This reverse method is rarely used in previous studies. According to the investigation of the test site, the lithology of the test site is mainly sodium-altered lava. Therefore, the RHT rock parameters used in this test are mainly obtained from the rock parameters obtained from the basic mechanical test of sodium-altered lava and the theoretical calculation. For the cutting effect under other lithology, it will be analyzed in the further test after encountering other lithology and obtaining the corresponding rock parameters in the subsequent test. The numerical model used in numerical simulation is a plate model with unit thickness, which is used to simulate the section cavity in cut blasting. In order to ensure the accuracy of the numerical simulation and be closer to the site, the particle spacing is set to 0.01 m. Therefore, there are more particles in the plane model, about 250,000 particles, so the calculation process is very longer. Due to the influence of the cyclone in the slot cavity, the UAV is scanned at a height of about 0.5 m. Therefore, the S_c-K_r curve is mainly fitted with 20 groups of data, The further experimental study is to overcome the problem of scanning accuracy, improve the scanning accuracy to 0.1 m, and further accurately fit the obtained curve.

7. Conclusions

In this paper, based on the lack of research under high in-situ stress conditions and the lack of criteria for determining the rock damage of the model instantiation, the SPH method and the RHT damage are used to numerically simulate the blasting section of the once-in-a-whole, and to study the cavity formation and the damage law of the cutting blasting under different rock clamping conditions by changing the applied stress state. The following conclusions are drawn from the study:

(1) Through the calculation of blast breaking strain mechanism and rock damage theory, combined with the embedded function in the RHT damage constitution, the connection between the critical parameters of rock blast damage and material plastic strain, ultimate strain and ontological in the RHT constitution is obtained. The parameters of sodic lava in the numerical simulation are derived based on the RHT constitution

parameters of variable sodic lava obtained from mechanical tests and theoretical calculations. The critical threshold of rock damage in the numerical simulation provides a more practical evaluation tool for the damage assessment of RHT constitution in numerical simulation.

(2) The simulation effect of cutting blasting under different rock clamping conditions shows that: rock clamping production inhibits the development and expansion of blast cracks, the stress applied in the numerical simulation affects the extent of fracture generation and development, and the blast produces mainly circular fractures, the larger the coefficient of rock clamping, the greater the blast inhibition effect. Furthermore, the fewer blast cracks and the smaller the rock damage circle.

(3) The rock clamping production in the test area was simulated and compared and inferred according to the field test scanning results to approach the rock clamping production in the field test area. The regression analysis was performed for the rock clamping production coefficient Kr for different depths of the slot cavity to obtain the relationship between Kr and the free surface depth Df, and the relationship between the slot cavity area Sc of the hollowing section and Kr, the mathematical relationship between Sc and Df is obtained, and the goodness of fit of the curve is about 0.94, which fully proves that the Sc-Df curve has high accuracy and provides a reference basis for the relevant design of subsequent deep burial projects.

(4) According to the comparison between the 3D scan data of field trenching and blasting and the numerical simulation results, the average fit of the numerical simulation is 94.5%, which indicates that the RHT parameters have certain accuracy and the numerical simulation has certain reliability.

Author Contributions: Conceptualization, methodology, software, validation, formal analysis, writing—original draft, writing—review and editing: B.S.; investigation, data curation: J.M.; supervision, software: Y.H.; supervision, methodology: H.L.; funding acquisition, project administration: Z.Z.; resources, investigation: J.W. All authors have read and agreed to the published version of the manuscript.

Funding: This research is supported by financial grants from the National Natural Science Foundation of China (52064025, 52164009, 52164010). The authors are very grateful to the financial contribution and convey their appreciation of the organization for supporting this basic research.

Institutional Review Board Statement: Not applicable.

Informed Consent Statement: Not applicable.

Data Availability Statement: Data sharing not applicable.

Conflicts of Interest: The authors declare no conflict of interest.

References

1. Wang, S.; Tang, Y.; Li, X.; Du, K. Analyses and predictions of rock cuttabilities under different confining stresses and rock properties based on rock *indentation* tests by conical pick. *Trans. Nonferrous Met. Soc.* **2021**, *31*, 1766–1783. [CrossRef]
2. Wang, S.; Li, X.; Yao, J.; Gong, F.; Li, X.; Du, K.; Tao, M.; Huang, L.; Du, S. Experimental investigation of rock breakage by a conical pick and its *application* to non-explosive mechanized mining in deep hard rock. *Int. J. Rock Mech. Min. Sci.* **2019**, *21*, 104063. [CrossRef]
3. Wang, S.; Sun, L.; Li, X.; Wang, S.; Du, K.; Li, X.; Feng, F. Experimental investigation of cuttability improvement for hard rock fragmentation using *conical* cutter. *Int. J. Geomech.* **2021**, *21*, 06020039. [CrossRef]
4. Wang, S.; Tang, Y.; Wang, S. Influence of brittleness and confining stress on rock cuttability based on rock indentation tests. *J. Cent. South Univ.* **2021**, *28*, 2786–2800. [CrossRef]
5. Song, Z.; Wang, Y.; Konietzky, H.; Cai, X. Mechanical behavior of marble exposed to freeze-thaw-fatigue loading. *Int. J. Rock Mech. Min. Sci.* **2021**, *138*, 104648. [CrossRef]
6. Yan, P.; Lu, W.; Li, H.; Chen, M.; Zhou, C. Influence of geo-stress on energy distribution of vibration induced by blasting excavation. *Explos. Shock. Waves* **2009**, *29*, 182–188.
7. Wei, C.; Zhu, W.; Bai, Y.; Niu, L. Numercial simulation on cutting seam cartridge blasting under different in-situ stress conditions. *Explos. Shock. Waves* **2016**, *36*, 161–169.

8. Yang, J.; Sun, W.; Yao, C.; Zhang, X. Mechanism of rock fragmentation by multi-hole blasting in highly-stressed rock masses. *Explos. Shock. Waves* **2020**, *40*, 118–127.
9. Yang, J.; Wu, Z.; Yao, C.; Jiang, S.; Jiang, Q. Influence of in-situ stress on blast-induced rock fracture and seismic waves. *J. Vib. Shock.* **2020**, *39*, 64–70.
10. Liu, Z.; He, C.; Fu, G. Numerical simulation and properties of in-situ stress field on roadway surrounding rock failure. *J. Heilongjiang Univ. Sci. Technol.* **2018**, *28*, 14–18.
11. Dai, J.; Qian, Q. Break blasting parameters for driving a roadway in rock with high residual stress. *Explos. Shock. Waves* **2007**, *27*, 272–277.
12. Tang, H.; Zhou, Y.; Liao, Y. Damage zone of surrounding rock of underground engineering under construction blasting. *J. Vib. Shock.* **2015**, *34*, 202–206.
13. Tao, M.; Li, X.; Wu, C. 3D numerical model for dynamic loading-induced multiple fracture zones around underground cavity faces. *Comput. Geotech.* **2013**, *54*, 33–45. [CrossRef]
14. Li, X.; Cao, W.; Zhou, Z.; Zou, Y. Influence of stress path on excavation unloading response. *Tunn. Undergr. Space Technol.* **2014**, *42*, 237–246. [CrossRef]
15. Li, X.; Cao, W.; Tao, M.; Zhou, Z.; Chen, Z. Influence of unloading disturbance on adjacent tunnels. *Int. J. Rock Mech. Min. Sci.* **2016**, *84*, 10–24. [CrossRef]
16. Cao, W.; Li, X.; Tao, M.; Zhou, Z. Vibrations induced by high initial stress release during underground excavations. *Tunn. Undergr. Space Technol.* **2016**, *53*, 78–95. [CrossRef]
17. Yang, D.; Li, H.; Xia, X.; Luo, C. Study of blasting-induced dynamic damage of tunnel surrounding rocks under high in-situ stress. *Rock Soil Mech.* **2014**, *35*, 1110–1116.
18. Luo, S.; Yan, P.; Lu, W.; Chen, M.; Wang, G. Research on the simulation of blasting damage and its mechanism of deep tunnel excavation. *Chin. J. Rock Mech. Eng.* **2021**, *40*, 2760–2772.
19. Chen, M.; Hu, Y.; Lu, W.; Yan, P.; Zhou, C. Numerical simulation of blasting excavation induced damage to deep tunnel. *Rock Soil Mech.* **2011**, *32*, 1531–1537.
20. Lu, W.; Yang, J.; Chen, M.; Zhou, C. Mechanism and equivalent numerical simulation of transient release of excavation load for deep tunnel. *Chin. J. Rock Mech. Eng.* **2011**, *30*, 1089–1096.
21. Li, Q.; Liu, K.; Li, Q.; Wang, Z.; Weng, L. Cutting parameter optimization for one-step shaft excavation technique based on parallel cutting method. *Trans. Nonferrous Met. Soc. China* **2018**, *28*, 1413–1423. [CrossRef]
22. Xiao, S.; Jiang, Y.; Liu, Z.; Su, L. Hard rock blasting energy distribution and fragmentation characteristics under high earth stress. *J. Vib. Shock.* **2018**, *37*, 143–149.
23. Wu, Z.; Zhang, C. Investigation of rock damage model and its mechanical behavior. *Chin. J. Rock Mech. Eng.* **1996**, *15*, 55–61.
24. Riedel, W.; Kawai, N.; Kondo, K.I. Numerical assessment for impact strength measurements in concrete materials. *Int. J. Impact Eng.* **2009**, *36*, 283–293. [CrossRef]
25. Brown, E.T.; Hoek, E. Technical note trends in relationships between measured in-situ stress and depth. *Int. J. Rock Mech. Min. Sci. Geomech. Abstr.* **1978**, *15*, 211–215. [CrossRef]
26. Fuchs, K.; Müller, B. World Stress Map of the Earth: A key to tectonic processes and technological applications. *Die Nat.* **2001**, *88*, 357–371. [CrossRef]

mathematics

Article

Reliability Analysis of High Concrete-Face Rockfill Dams and Study of Seismic Performance of Earthquake-Resistant Measures Based on Stochastic Dynamic Analysis

Zhuo Rong [1], Xiang Yu [2,3,*], Bin Xu [1,3] and Xueming Du [2]

1 Faculty of Infrastructure Engineering, School of Hydraulic Engineering, Dalian University of Technology, Dalian 116024, China; rongzhuo@mail.dlut.edu.cn (Z.R.); xubin@dlut.edu.cn (B.X.)
2 School of Water Conservancy Engineering, Zhengzhou University, Zhengzhou 450001, China; dxm2019@zzu.edu.cn
3 State Key Laboratory of Coastal and Offshore Engineering, Dalian University of Technology, Dalian 116024, China
* Correspondence: xiangyu@zzu.edu.cn

Citation: Rong, Z.; Yu, X.; Xu, B.; Du, X. Reliability Analysis of High Concrete-Face Rockfill Dams and Study of Seismic Performance of Earthquake-Resistant Measures Based on Stochastic Dynamic Analysis. *Mathematics* **2021**, *9*, 3124. https://doi.org/10.3390/math9233124

Academic Editor: Elena Benvenuti

Received: 19 October 2021
Accepted: 1 December 2021
Published: 4 December 2021

Abstract: The randomness of earthquake excitation has a significant impact on the seismic performance of high earth-rock dams. In this paper, the seismic performance of geosynthetic-reinforced soil structures (GRSS) of high concrete face rockfill dams (CFRDs) is evaluated from the stochastic perspective. Multiple groups of seismic ground motions are generated based on spectral expression-random function non-stationary model. Taking Gushui CFRD as an example, this study calculates the failure probability of each damage level of non-reinforce slopes and reinforce slopes based on generalized probability density evolution method (GPDEM) and reliability analysis is presented though multiple evaluation indicators. The result shows that GRSS can reduce the mild damage of CFRDs during earthquake and restrain the moderate and severe damage. The influence of vertical spacing and length of GRSS on the seismic performance is obtained, which provides a reference for the seismic design and risk analysis of CFRDs.

Keywords: high concrete face rockfill dam; geosynthetic-reinforced soil structures; generalized probability density evolution method; seismic performance; reliability analysis

1. Introduction

In recent years, the high earth-rock dams under construction or proposed in China are mainly distributed in the western regions, where earthquakes occur frequently. Once a dam breaks, it could cause immeasurable losses. According to domestic and international experimental results [1,2] and existing earthquake experiences [3,4] with earth-rock dams, the safety of earth-rock dams is closely related to the stability of the dam slope downstream.

Geosynthetic-reinforced soil structures (GRSS) perform well during strong earthquakes, in comparison with other earthquake-resistant measures, in maintaining the stability of slope rockfill [5,6]; therefore, the study of GRSS should receive greater attention. Li et al. [7] used the Newmark sliding block displacement method to evaluate the effect of reinforcement technology on dam crest rockfill, and Noorzad and Omidvar [8] performed a parametric analysis to study the effect of reinforcements on the seismic behavior of reinforced dams. The results showed that reinforcement measures can reduce dam settlement and maximum shear strain, but increase the maximum horizontal peak acceleration. Zhu et al. [9] analyzed the influence of various parameters of GRSS on seismic performance, based on the Fast Lagrangian Analysis of Continuum (FLAC) method. Yang [10] evaluated the seismic stability of reinforced earth-rock dams, based on the upper-bound theorem of limit analysis, and studied the influence of geogrid length on seismic performance. Glovatsky et al. [11] developed the theoretical foundations for the modeling and design of test stands in the study of volumetric models of dams of large channels made of reinforced soil, and evaluated the influence of reinforcement on the bearing capacity of earth-rock dams.

Indeed, variable associated seismic ground motions are filled with uncertainties and threaten the dynamic stability of GRSS [12]. However, the aforementioned studies were based on single ground motion, without considering the randomness of earthquake excitation, with respect to which it is difficult to calculate the probability of failure accurately, and which may even lead to calculation results that are significantly different from the facts. As reliability analysis theory has advanced, scholars have paid more and more attention to the role of this theory with respect to the seismic safety of dams. The Monte Carlo method, the generalized probability density method, the response surface method, and other methods have been used for risk analysis [13] and seismic safety evaluation [14,15] from the perspective of probability. However, relatively few studies have targeted the earthquake-resistant measures of high earth-rock dams from the perspective of random dynamics.

Based on the above, this paper adopted a seismic ground motion generation method and the generalized probability density evolution theory for random dynamic analysis of high concrete face rockfill dams (CFRDs), in considering the randomness of seismic ground motions. First, the finite element model of CFRDs was established, followed by the stochastic dynamic analysis of the dam non-reinforced slopes and reinforced slopes. Second, the seismic performance of the proposed measure was evaluated from the perspective of dynamic reliability. Third, the changes in the length and vertical spacing of GRSS were analyzed to provide a reference for the actual engineering design.

2. Reliability Analysis Method Based on Stochastic Dynamics

The reliability analysis of engineering has made great progress in the field of earthquake-resistant engineering research in recent years [16,17]. The system of reliability analysis of earthquake-resistant measures has four main steps: (1) multiple groups of random ground motion processes are generated, based on the non-stationary ground motion model; (2) the finite element model and the input the generated groups of random ground motions are utilized for batch computing; (3) the probability density function (PDF) and the cumulative distribution function (CDF) are obtained for dam safety performance indices, combined with the generalized probability density evolution theory; and (4) based on these steps, a probability model is constructed to analyze the reliability of a dam before and after the application of earthquake-resistant measures, so seismic performance may be evaluated.

2.1. Non-Stationary Ground Motion Model

In this paper, the spectral expression-random function non-stationary ground motion model [18] was established, based on the improved Clough-Penzien power spectrum model [19], and used for the generation of multiple groups of random ground motion.

The acceleration time series of the stochastic seismic ground motions are generated based on the spectral representation of the random function method of non-stationary stochastic processes [20]. This ground motion generation method has good applicability in the calculation considering the randomness of ground motions [21]. The random process of non-stationary ground motion acceleration with zero mean can be generated by the following formula:

$$\ddot{X}_g(t) = \sum_{k=1}^{N} \sqrt{2S_{\ddot{X}_g}(t,\omega_k)\Delta\omega}\,[\cos(\omega_k t)X_k + \sin(\omega_k t)Y_k], \tag{1}$$

where $\omega_k = k\Delta\omega(\Delta\omega = \omega/N)$. $\{X_k, Y_k\}$ $(k = 1,2,\ldots,N)$ are the standard orthogonal random variables with an interval frequency of $\Delta\omega$ = 0.15 rad/s, N is the number of the truncated items with N = 1600 here [18]. $S_{\ddot{X}_g}$ is the bilateral evolutionary power spectral density function, and the expression is as follows [18]:

$$S_{\ddot{X}_g}(t,\omega) = A^2(t)\frac{\omega_g^4(t) + 4\xi_g^2(t)\omega_g^2(t)\omega^2}{[\omega^2 - \omega_g^2(t)]^2 + 4\xi_g^2(t)\omega_g^2(t)\omega^2} \bullet \frac{\omega^4}{[\omega^2 - \omega_f^2(t)]^2 + 4\xi_f^2(t)\omega_f^2(t)\omega^2} \bullet S_0(t), \tag{2}$$

where $A(t)$ is the intensity modulation function and calculated as:

$$A(t) = \left[\frac{t}{c}\exp\left(1 - \frac{t}{c}\right)\right]^d, \tag{3}$$

where c is the average time of peak ground acceleration (PGA) emergence, d is the shape control index of $A(t)$. In this paper, c is taken as 4 s and d is taken as 2, according to [18]. In the evolutionary power spectral density function, the frequency modulation function can be determined by the following parameters:

$$\omega_g(t) = \omega_0 - a\frac{t}{T}, \; \xi_g(t) = \xi_0 + b\frac{t}{T}, \tag{4}$$

$$\omega_f(t) = 0.1\omega_g(t), \; \xi_f(t) = \xi_g(t), \tag{5}$$

where ω_0 and ξ_0 are the initial angular frequency and the initial damping ratio of the site soil; a and b are parameters determined on the basis of the field classification and seismic design categories; and T is the duration of the ground motion acceleration time history, which differs according to different sites. In this study, the site type used was I_1, and the parameter values ω_0, ξ_0, a, b, and T were 25 (rad/s), 0.45, 3.5, 0.3, and 15 (s), respectively, in accordance with the China Hydraulic Seismic Design Code (NB 35047-2015).

The spectral parameters $S_0(t)$ in Equation (2) reflecting the spectral intensity can be expressed as:

$$S_0(t) = \frac{\overline{a}_{max}^2}{\gamma^2\pi\omega_g(t)[2\xi_g(t) + 1/(2\xi_g(t))]}, \tag{6}$$

where $\overline{a}_{max}$ is the mean value of PGA with $\overline{a}_{max} = 0.340\ g$, which is the checking ground motion of Gushui CFRD; and γ is the equivalent peak factor, depended on the sort of seismic site, taken here to be 2.6 according to [18].

When $\omega = 0$, the following equation should be satisfied:

$$S_{\ddot{X}_g}(t, \omega_0) = S_{\ddot{X}_g}(t, 0) = 0. \tag{7}$$

In Equation (1), the variables $\{X_k, Y_k\}$ (k= $1, 2, \ldots, N$) are standard orthogonal random variables, which are uniquely determined by the orthogonal basis function constructed based on the idea of random function through mapping, and meet the following basic conditions:

$$E[X_k] = E[Y_k] = 0, \tag{8}$$

$$E[X_jY_k] = 0, \; E[X_jX_k] = E[Y_jY_k] = \delta_{jk}, \tag{9}$$

where E [•] represents mathematical expectation and δ_{jk} is the Kronecker delta.

The method of constructing standard orthogonal random variables is as follows:

Suppose $\overline{X}_n$ and $\overline{Y}_n$ (n= $1, 2, \ldots, N$) are two independent random variables respectively Θ_1 and Θ_2, then the random function can be recorded as:

$$\overline{X}_n = \text{cas}(n\Theta_1), \overline{Y}_n = \text{cas}(n\Theta_2), \tag{10}$$

where $\text{cas}(x) = \cos(x) + \sin(x)$ is the Hartley orthogonal basis function [22], basic random variables Θ_1 and Θ_2 are distributed uniformly and independent in the interval [0, 2π], which can usually be obtained by number-theoretic method. After certain deterministic mapping, they become the standard orthogonal random variables required by the Equation (1).

The mean square error of the above non-stationary ground motion acceleration process simulation can be expressed as:

$$\varepsilon(N) = 1 - \frac{\int_0^{\omega_u} \int_0^T S_{\ddot{X}_g}(t,\omega)\mathrm{d}t\mathrm{d}\omega}{\int_0^{\infty} \int_0^T S_{\ddot{X}_g}(t,\omega)\mathrm{d}t\mathrm{d}\omega}, \tag{11}$$

where $\omega_u = N\Delta\omega$ is the truncation frequency, and generally the mean square error of peak acceleration is limited to far less than 1.0 [20].

2.2. Generalized Probability Density Evolution Theory

The generalized probability density evolution method (GPDEM) was proposed by Li et al. [23]. Starting from random events, the theory combines the decouple system's physical equations, based on the principle of conservation of probability, to obtain the generalized probability density evolution equation and to establish the relationship between each physical quantity of interest in the project and in the dynamic system through constitutive relations and deformation coordination relations. Thus, the theory is expressed as a function of basic random variables, and then the probability density function of the studied physical quantity is solved by combining the initial conditions and the boundary conditions. In recent years, the theory has made good progress in the analysis of uncertainty reliability and in the application of large-scale nonlinear structures [24,25].

The motion equation of n degree of freedom system can be expressed as:

$$\overline{\mathbf{M}}(\mathbf{\Theta})\ddot{\mathbf{X}} + \mathbf{C}(\mathbf{\Theta})\dot{\mathbf{X}} + \mathbf{G}(\mathbf{\Theta},\mathbf{X}) = \mathbf{\Gamma}\mathbf{F}(\mathbf{\Theta},t), \tag{12}$$

where $\overline{\mathbf{M}}$, $\mathbf{C}$ are the $n \times n$ order mass and damping matrix, $\mathbf{G}\,(\cdot)$ is the linear or nonlinear restoring force vector, $\ddot{\mathbf{X}}$, $\dot{\mathbf{X}}$ and $\mathbf{X}$ are the acceleration, velocity and displacement vectors of the structural response respectively, the $\mathbf{\Gamma}$ is $n \times r$ order excitation influence matrix and the $\mathbf{F}(\mathbf{\Theta},t)$ is r order excitation vector. And n is the number of degrees of freedom of the system and r is the order of external excitation here.

For a general well posed dynamic system, the physical solution of Equation (12) exists, uniquely and continuously depends on the basic parameters, so the solution of Equation (12) can be expressed as:

$$\mathbf{X} = \mathbf{H}(\mathbf{\Theta},t). \tag{13}$$

The speed process can be expressed as:

$$\dot{\mathbf{X}} = \mathbf{h}(\mathbf{\Theta},t). \tag{14}$$

The information of other physical quantities $\mathbf{Z} = (\mathrm{Z}_1,\ldots,\mathrm{Z}_\mathrm{m})^\mathrm{T}$ in practical engineering can also be expressed as a function of basic random variables by establishing relations with $\ddot{\mathbf{X}}$ and $\dot{\mathbf{X}}$ through constitutive relations and deformation coordination relations:

$$\mathbf{Z} = \mathbf{H}_Z(\mathbf{\Theta},t). \tag{15}$$

The time change rate (speed) can be expressed as:

$$\dot{\mathbf{Z}} = \mathbf{h}_Z(\mathbf{\Theta},t). \tag{16}$$

Since Equation (16) itself can be considered as a random dynamic process, the randomness comes entirely from $\mathbf{\Theta}$. The extended random process ($\mathbf{Z}_\mathrm{t}$, $\mathbf{\Theta}$) is described according to the random event of probability conservation in the whole evolution process. The generalized probability density evolution equation can be obtained, considering the arbitrariness of $\mathbf{\Omega_\Theta}$ [26]:

$$\frac{\partial p_{\mathbf{Z\Theta}}(z,\boldsymbol{\theta},t)}{\partial t}+\sum_{l=1}^{m}\dot{Z}_l(\boldsymbol{\theta},t)\frac{\partial p_{\mathbf{Z\Theta}}(z,\boldsymbol{\theta},t)}{\partial z_l}=0, \tag{17}$$

where $p_{\mathbf{Z\Theta}}(z,\boldsymbol{\theta},t)$ refers to the joint PDF of $(\mathbf{Z},\mathbf{\Theta})$, in which the source random factors are completely described by $\mathbf{\Theta}$. $\mathbf{Z}$ refers to the physical quantity studied. The augmented system composed of $(\mathbf{Z},\mathbf{\Theta})$ is a conservative probability system, which follows the law of probability conservation.

It is worth pointing that the dimension m of this equation is the number of physical quantities studied. When only a certain response physical quantity is considered, the equation can further degenerate into a one-dimensional partial differential equation:

$$\frac{\partial p_{\mathbf{Z\Theta}}(z,\boldsymbol{\theta},t)}{\partial t}+\dot{Z}(\boldsymbol{\theta},t)\frac{\partial p_{\mathbf{Z\Theta}}(z,\boldsymbol{\theta},t)}{\partial z}=0. \tag{18}$$

Equation (18) has only partial differential for z and $\boldsymbol{t}$, while $\boldsymbol{\theta}$ is in the form of a parametric equation. Therefore, a series of deterministic values can be obtained, that is, for a given $\mathbf{\Theta}$ = θ_q $(q=1,2,\ldots,n_{sel})$, where q represents different ground motion processes, and n_{sel} = 89 in this paper. Derivative (velocity) $\dot{\mathbf{Z}}_j(\theta_q,t_m)(j=1,2,\ldots,m)$ of the required physical quantity and θ_q can be obtained by solving the physical equation. When the random parameters are determined by selecting points in the probability space, the partial differential equation of the random dynamic system is transformed into a set of deterministic dynamic equations. The engineering structure can be solved by various numerical simulation methods such as finite element method and finite difference method, and Equation (18) becomes the following series of equations:

$$\frac{\partial p_{\mathbf{Z\Theta}}(z,\boldsymbol{\theta}q,t)}{\partial t}+\sum_{j=1}^{m}\dot{Z}_l(\boldsymbol{\theta}q,t)\frac{\partial p_{\mathbf{Z\Theta}}(z,\boldsymbol{\theta}q,t)}{\partial zj}=0,q=1,2,\ldots,n_{sel}. \tag{19}$$

The initial condition of Equation (19) is:

$$\partial p_{\mathbf{Z\Theta}}(z,\boldsymbol{\theta}q,t)|_{t=t_0}=\delta(z-z_0)Pq. \tag{20}$$

The boundary conditions of Equation (19) are:

$$\partial p_{\mathbf{Z\Theta}}(z,\boldsymbol{\theta}q,t)\Big|_{z_j\to\pm\infty}=0,j=1,2,\ldots,m. \tag{21}$$

The discrete numerical solutions $p_{\mathbf{Z\Theta}}(z,\boldsymbol{\theta}q,t)$ can be obtained by bringing in the initial conditions and boundary conditions. The solutions $p_{\mathbf{Z}}(z,t)$ can be obtained by accumulating all the above discrete numerical solutions, and the result when m = 1 is as follows:

$$p_Z(z,t)=\sum_{q=1}^{n_{sel}}p_{\mathbf{Z\Theta}}(z,\boldsymbol{\theta}q,t). \tag{22}$$

2.3. Reliability Calculation Based on GPDEM

The dynamic reliability of engineering structures usually includes two kinds of problems: first exceedance probability failure and cumulative damage failure reliability, which can be obtained by constructing a virtual random process and solving the corresponding generalized probability density evolution equation [27].

Taking a random process $\mathbf{X}(\mathbf{\Theta},t)$ (random ground motions in this study) as an example, its extreme value can be expressed as follows [28]:

$$\mathbf{Y_X}=max(\mathbf{X}(\mathbf{\Theta},t),t\in[0,T]). \tag{23}$$

The extreme value or cumulative value depends on the source random vector $\mathbf{\Theta}$ and the random variables of the time period [0, T], so a virtual process can be constructed:

$$\mathbf{Z_X}(\tau) = \psi[\mathbf{Y_X}, \tau]. \tag{24}$$

Obviously, the condition Equation (25) can be met:

$$\mathbf{Z_X}(\tau)|_{\tau=0} = 0\,, \mathbf{Z_X}(\tau)|_{\tau=1} = \mathbf{Y_X}. \tag{25}$$

For Equation (24), the derivative of τ is:

$$\dot{\mathbf{Z}}_\mathbf{X} = \frac{\partial \mathbf{Z_X}}{\partial \tau} = \mathbf{W_X}(\mathbf{\Theta}, T). \tag{26}$$

Since the randomness of the virtual random process $\mathbf{Z_X}(\tau)$ comes entirely from Θ, $(\mathbf{Z_X}(\tau), \mathbf{\Theta})$ constitutes a probabilistic conservative system. According to the GPDEM, the joint PDF of $(\mathbf{Z_X}(\tau), \mathbf{\Theta})$, that is $p_{\mathbf{Z\Theta}}(z, \boldsymbol{\theta}, t)$, satisfies the following generalized probability density evolution equation:

$$\frac{\partial p_{\mathbf{Z\Theta}}(z, \boldsymbol{\theta}, t)}{\partial t} + \mathbf{W_X}(\boldsymbol{\theta}, T)\frac{\partial p_{\mathbf{Z\Theta}}(z, \boldsymbol{\theta}, t)}{\partial z} = 0. \tag{27}$$

The equation is solved to obtain the cumulative distribution curve of the studied physical quantity based on the solution method in Section 2.2, so as to obtain the transcendence probability corresponding to each value.

2.4. Calculation Indexes of Reliability Analysis of Concrete Face Rockfill Dam

At present, the finite element dynamic time history analysis method is used for the stability analysis of the dam body, and the indexes include cumulative time of Fs < 1.0 and cumulative slip. In this paper, the two indexes are used to judge the damage grade of the dam body.

2.4.1. Overall Stability Index

(1) Safety factor

The safety factor is the ratio of the maximum shear strength provided by the soil on the potential sliding surface to the actual shear stress generated by the external load. The pre-earthquake stress of the dam and the instantaneous dynamic stress during the earthquake are calculated by the finite element method. The stability of the dam is calculated based to the static and dynamic superposition results of the element based on Newmark method [29,30], and the safety factor is calculated with the following formula:

$$\mathbf{F_s} = \frac{\sum\limits_{i=1}^{n} (c_i + \sigma_i tan\varphi_i)l_i}{\sum\limits_{i=1}^{n} \tau_i l_i}, \tag{28}$$

where c_i and φ_i are the cohesion and internal friction angle of the soil in the ith cell respectively; l_i is the length of the slip arc through the ith cell; σ_i and τ_i are the normal stress and tangential stress on the slip arc surface of the ith cell, which are obtained by superposition of pre-earthquake stress and dynamic stress.

(2) Cumulative time of Fs < 1.0

The time cumulative time of Fs < 1.0 in the whole earthquake process is obtained by determining and judging the minimum safety factor at each time and cumulative addition. In this paper, the criteria for classifying the damage level according to cumulative time is taken as: it is considered that slight damage occurs when the cumulative time is greater than 0 s. The time of 0.5 s is the moderate failure limit, 1.5 s is the severe failure limit, and 2 s is the local non dam break limit according to [31].

2.4.2. Local Sliding Failure Index

In this paper, the cumulative slip is used as the index to evaluate the local slip failure degree of the dam body. Based on the existing research [32,33] and the China Hydraulic Seismic Design Code (NB 35047-2015), the evaluation criteria of the cumulative slip index used in this paper are as follows: when the slip begins, it is judged as mild failure; 20 cm is the moderate failure limit, 100 cm is the severe failure limit, and 150 cm is the local non dam break limit.

For any sliding arc, the sliding angular velocity of the slider around the center of the circle can be calculated by Equation (29) [30]:

$$\alpha(t) = \frac{\mathbf{M}}{\mathbf{I}}, \tag{29}$$

$$\mathbf{M} = [\sum_{i=1}^{n} \tau_i l_i - \sum_{i=1}^{n} (c_i + \sigma_i tan\varphi i)li]R, \tag{30}$$

where, $\mathbf{I}$ is the moment of inertia of the sliding body; $\alpha(t)$ is the sliding angular velocity of the sliding body after instantaneous instability; $\mathbf{M}$ is the rotational moment acting on the sliding body, and R is the slip arc radius.

When an instantaneous slip occurs in a slip arc at a certain time, the slip amount of the slip arc is:

$$D_i^k = R^k \iint \alpha_i^k dt. \tag{31}$$

Multiple instantaneous sliding may occur in the whole time period, and the cumulative sliding amount is:

$$D^k = \sum_{i=1}^{n} D_i^k. \tag{32}$$

Maximum slip of dam slope is the maximum cumulative slip of all possible slip arcs:

$$D_{max} = max(D^1, D^2, \ldots, D^k, \ldots, D^m). \tag{33}$$

2.5. Reliability Analysis

This paper mainly selects two characteristic physical quantities: safety factor cumulative time of Fs < 1.0 and cumulative slip to determine the stability of the dam body, and carries out reliability analysis by calculating the failure probability of different grades of the dam body. The state function of the structure can be expressed as:

$$U = R - S = g(X_1, X_2, \ldots, X_n), \tag{34}$$

where U is the structural state function, which is used to represent the safe state of the structure. At that time $U = 0$, it means that the structure reaches the failure limit state, when $U < 0$, it means that the structure is damaged, when $U > 0$, it means that the structure is safe; R indicates the comprehensive resistance of the structure (i.e., the allowable failure state of the structure in this paper), and S indicates the bearing effect of the structure (i.e., the calculated maximum value). If expressed as a function, R and S can be comprehensively expressed as functions of basic random variables (such as strength parameters, seismic load, displacement response, etc.).

The reliability analysis of the structures (i.e., the calculation formula of failure probability) is following:

$$P_f = P(Z < 0) = \int_{-\infty}^{0} f(Z)\mathrm{d}Z, \tag{35}$$

where P_f is failure probability. In this paper, it represents the probability that the calculated values of indicators exceed the allowable values for the specified failure level.

3. Example Analysis

3.1. Model Establishment

This paper has used the Gushui CFRD as an example [34]. The elevation of the dam crest is 2287 m; the elevation of the dam bottom is 2042 m; the dam height is 242 m (the wave wall is not considered); the upstream dam slope is 1:1.5; the downstream dam slope above the path is 1:1.6; and the slope below the path is 1:1.5. The width of the dam crest is 20 m and the length of the dam crest is 437 m. The cushion zone and the transition zone are set under the face slab. The width of the cushion zone is 4 m, the total thickness of the transition zone is 8 m, and the face slab thickness is 0.4 m~1.24 m. The face slab was poured in three phases to 2167 m, 2235 m, and 2285 m, respectively, and the water storage was stored in two phases to 2267 m (the designed normal water level). The model was densified at the upper part of the dam, where the vertical spacing was set at 1m, considering the dam size, the research purpose, and existing studies [35,36], so the number of model nodes was 8432 and the number of elements was 8277 (Figure 1) before the application of GRSS for the dam slope. The setting of earthquake-resistant measures is shown in Figure 2. The dam element was simulated by the quadrilateral isoparametric element, which is a continuous medium block isoparametric element. The non-thickness Goodman element was used for the indirect contact surface between the panel and the cushion zone, based on the assumption that the two contact surfaces were connected by countless tiny tangential and normal springs. The contact surface and adjacent contact surface elements only had force connected at the node. The above two element types are often used in the field of earth-rock dams [37], and the accuracy meets engineering needs. After filling and water storage, static, dynamic, and two-dimensional stability calculations were carried out. The hydrodynamic pressure on the panel was simulated by the added-mass method [38].

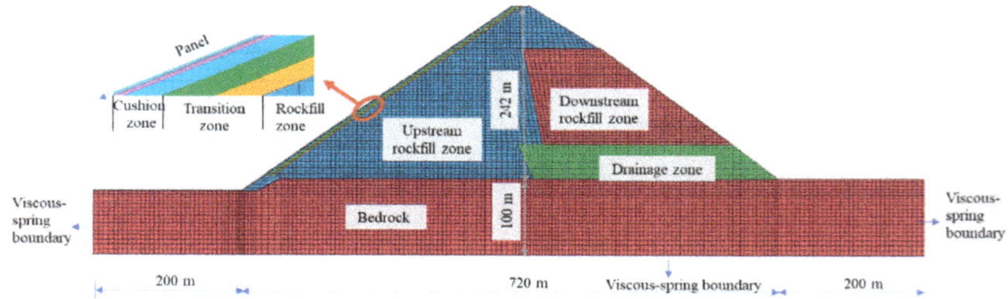

Figure 1. Finite element mesh of the dam.

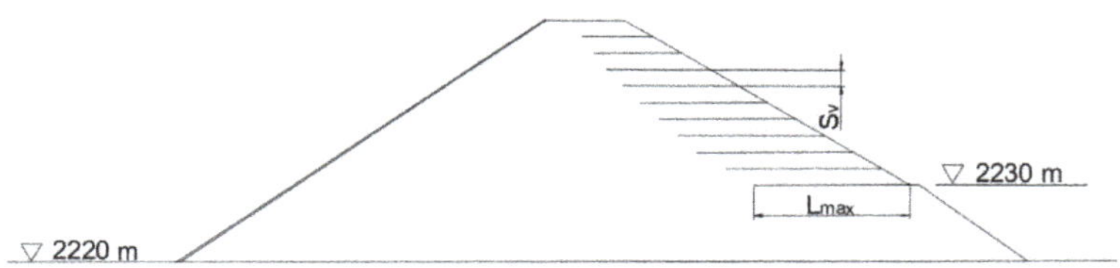

Figure 2. GRSS setting.

3.2. Constitutive Model and Material Parameters

In this paper, the static calculation of rockfill is simulated by Duncan-Chang E-B model [39]. Based on Duncan Chang E-μ, volume modulus B is used instead of Poisson's ratio ν_t as the calculation parameter in Duncan-Chang E-B model, solving the problem which is quite different from the actual situation in the hyperbolic assumption. The dynamic calculation is analyzed by equivalent linear analysis method based on equivalent visco-elastic model [40]. In this paper, the static, dynamic and stability calculation are based on GEODYNA, which has good performance in the calculation of large-scale nonlinear structures [41]. Based on the dam site conditions and specifications, the static and dynamic material parameters of dam rockfill are as show in Tables 1 and 2, comprehensively considering the existing studies [34,42–45]. The density of GRSS is taken as 2.6×10^3 kg/m^3, and the ultimate strength is 85 kN/m.

Table 1. Parameters of Duncan-Chang E~B model for static analysis.

Material	ρ (kg/m^3)	φ_0 (°)	Δ_φ (°)	n	R_f	K_b	m	K
Upstream rockfill	2214	55.5	11.3	1350	0.28	0.80	780	0.18
Downstream rockfill	2214	53.0	11.0	1000	0.26	0.79	700	0.16
Drainage zone	2214	55.0	12.2	1300	0.31	0.79	800	0.12
Transition material	2222	53.5	10.7	1250	0.31	0.78	720	0.16
Cushion material	2258	54.4	10.6	1200	0.30	0.75	680	0.15

Table 2. Parameters of equivalent visco-elastic model for dynamic analysis.

Material	K	*n*
Upstream rockfill	2660	0.444
Downstream rockfill	4997	0.298
Drainage zone	3115	0.396
Transition material	3828	0.345
Cushion material	5297	0.33

3.3. *Ground Motions Input*

Viscoelastic artificial boundary and equivalent node load are used for wave input [46,47] to realize ground motion input considering the radiation damping effect of infinite foundation and the influence of traveling wave effect. The horizontal PGA is 0.340 g. The vertical PGA is 2/3 of the horizontal acceleration.

This chapter uses the aforementioned spectral expression-random function non-stationary random ground motion model to generate the seismic input of 89 different action processes due to the difference of ground motions in the magnitude of ground motion frequency, the height of peak and valley value and the duration of fluctuation mode. When 89 sample acceleration time histories are generated, the error between the mean value of peak acceleration and the target value (zero) is 8.0%. The results show that the characteristics of the sample set are consistent with the target in the second-order numerical statistical sense [20]. Figure 3 shows the seismic acceleration time series information of the generated ground motion samples. Figure 3b,c show the mean and standard deviation of the generated ground motion samples. Figure 3e shows that the mean of response spectrum of 89 samples coincides with the specification spectrum in China Hydraulic Seismic Design Code (NB 35047-2015). It can be seen that the ground motions generated by the spectral expression-random function non-stationary ground motion model fit well with the target earthquake motion.

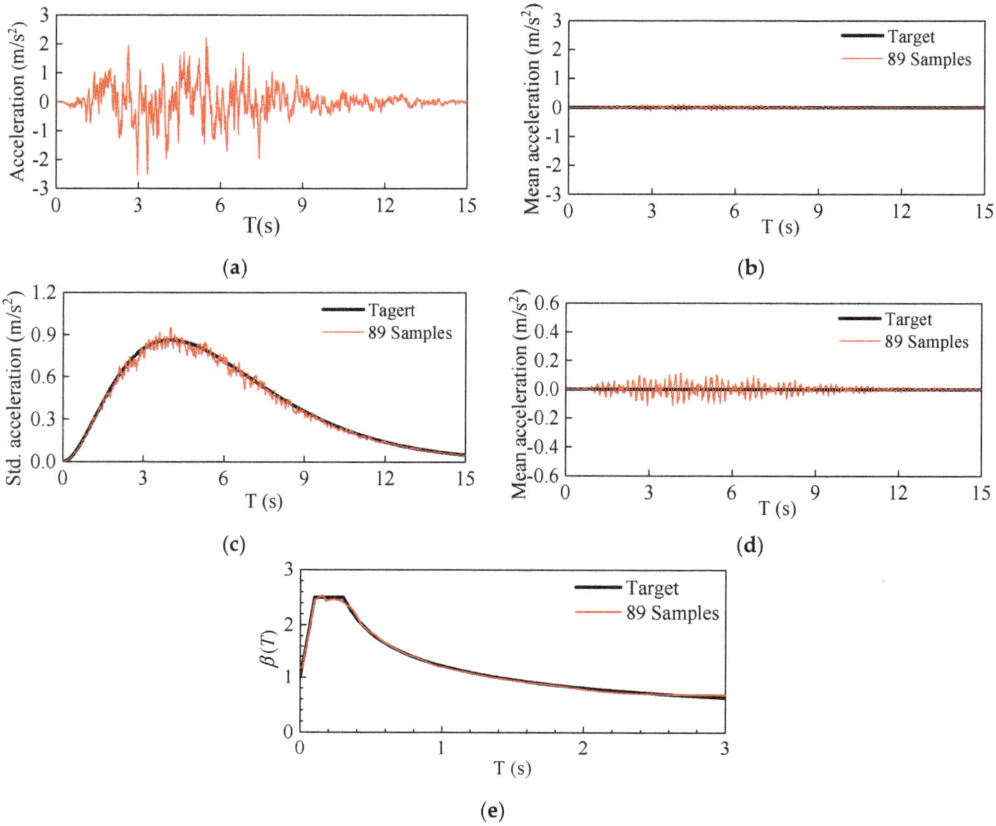

Figure 3. Comparison of the seismic acceleration time series between the samples and target: (**a**) Typical non-stationary seismic acceleration time series; (**b**) Mean acceleration time series; (**c**) Standard deviation acceleration time series; (**d**) Magnifying view of the mean acceleration time series; (**e**) Response spectrum.

3.4. Seismic Response Analysis

In order to study the influence of GRSS on the stability of dam slope during earthquakes, based on the existing research [9–11] and with reference to the common parameter settings in engineering [48], the vertical spacing of GRSS layers (S_v) was set at 4 m and the length (L_{max}) was set at 50 m, for comparison with the original dam slope as a typical work condition. The random dynamic response of the dam slope was analyzed, and the seismic performance of the proposed measure was evaluated from the perspective of reliability.

Eighty-nine groups of safety factor history curves were obtained by calculating the dynamic response of the high CFRD under 89 groups of random ground motions. The safety factor probability density surface (Figures 4b and 5b) was solved by the finite difference method (FDM) in TVD format [26] to reflect the transmission information of the safety factor in space and time, based on the GPDEM. The PDF of the safety factor fluctuated in time and space, and the safety factor gradually concentrated with increases in time. The comparison between the two figures showed that GRSS have no significant impact on the shape of PDF as a dam safety factor.

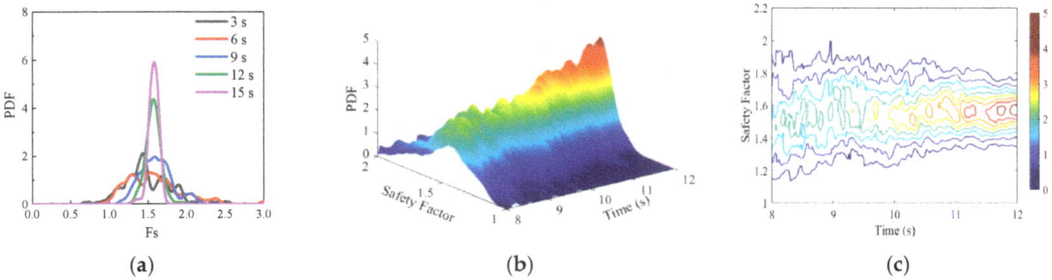

Figure 4. Probability density evolution information of safety factor before reinforcement: (**a**) Probability density function at typical time; (**b**) Probability density function evolution surface; (**c**) Probability density function contour.

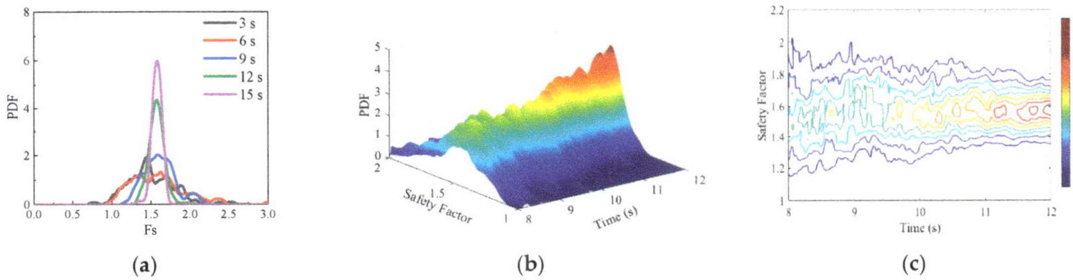

Figure 5. Probability density evolution information of safety factor after reinforcement: (**a**) Probability density function at typical time; (**b**) Probability density function evolution surface; (**c**) Probability density function contour.

Figure 6 shows the probability density information of the minimum safety factor before and after the application of earthquake-resistant measures, obtained by solving the equivalent extreme-value event. Before GRSS measures were taken, the average value of the minimum safety factor was 0.944 and the standard deviation was 0.131. After the measures were taken, the average value of the minimum safety factor increased to 1.003 and the standard deviation decreased to 0.105 (Figure 6a). The calculation results of the reinforced slopes showed that the minimum safety factor not only improved in value, but also reduced in dispersion, which indicated that GRSS can not only maintain the stability of a dam body when encountering earthquakes, but also improve the stability of the overall response of the dam body when encountering different ground motions. Similar conclusions were also reached in Figure 6b; that is, GRSS are conducive to improving the reliability of a dam under random ground motions.

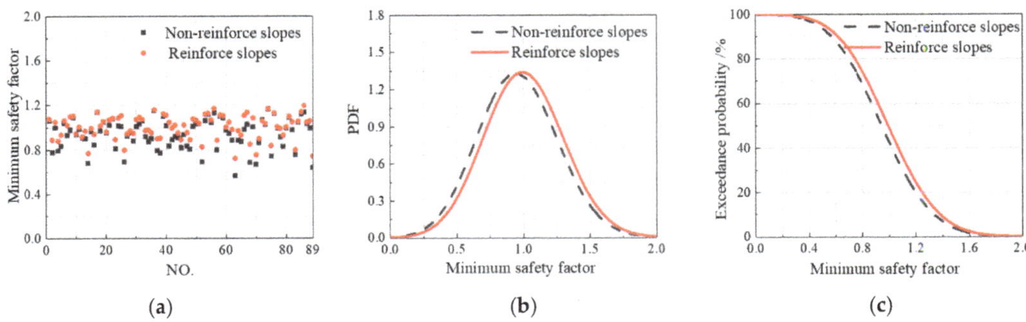

Figure 6. Discrete point distribution and probability information of minimum safety factor: (**a**) Discrete point distribution; (**b**) Probability density distribution function; (**c**) Exceedance probability.

3.5. Reliability Analysis of Dam

3.5.1. Overall Stability Analysis of Dam Body

In this section, the probability density function of the cumulative time of Fs < 1.0 was solved by constructing a virtual random process to analyze the stability of non-reinforced slopes and reinforced slopes, based on the 89 groups of random ground motions referred to in Section 3.3. Under different seismic ground motions, the maximum value of the cumulative time of the original dam body was 1.43 s, which was quite different from the minimum value of 0 s (Figure 7a); the seismic performance of the proposed measure in reducing cumulative time was inconsistent, indicating the need to analyze the seismic performance of earthquake-resistant measures from the stochastic perspective.

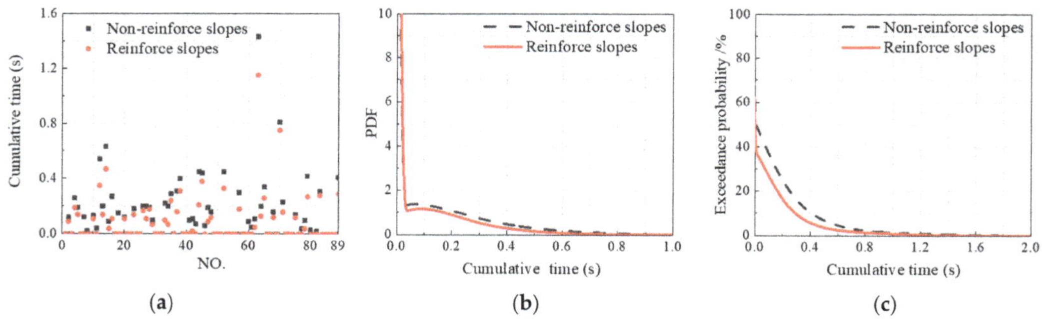

Figure 7. Discrete point distribution and probability information of cumulative time: (**a**) Discrete point distribution; (**b**) Probability density distribution function; (**c**) Exceedance probability.

According to the aforementioned standard, the probability of a mild failure of the original dam slopes was large, up to more than 50%, while it was not easy to cause moderate or severe damage, as GRSS have an obvious effect in maintaining the stability of the dam slope (Table 3).

Table 3. Relationship table of cumulative time-exceedance probability.

Exceedance Probability (%)		Non-Reinforce Slopes	Reinforce Slopes	Reduced Value	Relative Reduction Value
Cumulative time (s)	0	50.11	37.96	12.15	24.25
	0.2	24.92	16.55	8.37	33.59
	0.5	6.86	3.77	3.09	45.04
	1.0	1.36	0.94	0.42	30.88
	1.5	0.47	0.16	0.31	65.96
	2.0	0.06	-	0.06	100.00

3.5.2. Analysis of Local Sliding Failure of Dam Slope

Figure 8 and Table 4 show the probability information of cumulative slippage of dam slopes under random ground motions. According to the calculated results, smaller parts of ground motions will generate local sliding of a CFRD under proposed seismic ground motions, and the failure probability of moderate or severe sliding of the dam slope is small. From the perspective of reliability, the proposed measure has an obvious inhibitory effect on the local sliding of the dam slope.

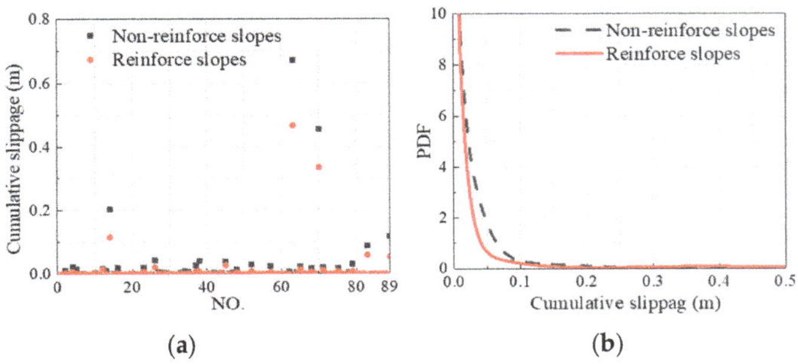

Figure 8. Discrete point distribution and probability information of cumulative slippage: (**a**) Discrete point distribution; (**b**) Probability density distribution function; (**c**) Exceedance probability.

Table 4. Relationship table of cumulative slippage-exceedance probability.

Exceedance Probability (%)		Non-Reinforce Slopes	Reinforce Slopes	Reduced Value	Relative Reduction Value
Cumulative slippage (m)	0	33.79	23.63	10.16	30.07
	0.2	2.88	2.23	0.65	22.57
	0.5	1.39	0.44	0.95	68.35
	1.0	-	-	-	-
	1.5	-	-	-	-

3.6. Analysis on Influenceing Factors of GRSS Performance

In this section, failure probability was calculated by adjusting the length and vertical spacing of reinforcement based on the 89 groups of random ground motions as described in Section 3.3. The influence of reinforcement-setting on the seismic performance of GRSS was considered, to find the optimal GRSS setting from the perspective of reliability, which can be used to provide a reference for the design of earthquake-resistant measures, considering both project costs and effects.

3.6.1. Impact on Maintaining Overall Stability

In order to analyze the influence of the length and vertical spacing of reinforcement on the stability of CFRD slopes, this section kept other factors unchanged, while different characteristic lengths (L_{max} = 40 m, 50 m, and 60 m) and characteristic vertical spacings (S_v = 2 m, 4 m, and 6 m) of GRSS were selected according to previous engineering experience and the research of previous literature [11,12]. The equivalent extreme values were calculated based on the GPDEM for the cumulative time of Fs < 1.0 (as shown in Figure 9). The results showed that with other factors unchanged, the reliability of CFRD increased with the increase in reinforcement length and decreased with the increase in reinforcement vertical spacing. The probability of mild failure of dam slope followed an order of 40.10%, 37.96%, and 37.80%, with the increases in the length of reinforcement. It can be seen that when the length increased to more than 50 m, the stability reliability of dam slope did not increase significantly. With the increase in vertical spacing, the probability of mild failure of dam slope was 30.73%, 37.96%, and 39.47%, respectively. Other failure conditions were similar, proving that the stability of dam slope decreases significantly when reinforcement spacing is expanded from 2 m to 4 m (Table 5).

Table 5 compares and analyzes the consistency risk of various working conditions based on performance. Taking working condition 1 (with GRSS vertical spacing of 2 m and length of 50 m) as an example, under random ground motion with a GPA of 0.340 g, the probability of mild failure, moderate failure, and severe failure of the dam slope followed an order of 30.73%, 2.69%, and 0.06%, respectively. The probability basically did not reach the situation of local dam-break, which implied that the overall stability reliability of the dam slope was high when the vertical spacing of the geogrid reinforcement was 2 m and the length was 50 m. Even if damage occurred, it was mild and easy to repair, and had little economic or social impact under condition 1. The analysis of other working conditions is similar. These findings can provide a reference for the reliability analysis of GRSS in engineering.

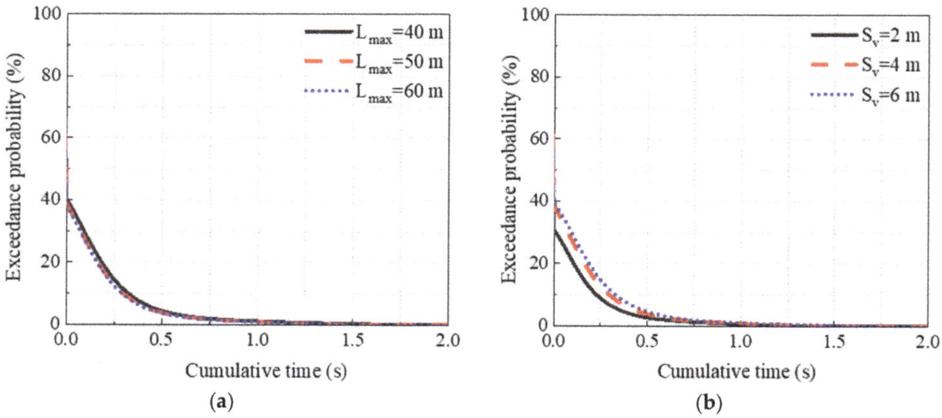

Figure 9. Cumulative time-exceedance probability curve under various working conditions: (**a**) Change of reinforcement length; (**b**) Change of reinforcement vertical spacing.

Table 5. Relationship table of cumulative time-exceedance probability under various working conditions.

Exceedance Probability (%)	S_v (m)	L_{max} (m)	Cumulative Time (s)				
			0	0.5	1.0	1.5	2.0
Condition 1	2	50	30.73	2.69	0.69	0.06	-
Condition 2	4	40	40.10	4.42	1.15	0.36	0.03
Condition 3	4	50	37.96	3.77	0.94	0.16	-
Condition 4	4	60	37.80	3.67	0.94	0.16	-
Condition 5	6	50	39.47	4.63	1.09	0.28	0.02

3.6.2. Influence on Restraining Local Sliding Failure

Similarly, in order to study the influence of adjusting the length and vertical spacing of GRSS on the local sliding of concrete-face rockfill dams subjected to earthquakes, different lengths and vertical spacings of geogrid reinforcement were adopted in this section, and the cumulative sliding index was then calculated. Taking mild failure as an example, with an increase in reinforcement length, the probability of mild failure of dam slope followed an order of 25.41%, 23.63%, and 23.07% under the random ground motions with GPA of 0.340 g. With increases in reinforcement vertical spacing, the probability was 18.05%, 23.63%, and 26.52%, respectively. Moderate failure was similar, indicating that the local sliding failure probability of CFRD decreases with an increase in reinforcement length and increases with an increase in reinforcement spacing (Figure 10 and Table 6).

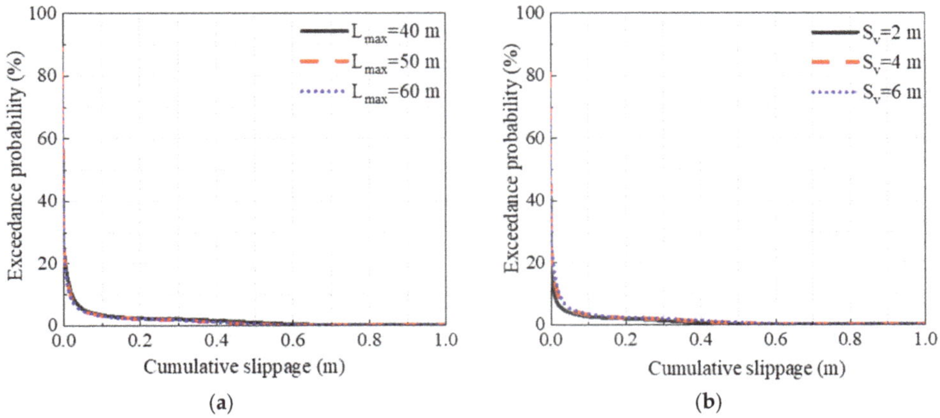

Figure 10. Cumulative slippage-exceedance probability curve under various working conditions: (**a**) Change of reinforcement length; (**b**) Change of reinforcement vertical spacing.

Table 6. Relationship table of cumulative slippage-exceedance probability under various working conditions.

Exceedance Probability (%)	S_V (m)	L_{max} (m)	Cumulative Slippage (m)				
			0	0.2	0.5	1.0	1.5
Condition 1	2	50	18.05	2.01	0.12	-	-
Condition 2	4	40	25.41	2.38	0.99	-	-
Condition 3	4	50	23.63	2.23	0.44	-	-
Condition 4	4	60	23.07	2.20	0.42	-	-
Condition 5	6	50	26.52	2.37	0.71	-	-

4. Conclusions

In this paper, the reliability of the seismic performance of GRSS was evaluated from the perspective of probability and the influence of factors on performance. The conclusions were as follows:

(1) A set of reliability analysis systems of high CFRDs was established by combining a non-stationary ground motion model, a generalized probability density evolution theory, and multi-evaluation indicators, providing a method for the evaluation of seismic performance of earthquake-resistant measures.
(2) GRSS can not only maintain the stability of a dam body during an earthquake; it can also improve the stability of the overall response of a dam under different ground motions.
(3) The dam slope of the Gushui CFRD is more susceptible to slight damage under random ground motions with a PGA of 0.340 g, with a certain probability of moderate damage and a lower probability of severe damage. The calculation results showed that the seismic performance of GRSS can inhibit mild damage of 20–30% of ground motion samples and avoid severe damage with a high probability.
(4) With an increase in geogrid reinforcement length and a decrease in vertical spacing, the seismic performance in maintaining dam slope stability increases due to strong constraints. However, the increase in reinforcement length has little influence on the performance of GRSS. When reinforcement vertical spacing is reduced from 4 m to 2 m, the reduction in the earthquake-resistant effect of GRSS is relatively obvious.

This is conducive to actual engineering design, considering economics and seismic performance.

Author Contributions: Data curation, X.D.; Methodology, B.X.; Writing—original draft, Z.R.; Writing—review & editing, X.Y. All authors have read and agreed to the published version of the manuscript.

Funding: This research was supported by the Natural Science Foundation of China (No. 51809034, 51779034), Open Found of State Key Laboratory of Coastal and Offshore Engineering, Dalian University of Technology (No. LP2014), China Postdoctoral Science Foundation (No. 2021M692938) and Henan Postdoctoral Foundation (No. 202002022).

Data Availability Statement: All data, models, and code generated or used during the study appear in the submitted article.

Conflicts of Interest: The authors declare no conflict of interest.

References

1. Chen, S.; Fu, Z.; Wei, K.; Han, H. Seismic Responses of High Concrete Face Rockfill Dams: A Case Study. *Water Sci. Eng.* **2016**, *9*, 195–204. [CrossRef]
2. Zhou, Y.; Zhang, Y.; Pang, R.; Xu, B. Seismic Fragility Analysis of High Concrete Faced Rockfill Dams Based on Plastic Failure with Support Vector Machine. *Soil Dyn. Earthq. Eng.* **2021**, *144*, 106587. [CrossRef]
3. Zhang, J.M.; Yang, Z.; Gao, X.; Zhang, J. Geotechnical Aspects and Seismic Damage of the 156-m-High Zipingpu Concrete-Faced Rockfill Dam Following the Ms 8.0 Wenchuan Earthquake. *Soil Dyn. Earthq. Eng.* **2015**, *76*, 145–156. [CrossRef]
4. Liu, H.; Chen, Y.; Yu, T.; Yang, G. Seismic Analysis of the Zipingpu Concrete-Faced Rockfill Dam Response to the 2008 Wenchuan, China, Earthquake. *J. Perform. Constr. Facil.* **2015**, *29*, 04014129. [CrossRef]
5. Ling, H.I.; Leshchinsky, D.; Perry, E.B. Seismic Design and Performance of Geosynthetic-Reinforced Soil Structures. *Géotechnique* **1997**, *47*, 933–952. [CrossRef]
6. Ling, H.I.; Leshchinsky, D. Effects of Vertical Acceleration on Seismic Design of Geosynthetic-Reinforced Soil Structures. *Géotechnique* **1998**, *48*, 347–373. [CrossRef]
7. Li, H.J.; Chi, S.C.; Li, G. Seismic Stability Analysis of Reinforced Slope of High Core Rockfill Dam. *Chin. J. Geotech. Eng.* **2007**, *29*, 1881–1887. (In Chinese) [CrossRef]
8. Noorzad, R.; Omidvar, M. Seismic Displacement Analysis of Embankment Dams with Reinforced Cohesive Shell. *Soil Dyn. Earthq. Eng.* **2010**, *30*, 1149–1157. [CrossRef]
9. Zhu, Y.; Kong, X.; Zou, D.; Zhu, S. Dynamic Elastoplastic Analysis of Geogrid Reinforced High Earth Rockfill Dam. *J. Hydraul. Eng.* **2012**, *43*, 1478–1486. (In Chinese)
10. Yang, X.; Chi, S.; Lu, X. Upper Limit Analysis of Seismic Stability of Reinforced Earth Rock Dam Slope. *J. Hydraul. Eng.* **2014**, *45*, 304–311. (In Chinese) [CrossRef]
11. Glovatsky, O.; Hamdamov, B.; Bekchanov, F.; Saparov, A. Strengthening Technology and Modeling of Dams from Reinforced Soil. *IOP Conf. Ser. Mater. Sci. Eng.* **2021**, *1030*, 012155. [CrossRef]
12. Peng, M.; Sun, R.; Chen, J.-F.; Zhang, L.-M.; Yu, S.-B. Stochastic Seismic Analysis of Geosynthetic-Reinforced Soil Slopes Using the Probability Density Evolution Method. *Comput. Geotech.* **2021**, *140*, 104485. [CrossRef]
13. Hariri-Ardebili, M.A. Risk, Reliability, Resilience (R 3) and beyond in Dam Engineering: A State-of-the-Art Review. *Int. J. Disaster Risk Reduct.* **2018**, *31*, 806–831. [CrossRef]
14. Kartal, M.E.; Bayraktar, A.; Başağa, H.B. Seismic Failure Probability of Concrete Slab on CFR Dams with Welded and Friction Contacts by Response Surface Method. *Soil Dyn. Earthq. Eng.* **2010**, *30*, 1383–1399. [CrossRef]
15. Li, Y.; Tang, W.; Wen, L.; Wang, J. Study on Seismic Failure Probability of High Earth-Rock Dam Considering Dam Body Deformation and Slope Stability. *Eur. J. Environ. Civ. Eng.* **2020**, 1–15. [CrossRef]
16. Zai, D.; Pang, R.; Xu, B.; Fan, Q.; Jing, M. Slope System Stability Reliability Analysis with Multi-Parameters Using Generalized Probability Density Evolution Method. *Bull. Eng. Geol. Environ.* **2021**, *80*, 8419–8431. [CrossRef]
17. Pang, R.; Xu, B.; Zhou, Y.; Song, L. Seismic Time-History Response and System Reliability Analysis of Slopes Considering Uncertainty of Multi-Parameters and Earthquake Excitations. *Comput. Geotech.* **2021**, *136*, 104245. [CrossRef]
18. Pang, R.; Xu, B.; Zou, D.; Kong, X. Stochastic Seismic Performance Assessment of High CFRDs Based on Generalized Probability Density Evolution Method. *Comput. Geotech.* **2018**, *97*, 233–245. [CrossRef]
19. Deodatis, G. Non-Stationary Stochastic Vector Processes: Seismic Ground Motion Applications. *Probabilistic Eng. Mech.* **1996**, *11*, 149–167. [CrossRef]
20. Liu, Z.; Liu, W.; Peng, Y. Random Function Based Spectral Representation of Stationary and Non-Stationary Stochastic Processes. *Probabilistic Eng. Mech.* **2016**, *45*, 115–126. [CrossRef]
21. Zhou, Y.; Xu, B.; Pang, R.; Zou, D.; Kong, X. Stochastic Seismic Response and Stability Reliability Analysis of a Vertical Retaining Wall in Front of the Pumping Station of a Nuclear Power Plant Using the Probability Density Evolution Method. *Nuclear Eng. Design* **2018**, *334*, 110–120. [CrossRef]

22. Jacobson, R.A.; Bracewell, R.N. The Hartley Transform. *Math. Comput.* **1987**, *48*, 847. [CrossRef]
23. Li, J.; Chen, J.B. Probability Density Evolution Method for Dynamic Response Analysis of Structures with Uncertain Parameters. *Comput. Mech.* **2004**, *34*, 400–409. [CrossRef]
24. Li, J.; Chen, J. The Number Theoretical Method in Response Analysis of Nonlinear Stochastic Structures. *Comput. Mech.* **2007**, *39*, 693–708. [CrossRef]
25. Li, J. Probability Density Evolution Method: Background, Significance and Recent Developments. *Probabilistic Eng. Mech.* **2016**, *44*, 111–117. [CrossRef]
26. Li, J.; Chen, J. The Principle of Preservation of Probability and the Generalized Density Evolution Equation. *Struct. Saf.* **2008**, *30*, 65–77. [CrossRef]
27. Pang, R.; Xu, B.; Kong, X.; Zhou, Y.; Zou, D. Seismic Performance Evaluation of High CFRD Slopes Subjected to Near-Fault Ground Motions Based on Generalized Probability Density Evolution Method. *Eng. Geol.* **2018**, *246*, 391–401. [CrossRef]
28. Li, J.; Chen, J.; Fan, W. The Equivalent Extreme-Value Event and Evaluation of the Structural System Reliability. *Struct. Saf.* **2007**, *29*, 112–131. [CrossRef]
29. Newmark, N.M. Effects of Earthquakes on Dams and Embankments. *Géotechnique* **1965**, *15*, 139–160. [CrossRef]
30. Ling, H.I.; Leshchinsky, D.; Mohri, Y. Soil Slopes under Combined Horizontal and Vertical Seismic Accelerations. *Earthq. Engng. Struct. Dyn.* **1997**, *26*, 1231–1241. [CrossRef]
31. Softening of Rockfill Based on Generalized Probability Density Evolution Method. *Soil Dyn. Earthq. Eng.* **2018**, *107*, 96–107. [CrossRef]
32. Ozkan, M.Y. A Review of Considerations on Seismic Safety of Embankments and Earth and Rock-fill Dams. *Soil Dyn. Earthq. Eng.* **1998**, *17*, 439–458. [CrossRef]
33. Darbre, G.R. Swiss Guidelines for the Earthquake Safety of Dams. In Proceedings of the 13th WCEE 2004, Vancouver, BC, Canada, 1–6 August 2004.
34. Feng, Y.L.; Chen, Y.; Lei, H.J. Research Results on Safety Index and Design Engineering Measures of Gushui High Face Rockfill Dam. In *Technology for Earth-Rockfill Dam*; China Electric Power Press: Beijing, China, 2015. (In Chinese)
35. Kong, X.; Zhou, Y.; Zou, D. Numerical analysis of dislocations of the face slabs of the Zipingpu Concrete Faced Rockfill Dam during the Wenchuan earthquake. *Earthq. Eng. Eng. Vib.* **2011**, *10*, 9. [CrossRef]
36. Zhu, Y.; Peng, N.; Zuo, C.; Dong, Y. Study on Presettlement Control Measures for High Concrete-Faced Rockfill Dams. *Eur. J. Environ. Civil Eng.* **2021**, 1–14. [CrossRef]
37. Li, Y.; Pang, R.; Xu, B.; Wang, X.; Fan, Q.; Jiang, F. GPDEM-Based Stochastic Seismic Response Analysis of High Concrete-Faced Rockfill Dam with Spatial Variability of Rockfill Properties Based on Plastic Deformation. *Comput. Geotech.* **2021**, *139*, 104416. [CrossRef]
38. Westergaard, H.M. Water Pressures on Dams During Earthquakes. *Trans. Asce* **1933**, *98*, 418–432. [CrossRef]
39. Duncan, J.M.; Byrne, P.; Wong, K.S. Strength, Stress-strain and Bulk Modulus Parameters for Finite Element Analysis of Stress and Movements in Soil Masses. *J. Consult. Clin. Psychol.* **1981**, *49*, 67–554.
40. Clough, G.W.; Duncan, J.M. Finite Element Analyses of Retaining Wall Behavior. *Soil Mech. Found. Div. J.* **1971**, *97*, 1657–1673. [CrossRef]
41. Pang, R.; Xu, B.; Zhou, Y.; Zhang, X.; Wang, X. Fragility Analysis of High CFRDs Subjected to Mainshock-Aftershock Sequences Based on Plastic Failure. *Eng. Struct.* **2020**, *206*, 110152. [CrossRef]
42. Wang, S.; Tang, Y.; Li, X.; Du, K. Analyses and Predictions of Rock Cuttabilities under Different Confining Stresses and Rock Properties Based on Rock Indentation Tests by Conical Pick. *Trans. Nonferrous Met. Soc. China* **2021**, *31*, 1766–1783. [CrossRef]
43. Wang, S.; Tang, Y.; Wang, S. Influence of Brittleness and Confining Stress on Rock Cuttability Based on Rock Indentation Tests. *J. Cent. South Univ.* **2021**, *28*, 2786–2800. [CrossRef]
44. Cai, X.; Cheng, C.; Zhou, Z.; Konietzky, H.; Song, Z.; Wang, S. Rock mass watering for rock-burst prevention: Some thoughts on the mechanisms deduced from laboratory results. *Bull. Eng. Geol. Environ.* **2021**, *80*, 8725–8743. [CrossRef]
45. Xu, B.; Pang, R.; Zhou, Y. Verification of Stochastic Seismic Analysis Method and Seismic Performance Evaluation Based on Multi-Indices for High CFRDs. *Eng. Geol.* **2020**, *264*, 105412. [CrossRef]
46. Jingbo, L.; Yandong, L. A Direct Method for Analysis of Dynamic Soil-Structure Interaction Based on Interface Idea. In *Developments in Geotechnical Engineering*; Elsevier: Amsterdam, The Netherlands, 1998; Volume 83, pp. 261–276, ISBN 978-0-444-50035-9.
47. Zou, D.; Han, H.; Liu, J.; Yang, D.; Kong, X. Seismic Failure Analysis for a High Concrete Face Rockfill Dam Subjected to Near-Fault Pulse-like Ground Motions. *Soil Dyn. Earthq. Eng.* **2017**, *98*, 235–243. [CrossRef]
48. Yang, G.; Li, H.; Yu, Y.; Lv, H. Study on Seismic Measures of Nuozhadu High Core Rockfill Dam. *J. Hydroelectr. Eng.* **2008**, *27*, 89–93. (In Chinese)

mathematics

Article

Rockburst Interpretation by a Data-Driven Approach: A Comparative Study

Yuantian Sun, Guichen Li * and Sen Yang

School of Mines, Key Laboratory of Deep Coal Resource Mining, Ministry of Education of China, China University of Mining and Technology, Xuzhou 221116, China; yuantiansun@cumt.edu.cn (Y.S.); TS20020064A31TM@cumt.edu.cn (S.Y.)
* Correspondence: liguichen@cumt.edu.cn; Tel.: +86-158-0521-5566

Abstract: Accurately evaluating rockburst intensity has attracted much attention in these recent years, as it can guide the design of engineering in deep underground conditions and avoid injury to people. In this study, a new ensemble classifier combining a random forest classifier (RF) and beetle antennae search algorithm (BAS) has been designed and applied to improve the accuracy of rockburst classification. A large dataset was collected from across the world to achieve a comprehensive representation, in which five key influencing factors were selected as the input variables, and the rockburst intensity was selected as the output. The proposed model BAS-RF was then validated by the dataset. The results show that BAS could tune the hyperparameters of RF efficiently, and the optimum model exhibited a high performance on an independent test set of rockburst data and new engineering projects. According to the ensemble RF-BAS model, the feature importance was calculated. Furthermore, the accuracy of the proposed model on rockburst prediction was higher than the conventional machine learning models and empirical models, which means that the proposed model is efficient and accurate.

Keywords: rockburst classification; data-driven approach; random forest; beetle antennae search algorithm

Citation: Sun, Y.; Li, G.; Yang, S. Rockburst Interpretation by a Data-Driven Approach: A Comparative Study. *Mathematics* **2021**, *9*, 2965. https://doi.org/10.3390/math9222965

Academic Editor: Mario Versaci

Received: 26 September 2021
Accepted: 18 November 2021
Published: 20 November 2021

1. Introduction

Rock stability in deep underground conditions is seriously affected by rockburst, which still attracts a lot of attention nowadays [1,2]. In civil engineering and mining engineering, rockburst events normally occur suddenly, causing a loss of money in working facilities. Accurately evaluating the rockburst intensity has been a significant task as it can be a guideline in this area and guide managers to design carefully [3,4].

Rockburst cases occur in different conditions, such as tunneling and mining [5–7]. For instance, in the deep traffic tunnel in China, there are different grades of rockburst, which have caused different types of damage to the tunnel. Slight rockburst causes some cracks in the concrete in the tunnel face, and moderate rockburst affects the arc cavity pits, with depths of about 1 m, while intense rockburst affects the arc and wedge-shaped pits with depths of about 2 m, and the extremely intense rockburst almost destroyed the working condition, causing the depth of the pits to be about 3 m. Therefore, classifying and predicting the rockburst intensity plays a significant role in working safety.

Nowadays, the mechanism of rockburst is still not clear, but the basic laws of it are known as instantaneous slip and instantaneous fracturing. To control the rockburst, different methods have been proposed, such as temporary and permanent rock support systems; however, these approaches are not efficient as the rockburst intensity is difficult to know properly. Thus, some monitoring methods, such as a microseismic monitoring system, were applied to record and analyze the rockburst events [8]. The microseismic monitoring system records the rockburst intensity after the rockburst events, and it cannot predict the rockburst in advance. Hence, estimating and predicting rockburst intensity before its occurrence is of importance. Different models have been proposed, such as stress criteria, including the Barton, Hoek and Brown, Hou, Russenes, and Turchaninov criteria. Furthermore, the existing prediction approaches can be regarded as short-term and long-term predictions. In short-term predictions, the rockburst occurrence is based on in-situ site testes; however, the long-term prediction is basically according to the fundamental methods, such as strength theory and energy theory, which are similar to simulation, machine learning, and empirical knowledge methods [9].

Due to the uncertainties of rockburst and the unclear mechanism of occurrence, a curtained model or method is not suitable for the accurate prediction of rockburst. The method should consider more influencing factors related to rockburst occurrence, with random, fuzzy, or even both mechanisms, and thus, the artificial intelligence method can perfectly solve the problem [10,11]. For instance, there are various machine learning methods for predicting long-term rockburst hazards, such as support vector machines, artificial neural networks, and decision trees. The previous studies are summarized in Table 1. It can be noted that the prediction accuracy of rockburst intensity is affected by the number of data and different machine learning algorithms. Therefore, developing a high-performance and less-time-consuming ensemble classifier for the larger dataset is quite important.

Table 1. Previous studies on rockburst prediction with different machine learning methods.

Algorithms	Accuracy (%)	Data	References
SVM	100	16	Zhao et al. [12]
	93.8	45	Zhu et al. [13]
	51.7–67.2	246	Zhou et al. [14]
ANFIS	66.5–95.6	174	Adoko et al. [15]
ANN	72.2	18	Chen et al. [16]
	100	19	Xiao et al. [17]
	100	10	Feng et al. [10]
	85.2	134	Faradonbeh et al. [18]
CM	90–94.1	164	Liu et al. [19]
	76.4–82	209	Zhou et al. [20]
	71–76	246	Zhao et al. [14]
LR	80.2–90.9	135	Li et al. [21]
	88.3	188	Afraei et al. [22]
BN	91.7	135	Li et al. [23]
	53.9–65.8	246	Lin et al. [24]
KNN	53.2–67.2	246	Zhou et al. [14]
	50–65.9	246	Lin et al. [24]
DT	81.5	134	Faradonbeh et al. [18]
	73–93	132	Pu et al. [25]
	89.2–90.2	174	Ghasemi al. [26]

Note: SVM, Support vector machines; ANFIS, adaptive neuron fuzzy inference system; ANN, Artificial neural network; LR, Logistic regression; CM, Cloud model; BN, Bayesian network; KNN, k-nearest neighbors; DT, Decision tree.

Random forest (RF) has been applied in rockburst classification [27]. However, the relevant studies are fewer [14,24,28], by which their accuracy is limited by the hyperparameters, i.e., the number of the trees and the minimum leaf node. To optimize the structure of RF, there are some global optimization algorithms, such as the firefly algorithm (FA) and particle swarm optimization (PSO). However, these algorithms are time-consuming, and therefore, a new global algorithm should be proposed. Beetle Antennae Search (BAS), is a biologically inspired, intelligent optimization algorithm, which is inspired by the foraging principle of longicorn beetles. Furthermore, it has been used for tuning the hyper-parameters of ML algorithms in recent years.

This research aims to develop a machine learning-based model to study rockburst classification. The BAS algorithm was employed to tune the hyper-parameters of the RF algorithm. The performance of the ensemble BAS-RF model has been compared with other machine learning algorithms: the support vector machine, k-nearest neighbors, and decision tree algorithms. Furthermore, the BAS-RF has been tested against empirical criteria as well as previously published RF models, which were developed to address the rockburst problem.

2. Dataset Preparation

A total of 279 cases of rockburst events reported in the literature were collected to build a dataset [14,26,29–32]. The dataset included five influencing variables, with the buried depth of opening (H), the maximum tangential stress of the excavation boundary (σ_θ), the uniaxial compressive strength of rock (σ_c), the tensile rock strength (σ_t), and the elastic energy index (W_{et}) as input parameters and rockburst intensity as the output. These input variables are commonly applied in rockburst classification and can provide fundamental understandings about rockburst occurrence in underground conditions. According to rock failure properties, the output parameter, i.e., rockburst intensity, contains four different classes, namely none, light, moderate, and strong. The frequency of each input parameter is depicted in Figure 1. The statistics of the input parameters are summarized in Table 2.

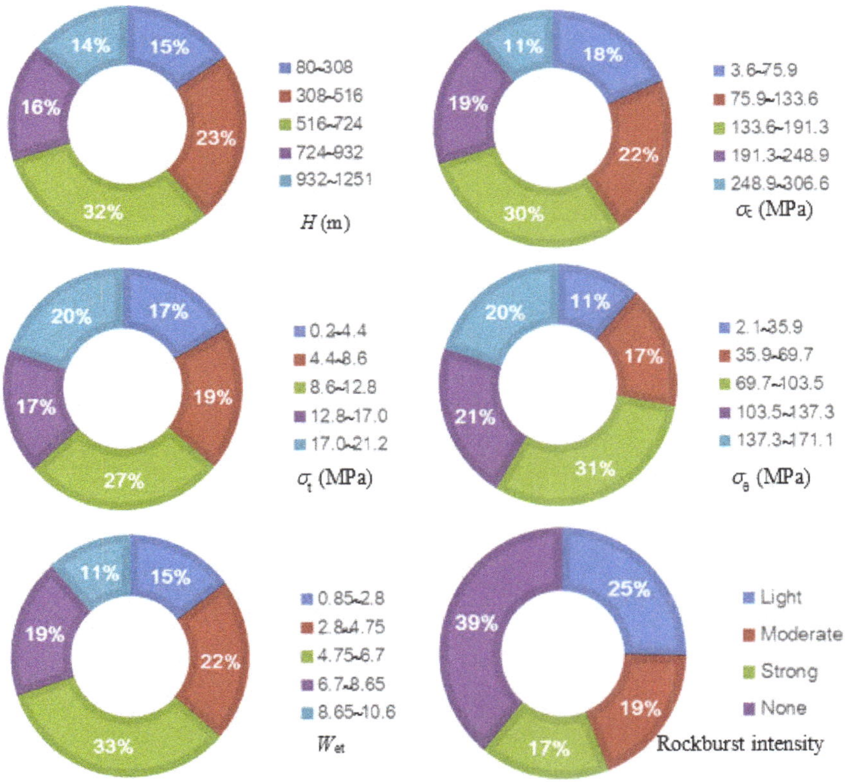

Figure 1. The statistics of input and output in the rockburst dataset.

Table 2. The collected input variables.

Parameters	Min	Max	Mean	Standard Deviation
H (m)	80	1251	682.2	291.4
σ_c (MPa)	3.6	306.6	118.9	69.8
σ_t (MPa)	0.2	21.2	8.6	6.1
σ_θ (MPa)	2.1	171.1	63.4	42.5
W_{et}	0.85	10.57	5.2	3.4

3. Algorithm Background and Ensemble Model

3.1. Algorithms Description

3.1.1. Decision Tree and Random Forest

The Decision Tree (DT) and Random Forest (RF) both have tree structures. In contrast to the DT, the random forest uses the method of majority votes. The normal structure of DT and RF is shown in Figure 2.

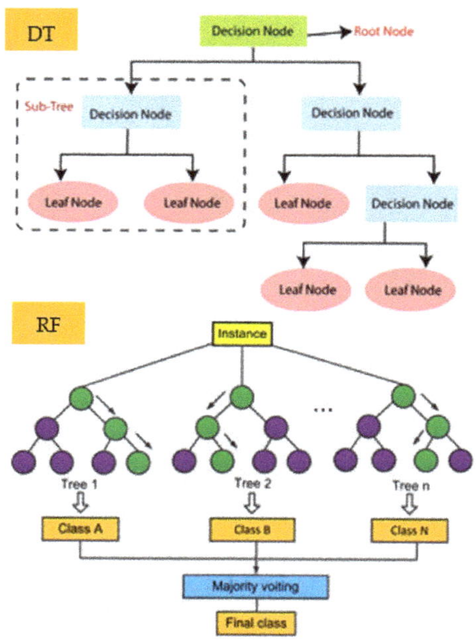

Figure 2. The structure of DT and RF.

The C 4.5 algorithm in this study was applied for the attribute selection process, which can be expressed as follows:

$$GainRatio\ (S, A) = \frac{Gain\ (S, A)}{SplitInfo\ (A)} \quad (1)$$

where S is the training set; A is the attribute; $SplitInfo\ (A)$ is given by

$$SplitInfo\ (A) = \sum_{v \epsilon Domain\ (A)} \frac{|S_v^A|}{|S|} \cdot \log_2 \frac{|S_v^A|}{|S|} \quad (2)$$

The necessary steps are (1) selecting random K data points from the training set, (2) building the decision trees associated with the selected data points, (3) choosing the number of decision trees, (4) repeating steps 1 and 2, (5) finding the predictions of each decision tree for new data points, and assigning new data points to the category having the majority of votes.

3.1.2. K-Nearest Neighbor

The K-Nearest Neighbor (KNN) is a non-parametric and lazy learning algorithm. K is the number of nearest neighbors. The number of neighbors is the core deciding factor. There are some basic steps, i.e., calculate the distance, find the closest neighbors, and vote for labels. The structure of KNN is depicted in Figure 3.

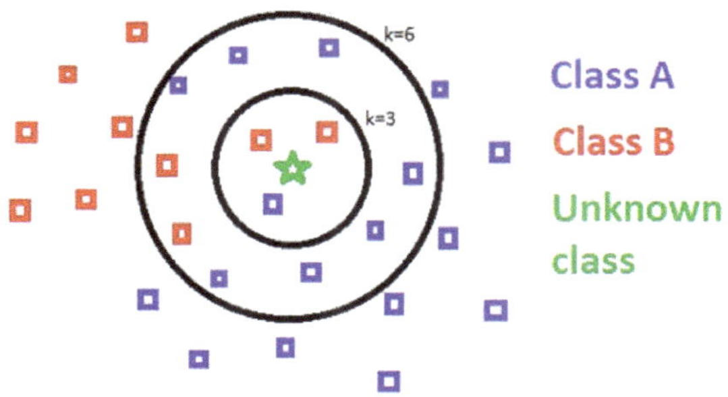

Figure 3. The structure of KNN.

3.1.3. Support Vector Machines

Support Vector Machines (SVM) are considered to be a classification approach by constructing a hyperplane in a multidimensional space to separate different classes. They include the following steps: generate hyperplanes and select the right hyperplane with the maximum segregation. The structure of SVM is shown in Figure 4.

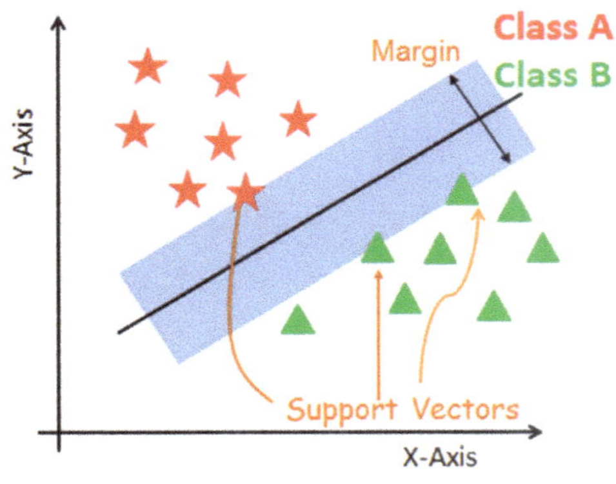

Figure 4. The structure of SVM.

3.1.4. Beetle Antennae Search Algorithm

The Beetle Antennae Search algorithm (BAS) is an intelligent optimization algorithm, proposed by Jiang et al. in 2017. Different from other bionic algorithms, the Beetle Antenna Search algorithm is a monomer search algorithm with the advantages of a simple principle, fewer parameters, and less computation. It has great advantages in dealing with low-dimensional optimization objectives, such as low time complexity and strong searchability. The flow chart of BAS is given in Figure 5. In this study, the iteration of BAS was set as 50, and the step factor was set as 0.95. All algorithms were developed by Matlab software.

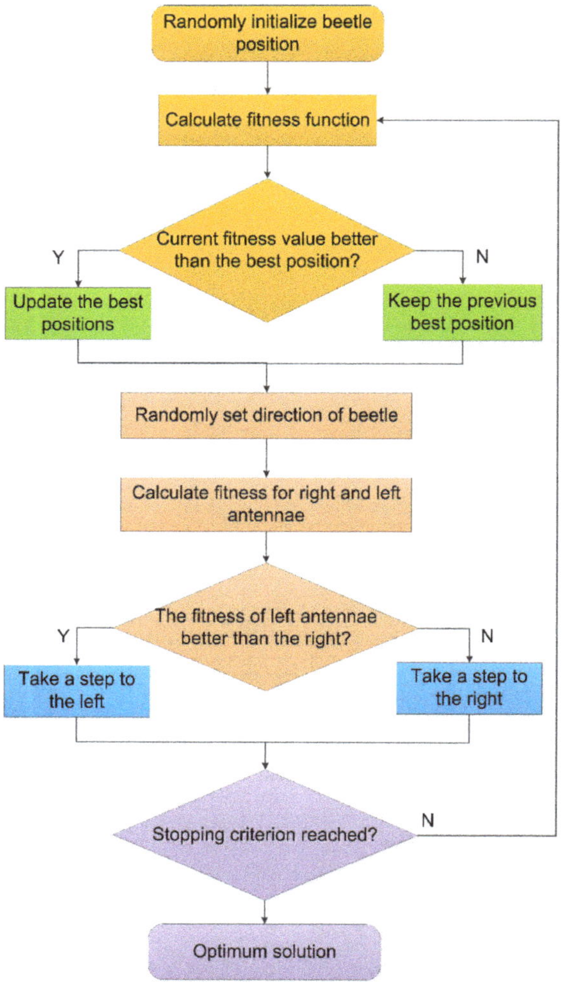

Figure 5. The flowchart of BAS.

3.2. *The Methodology of Ensemble RF-BAS Model*

There are several procedures for constructing an ensemble model.

Step 1: Splitting the dataset into a train dataset and test dataset, and normally, the proportion is 70% and 30%, respectively. It should be pointed out that due to the rockburst intensity being classed into four classes, the train and test dataset should also be divided into four subsets accordingly.

Step 2: Initialing the parameters of BAS, i.e., the beetle's position in the space, in which the dimension of the position vector is the number of hyperparameters of the algorithm.

Step 3: Training the model and calculating the fitness value on the remaining subset of the training set.

Step 4: The BAS will tune the hyper-parameters by decreasing the fitness value. When the iteration of 50 is reached, the optimal hyperparameters can be found.

Step 5: The above process is repeated five times, and it can be called a fivefold cross-validation (CV) (shown in Figure 6). The full procedure is depicted in Figure 7.

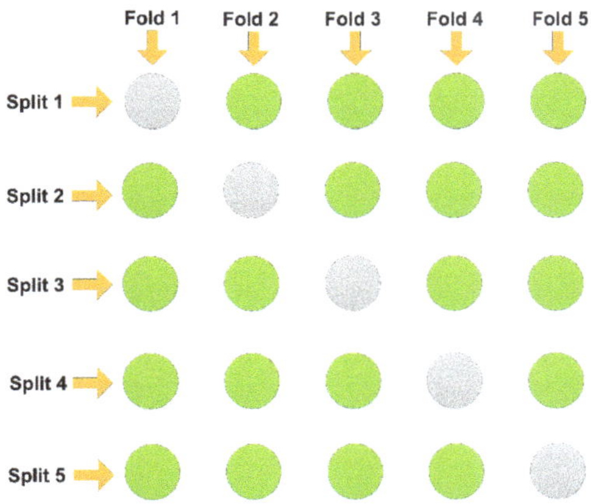

Figure 6. Fivefold cross-validation (CV).

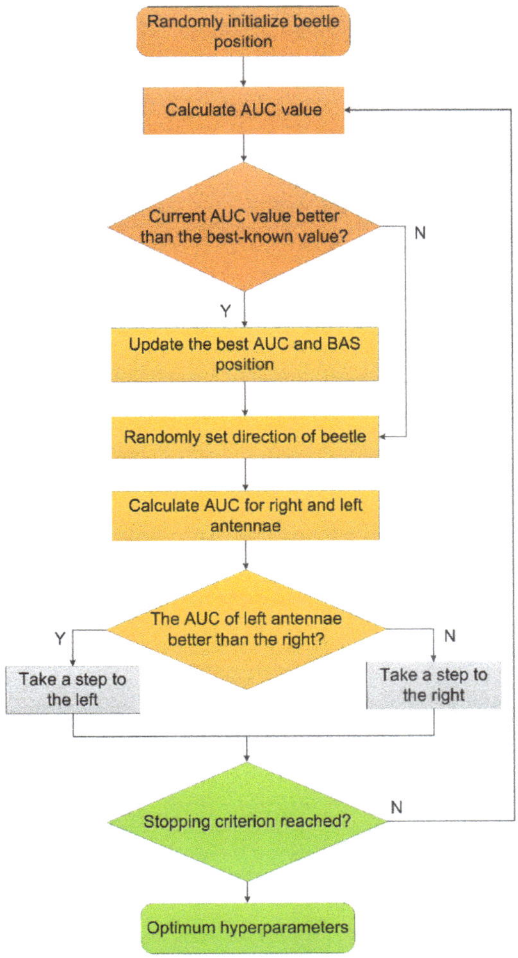

Figure 7. The procedures of hyperparameter tuning of RF by BAS.

3.3. Performance Evaluation Methods

In this study, we applied the classical methods for model evaluation. The receiver operating characteristic (ROC) curve and the AUC curve (the area under the ROC) were used in the evaluation of rockburst classification. The horizontal axis is the false positive rate (FPR); however, the vertical axis represents the true positive rate (TPR) in the ROC curve.

4. Results

4.1. Hyper-Parameter Tuning

In this procedure, AUC was set as the objective function, and the hyper-parameters of RF, i.e., (the number of the trees and the minimum required samples at a leaf node) were tuned by BAS. Then, in the test process, four BAS-RF models were used to classify the unknown samples. The AUC values and convergence showed different patterns with different classes (given in Figure 8). With the increase of iteration, the average AUC values increased sharply before five iterations, meaning that the BAS could tune the hyper-parameters quickly and effectively. The hyper-parameters of RF were given in Table 3.

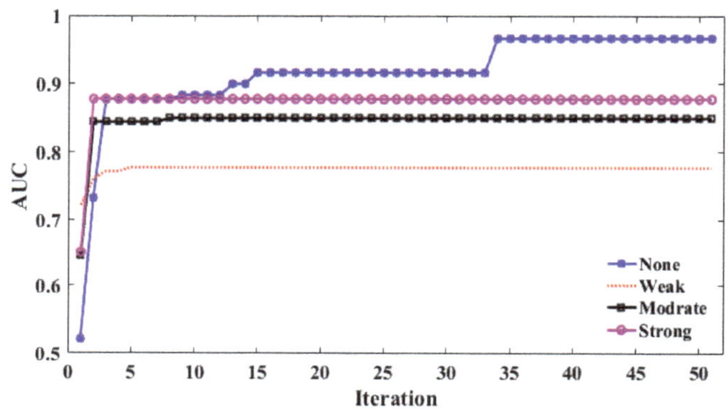

Figure 8. The evolution of AUC values with different dataset classes by BAS tuning.

Table 3. The optimum hyperparameters in each class.

Number of Input Variables	Hyperparameter	Definition	Scope	Class			
				None	Light	Moderate	Strong
5	tree_num	The number of the trees	2–100	42	29	34	17
	min_sample_leaf	The minimum required samples at a leaf node	1–10	1	1	2	1

4.2. Validation of BAS-RF

In the testing dataset, the proposed BAS-RF model was applied to validate the accuracy on that dataset. The final results are given in Table 4. As can be seen, the accuracy was over 0.90, which means that the proposed model could be used for a new dataset.

Table 4. The confusion matrix of the proposed model on test validation.

Rockburst	Actual	Predicted				Percentage Correct	Accuracy
		None	Light	Moderate	Strong		
None	35	32	1	1	1	0.91	0.92
Light	20	1	19	0	0	0.95	
Moderate	15	0	1	13	1	0.86	
Strong	14	0	1	0	13	0.92	

4.3. The Rank of Influencing Variables

The ranking influence of each input variable on the rockburst is depicted in Figure 9. W_{et} was the most important variable influencing rockburst intensity, followed by σ_θ, H, σ_c, and σ_t. The results indicate that more attention should be given to W_{et}, σ_θ, and H in engineering projects. Although some parameters, i.e., σ_c, and σ_t, have a lower influence on rockburst intensity, they should still be taken into account when analyzing rockburst events.

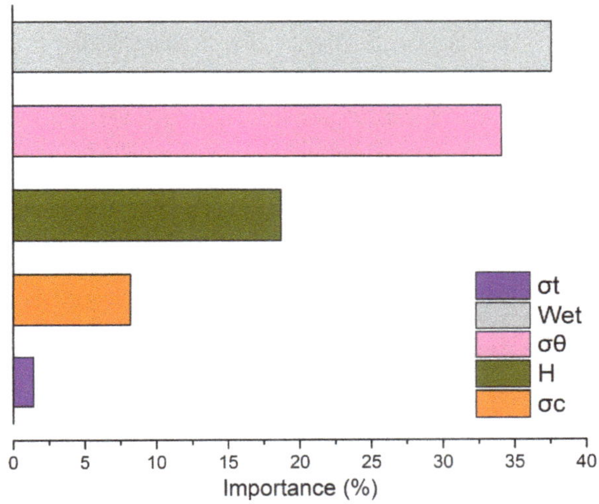

Figure 9. The relative importance of variables.

5. Discussion

5.1. Comparison of the BAS-RF with Baseline Models

The performance of the BAS-RF model was evaluated with SVM, DT, and KNN machine learning models. The BAS-RF model was the most accurate model, having an accuracy of 0.92 (Table 5). The DT, SVM, and KNN models had accuracies of 0.84, 0.76, and 0.71, respectively. The comparison analysis confirmed that the proposed BAS-RF model achieved a better performance than the other machine learning classifiers. Furthermore, the conventional RF models on rockburst assessment in previous studies were compared with BAS-RF; the accuracy performance of BAS-RF was higher than existing RF models. Few studies have already applied the conventional RF model for rockburst assessment. In this section, the proposed ensemble classifier BAS-RF was compared with the findings of the previous studies. Zhou et al. (2016) compared the performance of 10 machine learning algorithms to analyze rockburst events. They used 246 cases and considered seven input variables. Lin et al. (2018) investigated rockburst events using machine learning models. They investigated 246 rockburst cases, considering six input variables. The accuracy performances of the RF model developed by Zhou et al. (2016) and Lin et al. (2018) were 0.73 and 0.61, respectively. The BAS-RF model performed much better compared to the existing RF models. The model was developed using a larger dataset, and thus, it can be applied over a wider range of conditions. Although both models developed by Zhou et al. (2016) and Lin et al. (2018), respectively, considered seven and six input variables, they still led to a lower prediction accuracy. Furthermore, we compared the results (summarized in Table 5) with conventional empirical models, such as the rock brittleness coefficient criterion, burst proneness index, and Russenes criterion. The BAS-RF model performed better than the empirical models.

Table 5. Classification accuracy of ensemble classifiers and baseline models.

ML Models					Empirical Models			
BAS-RF	SVM	DT	KNN	Conventional RF	Rock Brittleness Coefficient Criterion	Elastic Energy Index	Russenes Criterion	Burst Proneness Index
0.92	0.76	0.84	0.71	0.73, Zhou et al. (2016) [14]; 0.61, Lin et al. (2018) [24]	0.32, Wang et al. (1998) [33]	0.41, Kidybinski (1981) [34]	0.36, Russenes (1974) [35]	0.21, Singh (1989) [36]

The TPR (True Positive Rate) and AUC values calculated for all classifiers shown in Figure 10 indicate that the ensemble classifier BAS-RF provided the most accurate classification. The ensemble BAS-RF led to an AUC value of 0.95, followed by DT, SVM, and KNN. The AUC values of DT, SVM and KNN were 0.82, 0.81, and 0.7, respectively.

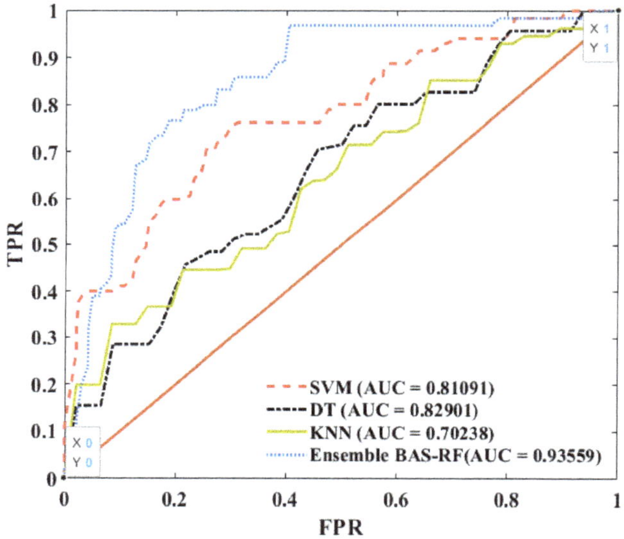

Figure 10. ROC curve of the proposed ensemble BAS-RF, SVM, DT, KNN.

5.2. Cases Application

Eight rockburst events in four different tunnel and mining projects were predicted by the BAS-RF model. The field data were collected from available literature, including the Calling tunnel, Dongguashan mine [37], Duoxiongla tunnel [38], and Daxiangling tunnel [30]. The prediction outcomes are summarized in Table 6, which indicated that the rockburst intensity for all cases was predicted correctly. The results of this study confirm that the BAS-RF model is a robust alternative tool for the rockburst assessment, and it can be successfully applied in various geotechnical engineering projects.

Table 6. Engineering application of the proposed BAS-RF model.

No.	H (m)	σ_θ (MPa)	σ_c (MPa)	σ_t (MPa)	W_{et}	Actual	Predicted
1 [37]	768	32.8	160	6.6	4.6	Light	Light
2 [37]	768	50.9	160	7.5	5.3	Moderate	Moderate
3 [37]	730	105.5	190.3	17.1	4.0	Moderate	Moderate
4 [38]	700	87.3	137.7	9.62	7.14	Strong	Strong
5 [38]	700	87.3	94.4	9.16	3.57	Light	Light
6 [30]	808	45.6	114	2.3	4.7	Moderate	Moderate
7 [30]	362	25.6	59.7	1.3	1.7	None	None
8 [30]	981	57.2	80.6	2.5	5.5	Strong	Strong

6. Summary and Conclusions

A novel ensemble classifier combining the random forest (RF) and Beetle Antennae search algorithm (BAS) was proposed to classify rockburst intensity in underground projects. The BAS algorithm was applied to tune hyperparameters of the RF. The performance of the proposed model (BAS-RF) was evaluated by its accuracy, precision, and recall criteria. Additionally, the ROC curve and AUC values were used to assess the rockburst intensity. The conclusions can be summarized as follows:

- The BAS algorithm could tune hyperparameters of the RF model effectively, leading to a satisfactory performance of the BAS-RF model in rockburst classifications.
- The BAS-RF model performed much better compared to the other classifier. The BAS-RF was the most accurate model, followed by DT, SVM, and KNN models.
- Analyzing the relative importance of input variables based on the BAS-RF model demonstrated that Wet has a substantial influence on rockburst.
- The BAS-RF model provided the most accurate classification as compared to the existing RF model as well as the empirical criteria.
- We successfully applied the BAS-RF model for predicting rockburst events in new projects. The proposed model had a high generalization ability, which facilitates its future application in rockburst intensity assessments.
- It should be pointed out that the generalization could have been improved if we had used a large dataset to train the model.

Author Contributions: Conceptualization by Y.S., G.L. and S.Y.; writing—original draft preparation, Y.S. and G.L.; writing—review and editing, G.L. and S.Y.; supervision, G.L.; funding acquisition, G.L. All authors have read and agreed to the published version of the manuscript.

Funding: This research was supported by the projects of "the Fundamental Research Funds for the Central Universities (2020ZDPY0221, 2021QN1003)", "National Natural Science Foundation of China (52104106, 52174089)", and "Basic Research Program of Xuzhou (KC21017)".

Institutional Review Board Statement: Not applicable.

Informed Consent Statement: Not applicable.

Data Availability Statement: The data presented in this study are available on request from the corresponding author. The data are not publicly available due to privacy.

Acknowledgments: Special thanks to Zuqi Wang for her encouragements.

Conflicts of Interest: The authors declare no conflict of interest.

References

1. Kaiser, P.K.; McCreath, D.R.; Tannant, D.D. *Rockburst Support Handbook*; Geomechanics Research Centre, Laurentian University of Canada: Sudbury, ON, USA, 1996.
2. Ortlepp, W. RaSiM Comes of Age—A Review of the contribution to the understanding and control of mine rockbursts. In Proceedings of the Sixth International Symposium on Rockburst and Seismicity in Mines, Perth, Australia, 9–11 March 2005; pp. 3–20.
3. Cai, M. Principles of rock support in burst-prone ground. *Tunn. Undergr. Space Technol.* **2013**, *36*, 46–56. [CrossRef]

4. Kie, T.T. Rockbursts, case records, theory and control. In Proceedings of the International Symposium on Engineering in Complex Rock Formations, Beijing, China, 3–7 November 1988; Elsevier BV: Cham, Switzerland, 1988; pp. 32–47.
5. Zhou, J.; Li, X.; Mitri, H.S. Evaluation method of rockburst: State-of-the-art literature review. *Tunn. Undergr. Space Technol.* **2018**, *81*, 632–659. [CrossRef]
6. He, M.; Ren, F.; Liu, D. Rockburst mechanism research and its control. *Int. J. Min. Sci. Technol.* **2018**, *28*, 829–837. [CrossRef]
7. Dehghan, S.; Shahriar, K.; Maarefvand, P.; Goshtasbi, K. 3-D modeling of rock burst in pillar No. 19 of Fetr6 chromite mine. *Int. J. Min. Sci. Technol.* **2013**, *23*, 231–236. [CrossRef]
8. Lu, C.-P.; Dou, L.-M.; Liu, B.; Xie, Y.-S.; Liu, H.-S. Microseismic low-frequency precursor effect of bursting failure of coal and rock. *J. Appl. Geophys.* **2012**, *79*, 55–63. [CrossRef]
9. Wang, S.; Li, X.; Yao, J.; Gong, F.; Li, X.; Du, K.; Tao, M.; Huang, L.; Du, S. Experimental investigation of rock breakage by a conical pick and its application to non-explosive mechanized mining in deep hard rock. *Int. J. Rock Mech. Min. Sci.* **2019**, *122*, 104063. [CrossRef]
10. Wang, S.; Sun, L.; Li, X.; Wang, S.; Du, K.; Li, X.; Feng, F. Experimental investigation of cuttability improvement for hard rock fragmentation using conical cutter. *Int. J. Geomech.* **2021**, *21*, 6020039. [CrossRef]
11. Sun, Y.; Li, G.; Zhang, N.; Chang, Q.; Xu, J.; Zhang, J. Development of ensemble learning models to evaluate the strength of coal-grout materials. *Int. J. Min. Sci. Technol.* **2021**, *31*, 153–162. [CrossRef]
12. Zhao, H.-B. Classification of rockburst using support vector machine. *Rock Soil Mech.* **2005**, *26*, 642–644.
13. Zhu, Y.H.; Liu, X.R.; Zhou, J.P. Rockburst prediction analysis based on v-SVR algorithm. *J. China Coal Soc.* **2008**, *33*, 277–281.
14. Zhou, J.; Li, X.; Mitri, H.S. Classification of Rockburst in Underground Projects: Comparison of Ten Supervised Learning Methods. *J. Comput. Civ. Eng.* **2016**, *30*, 04016003. [CrossRef]
15. Adoko, A.C.; Gokceoglu, C.; Wu, L.; Zuo, Q.J. Knowledge-based and data-driven fuzzy modeling for rockburst prediction. *Int. J. Rock Mech. Min. Sci.* **2013**, *61*, 86–95. [CrossRef]
16. Haijun, L.C.; Dexin, S.N. Prediction of rockburst by artificial neural network. *Chin. J. Rock Mech. Eng.* **2003**, *22*, 762.
17. Guo, L.; Li, X.; Yan, X.; Xiong, L. Rock Burst Prediction Methods Based on BP Network Theory. *Ind. Saf. Dust Control* **2005**, *10*.
18. Faradonbeh, R.S.; Taheri, A. Long-term prediction of rockburst hazard in deep underground openings using three robust data mining techniques. *Eng. Comput.* **2019**, *35*, 659–675. [CrossRef]
19. Liu, Z.; Shao, J.; Xu, W.; Meng, Y. Prediction of rock burst classification using the technique of cloud models with attribution weight. *Nat. Hazards* **2013**, *68*, 549–568. [CrossRef]
20. Zhou, K.-P.; Lin, Y.; Deng, H.-W.; Li, J.; Liu, C.-J. Prediction of rock burst classification using cloud model with entropy weight. *Trans. Nonferrous Met. Soc. China* **2016**, *26*, 1995–2002. [CrossRef]
21. Li, N.; Jimenez, R. A logistic regression classifier for long-term probabilistic prediction of rock burst hazard. *Nat. Hazards* **2017**, *90*, 197–215. [CrossRef]
22. Afraei, S.; Shahriar, K.; Madani, S.H. Statistical assessment of rock burst potential and contributions of considered predictor variables in the task. *Tunn. Undergr. Space Technol.* **2018**, *72*, 250–271. [CrossRef]
23. Li, N.; Feng, X.; Jimenez, R. Predicting rock burst hazard with incomplete data using Bayesian networks. *Tunn. Undergr. Space Technol.* **2017**, *61*, 61–70. [CrossRef]
24. Sun, Y.; Li, G.; Zhang, J. Developing Hybrid Machine Learning Models for Estimating the Unconfined Compressive Strength of Jet Grouting Composite: A Comparative Study. *Appl. Sci* **2020**, *10*, 1612. [CrossRef]
25. Pu, Y.; Apel, D.B.; Lingga, B. Rockburst prediction in kimberlite using decision tree with incomplete data. *J. Sustain. Min.* **2018**, *17*, 158–165. [CrossRef]
26. Sun, Y.; Li, G.; Zhang, J.; Huang, J. Rockburst intensity evaluation by a novel systematic and evolved approach: Machine learning booster and application. *Bull. Eng. Geol. Environ.* **2021**, *80*, 8385–8395. [CrossRef]
27. Breiman, L. Random forests. *Mach. Learn.* **2001**, *45*, 5–32. [CrossRef]
28. Dong, L.-J.; Li, X.-B.; Peng, K. Prediction of rockburst classification using Random Forest. *Trans. Nonferrous Met. Soc. China* **2013**, *23*, 472–477. [CrossRef]
29. Afraei, S.; Shahriar, K.; Madani, S.H. Developing intelligent classification models for rock burst prediction after recognizing significant predictor variables, Section 2: Designing classifiers. *Tunn. Undergr. Space Technol.* **2018**, *84*, 522–537. [CrossRef]
30. Long, L.; Chen, J. Fuzzy Comprehensive Assessment Method Adopted to Predict Rock Burst in Daxiangling Tunne. *Xiandai Suidao Jishu* **2010**, *47*, 23–27.
31. Li, N.; Jimenez, R.; Feng, X. The Influence of Bayesian Networks Structure on Rock Burst Hazard Prediction with Incomplete Data. *Procedia Eng.* **2017**, *191*, 206–214. [CrossRef]
32. Zhou, J.; Li, X.; Shi, X. Long-term prediction model of rockburst in underground openings using heuristic algorithms and support vector machines. *Saf. Sci.* **2012**, *50*, 629–644. [CrossRef]
33. Lee, P.; Tsui, Y.; Tham, L.; Wang, Y.; Li, W. Method of fuzzy comprehensive evaluations for rockburst prediction. *Chin. J. Rock Mech. Eng.* **1998**, *17*, 493–501.
34. Kidybiński, A. Bursting liability indices of coal. *Int. J. Rock Mech. Min. Sci. Géoméch. Abstr.* **1981**, *18*, 295–304. [CrossRef]
35. Russenes, B.F. *Analysis of Rock Spalling for Tunnels in Steep Valley Sides*; Norwegian Institute of Technology: Trondheim, Norway, 1974; Volume 247.
36. Singh, S. Classification of mine workings according to their rockburst proneness. *Min. Sci. Technol.* **1989**, *8*, 253–262. [CrossRef]

37. Jia, Y.; Lv, Q.; Shang, Y. Rockburst prediction using particle swarm optimization algorithm and general regression neural network. *Chin. J. Rock Mech. Eng.* **2013**, *32*, 343–348.
38. Tang, Z.; Xu, Q. Rock burst prediction based on nine machine learning algorithms. *Chin. J. Rock. Mech Eng.* **2020**, *161*.

mathematics

Article

Diffusion Model of Parallel Plate Crack Grouting Based on Foaming Expansion Characteristics of Polymer Slurry

Jiasen Liang [1], Shaokun Ma [2,*] and Xueming Du [1,*]

[1] School of Water Conservancy Science and Engineering, Zhengzhou University, Zhengzhou 450001, China; liangjiasen888@163.com
[2] School of Civil Engineering, Guangxi University, Nanning 530004, China
* Correspondence: mashaokun@sina.com (S.M.); 2007-dxm@163.com (X.D.)

Abstract: Polymers as a new chemical grouting material have been widely used in fractured rock mass; however, the understanding of polymer diffusion characteristics still needs to be further improved. In order to study the diffusion mechanism of foamed polymer slurry in rock fissures, the radial diffusion model of polymer single crack grouting is derived in consideration of the factors such as grouting volume, crack width and expansion rate. The influence of different factors on slurry diffusion radius, diffusion pressure and flow rate is analyzed, The diffusion model is verified by finite element numerical simulation. The findings show that (1) The results of slurry diffusion radius, pressure and velocity distribution at different times under different working conditions in the present model are in good agreement with the analytical solution; (2) The diffusion pressure is directly proportional to the grouting volume and expansion multiple, and inversely proportional to the crack width. In addition, diffusion pressure decreases with the increase of diffusion distance, and the pressure at the corresponding distance increases slowly with time, and finally tends to be stable; (3) For the same section, the radial velocity decreases slowly with the increase of time; for different sections, the flow velocity increases sharply with the increase of the distance between the section and the central axis of the grouting hole.

Keywords: polymer; diffusion model; crack; expansion ratio; grouting amount

Citation: Liang, J.; Ma, S.; Du, X. Diffusion Model of Parallel Plate Crack Grouting Based on Foaming Expansion Characteristics of Polymer Slurry. *Mathematics* **2021**, *9*, 2907. https://doi.org/10.3390/math9222907

Academic Editor: Aleksandr Rakhmangulov

Received: 10 October 2021
Accepted: 14 November 2021
Published: 15 November 2021

1. Introduction

In underground construction, grouting is one of the effective technical means to improve rock mechanical properties and block groundwater [1–6]. Grouting refers to injecting the slurry with cementitious capacity into the cracks, voids or cavities in the rock layer (or soil layer) through grouting drilling holes or grouting pipes, which can drive away the water and air in the cracks, voids or cavities, cement the original loose soil particles or cracks into a whole, and form a "stone body" with new structure, high strength and strong waterproof and impermeability, so as to improve the performance of the rock (soil) layer [7]. In recent years, many new grouting materials and equipment have been come into use, which greatly improves the effectiveness of grouting plugging. With the advantages of safety and environmental protection, fast response, high expansion rate, impermeability and durability, polymer materials have become grouting materials with excellent comprehensive performance [8], and are widely used in foundation reinforcement, dam seepage prevention, road maintenance, etc.

The diffusion mechanism of slurry in rock and soil fractures has always been the focus of research. Many scholars have studied the migration law of slurry in fractured rock mass. Generally, the fractured rock mass is simplified as a parallel plate model [9–12]. According to different slurry constitutive equations, the diffusion of slurry in rock fractures is studied by the analytical method, and the slurry flow and pressure distribution equations are established to explore the flow law of slurry in fractured rock mass. Some scholars regard the slurry as Newtonian fluid and deduce the fracture grouting diffusion model. For instance,

Zhang et al. [13] established a two-dimensional slurry flow model in fractures considering fracture roughness and groundwater viscous resistance, and Li et al. [14] deduced an ideal self-expanding slurry single fracture diffusion model; Furthermore, other scholars regard the slurry as Bingham fluid and deduce the crack grouting diffusion model. For example, Gustafson et al. [15] proposed the analytical model for calculating the penetration length of silica sol grouting. Liu et al. [16] proposed the grouting numerical model of water-rich broken sandstone for ordinary portland cement 42.5 (PO. 42.5), aluminum sulfate cement 42.5 (sac. 42.5) and self-developed cement-based composite grouting material (CGM). And a crack grouting diffusion model considering the time-varying viscosity of cement slurry has been proposed [17].

In the diffusion of one-dimensional flow model, Amadei and Savage [18] proposed a one-dimensional flow model of Bingham slurry in the channel between parallel walls, and analyzed the influence of pressure gradients on slurry flow characteristics; Luo et al. [19] deduced the flow equation of Bingham slurry in one-dimensional inclined single fracture, and analyzed the effects of fracture inclination, viscosity and other factors on the velocity of slurry. In the two-dimensional radial diffusion model, Tani et al. [9] studied the radial diffusion law of cement slurry diffusing in the cracks between parallel plates; while Zhan et al. [20,21] established a hydrodynamic grouting diffusion model of single fracture and analyzed the influence of water flow velocity on the slurry diffusion range. Zhang et al. [22] proposed a theoretical model of horizontal fracture grouting diffusion considering the temporal and spatial variation of slurry viscosity under the condition of a constant grouting rate.

In general, great progress has been made in the research on crack grouting mechanism at home and abroad; however, the existing models take constant density slurry such as sodium silicate, cement slurry and ordinary chemical slurry as the objects, and there is no research on the crack grouting diffusion model of foamed polymer slurry. Two component foamed polyurethane generally has faster reaction speeds and larger expansion rates. According to the added foaming dose, it can expand 10–30 times in 6–30 s [8]. Due to the expansibility of polymer slurry and the void structure characteristics of injected medium, the diffusion characteristics of polymers are more complex, which also brings great difficulties to the selection of polymer grouting parameters in engineering practice [23].

In this paper, based on the foaming expansion characteristics of polymer slurry and the theory of viscous hydrodynamics, the radial diffusion model of foaming polymer in single crack considering the self-expansion characteristics of polymer is theoretically deduced. In addition, the numerical simulation of grouting diffusion in parallel plate cracks is established by using finite element software (Comsol Multiphysics), which verifies the accuracy of the present analysis model. The diffusion model can fully consider the influence of grouting volume, crack opening, radial distance, time, expansion ratio and other factors on diffusion characteristics of slurry, and has a certain reference value for the theoretical research of foam polymer slurry crack grouting.

2. Grout Diffusion Model

2.1. Density Model

Here, three kinds of polymer grouting materials (10 times expansion rate, 20 times expansion rate and 30 times expansion rate) are taken as the research object. First, the ex-factory expansion rate of the polymer is calibrated and verified. A self-made reaction vessel for measuring the expansion rate of polymers was made according to the requirement of the test (Figure 1), which is mainly composed of a plexiglass cylinder with a bottom. Meanwhile, in order to observe the change of expansion volume, a scale is pasted on the reaction cylinder, and a high-definition camera(HD) is used to capture the expansion rise height of the polymer at different times. The two-component polymer grouting materials are mixed according to the requirements of the same quality and poured into the reaction vessel for reaction expansion at room temperature.

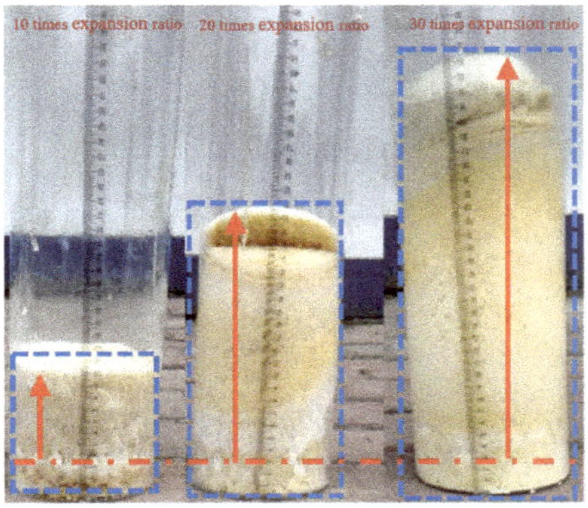

Figure 1. Polymer consolidation after reaction.

Figures 1 and 2, respectively, show the morphology photos of the consolidated body of three polymers in the reaction cylinder and the comparison diagram of slurry height before and after the reaction. It can be seen from the figure that the three polymer grouting materials have expanded to a certain extent after reaction. In particular, it can be clearly seen from Figure 2 that the three polymers have expanded from the original height of 2.5 cm to 11.5 cm, 21 cm and 31.5 cm, respectively. In addition, the heights of consolidated polymers with 20 and 30 times the expansion ratio are 1.83 and 2.74 times, respectively. It can be proved that the ex-factory expansion ratio of the three polymers is reliable within the allowable range of test error.

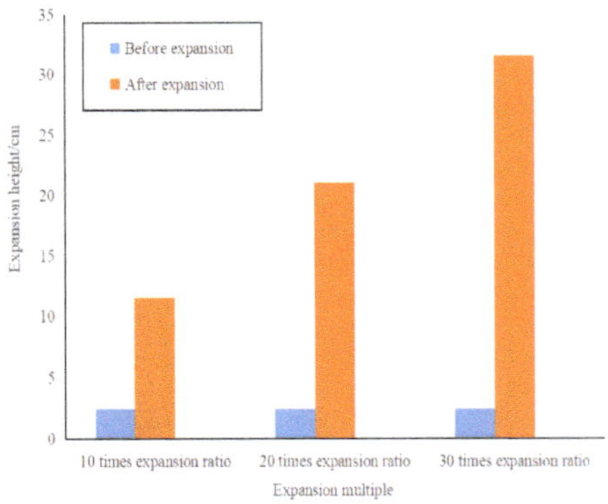

Figure 2. Comparison of slurry height before and after reaction.

The density integrity relationship is obtained by the nonlinear fitting of the experimental results.

$$\rho = Ae^{-Bt} + C \quad (1)$$

where A = 1087.297, 1115.818, 1144.339, B = 1/4.87, 1/5.48, 1/5.92, C = 85.563, 57.042, 20.521 represent polymer grouting materials with 10, 20 and 30 times expansion rate respectively, t is the time (s), ρ is a slurry density (kg/m^3) for a certain hour.

2.2. Basic Hypothesis

Based on the existing derivation method of fracture diffusion model, the following assumptions were made [24,25]: (1) the slurry is a homogeneous isotropic fluid; (2) There is no slip boundary at the upper and lower surfaces of the crack, that is, the slurry velocity at the contact with the wall is 0; (3) The slurry is in a laminar flow during the diffusion process; (4) The slurry is a Newtonian fluid, and its viscosity and flow pattern remain unchanged in the grouting process; (5) The fracture wall is rigid, and there is no deformation under the pressure of the slurry; (6) The groutability of the slurry is good, and there is no blockage, and the movement of the slurry in the fracture model is a full plane radiation; (7) The crack wall has no adsorption effect on the slurry, and there is no precipitation during the movement of the slurry.

2.3. Diffusion Model Derived

The schematic diagram of single crack grouting [14] is shown in Figure 3, and its opening of crack is h, and the radius of grouting hole is R_0. It is assumed that the polymer slurry will flow radially around the grouting hole between the upper and lower crack surfaces; and ignoring the grouting pressure in the grouting hole, the slurry will flow in the crack completely by virtue of the volume expansion mechanism.

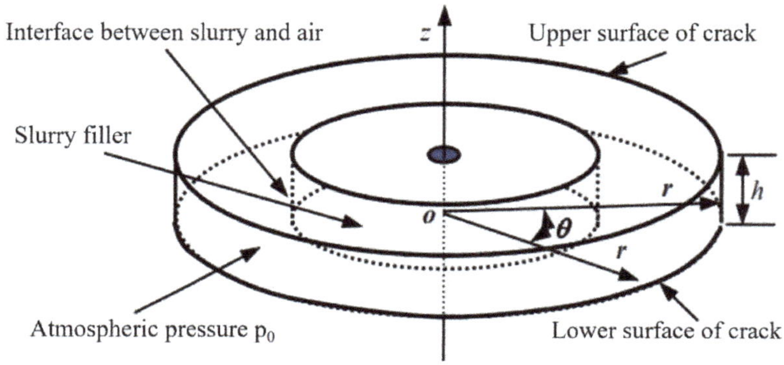

Figure 3. Schematic diagram of self-expanded slurry diffusion in flat single fracture.

Taking any fluid unit from the basin (see Figure 4), the external force on the fluid element are a normal stress p and the shear stressτ, and there is no shear resistance between the radial planes perpendicular to the crack plane. If the influence of velocity change is not considered, the sum of the respective force along the direction of the center radial axial axis of the single element should be equal to zero [26].

$$pr\Delta\theta\Delta z-\left(p+\frac{dp}{dr}\Delta r\right)(r+\Delta r)\Delta\theta\Delta r+\left(p+\frac{dp}{dr}\frac{\Delta r}{2}\right)\Delta\theta\Delta r\Delta z+\left(\frac{d\tau}{dz}\Delta z\right)\frac{(2r+\Delta r)\Delta\theta}{2}\Delta r=0 \quad (2)$$

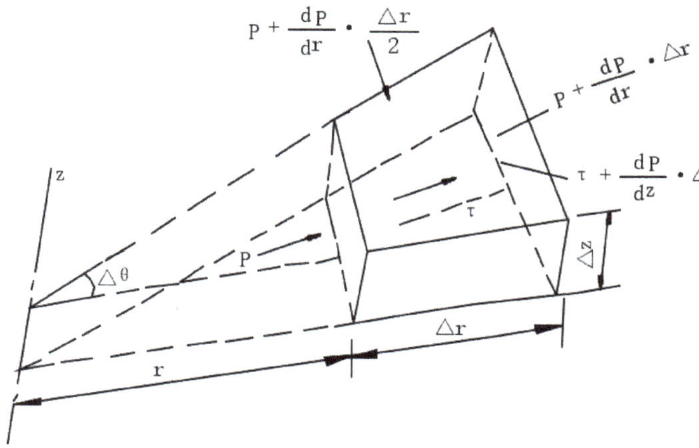

Figure 4. Determination of slurry unit.

High-order trace elements are omitted, it can be expressed by

$$\frac{dp}{dr} - \frac{d\tau}{dz} = 0 \tag{3}$$

According to Newton's law of frictional resistance, the shear force can be expressed by

$$\tau = \eta \frac{du}{dz} \tag{4}$$

Integrating from Equations (3) and (4):

$$\eta \frac{\partial u_r}{\partial z} = \frac{\partial p}{\partial r}(z + C_1) \tag{5}$$

By substituting the boundary condition $\frac{\partial u_r}{\partial z}\Big|_{z=0} = 0$ into the above formula, the following is obtained:

$$\eta \frac{\partial u_r}{\partial z} = \frac{\partial p}{\partial r} z \tag{6}$$

Then we integrate Z and substitute the boundary conditions $\frac{\partial u_r}{\partial z}\Big|_{z=0} = 0$ into the above formula:

$$u_r = \frac{\partial p}{2\eta \partial r}\left(z^2 + C_2\right) \tag{7}$$

Substituting the boundary conditions $u_r|_{z=\pm\frac{h}{2}} = 0$ into the above formula, it is rewritten as [17]:

$$u_r = \frac{\partial p}{2\eta \partial r}\left(z^2 - \frac{h^2}{4}\right) \tag{8}$$

where: u_r is slurry viscosity.

Then the average flow velocity on the cross section of slurry crack is:

$$\overline{u_r} = \frac{1}{h}\int_{-\frac{h}{2}}^{\frac{h}{2}} u_r dz = -\frac{h^2}{12\eta}\frac{\partial p}{\partial r} \tag{9}$$

Let the diffusion radius of slurry at time t be Rt. For any slurry within the range of $R \le R_t$, let the increment of filling range after volume expansion within $\triangle t$ be $\triangle R$. Due to the low of conservation of mass, it can be expressed by [14]:

$$\pi r^2 h\rho(t) = \pi(r + \Delta r)^2 h\rho(t + \Delta t) \tag{10}$$

where: $\rho(t) = Ae^{-Bt} + C$

After sorting:

$$\frac{(r + \Delta r)^2}{r^2} = \frac{Ae^{-Bt} + C}{Ae^{-B(t+\Delta t)} + C} \tag{11}$$

Further, it can be obtained that:

$$1 + \frac{\Delta r}{r} = \sqrt{1 + \frac{A(1 - e^{-B\Delta t})}{Ae^{-B\Delta t} + Ce^{Bt}}} \tag{12}$$

when $\Delta t \to 0$, $\frac{A\left(1-e^{-B\Delta t}\right)}{Ae^{-B\Delta t}+Ce^{Bt}} \to 0$.

Ordering $x = \frac{A\left(1-e^{-B\Delta t}\right)}{Ae^{-B\Delta t}+Ce^{Bt}}$, $\sqrt{1+x}$ Taylor expansion is made for $\sqrt{1+x}$ when $x_0 = 0$, the first two terms are $\sqrt{1+x} \approx 1 + \frac{1}{2}x$

$$\sqrt{1+\frac{A(1-e^{-B\Delta t})}{Ae^{-B\Delta t}+C_1e^{Bt}}} \approx 1+\frac{A(1-e^{-B\Delta t})}{2\left(Ae^{-B\Delta t}+Ce^{Bt}\right)} \tag{13}$$

Substituting Equation (13) into Equation (12), it can be obtained,

$$\frac{\Delta r}{r}=\frac{A(1-e^{-B\Delta t})}{2\left(Ae^{-B\Delta t}+Ce^{Bt}\right)} \tag{14}$$

when $\Delta t \to 0,\ 1-e^{-B\Delta t} \to 0$, and $1-e^{-B\Delta t} \to B\Delta t$, then $\frac{\Delta r}{r}=\frac{AB\Delta t}{2\left(Ae^{-B\Delta t}+Ce^{Bt}\right)}$.

$$1-e^{-B\Delta t} \sim B\Delta t \tag{15}$$

After sorting:

$$\frac{\Delta r}{\Delta t}=\frac{ABr}{2\left(Ae^{-B\Delta t}+Ce^{Bt}\right)} \tag{16}$$

when $\Delta t \to 0,\ \Delta r \to 0$, Equation (16) can be rewritten into differential form:

$$\frac{dr}{dt}=\frac{rAB}{2\left(A+Ce^{Bt}\right)} \tag{17}$$

If the radial average velocity is equal to the change rate of radius with time, we have

$$\overline{u_r}=\frac{dr}{dt}=\frac{rAB}{2\left(A+Ce^{Bt}\right)} \tag{18}$$

Combining Equations (9) and (18), we have

$$\frac{dp}{dr}=-\frac{6r\eta AB}{b^2\left(A+Ce^{Bt}\right)} \tag{19}$$

Substituting Equation (19) into Equation (8), we have

$$u_r=-\frac{3r^2AB}{h^2\left(A+Ce^{Bt}\right)}\left(\frac{z^2}{h^2}-\frac{1}{4}\right) \tag{20}$$

Integrating equation (19) with R, we have

$$p=-\frac{3r^2\eta AB}{h^2\left(A+Ce^{Bt}\right)}+C \tag{21}$$

Assuming that the radius of the initially injected slurry is R_0, since the total mass of the slurry remains unchanged during the slurry diffusion process, it can be expressed at the time t,

$$\pi R_0^2 h\rho_0=\pi R_t^2 h\rho(t) \tag{22}$$

The diffusion radius of slurry at time t can be written as

$$R_t=R_0\sqrt{\frac{A+C}{Ae^{-Bt}+C}} \tag{23}$$

Substituting Equation (23) into Equation (18), it can be obtained that the average velocity at the interface between slurry and air at time t,

$$\overline{u_r} = \frac{dr}{dt} = \frac{R_0AB}{2(A + C_1e^{Bt})}\sqrt{\frac{A + C}{Ae^{-Bt} + C}} \tag{24}$$

when t = 0 s, the slurry pressure is the same as the atmospheric pressure P_0 at the interface between slurry and air. According to Formula (21), we have

$$p_0 = -\frac{3R_0^2\eta AB}{h^2\left(A + Ce^{B\times 0}\right)} + C \tag{25}$$

where, $C = p_0 + \frac{3\eta R_0^2 AB}{h^2(A+C)} = p_0 + \frac{3\eta R_0^2 AB}{h^2(A+C)}$

Substituting it into Equation (21), we have

$$p = -\frac{3r^2\eta AB}{h^2\left(A + Ce^{Bt}\right)} + p_0 + \frac{3\eta R_0^2 AB}{h^2(A + C)} \tag{26}$$

where $0 \leq r \leq R_t$.

3. Numerical Simulation of Polymer Slurry Diffusion in Parallel Plate Cracks

3.1. Governing Equations

In order to verify the correctness of the numerical method, the numerical simulation of polymer slurry parallel plate crack grouting diffusion was carried out. Under the assumption of ideal mixing and rapid reaction, the polymer is regarded as a continuum with the characteristics of compressible Newtonian fluid. The growth of bubbles causes the dependence of polymer density on time. The diffusion and flow of polymers in the crack follow the mass conservation equation and momentum conservation equation. Ignoring the tension of the surface, the diffusion velocity and pressure can be expressed as:

$$\frac{\partial \rho}{\partial t} + \nabla \cdot (\rho U) = 0 \tag{27}$$

$$\frac{\partial \rho U}{\partial t} + \nabla \cdot (\rho UU) = -\nabla p + \eta \nabla^2 U + \frac{1}{3}\eta \nabla(\nabla \cdot U) + \rho gb \tag{28}$$

where ρ is the mixing density; U is the velocity vector; t is the time; P is the pressure; η is shear viscosity; G is the acceleration of gravity.

3.2. Level Set Method

The level set method tracks the interface position by solving the transport equation of the level set function, that is, by tracking the level set function φ to determine the interface of fluid. For convective transport, the velocity vector can be calculated by the Navier Stokes equation.

$$\frac{\partial \varphi}{\partial t} + U \cdot \nabla \varphi = \gamma \nabla \cdot \left(\varepsilon \nabla \varphi - \varphi(1 - \varphi)\frac{\nabla \varphi}{|\nabla \varphi|}\right) \tag{29}$$

where γ and ε is a reinitialization parameter. In this paper ε is taken as the maximum element size in the domain. γ is equal to 1.

Within a given value range of level set function, the fluid characteristics transition smoothly from liquid to gas. The level set function changes between 0 and 1, and it is expressed as 0 or 1 in the two fluids. Specifically, it is 0 in the liquid phase and 1 in the gas phase. On the interface between liquid and gas, the corresponding level set function value $\varphi = 0.5$. The density can be expressed by the level set function

$$\rho = \rho_1 + \varphi(\rho_2 - \rho_1) \tag{30}$$

The dynamic viscosity can be expressed by

$$\eta = \eta_1 + \varphi(\eta_2 - \eta_1) \tag{31}$$

where ρ_1 = density of polymer; ρ_2 = density of air; μ_1 = viscosity of polymer; and μ_2 = viscosity of air.

3.3. Numerical Realization (Boundary and Initial Conditions)

The self-expansion of polymer in a single crack with a radius of 1.3 m and a crack opening of 6mm is analyzed using a parallel plate model, as shown in Figure 5. It is assumed that the static pressure grouting stage has been completed, so some parts of the crack at t = 0 s will initially be filled with unexpanded polymer grouting. Taking the circle with radius r_0 as the initial shape of diffusion and assuming that the static pressure grouting stage has been completed, the grouting amount is the quality required to diffuse to the corresponding radius r_0, and R_0 is calculated by

$$R_0 = \sqrt{\frac{Q}{\pi h \rho_0}} \tag{32}$$

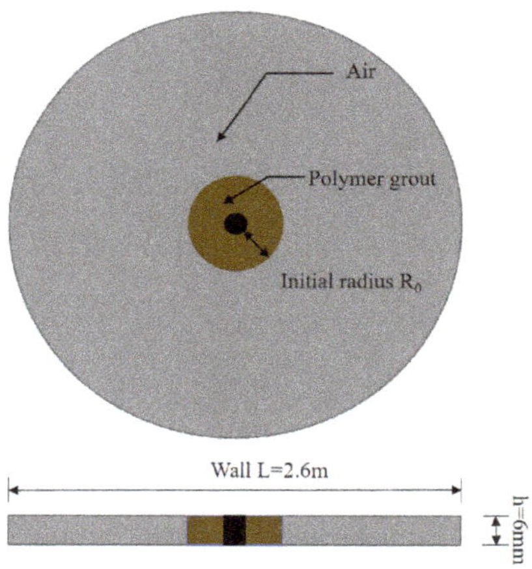

Figure 5. Schematic diagram of numerical model of polymer grouting in crack.

In this paper, 3 types of polymer material are selected, and the grouting amount is 125 g, 250 g and 370 g. The initial density of slurry ρ_0 is 1172.86 kg/m^3. The initial radius R_0 calculated by Formula (32) is 0.11 m, 0.15 m and 0.18 m, respectively.

The UDFs function (user-defined functions) is used to edit the user-defined function of the attribute change of slurry fluid, so as to realize the real-time adjustment of density parameters in the process of slurry expansion.

3.4. Numerical Verification of Diffusion Model

The reliability of the grouting model is verified by comparing the numerical solution with the analytical solution. Figure 6 shows the theoretical curve and simulation curve of diffusion radius with time under different grouting quantities. It can be seen from the figure that the curve obtained by the present model is consistent with that obtained by numerical simulation., and the diffusion range of slurry gradually increases with time and tends to be stable at 30 s. In addition, it is observed that the time-consuming of slurry expansion and the diffusion stage is not related to the grouting quantity. When the grouting quantity is 125 g, 250 g and 170 g, the slurry stops diffusion at about 30 s. At this time, the slurry stops expanding and reaches the maximum diffusion range, which is the effective diffusion area of the slurry. It can also be seen from the figure that the effective diffusion area of the slurry is directly proportional to the grouting volume.

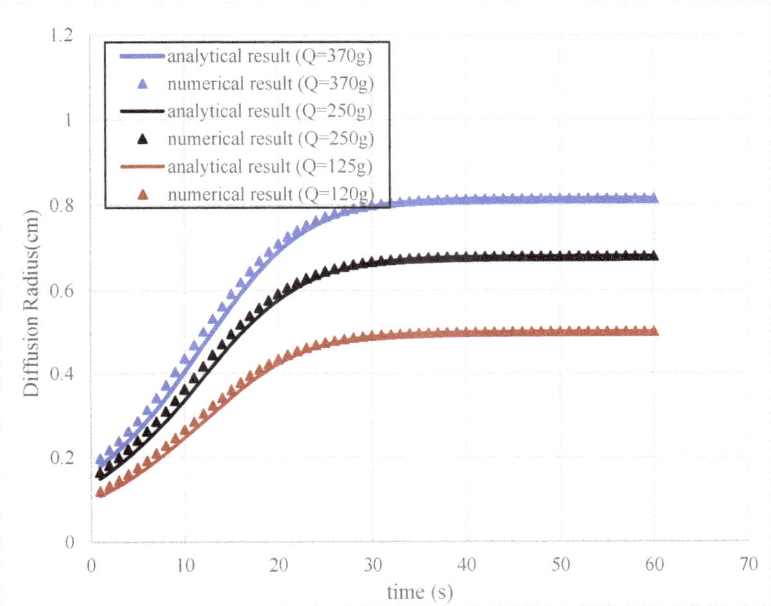

Figure 6. Variation curve of diffusion radius with time under different grouting amounts.

Figure 7 shows the simulation results of the change of slurry volume fraction with time. It can be seen from the figure that the slurry diffusion form of single crack grouting is circular, and gradually diffuses outward in concentric circular form with time, finally reaching the effective diffusion area of slurry, which is consistent with the diffusion form assumed by the present theoretical model.

Table 1 shows the simulation results and analytical results of the pressure distribution with time at 2 cm away from the grouting hole when the grouting amount is 250 g. It can be seen from the Table 1 that the diffusion pressure gradually increases with time and finally tends to be stable. The maximum relative error between the numerical solution and the analytical solution is 0.9%, and the average relative error is 0.28%.

Table 1. Simulation results and analytical results of pressure distribution at different times.

Position r (cm)	Time t (s)	Diffusion Pressure (KPa)		Relative Error
		Numerical Solution	Theoretical Solution	
r = 2	0	0.3554	0.3589	0.9%
	10	0.3591	0.3604	0.4%
	20	0.3634	0.3636	0.1%
	30	0.3652	0.3650	0.1%
	40	0.36581	0.3653	0.1%
	50	0.3662	0.3653	0.2%
	60	0.3662	0.3653	0.2%

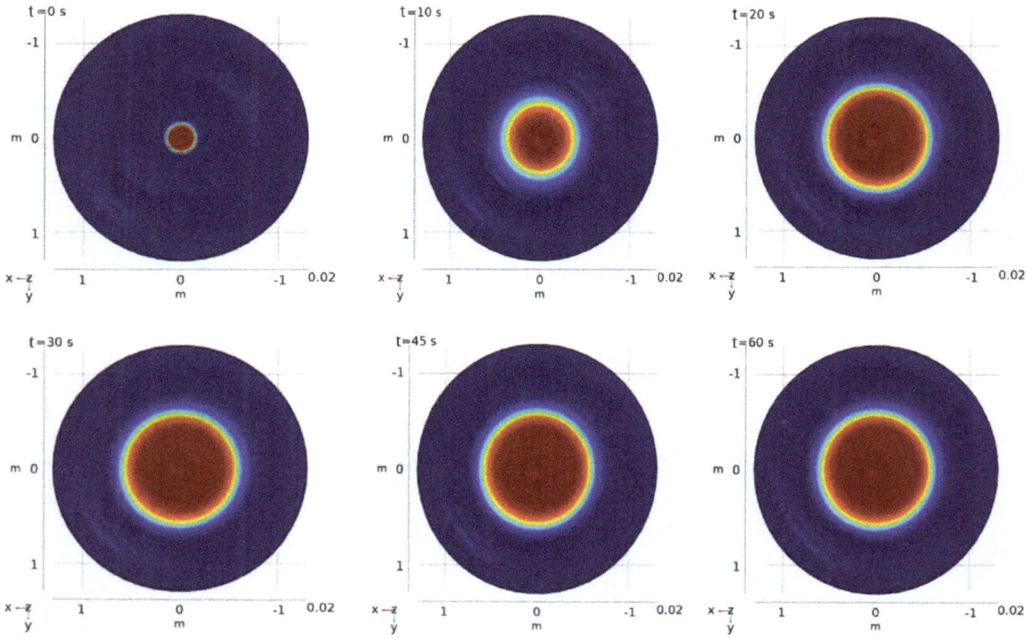

Figure 7. Simulation results of slurry volume fraction varying with time at different times.

Table 2 shows the simulation results and analytical results of pressure radial distribution with distance when the grouting amount is 250 g and the time is 30 s. It can be seen in Table 2 that the diffusion pressure gradually decreases with the increase of radial distance. The maximum relative error between the numerical solution and the analytical solution is 1.4%, and the average relative error is 1%. It can be seen that the theoretical solutions in Tables 1 and 2 are in good agreement with the numerical solutions.

Table 2. Simulation results and analytical results of radial pressure distribution at different positions.

Time t (s)	Position r (cm)	Diffusion Pressure (KPa)		Relative Error
		Numerical Solution	Theoretical Solution	
t = 30	2	0.3630	0.3651	0.5%
	4	0.3592	0.3643	0.5%
	6	0.3579	0.3629	1.3%
	8	0.3560	0.3610	1.0%
	10	0.3535	0.3585	1.3%
	12	0.3506	0.3556	1.0%
	14	0.3470	0.3520	1.4%

4. Analysis of Slurry Diffusion Characteristics

4.1. Analysis of Slurry Pressure Field

Figures 8 and 9 show the temporal and spatial distribution characteristics of the slurry pressure field. Figure 8 shows the variation curve of pressure with diffusion distance at different times. It can be seen from the figure that the farther away from the grouting hole, the smaller the diffusion pressure. Figure 9 shows the variation curve of pressure with time at different distances. It can be seen from the figure that the pressure at different positions increases slowly with time and finally ends up being stable, and the pressure values at different positions are basically the same.

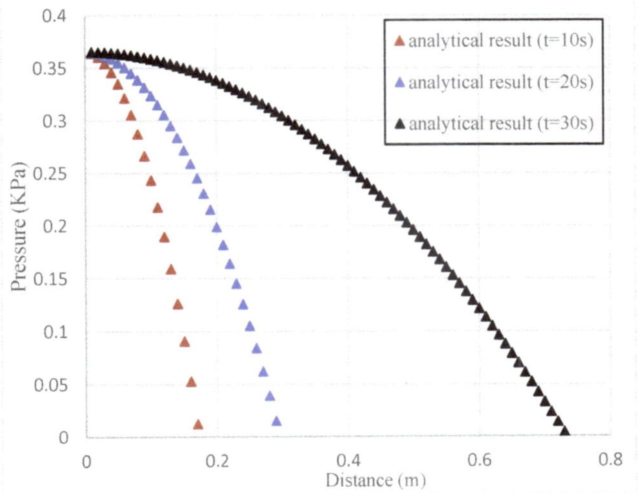

Figure 8. Variation curve of pressure with diffusion distance at different times.

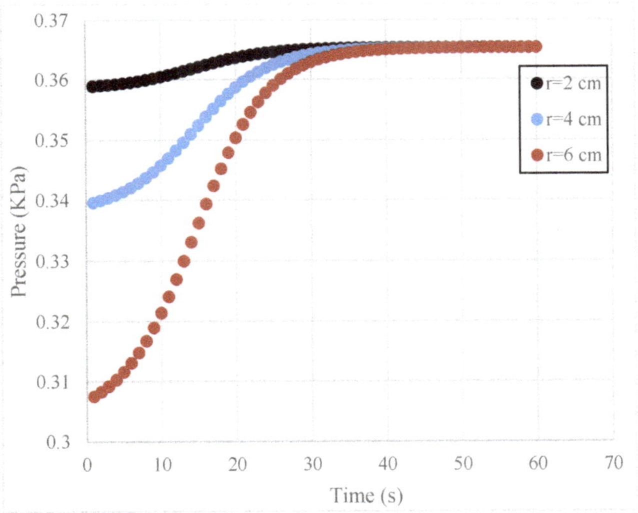

Figure 9. Pressure versus time curves at different distances.

Figure 10 shows the variation curve of pressure with time under different grouting quantities at the radius r = 2 cm. It can be seen from the figure that the pressure increases slowly with time, and finally the pressure tends to be stable. Figure 11 shows the relationship curve between the maximum pressure and the grouting amount. It can be seen from Figures 10 and 11 that the diffusion pressure is directly proportional to the grouting amount. Figure 12 shows the change of pressure with time under different crack opening, and Figure 13 shows the relationship curve between the maximum pressure and crack opening. It can be seen from Figures 12 and 13 that the diffusion pressure is inversely proportional to the grouting volume, and the diffusion pressure is very sensitive to the change in the crack opening. When the crack decreases from 4 mm to 2 mm, the pressure increases rapidly, the crack opening decreases from 8 mm to 2 mm, and the maximum pressure value increases by a factor of 20. Figure 14 shows the change curve of pressure with time under different expansion ratios, and Figure 15 shows the relationship curve between the maximum pressure and expansion ratio. It can be seen from Figures 14 and 15 that the diffusion pressure is directly proportional to the grouting amount. In general, the grouting volume, the crack opening and the expansion ratio are three key factors affecting the slurry pressure field.

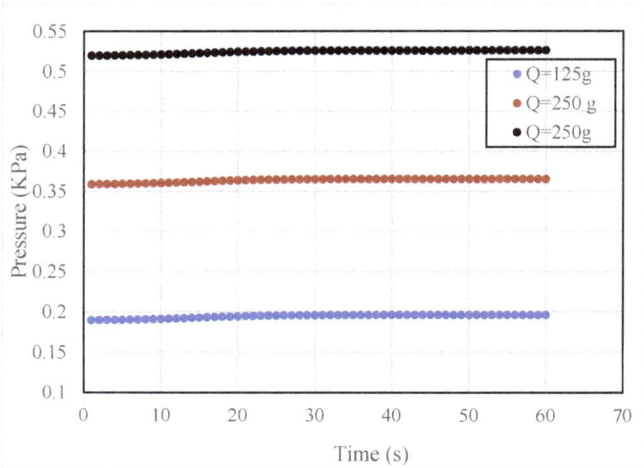

Figure 10. Variation curve of pressure with time under different grouting amount.

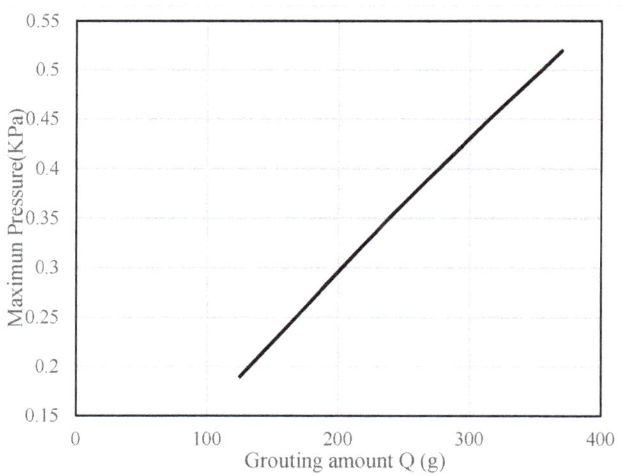

Figure 11. Relation curve between maximum pressure and grouting volume.

4.2. *Analysis of Slurry Diffusion Flow Field*

Table 3 shows the flow velocity values at different positions on the fracture section when t = 20 s, which can be described as an intuitive three-dimensional distribution of flow velocity, as shown in Figure 16. It can be seen from Table 3 and Figure 16 that the farther the fracture section is from the grouting hole and the fracture wall, the greater the diffusion pressure is. Figures 17 and 18 show the velocity on temporal and spatial distribution characteristics of slurry. Figure 17 shows the velocity variation of slurry with radial distance at different times. It can be seen from the Figure 17 that the slurry velocity increases along the radial direction, and the velocity reaches the maximum at the interface between slurry and air. Figure 18 shows the velocity variation curve of slurry along the z-axis direction at different times. It can be seen from Figure 18 that the slurry velocity presents a parabolic distribution in the z-axis direction, and the velocity reaches the maximum at the midpoint of the fracture (z = 0) and reaches the minimum at the fracture wall, and the velocity is zero. It can also be seen from Figures 17 and 18 that the slurry flow rate gradually decreases with time and finally becomes zero. Figures 19 and 20 are the time-varying curves of flow velocity at the interface between slurry and air under different grouting amounts and expansion ratios. It can be seen from Figures 19 and 20 that the greater the grouting amount and expansion ratio, the greater the interface flow velocity; the grouting amount and expansion ratio are the key factors affecting the flow velocity.

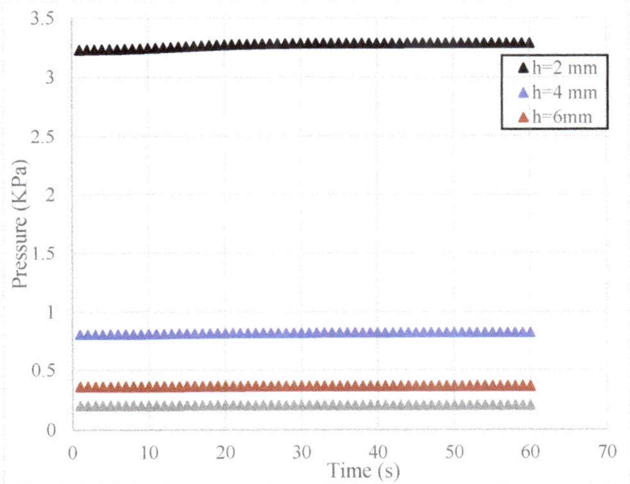

Figure 12. Variation curve of pressure with time under different fracture opening.

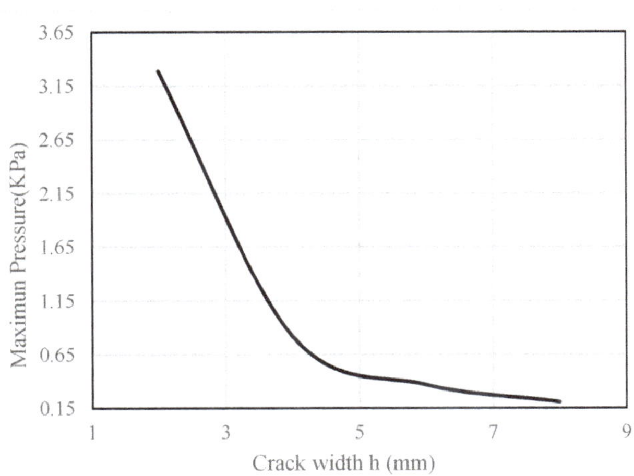

Figure 13. Relation curve between maximum pressure and fracture opening.

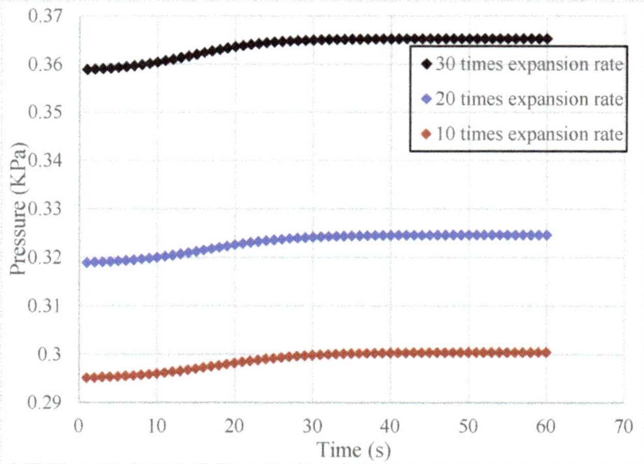

Figure 14. Pressure versus time curves at different expansion ratios.

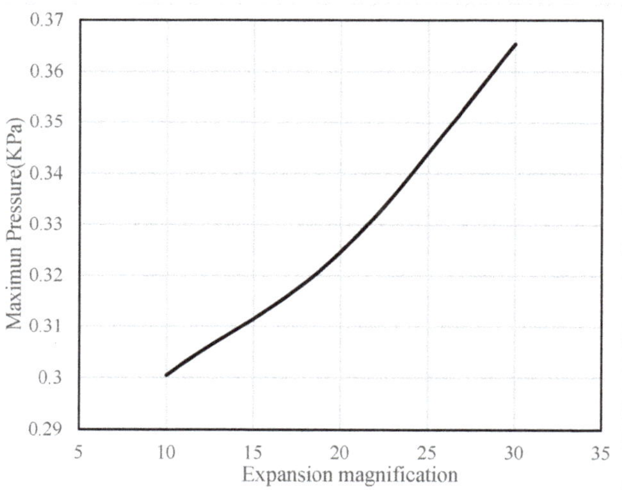

Figure 15. Relation curve between maximum pressure and expansion ratio.

Table 3. Velocity values at different positions on the fracture section.

Position(x)/m	z/m						
	−0.0030	**−0.0020**	**−0.0010**	**0.0000**	**0.0010**	**0.0020**	**0.0030**
0.1653	0	0.0158	0.0253	0.0285	0.0253	0.0158	0
0.1820	0	0.0192	0.0307	0.0345	0.0307	0.0192	0
0.2000	0	0.0232	0.0370	0.0417	0.0370	0.0232	0
0.2195	0	0.0279	0.0446	0.0502	0.0446	0.0279	0
0.2404	0	0.0334	0.0535	0.0602	0.0535	0.0334	0
0.2627	0	0.0399	0.0639	0.0719	0.0639	0.0399	0
0.2862	0	0.0474	0.0759	0.0854	0.0759	0.0474	0
0.3110	0	0.0560	0.0896	0.1008	0.0896	0.0560	0
0.3367	0	0.0656	0.1050	0.1181	0.1050	0.0656	0
0.3631	0	0.0763	0.1221	0.1373	0.1221	0.0763	0
0.3899	0	0.0880	0.1408	0.1584	0.1408	0.0880	0

Table 3. *Cont.*

Position(x)/m	z/m						
	−0.0030	−0.0020	−0.0010	0.0000	0.0010	0.0020	0.0030
0.4167	0	0.1005	0.1608	0.1809	0.1608	0.1005	0
0.4432	0	0.1137	0.1819	0.2046	0.1819	0.1137	0
0.4689	0	0.1273	0.2037	0.2291	0.2037	0.1273	0
0.4936	0	0.1410	0.2256	0.2538	0.2256	0.1410	0
0.5168	0	0.1546	0.2474	0.2783	0.2474	0.1546	0
0.5384	0	0.1678	0.2684	0.3020	0.2684	0.1678	0
0.5581	0	0.1803	0.2884	0.3245	0.2884	0.1803	0
0.5758	0	0.1919	0.3071	0.3455	0.3071	0.1919	0
0.5916	0	0.2026	0.3241	0.3646	0.3241	0.2026	0

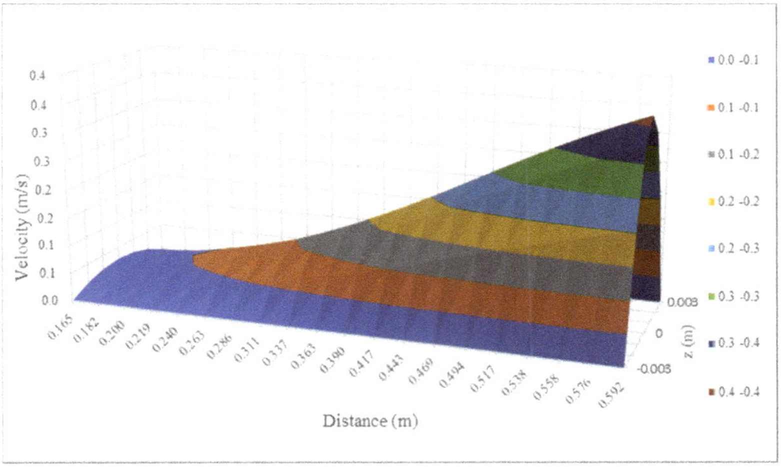

Figure 16. Three -dimensional velocity distribution.

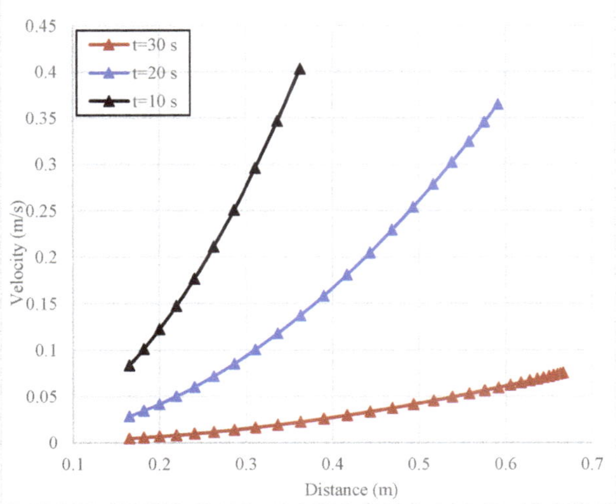

Figure 17. Variation curve of slurry velocity with radial distance at different time.

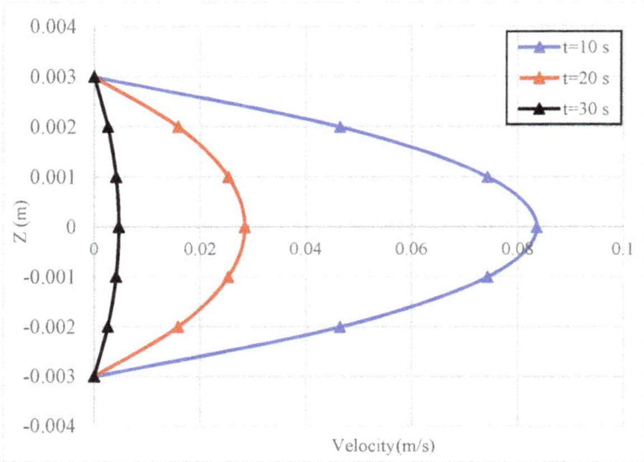

Figure 18. Variation curve of slurry velocity along z-axis at different times.

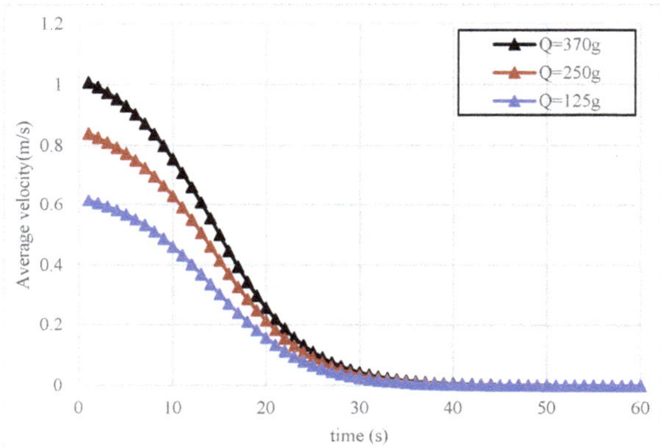

Figure 19. Variation curve of velocity at the interface between slurry and air with time under different grouting amount.

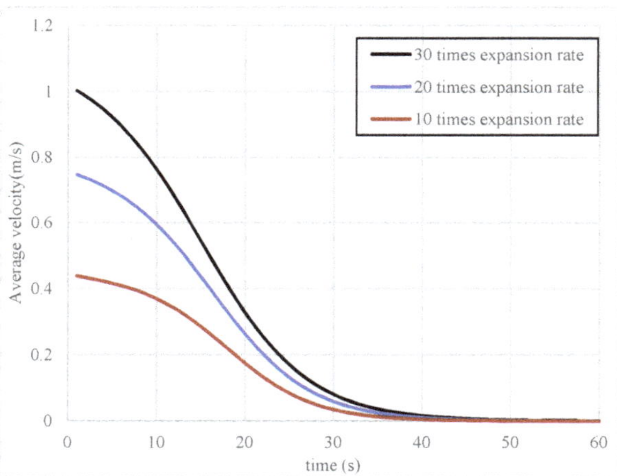

Figure 20. Variation curve of velocity at the interface between slurry and air with time under different expansion ratios.

5. Conclusions

(1) In this paper, the effects of grouting volume, crack opening, radial distance, time, expansion ratio and other factors were fully considered. The radial diffusion model of polymer grouting in a single crack was theoretically deduced, and the theoretical model was verified by the numerical simulation of slurry in parallel plate cracks The variation laws of slurry diffusion radius and pressure at different times are basically consistent with the analytical solution,

(2) The amount of grouting, the expansion ratio and the crack opening have a great influence on the diffusion law of polymer crack grouting. The diffusion pressure is directly proportional to the grouting amount and expansion ratio, and inversely proportional to the crack opening. In addition, the greater the grouting amount and expansion ratio, the greater the interfacial velocity.

(3) The diffusion pressure is directly proportional to the grouting volume and expansion multiple, and inversely proportional to the crack width. In addition, diffusion pressure decreases with the increase of diffusion distance, and the pressure at the corresponding distance increases slowly with time, eventually reaching stability.

Author Contributions: Conceptualization, J.L. and X.D.; methodology, J.L.; software, J.L.; validation, J.L., X.D. and S.M.; formal analysis, J.L.; investigation, J.L.; resources, J.L.; data curation, J.L.; writing—original draft preparation, J.L.; writing—review and editing, J.L.; visualization, J.L.; supervision, J.L.; project administration, J.L.; funding acquisition, X.D. All authors have read and agreed to the published version of the manuscript.

Funding: This work was supported by the National Natural Science Foundation of China (No. 52008379, 51678536, 41404096), the Postdoctoral research projects in Henan Province (201901012), Supported by the Systematic Project of Guangxi Key Laboratory of Disaster Prevention and Engineering Safety (2020ZDK001) and the Open Research Fund of Guangxi Key Laboratory of Water Engineering Materials and Structures, Guangxi institute of water resources research(GXHRI-WEMS-2020-12); the National Key Research and Development Program of China (No.2017YFC1501204), the National Basic Research Program of China (No.2015CB060200), the Program for Science and Technology Innovation Talents in Universities of Henan Province (Grant No. 19HASTIT043), the Outstanding Young Talent Research Fund of Zhengzhou University (1621323001) and the Program for Innovative Research Team (in Science and Technology) in University of Henan Province (18IRTSTHN007).

Institutional Review Board Statement: Not applicable.

Informed Consent Statement: Not applicable.

Data Availability Statement: The study did not report data.

Acknowledgments: The first author would like to thank the Chinese Scholarship Council for financial support toward his joint at the University of Newcastle, Australia. We would also like to acknowledge the reviewers for their invaluable comments.

Conflicts of Interest: The authors declare that they have no known competing financial interests or personal relationships that could have appeared to influence the work reported in this paper.

References

1. Du. X.M.; Fang, H.Y.; Wang, S.Y.; Xue, B.H. Experimental and practical investigation of the sealing efficiency of cement grouting in tortuous fractures with flowing water. *Tunn. Undergr. Space Technol.* **2021**, *108*, 103693. [CrossRef]
2. Algin, H.M. Optimised design of jet-grouted raft using response surface method. *Comput. Geotech.* **2016**, *74*, 56–73 [CrossRef]
3. Maghous, S.; Saada, Z.; Dormieux, L.; Canou, J.; Dupla, J.C. A model for in situ grouting with account for particle filtration. *Comput. Geotech.* **2006**, *34*, 164–174. [CrossRef]
4. Wang, Z.F.; Shen, S.L.; Modoni, G. Enhancing discharge of spoil to mitigate disturbance induced by horizontal jet grouting in clayey soil: Theoretical model and application. *Comput. Geotech.* **2019**, *111*, 222–228. [CrossRef]
5. Wang, S.F.; Tang, Y.; LI, X.B.; Du, K. Analyses and predictions of rock cuttabilities under different confining stresses and rock properties based on rock indentation tests by conical pick. *Trans. Nonferrous Met. Soc. China* **2021**, *31*, 1766–1783. [CrossRef]
6. Wang, S.F.; Tang, Y.; Wang, S.Y. Influence of brittleness and confining stress on rockcuttability based on rock indentation tests. *J. Cent. South Univ.* **2021**, *28*, 2786–2800. [CrossRef]

7. Editorial Committee of Foundation Treatment Manual. *Foundation Treatment Manual*; China Construction Industry Press: Beijing, China, 1988; pp. 283–287, 331–371.
8. Shi, M.S. Research on Polymer Grouting Material Properties and Directional Fracturing Grouting Mechanism for Dykes and Dams. Ph.D. Thesis, Dalian University of Technology, Dalian, China, 2011.
9. El Tani, M. Grouting rock fractures with cement grout. *Rock Mech. Rock Eng.* **2012**, *45*, 547–561. [CrossRef]
10. El Tani, M.; Stille, H. Grout spread and injection period of silica solution and cement mix in rock fractures. *Rock Mech. Rock Eng.* **2017**, *50*, 2365–2380. [CrossRef]
11. Yang, M.J.; Yue, Z.Q.; Lee, P.K.; Su, B.; Tham, L.G. Prediction of grout penetration in fractured rocks by numerical simulation. *Can. Geotech. J.* **2002**, *39*, 1384–1394. [CrossRef]
12. Zhang, Q.S.; Zhang, L.Z.; Liu, R.T. Grouting mechanism of quick setting slurry in rock fissure with consideration of viscosity variation with space. *Tunn. Undergr. Space Technol.* **2017**, *70*, 262–273. [CrossRef]
13. Zhang, L.H. Seepage Mechanism and Mechanics of Geotechnical Grouting. Ph.D. Thesis, Northern Jiaotong University, Beijing, China, 1996.
14. Li, X.L.; Jin, D.; Wang, F.M.; Zhong, Y.H.; Zhang, B. Diffusion model of an ideal expansible grout in single fracture. *Chin. J. Rock Mech. Eng.* **2018**, *37*, 37–1207.
15. Funehag, J.; Gustafson, G. Design of grouting with silica sol in hard rock-new methods for calculation of penetration length. part I. *Tunn. Undergr. Space Technol.* **2008**, *23*, 1–8. [CrossRef]
16. Li, S.F.; Li, S.C.; Liu, R.T. Grouting reinforcement experiment for water-rich broken rock mass. *Chin. J. Rock Mech. Eng.* **2017**, *36*, 198–207.
17. Ruan, W.J. Spreading model of grouting in rock mass fissures based on time-dependent behavior of viscosity of cement-based grouts. *Chin. J. Rock Mech. Eng.* **2005**, *24*, 2709–2714.
18. Amadei, B.; Savage, W.Z. An analytical solution for transient flow of Bingham viscoplastic materials in rock fractures. *Int. J. Rock Mech. Min. Sci.* **2001**, *38*, 285–296. [CrossRef]
19. Luo, P.P.; Li, Z.P.; Fan, B. Theoretical study on flow model for tilted single fracture Binghamian grout. *J. Shandong Univ. Sci. Technol.* **2010**, *29*, 43–47.
20. Zhan, K.Y.; Sui, W.H.; Gao, Y. A model for grouting into single fracture with flowing water. *Rock Soil Mech.* **2011**, *32*, 1659–1663.
21. Tian, M.X. A model for grouting into fracture with flowing water based on time-dependent behavior of viscosity. *J. Chongqing Jiaotong Univ. Nat. Sci.* **2011**, *30*, 536–537.
22. Zhang, Q.S.; Zhang, L.Z.; Zhang, X. Grouting diffusion in a horizontal crack considering temporal and spatial variation of viscosity. *Chin. J. Rock Mech. Eng.* **2015**, *34*, 1198–1210.
23. Guo, C.C.; Wang, F.M. Mechanism Study on the Construction of Ultra-Thin Antiseepage Wall by Polymer Injection. *J. Mater. Civ. Eng.* **2012**, *24*, 1183–1192.
24. Zheng, C.C. Simulation study on Grouting of Fractured Rock Mass. Ph.D. Thesis, Central South University of Technology, Changsha, China, 1999.
25. Ruan, W.J. Study on Basic Properties of Grout and Grouting Diffusion in Rock Cracks. Ph.D. Thesis, Jilin University, Changchun, China, 2003.
26. Zheng, Y.H. Research on Grouting and Grouting Control Method for Fractured Rock Mass. Ph.D. Thesis, Jilin University, Changchun, China, 2005.

mathematics

Article

Variable Weight Matter–Element Extension Model for the Stability Classification of Slope Rock Mass

Shan Yang *, Zitong Xu and Kaijun Su

School of Resources and Safety Engineering, Central South University, Changsha 410083, China; xuzitong@csu.edu.cn (Z.X.); 205512132@csu.edu.cn (K.S.)
* Correspondence: yangshan@csu.edu.cn

Abstract: The slope stability in an open-pit mine is closely related to the production safety and economic benefit of the mine. As a result of the increase in the number and scale of mine slopes, slope instability is frequently encountered in mines. Therefore, it is of scientific and social significance to strengthen the study of the stability of the slope rock mass. To accurately classify the stability of the slope rock mass in an open-pit mine, a new stability evaluation model of the slope rock mass was established based on variable weight and matter–element extension theory. First, based on the main evaluation indexes of geology, the environment, and engineering, the stability evaluation index system of the slope rock mass was constructed using the corresponding classification criteria of the evaluation index. Second, the constant weight of the evaluation index value was calculated using extremum entropy theory, and variable weight theory was used to optimize the constant weight to obtain the variable weight of the evaluation index value. Based on matter–element extension theory, the comprehensive correlation between the upper and lower limit indexes in the classification criteria and each classification was calculated, in addition to the comprehensive correlation between the rock mass indexes and the stability grade of each slope. Finally, the grade variable method was used to calculate the grade variable interval corresponding to the classification criteria of the evaluation index and the grade variable value of each slope rock mass, so as to determine the stability grade of the slope rock. The comparison results showed that the classification results of the proposed model are in line with engineering practice, and more accurate than those of the hierarchical-extension model and the multi-level unascertained measure-set pair analysis model.

Keywords: mine slope; stability classification of rock mass; extremum entropy; variable weight theory; matter–element extension; grade variable

Citation: Yang, S.; Xu, Z.; Su, K. Variable Weight Matter–Element Extension Model for the Stability Classification of Slope Rock Mass. *Mathematics* **2021**, *9*, 2807. https://doi.org/10.3390/math9212807

Academic Editor: Mario Versaci

Received: 17 September 2021
Accepted: 26 October 2021
Published: 4 November 2021

1. Introduction

In the mining of an open pit, the stability of the slope rock mass has a significant impact on mining design, intensity, and safety. As a result of the gradual depletion of surface mineral resources, underground mining has been employed instead of open-pit mining in large mines [1,2]. Therefore, it is necessary to study the stability of open-pit slopes. The slope rock mass of open-pit mines is significantly affected by many factors, such as weathering, in situ stress, groundwater, and blasting vibration. There is an urgent need for the accurate determination of the stability grade of the slope rock mass. As a result of the progress of scientific theory and method, the evaluation of the stability of the slope rock mass in open-pit mines has developed from empirical judgment, theoretical analysis, and qualitative evaluation of a single index, to a comprehensive evaluation based on an index system [3–5]. In general, three main methods are used to evaluate the stability of the slope rock mass in open-pit mines: (1) solid modeling and numerical simulation methods. By analyzing the influence of the distribution of structural planes (such as joints and fissures) in the slope rock mass on rock anisotropy, Shi Wenhao, Yang Tianhong et al. [6] evaluated the stability of the slope rock mass in an open-pit using a 3D solid modeling method. To determine the stability of the open-pit slope, Zhao Haijun, Ma Fengshan et al. [7] analyzed the surrounding rock of the underground stope, the mechanical environment of the slope, and the movement and deformation of the rock mass using a GPS monitoring network and a 3D numerical simulation. (2) Geological survey and refined analysis. Through a detailed geological survey of the quality components and size of the slope, Du Shigui [8] conducted a refined statistical analysis of the survey results, and finally determined the stability of the slope rock mass based on the cybernetics of the rock mass structure. Neil Bara, Michael Kostadinovskia et al. [9] described the rapid and robust process utilized at BHP Limited for appraising a slope failure at an iron ore mine site in the Pilbara region of Western Australia, using a combination of UAV photogrammetry and 3D slope stability models in less than a shift (i.e., less than 12 h). Fehmi Arikan, Fatih Yoleri et al. [10] conducted a geotechnical assessment of slope stability and collected geological data from sources such as geologic reconnaissance, core logging, topographical surveys, and geomechanical laboratory testing data; in addition, kinematical and two-dimensional limit equilibrium back analyses were performed. (3) Qualitative evaluation and analysis based on the evaluation index system. Wang Xinmin, Kangqian et al. [11] evaluated the stability of the open-pit slope rock mass via the construction of an analytic hierarchy process-extension model. Zhang Xu, Zhou Shaowu et al. [12] evaluated the slope and excavation stability of the open-pit mines by building an entropy weight-set evaluation model. Huangdan and Shi Xiuzhi et al. [13] evaluated the stability of the slope rock mass by constructing a multi-level unascertained measure-set pair analysis model. Liu Leilei, Zhang Shaohe et al. [14] evaluated the stability of the slope rock mass by constructing an AHP-ideal point model. Bar N and Barton N. [15] discussed the applicability of the Q-slope method to slopes ranging from less than 5 m to more than 250 m in height, in both civil and mining engineering projects. Pastor, J.L., Riquelme, A.J., et al. [16] used SMRTool, an open-source software package, to derive a complete and detailed definition of the angular relationship between discontinuity and slope, and clarified the evaluation of SMR parameters.

Among the three methods mentioned above, the first is highly theoretical and accurate for the stability classification of slope rock masses with less complex environmental conditions. Although the second method yields accurate evaluation results, significant amounts of manpower and material resources are required in the field survey. Hence, it is less used in the classification of general slope rock stability. The third method, which is based on a mathematical model and an evaluation index system, has good generality and can be used in the stability classifications of various rock masses. However, the construction of the mathematical model and the selection of an index system need to be further improved. Based on the existing research, an evaluation index system for the stability classification of the slope rock mass in open-pit mines was established in this study. First, the index weight was calculated using variable weight theory to address the unreasonable index weighting caused by ignoring the index change in single- or multi-method weighting. Second, based on the matter–element extension model and the grade variable method, the stability of the slope rock mass in the open-pit mine was evaluated, so as to improve the accuracy of the matter–element extension model in the stability classification of the slope rock mass. Finally, a new classification model of slope rock mass stability was constructed.

2. Basic Principles of the Matter–Element Extension Model

The matter–element extension model is a mathematical model based on matter–element theory and extension mathematics. In this model, the matter element is taken as the basic element to describe objects. The matter element is expressed as the ordered triple $\boldsymbol{R} = (S,y,v)$, where S represents the objects; y represents the feature of objects; and v represents quantities of S about y. S, y, and v are called the three elements of the matter element [17]. Based on extension set theory and decision-making theory, matter–element transformation and the correlation function are used as tools. The extension engineering method can be used to solve the application problems in the fields of management, control, and engineering [18].

2.1. Matter Elements, Classical Domains, and Nodal Domains

If the evaluation object S contains a feature y expressed by v, then S, y, and v constitute the ordered triple $\boldsymbol{R} = (S,y,v)$, and are called matter elements [19]. If the evaluation object S has n features, the corresponding values of features $y_1, y_2 \ldots y_n$ are $v_1, v_2 \ldots v_n$, respectively. Matter elements describing the evaluation object S are recorded as $\boldsymbol{R}$:

$$\boldsymbol{R} = (S, y_i, v_i) = \begin{bmatrix} S & y_1 & v_1 \\ & y_2 & v_2 \\ & \cdots & \cdots \\ & y_n & v_n \end{bmatrix} \tag{1}$$

The classical domain of evaluation object S about grade j is recorded as $\boldsymbol{R}_j$:

$$\boldsymbol{R}_j = (S_j, y_i, V_{ji}) = \begin{bmatrix} S_j & y_1 & [a_{j1}, b_{j1}] \\ & y_2 & [a_{j2}, b_{j2}] \\ & \cdots & \cdots \\ & y_n & [a_{jn}, b_{jn}] \end{bmatrix} \tag{2}$$

The feature section of the evaluation object S is recorded as $\boldsymbol{R}_0$:

$$\boldsymbol{R}_0 = (S_0, y_i, V_{0i}) = \begin{bmatrix} S_j & y_1 & [a_{01}, b_{01}] \\ & y_2 & [a_{02}, b_{02}] \\ & \cdots & \cdots \\ & y_n & [a_{0n}, b_{0n}] \end{bmatrix} \tag{3}$$

where S is the evaluation object; v_i is the eigenvalue of the evaluation object; S_j is the evaluation object corresponding to the grade j, j = 1, 2, … , m; y_i is the eigenvalue of the evaluation object i, i = 1, 2, … , n; V_{ji} is the eigenvalue range of S_j corresponding to y_i, $V_{ij} = [a_{ij}, b_{ij}]$; S_0 is the evaluation object corresponding to all levels; and V_{0i} is the eigenvalue range of S_0 corresponding to y_i, $V_{0i} = [a_{0i}, b_{0i}]$.

2.2. Extension Correlation Functions

According to the extension set theory and the definition of extension distance [20], the extension distance equation of the feature y_i of the evaluation object S with respect to the stability grade j is expressed as follows:

$$\begin{cases} \rho(v_i, V_{ji}) = \left| v_i - \frac{a_{ji}+b_{ji}}{2} \right| - \frac{b_{ji}-a_{ji}}{2} \\ \rho(v_i, V_{0i}) = \left| v_i - \frac{a_{0i}+b_{0i}}{2} \right| - \frac{b_{0i}-a_{0i}}{2} \end{cases} \tag{4}$$

where v_i is the feature i of the object to be evaluated, i = 1, 2, … , n; V_{ji} is the value range of S_j for the feature y_i, $V_{ij} = [a_{ij}, b_{ij}]$; V_{0i} is the value range of S_0 for the feature y_i, $V_{0i} = [a_{0i}, b_{0i}]$; $\rho(v_i, V_{ji})$ is the extension distance of the feature y_i for the j-level classical domain; and $\rho(v_i, V_{ji})$ is the extension distance of the feature y_i for node region; i = 1, 2, n; n = 1, 2, … m.

If $v_i \in V_{0i}$, the specified correlation function [21] is expressed as follows:

$$S_j(v_i) = \begin{cases} \frac{\rho(v_i, V_{ji})}{\rho(v_i, V_{0i}) - \rho(v_i, V_{ji})} & \rho(v_i, V_{0i}) - \rho(v_i, V_{ji}) \neq 0, v_i \in V_{0i} \\ -\rho(v_i, V_{ji}) + 1 & \rho(v_i, V_{0i}) - \rho(v_i, V_{ji}) = 0, v_i \in V_{ij} \\ 0 & \rho(v_i, V_{0i}) - \rho(v_i, V_{ji}) = 0, v_i \notin V_{ij}, v_i \in V_{0i} \end{cases} \tag{5}$$

where $S_j\,(v_i)$ is the single index correlation degree of the feature y_i of the evaluation object S with respect to the grade j, i = 1, 2, … n; j = 1, 2, … , m.

Combining with the feature weight vector W of the evaluation object, the calculation expression of the comprehensive correlation [22] of the evaluation object S with respect to the grade j is as follows:

$$S_j(V_j) = \sum_{i=1}^{n} w_i S_j(v_i) \tag{6}$$

where w_i is the weight coefficient of the evaluation object feature y_i; $S_j\,(V_j)$ is the comprehensive correlation between the evaluation object S and the grade j; $S_j\,(v_i)$ is the single index correlation of the evaluation object S feature y_i with respect to the grade j; $i = 1, 2, \ldots, n$.

2.3. Level Variable Method

According to the level variable method [23–25], the calculation expression of the level variable k is obtained as follows:

$$P_j(V_j) = \frac{S_j(V_j) - B_s}{A_s - B_s} \tag{7}$$

$$k = \frac{\sum_{j=1}^{m} j P_j(V_j)}{\sum_{j=1}^{m} P_j(V_j)} \tag{8}$$

where $S_j\,(V_j)$ is the comprehensive correlation of the evaluation object S and the grade j; $P_j\,(V_j)$ is the standardized value of the comprehensive correlation; k is the eigenvalue of the stability grade variable of the evaluation object S; $B_s = \min\{S_j\,(V_j)\}$, $A_s = \max\{S_j\,(V_j)\}$; and $j = 1, 2, \ldots, m$.

3. Extreme Entropy Weighting and Variable Weighting Theory

3.1. Principle of Extreme Entropy Weighting

Extremum entropy method can be performed by the following two steps: (1) process the eigenvalues of the evaluation object without dimension to obtain the identical type of eigenvalue, and (2) determine the feature weight of the evaluation object. The previous research has proved that the extremum entropy method has the best performance compared with other entropy methods, and is also called the optimal entropy method [26]. In this study, extreme entropy is used to determine the feature weights of evaluation objects via the following steps.

1. The eigenvalue x_{ij} of the evaluation object X_i is obtained by extremum method and transformed into a dimensionless value v_{ij}. If the feature x_{ij} of the evaluation object x_i belongs to the positive type, it is processed by Equation (9). If the feature x_{ij} of the evaluation object x_i belongs to the reverse type, it is processed by Equation (10) [27]:

$$v_{ij} = \begin{cases} 1 & x_{ij} \geq m_j \\ \frac{x_{ij}-m_j}{M_j-m_j} & x_{ij} \in (m_j, M_j) \\ 0 & x_{ij} \leq M_j \end{cases} \tag{9}$$

$$v_{ij} = \begin{cases} 1 & x_{ij} \leq m_j \\ \frac{M_j-x_{ij}}{M_j-m_j} & x_{ij} \in (m_j, M_j) \\ 0 & x_{ij} \geq M_j \end{cases} \tag{10}$$

where x_{ij} is the eigenvalue of X_i, M_j is the maximum value of x_{ij}; m_j is the minimum value of x_{ij}; v_{ij} is the dimensionless value of x_{ij}, $i = 1, 2, \ldots, n$; $j = 1, 2, \ldots, m$.

2. The dimensionless v_{ij} of the evaluation object X_i is normalized:

$$r_{ij} = \frac{v_{ij}}{\sum\limits_{i=1}^{n} v_{ij}} \tag{11}$$

where v_{ij} is the dimensionless value of the feature x_{ij}; r_{ij} is the normalized value of the feature x_{ij}; and n is the total number of objects to be evaluated.

3. The feature information entropy of evaluation object X_i is calculated as follows:

$$\begin{cases} p_{ij} = 0 & r_{\cdot j} = 0 \\ p_{ij} = r_{ij}\ln(r_{ij}) & r_{\cdot j} \neq 0 \\ e_j = -\frac{1}{\ln(n)}\sum\limits_{i=1}^{n} p_{ij} \end{cases} \tag{12}$$

where e_j is the feature information entropy of the object X_i to be evaluated; r_{ij} is the normalized value of the feature x_{ij}, $i = 1, 2, \ldots, n$; $j = 1, 2, \ldots, m$; and $ej \in [0,1]$.

4. The feature weight of evaluation object X_i is calculated as follows:

$$w_j = \frac{1 - e_j}{m - \sum\limits_{j=1}^{m} e_j} \tag{13}$$

where w_{0j} is the weight of the feature x_{ij}; e_j is the feature information entropy of the feature x_{ij}; and m is the total number of features to be evaluated.

The constant weight vector is $W_0 = (w_{01}, w_{02}, \ldots, w_{0m})$.

3.2. Basic Theory of Variable Weight

As a result of the fixed weight value in the constant weight empowerment, the relative importance of each feature of the evaluation object is only reflected, while the impact of the eigenvalue change of the evaluation object on the feature weight is ignored [28]. For this reason, Wang Peizhuang, Li Hongxing et al. proposed and improved the variable weight theory [29]. According to variable weight theory, the constant weight of the evaluation object can be optimized by constructing the variable weight vector to obtain the variable weight of the evaluation object. According to the axiomatic system of the variable weight vector, the variable weight vector is defined as follows [30]:

In the following mapping **P**: [0,1]m → [0,1]m; X → P(X) = (P_1(X); and P_2(X), … , P_m(X)); then **P** is called eigenvariable weight vector. If **P** is satisfied by (1) punitiality, $x_i \geq x_j \Rightarrow P_i(X) \leq P_j(X)$; (2) continuity, $P_i(X)$ is continuous for each variable ($i = 1, 2, \ldots, n$) for any constant weight vector; (3) for any constant weight vector $W_0 = (w_{01}, w_{02}, \ldots, w_{0m})$; then Equation (14) is satisfied by ① polarity, ($w_{01} + w_{02} + \ldots + w_{0m} = 1$); ② continuity, w_{0j} ($x_1, x_2, \ldots, x_m$) is continuous with respect to each variable x_j ($j = 1, 2, \ldots, m$); ③ punitiveness, w_{0j} ($x_1, x_2, \ldots, x_m$) is reduced with respect to the variable x_j ($j = 1, 2, \ldots, m$). Then, **P** is called penalty contingency vector.

$$\mathbf{W}(\mathbf{X}) = \frac{(w_{01}p_1(X), w_{02}p_2(X), \cdots, w_{0m}p_m(X))}{\sum\limits_{j=1}^{m}(w_{0j}P_j(X))} = \frac{\mathbf{W}_0\mathbf{P}(\mathbf{X})}{\sum\limits_{j=1}^{m}(w_{0j}P_j(X))} \tag{14}$$

where $\mathbf{W}_0\mathbf{P}(\mathbf{X}) = (w_{01}p_1(X), w_{02}p_2(X), \cdots, w_{0m}p_m(X))$ is the Hardarmard product [31], and $\mathbf{W}(\mathbf{X})$ is the variable weight vector of the evaluation object X.

Subsequently, the definition changes to the following: (1) punitiveness $x_i \geq x_j \Rightarrow P_i(X) \leq P_j(X)$ is changed to incentive $x_i \geq x_j \Rightarrow P_i(X) \geq P_j(X)$; (3) ③ punitiveness $\mathbf{W}_0 = (w_{01}, w_{02}, \ldots, w_{0m})$ with a single reduction of ($j = 1, 2, \ldots, m$). Regarding the variable x_j, it is changed into incentive $w_{0j}(x_1, x_2, \ldots, x_m)$ with the single increase ($j = 1, 2, \ldots, m$) regarding the variable x_j. Then, $\boldsymbol{P}$ is called incentive contingency vector.

The eigenvariable weight vector $P_j(X)$ is essentially a gradient vector of the m-dimensional real function $B(x)$ (also known as equilibrium function) [32]. Its calculation formula is as follows:

$$P_j(X) = P_j(x_1, x_2, \cdots, x_m) = \frac{\partial B(x)}{\partial x_j} \tag{15}$$

According to Equations (14) and (15), the variable weight vector $\boldsymbol{W}(\boldsymbol{X})$ of the evaluation object X can be obtained.

4. Establishment of Variable Weight Matter–Element Extension Model for Slope Rock Mass Stability Classification

There are many factors affecting the stability of slope rock masses. Establishing a scientific and reasonable evaluation index system is the premise for the accurate evaluation of the slope rock mass stability. The safety evaluation of the slope stability is a dynamic system engineering. The establishment of an evaluation index system is the basic work of evaluation, and the rationality of the evaluation index system directly affects the accuracy of evaluation results. The principle of selecting evaluation indicators is to reflect the most important and comprehensive information with least indicators. Referring to the engineering rock mass classification standard [33,34], the hydroelectric engineering geological survey standard [35], and other researches on the classification criteria of the slope stability and safety evaluation indexes [11,13,36], the geological, environmental, and engineering conditions of the slope rock mass are considered comprehensively in this study. A classification evaluation index system of the slope rock mass stability is constructed by evaluation indexes, such as uniaxial compressive strength; elastic modulus; Poisson's ratio; structural features; cohesion; internal friction angle; daily maximum rainfall; maximum in situ stress; groundwater state; slope gradient; slope height; and rock acoustic velocity, as shown in Figure 1.

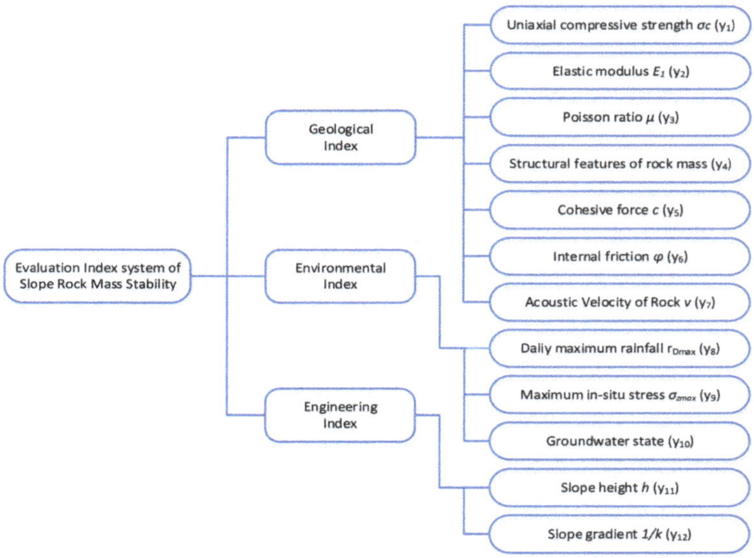

Figure 1. Evaluation index system of the slope rock mass stability.

In this study, the classification criteria of the slope rock mass stability evaluation indexes in references [11,13] are employed, as shown in Table 1. Equations (9) and (10) are used to normalize the values in the classification criteria of evaluation indexes, and the classical domain R_j and the nodal domain R_0 of the classification criteria of the slope rock mass stability are obtained.

Table 1. Evaluation index classification criteria of the slope rock mass stability.

Grade	σ_c y_1/MPa	E_1 y_2/GPa	μ y_3	y_4/%	c y_5/MPa	φ y_6/(°)
Level I (extremely stable)	[150,200]	[33.0,60.0]	[0,0.20]	[90,100]	[2.10,8.00]	[60,90]
Level II (stable)	[125,150)	[20.0,33.0)	[0.20,0.25)	[75,90)	[1.50,2.10)	[50,60)
Level III (basically stable)	[90,125)	[6.0,20.0)	[0.25,0.30)	[50,75)	[0.70,1.50)	[39,50)
Level IV (unstable)	[40,90)	[1.3,6.0)	[0.30,0.35)	[30,50)	[0.20,0.70)	[27,39)
Level V (extremely unstable)	[10,40)	[0,1.3)	[0.35,0.50)	[0,30)	[0.05,0.20)	[0,27)
Grade	v y_7/km·s^{-1}	r_{Dmax} y_8/mm	σ_{Zmax} y_9/MPa	y_{10}/L·(min.10m)$^{-1}$	h y_{11}/m	$1/k$ y_{12}/(°)
Level I (extremely stable)	[5.0,7.5]	[0,20]	[0,2]	[0,25]	[0,30]	[0,10]
Level II (stable)	[4.0,5.0)	(20,40]	(2,8]	(25,50]	(30,45]	(10,20]
Level III (basically stable)	[2.5,4.0)	(40,60]	(8,14]	(50,100]	(45,60]	(20,40]
Level IV (unstable)	[2.0,2.5)	(60,100]	(14,20]	(100,125]	(60,80]	(40,60]
Level V (extremely unstable)	[0,2.0)	(100,150]	(20,25]	(125,150]	(80,100]	(60,80]

$$R_j = \begin{bmatrix} & S(5) & S(4) & S(3) & S(2) & S(1) \\ y_1 & [0,0.16) & [0.16,0.42) & [0.42,0.61) & [0.61,0.74) & [0.74,1.00] \\ y_2 & [0,0.02) & [0.02,0.10) & [0.10,0.33) & [0.33,0.55) & [0.55,1.00] \\ y_3 & [0,0.30) & [0.30,0.40) & [0.40,0.50) & [0.50,0.60) & [0.60,1.00] \\ y_4 & [0,0.30) & [0.30,0.50) & [0.50,0.75) & [0.75,0.90) & [0.90,1.00] \\ y_5 & [0,0.02) & [0.02,0.08) & [0.08,0.18) & [0.18,0.26) & [0.26,1.00] \\ y_6 & [0,0.30) & [0.30,0.43) & [0.43,0.56) & [0.56,0.67) & [0.67,1.00] \\ y_7 & [0,0.27) & [0.27,0.33) & [0.33,0.53) & [0.53,0.67) & [0.67,1.00] \\ y_8 & [0,0.33) & [0.33,0.60) & [0.60,0.73) & [0.73,0.87) & [0.87,1.00] \\ y_9 & [0,0.20) & [0.20,0.44) & [0.44,0.68) & [0.68,0.92) & [0.92,1.00] \\ y_{10} & [0,0.17) & [0.17,0.33) & [0.33,0.67) & [0.67,0.83) & [0.83,1.00] \\ y_{11} & [0,0.20) & [0.20,0.40) & [0.40,0.55) & [0.55,0.70) & [0.70,1.00] \\ y_{12} & [0,0.25) & [0.25,0.50) & [0.50,0.75) & [0.75,0.88) & [0.88,1.00] \end{bmatrix} \quad (16)$$

$$R_0 = \begin{bmatrix} S & y_1 & [0,1.00] \\ & y_2 & [0,1.00] \\ & y_3 & [0,1.00] \\ & y_4 & [0,1.00] \\ & y_5 & [0,1.00] \\ & y_6 & [0,1.00] \\ & y_7 & [0,1.00] \\ & y_8 & [0,1.00] \\ & y_9 & [0,1.00] \\ & y_{10} & [0,1.00] \\ & y_{11} & [0,1.00] \\ & y_{12} & [0,1.00] \end{bmatrix} \quad (17)$$

The classical domain R_j expresses the variation range of the standardized index values of the slope rock mass stability evaluation index in each stability grade, and the joint domain R_0 expresses the entire range of the standardized index values of the slope rock mass stability classification. Figure 2 shows the calculation process of the rock mass stability classification evaluation using the variable weight matter–element extension model.

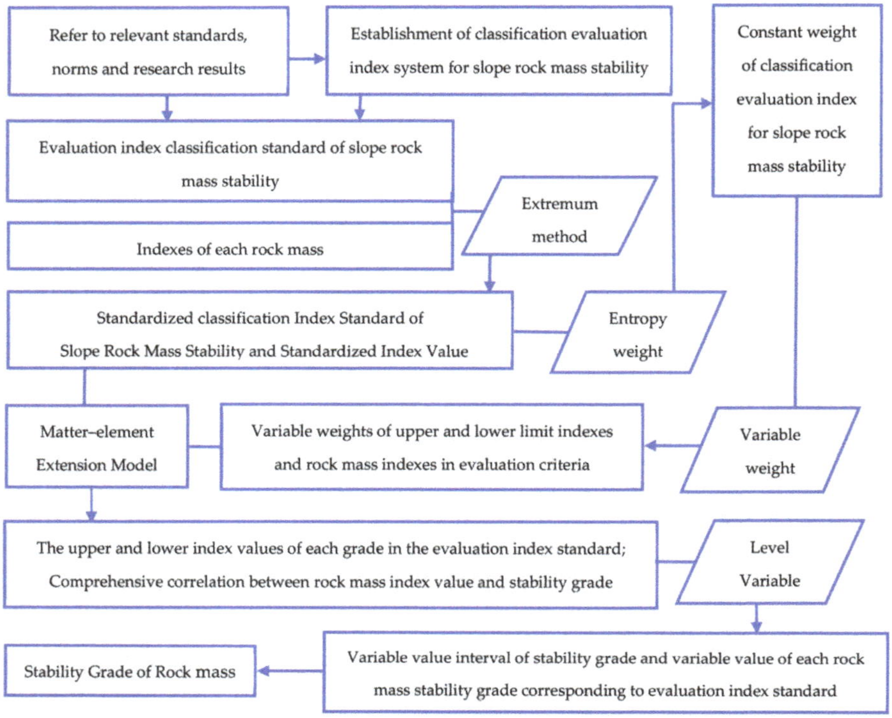

Figure 2. Evaluation process of the variable weight matter–element extension model for the classification of the slope rock mass stability.

5. Case Study in a Mining Project

According to the previous studies [13], the slope rock mass of an open-pit copper mine was formed in 2008. To date, only a part of the slope has been maintained, and the entire slope has remained stable. The evaluation indexes of the slope rock mass were measured and presented in the first four groups of data in Table 2. The data in the fifth group of Table 2 are the slope data of an open-pit mine, presented in previous studies [11]. The actual situation of the slope rock mass was extremely stable. Based on the measured data of slope rock masses in these two mines, the stability of the slope rock masses was evaluated by the proposed model to verify its validity.

According to Equations (9) and (10), the dimensionless classification index of the slope rock mass is processed. Table 3 presents the dimensionless unified classification indexes. The indexes of five rock mass samples in Table 3 are compared with the evaluation indexes in the classification criteria of the slope rock mass stability, as shown in mboxfigfig:mathematics-1402239-f003. The solid broken line is connected by the upper limit indexes of each stability grade in the standardized classification criteria of the slope rock mass stability evaluation indexes, and the virtual broken line is connected by the indexes of each rock mass sample after standardization. This method allows the stability of each rock mass sample to be obtained intuitively using the single index from the distribution of each turning point in the broken line. As shown in Figure 3, the distribution law of rock mass sample indexes in (a), (b), (c), and (d) is essentially the same, while the distribution law of the rock mass sample indexes in (e) is quite different. It indicates that there is a great difference between the two slopes, which is helpful to verify the accuracy of the proposed model.

Table 2. Evaluation indexes of the slope rock mass stability.

Rock	Measured Values of Geological Indexes						
	σ_c y_1/MPa	E_1 y_2/GPa	μ y_3	y_4/%	c y_5/MPa	φ y_6/°	v y_7/km·s^{-1}
X_1	52.60	2.3	0.27	90	17.80	31.8	3.700
X_2	53.00	2.0	0.21	85	17.80	35.6	3.789
X_3	61.60	1.9	0.18	90	23.10	24.6	3.847
X_4	60.04	1.9	0.19	92	23.10	24.6	3.896
X_5	28.97	25.7	0.22	57	5.08	52.0	3.200

Rock	Measured Values of Environmental Indexes			Measured Values of Engineering Indexes	
	r_{Dmax} y_8/mm	σ_{Zmax} y_9/MPa	y_{10}/L·(min·10 m)$^{-1}$	h y_{11}/m	$1/k$ y_{12}/°
X_1	6.06	6.18	10	48	60
X_2	6.06	8.73	10	67	44
X_3	6.06	9.05	9	76	38
X_4	6.06	10.18	9.5	45	54
X_5	12.00	15.32	35	12	36

Table 3. Normalized values of the evaluation indexes for the slope rock mass stability.

Rock Sample	y_1	y_2	y_3	y_4	y_5	y_6	y_7	y_8	y_9	y_{10}	y_{11}	y_{12}
X_1	0.22	0.04	0.46	0.90	1.00	0.35	0.49	0.96	0.75	0.93	0.52	0.25
X_2	0.23	0.03	0.58	0.85	1.00	0.40	0.51	0.96	0.65	0.93	0.33	0.45
X_3	0.27	0.03	0.64	0.90	1.00	0.27	0.51	0.96	0.64	0.94	0.24	0.53
X_4	0.26	0.03	0.62	0.92	1.00	0.27	0.52	0.96	0.59	0.94	0.55	0.33
X_5	0.10	0.43	0.56	0.57	0.63	0.58	0.43	0.92	0.39	0.77	0.88	0.55

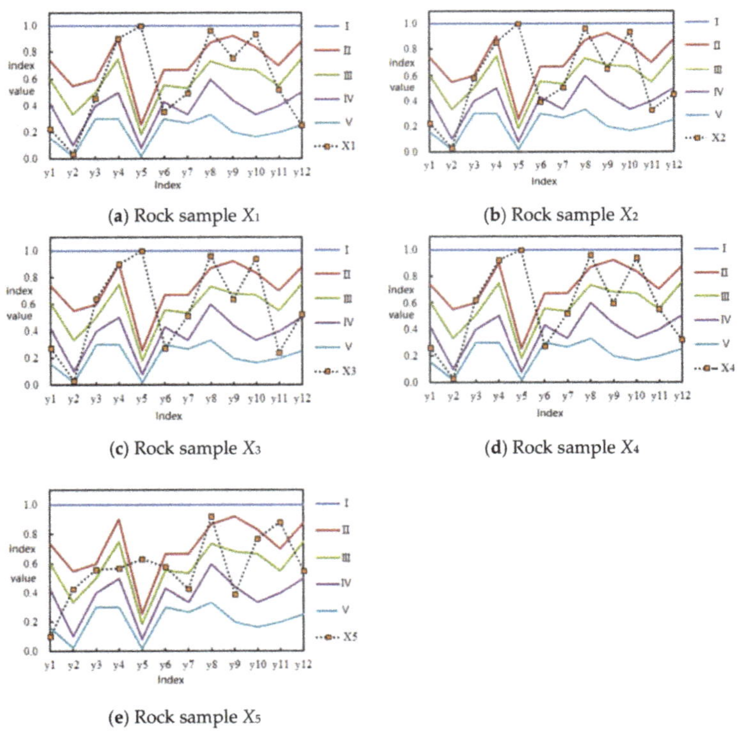

Figure 3. Distribution of the evaluation index values for each slope rock mass.

In this study, the feature variable weight vector $P_j(X)$ is constructed by using the full excitation feature variable weight function [37]:

$$P_j(X) = x_j^{\frac{1}{mw_{0j}}} \tag{18}$$

where $P_j(X)$ is the feature variable weight vector element; x_j is the standardized evaluation object index; m is the number of evaluation indicators, and m = 2; w_{0j} is the constant weight vector element.

Combining the theory of extreme entropy weighting and variable weight, the constant weight of the slope rock mass stability evaluation index and the variable weight of rock mass to be evaluated are calculated using Equations (9)–(15) and (18), as shown in Table 4. As shown in Figure 4, the constant weight of the evaluation index reflects the relative importance of each evaluation index and the overall trend of the index weight. The influence of different index values on the weight is considered in the variable weight. Therefore, when the value of the same index is different, the weight will change. The variation law shows that when the index value is relatively good, the index weight will be greater; when the index value is relatively poor, the index weight will be smaller.

Table 4. Evaluation index weights.

Index		y_1	y_2	y_3	y_4	y_5	y_6	y_7	y_8	y_9	y_{10}	y_{11}	y_{12}
Constant weight		0.107	0.254	0.048	0.048	0.129	0.063	0.053	0.045	0.057	0.060	0.070	0.066
Weighted variable	X_1	0.065	0.169	0.025	0.078	0.252	0.031	0.034	0.081	0.074	0.107	0.062	0.022
	X_2	0.066	0.164	0.037	0.072	0.255	0.037	0.036	0.082	0.060	0.108	0.036	0.047
	X_3	0.076	0.160	0.044	0.079	0.253	0.022	0.036	0.081	0.058	0.109	0.025	0.057
	X_4	0.073	0.158	0.041	0.081	0.251	0.022	0.037	0.080	0.052	0.107	0.066	0.031
	X_5	0.031	0.336	0.031	0.032	0.168	0.054	0.024	0.067	0.025	0.073	0.104	0.054

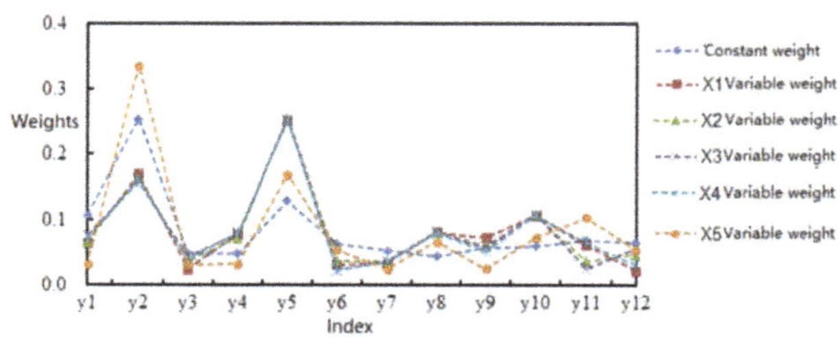

Figure 4. Comparison between the constant weights and variable weights of the evaluation index for each slope rock mass (the quantity of X is associated with the slope rock mass).

According to Equations (1)–(5), (16), and (17), the single index correlation of the slope rock mass sample X_1 with respect to each stability grade is calculated, as shown in Table 5. Similarly, the upper and lower limit index values of each grade in the classification evaluation index standard of the slope rock mass stability are calculated, as well as the single index correlation of the index values of the slope rock mass samples X_2, X_3, X_4, and X_5, with respect to each stability grade. Subsequently, the comprehensive correlation is calculated using Equation (6). Finally, the slope rock mass stability is calculated using Equations (7) and (8), and the upper and lower limits of each grade and the eigenvalues of the corresponding grade variables of the slope rock mass are also obtained, as is shown in Tables 6 and 7. Based on Table 6, the corresponding stability grade of each slope rock mass sample can be determined according to the variable eigenvalue.

Table 5. Single index correlation of the slope rock mass X_1.

Index	Single Index Correlation											
	y_1	y_2	y_3	y_4	y_5	y_6	y_7	y_8	y_9	y_{10}	y_{11}	y_{12}
I	−0.696	−0.931	−0.233	0.000	1.000	−0.470	−0.260	1.040	−0.403	1.067	−0.273	−0.714
II	−0.630	−0.886	−0.080	0.000	0.000	−0.364	−0.075	−0.699	0.420	−0.599	−0.059	−0.667
III	−0.468	−0.620	0.095	−0.600	0.000	−0.185	0.088	−0.850	−0.228	−0.799	0.067	−0.500
IV	0.418	0.810	−0.115	−0.800	0.000	0.177	−0.245	−0.900	−0.559	−0.900	−0.200	0.000
V	−0.228	−0.309	−0.258	−0.857	0.000	−0.131	−0.315	−0.940	−0.691	−0.920	−0.400	0.000

Table 6. Grade variable intervals and the slope rock mass stability grades.

Grade	I	II	III	IV	V
Level Variable k	[1.00, 1.86)	[1.86, 2.43)	[2.43, 3.58)	[3.58, 4.27)	[4.27, 5]

Table 7. Evaluation results of the slope rock mass stability.

Sample	Comprehensive Correlation					Level Variable k	The Proposed Model	Hierarchical-Extension Model [5]	Multi-Level Unascertained Measure-Set Pair Analysis Model [7]	Actual Situation
	$S(1)$	$S(2)$	$S(3)$	$S(4)$	$S(5)$					
X_1	0.155	−0.315	−0.361	−0.127	−0.405	2.04	Level II	Level II	Level III	Level II
X_2	0.144	−0.323	−0.337	−0.123	−0.384	2.05	Level II	Level I	Level II (near Level III)	Level II
X_3	0.172	−0.368	−0.359	−0.155	−0.392	1.93	Level II (near Level I)	Level II	Level III	Level II
X_4	0.201	−0.364	−0.359	−0.165	−0.394	1.88	Level II (near Level I)	Level I	Level II	Level II
X_5	0.289	−0.111	−0.312	−0.475	−0.508	1.62	Level I	Level I	Level II	Level I

To prove the validity of the evaluation results, under the same evaluation index system, the evaluation results are obtained by using the hierarchical extension model, the multi-level unascertained measure-set pair analysis model, and the actual stability of the project, as shown in Table 7. A comparison of the results shows that the classification results of the slope rock mass obtained by the proposed model are consistent with the actual situation of the mine slope. Moreover, the evaluation results of the proposed model are more accurate than those of the hierarchical extension model and the multi-level unascertained measure-set pair analysis model.

6. Conclusions

(1) The evaluation index dimension was unified using the extremum method, and the objective constant weight of the evaluation index (namely, uniaxial compressive strength; elastic modulus; Poisson's ratio; structural features; cohesion force; internal friction; and daily maximum rainfall) was calculated using the entropy weight method. The constant weight reflects the relative importance of the evaluation indexes. On this basis, the variable weight theory was introduced to fully consider the influence of the value difference of the classification evaluation index on the index weight, and the excitation feature variable weight function was used to calculate the weighting of the evaluation index of each rock mass, so that the weighting of the evaluation index was more reasonable.

(2) By applying the matter–element extension model and grade-variable method, the variable interval corresponding to the evaluation index standard of the stability grade of the slope rock mass and the variable value of the stability grade of each rock mass were calculated, and the stability grade of each rock mass was obtained.

The evaluation results presented in this study are consistent with the engineering practice and are more accurate than those of the hierarchical extension model and the multi-level unascertained measure-set pair analysis model.

(3) The variable value of the slope rock mass stability grade was obtained by the integrated information of comprehensive correlation between the evaluation index value of the slope rock mass stability and each stability grade. The accuracy of the extension model, in the classification of the slope rock mass stability, can be improved by the classification of the slope rock mass stability through the interval of the variable values of the grades corresponding to the evaluation index standard and the variable values for each slope rock mass stability grade.

Author Contributions: Conceptualization, S.Y.; methodology, S.Y.; data curation, Z.X. and K.S.; formal analysis, Z.X. and K.S.; validation, S.Y.; resources, S.Y. and Z.X.; writing—original draft preparation, Z.X.; writing—review and editing, Z.X.; project administration, S.Y. All authors have read and agreed to the published version of the manuscript.

Funding: This research was funded by the National Natural Science Foundation Project of China under Grant No. 72088101 and No. 51404305, and the Innovation Fund Project of Central South University under Grant No. 2021zzts0283.

Institutional Review Board Statement: Not applicable.

Informed Consent Statement: Not applicable.

Data Availability Statement: Data is contained within the article.

Acknowledgments: The authors would like to express their thanks to the National Natural Science Foundation and Innovation Fund Project of Central South University.

Conflicts of Interest: The authors declare no conflict of interest.

References

1. Wang, S.H.; Li, X.H.; Yao, J.R.; Gong, F.Q.; Li, X.; Du, K.; Tao, M.; Huang, L.Q.; Du, S.L. Experimental investigation of rock breakage by a conical pick and its application to non-explosive mechanized mining in deep hard rock. *Int. J. Rock Mech. Min. Sci.* **2019**, *122*, 104063. [CrossRef]
2. Wang, S.F.; Sun, L.C.; Li, X.B.; Wang, S.Y.; Du, K.; Li, X.; Feng, F. Experimental investigation of cuttability improvement for hard rock fragmentation using conical cutter. *Int. J. Geomech.* **2021**, *21*, 06020039. [CrossRef]
3. Yang, T.H.; Zhang, F.C.; Yu, Q.L.; Cai, M.F.; Li, H.Z. Research situation of open-pit mining high and steep slope stability and its developing trend. *Rock Soil Mech.* **2011**, *32*, 1437–1452.
4. Wang, S.F.; Tang, Y.; Li, X.B.; Du, K. Analyses and predictions of rock cuttabilities under different confining stresses and rock properties based on rock indentation tests by conical pick. *Trans. Nonferrous Met. Soc. China* **2021**, *31*, 1766–1783. [CrossRef]
5. Wang, S.F.; Tang, Y.; Wang, S.Y. Influence of brittleness and confining stress on rock cuttability based on rock indentation tests. *J. Cent. South Univ.* **2021**, *28*, 2786–2800. [CrossRef]
6. Shi, W.H.; Yang, T.H.; Wang, P.T.; Hu, G.J.; Xiao, P. Anisotropy analysis method for stability of open-pit slope rock mass and its application. *Chin. J. Geotech. Eng.* **2014**, *36*, 1924–1933.
7. Zhao, H.J.; Ma, F.S.; Guo, J.; Wu, Z.Q.; Zhang, Y.L. The influence of open pit to underground mining on the stability of slope rock mass. *J. Coal Mine* **2011**, *36*, 1635–1641.
8. Du, S.G.; Yong, R.; Chen, J.Q.; Chen, J.Q.; Xia, C.C.; Li, G.P.; Liu, W.L.; Liu, Y.M.; Liu, H. Graded analysis for slope stability assessment of large open-pit mines. *Chin. J. Rock Mech. Eng.* **2017**, *36*, 2601–2611.
9. Bar, N.; Kostadinovski, M.; Tucker, M.; Byng, G.; Rachmatullah, R.; Maldonado, A.; Pötsch, M.; Gaich, A.; McQuillan, A.; Yacoub, T. Rapid and robust slope failure appraisal using aerial photogrammetry and 3D slope stability models. *Int. J. Min. Sci. Technol.* **2020**, *30*, 651–658. [CrossRef]
10. Arikan, F.; Yoleri, F.; Sezer, S.; Caglan, D.; Biliyul, B. Geotechnical assessments of the stability of slopes at the Cakmakkaya and Damar open pit mines (Turkey): A case study. *Environ. Earth Sci.* **2010**, *61*, 741–755. [CrossRef]
11. Wang, X.; Kang, Q.; Qin, J.; Zhang, Q.; Wang, S. Application of AHP-extenics model to safety evaluation of rock slope stability. *J. Cent. South Univ. Sci. Technol.* **2013**, *44*, 2455–2462.
12. Zhang, X.; Zhou, S.W.; Lin, P.; Tan, Z.; Chen, Z.; Jiang, S. Slope stability evaluation based on entropy coefficient-set pair analysis. *Chin. J. Rock Mech. Eng.* **2018**, *37*, 3400–3410.
13. Huang, D.; Shi, X.Z.; Qiu, X.Y.; Gou, Y. Stability gradation of rock slopes based on multilevel uncertainty measure-set pair analysis theory. *J. Cent. South Univ. Sci. Technol.* **2017**, *48*, 1057–1064.

14. Liu, L.L.; Zhang, S.H.; Liu, L.M. Model and application of AHP and ideal point method based on stability gradation of rock slope. *Chin. J. Cent. South Univ. Sci. Technol.* **2014**, *10*, 3499–3504.
15. Bar, N.; Barton, N. The Q-slope method for rock slope engineering. *Rock Mech. Rock Eng.* **2017**, *50*, 3307–3322. [CrossRef]
16. Pastor, J.L.; Riquelme, A.J.; Tomás, R.; Cano, M. Clarification of the slope mass rating parameters assisted by SMRTool, an open-source software. *Bull. Eng. Geol. Environ.* **2019**, *78*, 6131–6142. [CrossRef]
17. Cai, W. Introduction of Extenics. *Syst. Eng. Theory Pract.* **1998**, *18*, 76–84.
18. Cai, W. Extension theory and its application. *Chin. Sci. Bull.* **1999**, *44*, 673–682. [CrossRef]
19. Xu, H.J.; Zhao, B.F.; Zhou, Y.; Liu, S.X. Evaluation on water disaster from roof strata based on the entropy-weight and matter-element extension model. *J. Min. Saf. Eng.* **2018**, *35*, 112–117.
20. Cai, W.; Yang, C.Y. Basic theory and methodology on extenics. *Chin. Sci. Bull.* **2013**, *58*, 1190–1199.
21. Yang, C.Y.; Cai, W. Recent research progress in dependent functions in extension sets. *J. Guangdong Univ. Technol.* **2012**, *29*, 7–14.
22. Shi, Z.P.; Shan, T.H.; Liu, W.F.; Zhang, X.P. Comprehensive evaluation of power network operation risk based on matter-element extensible model. *Power Syst. Technol.* **2015**, *39*, 3233–3239.
23. Chen, S.Y. *Theory and Model of Variable Fuzzy Sets and Its Application*; Dalian University of Technology Press: Dalian, China, 2009.
24. Li, Z.Y.; Rong, W.Y.; Chen, Z.D. Evaluation of railway luggage and parcel transportation safety based on variable fuzzy sets theory. *China Saf. Sci. J.* **2018**, *28*, 186–190.
25. Hu, B.Q.; Zhang, X. Improvement and application of extension evaluation method. *J. Wuhan Univ.* **2003**, *36*, 79–84.
26. Zhu, X.A.; Wei, G.D. Discussion on the good standard of dimensionless method in entropy method. *Stat. Decis.* **2015**, *31*, 12–15.
27. Wang, W.; Luo, Z.Q.; Xiong, L.X.; Jia, N. Research of goaf stability evaluation based on improved matter-element extension model. *J. Saf. Environ.* **2015**, *15*, 21–25. [CrossRef]
28. Tang, X.W.; Zhou, Z.F.; Shi, Y. The variable weighted functions of combined forecasting. *Comput. Math. Appl.* **2003**, *45*, 723–730. [CrossRef]
29. Wang, P.Z.; Li, H.X. *Fuzzy System Theory and Fuzzy Computer*; Science Publishing Company of Beijing: Beijing, China, 1996.
30. Li, D.Q.; Li, H.X. The properties and construction of state variable weight vectors. *J. Beijing Norm. Univ. Nat. Sci.* **2002**, *38*, 455–461.
31. Atanassov, K.T. Operators over interval-valued intuitionistic fuzzy sets. *Fuzzy Sets Syst.* **1994**, *64*, 159–174. [CrossRef]
32. Li, H.X. Factor spaces and mathematical frame of knowledge represention (VIII). *Fuzzy Syst. Math.* **1995**, *9*, 1–9.
33. GB/T 50218-2014, Engineering Rock Mass Classification Standard. Available online: https://www.chinesestandard.net/PDF/BOOK.aspx/GBT50218-2014 (accessed on 10 August 2021).
34. GB/T 50218-1994, Engineering Rock Mass Classification Standard. Available online: https://www.chinesestandard.net/PDF/BOOK.aspx/GB50218-1994 (accessed on 10 August 2021).
35. GB 50487-2008, Code for Engineering Geological Investigation of Water Resources and Hydropower. Available online: https://www.chinesestandard.net/PDF/BOOK.aspx/GB50487-2008 (accessed on 10 August 2021).
36. Kang, Z.Q.; Zhou, H.; Feng, X.T.; Yang, C.X. Evaluation of high rock slope quality based on theory of extenics. *J. Northeast. Univ. Nat. Sci.* **2007**, *28*, 1770–1774.
37. Mo, G.L.; Zhang, W.G.; Liu, Y.J. Construction of Variable Spatial Weight and Analysis of Spatial Effect. *J. Syst. Manag.* **2018**, *27*, 219–229.

mathematics MDPI

Article

A Numerical Investigation to Determine the *p–y* Curves of Laterally Loaded Piles

Kexin Yin [1,*], Lianghui Li [2,*] and Eugenia Di Filippo [1,3]

1 Institut de Recherche en Génie Civil et Mécanique (GeM), Ecole Centrale de Nantes, UMR 6183 CNRS, 1 Rue de la Noë, CEDEX 3, 44321 Nantes, France; eugenia-difilippo@libero.it
2 School of Energy and Mining Engineering, China University of Mining and Technology-Beijing, Beijing 100083, China
3 Department of Civil and Environmental Engineering, University of Perugia, via G. Duranti, 93-06125 Perugia, Italy
* Correspondence: kexin.yin.research@gmail.com (K.Y.); lilianghui@cumt.cn (L.L.)

Abstract: This paper focuses on a numerical approach to finding the *p–y* curves for laterally loaded piles. The Drucker–Prager plastic model is employed and implemented within a finite element MATLAB code. The pre- and post-processing code for Gmsh and related numerical tools are established as well. The *p–y* curve results from this new approach have been validated and compared to the typical design equations of API (American Petroleum Institute) and Matlock. The validation reveals that the code leads to lower *p–y* curves than the API and Matlock equations when the horizontal displacement is less than 0.35 times the diameter of the pile (B). A sensitivity analysis of the number of elements and the interface thickness is presented. The results indicate that the obtained *p–y* curves are independent of the two factors. Finally, the influence of clay content on the *p–y* behavior is investigated by the implemented MATLAB code. When $y < 0.15B$, the same lateral capacity values are resulted at clay contents of 27.5% and 55%, and they are higher than the ones for 0% clay content. The *p–y* curves show a decreasing trend with increasing clay content after $y > 0.15B$.

Keywords: laterally loaded pile; *p–y* curve; soil-pile interface; Drucker–Prager model

Citation: Yin, K.; Li, L.; Di Filippo, E. A Numerical Investigation to Determine the *p–y* Curves of Laterally Loaded Piles. *Mathematics* **2021**, *9*, 2783. https://doi.org/10.3390/math9212783

Academic Editors: Ioannis K. Argyros and Clemente Cesarano

Received: 10 August 2021
Accepted: 29 October 2021
Published: 2 November 2021

1. Introduction

Pile foundation is one of the most commonly used foundation types in complicated site conditions. Piles in port engineering, offshore oil platforms, and wind turbines are highly subjected to lateral loads. At the engineering scale, one of the main aspects that we want to focus on is the response of piles when they are subjected to lateral loading. This problem can be typically investigated by tracing the so-called *p–y* curves, which represent the relationship between the lateral displacement (y) and the lateral force (p) for a generic transversal section of the pile. The *p–y* approach has been widely employed to design laterally loaded piles [1–6]. The development of computer technology has made it possible to study complicated engineering problems by using numerical methods [4–11]. It is more flexible and less time consuming to analyze the *p–y* characteristics by using the finite element method than the full-scale in-situ tests.

This paper aims to construct *p–y* curves that take into account the properties of the interface between the soil and the concrete pile as characterized via the experimental campaign of interface direct shear test in the laboratory. To do this, the Drucker–Prager constitutive model is adopted to describe irreversible (plastic) strains occurring at the interface between the surrounding soil and the concrete pile. The Drucker–Prager constitutive law is then implemented into a two-dimensional finite element (FE) code within a MATLAB environment that was initially developed by Bonnet and Frangi [12] and Bonnet et al. [13]. A 2-D problem describing the effect of a transversal displacement applied to a pile section has been simulated. The resultant transversal force acting on the pile is found to be a function of the applied displacement. The obtained *p–y* curves are validated and compared with those in the literature relative to different design methods. A sensitivity analysis with respect to the number of elements and the thickness of the interface is also presented to check whether they affect the *p–y* results. Finally, since there is a lack of literature on the *p–y* response of piles in clay-rich soils, the *p–y* curves of a concrete pile in soil with different percentages of clay are provided to see the effect of the varying amounts of clay present in the soil.

2. *p–y* Method and Code Development

2.1. *p–y* Method

When a pile is subjected to horizontal loading, the lateral bearing capacity can be expressed as [14–16]:

$$p_u = N_p s_u B \tag{1}$$

where p_u is the lateral bearing capacity, N_p is the lateral bearing capacity factor, s_u is the undrained shear strength of the soil, and B is the pile diameter. The soil shear strength s_u depends on soil properties, and N_p is related to the failure mechanism of the soil–pile interface. Equation (1) can be used to normalize the *p–y* curves in the next section.

The concept of the *p–y* curve was firstly proposed by McClelland and Focht [17], and developed by others [18]. The *p–y* model, also known as the non-linear Winkler spring model, is commonly used in the design of piles under lateral loading because of its simplicity and low computational cost [5,6,19,20]. In the *p–y* model, the pile is simplified as a beam while the soil–pile interface is treated as a set of 1-D, non-linear springs [2,19–22]. As mentioned above, p is the resistance force of the soil per unit of pile length, and y is the local pile deflection caused by the horizontal loading [14,20,23]. Since the *p–y* concept was introduced, several *p–y* curves were proposed based on different influence factors by previous researchers. For example, the Matlock *p–y* curve, API *p–y* curve, Jeanjean *p–y* curve, and Zhang *p–y* curve [14,16,24,25]. Three typical *p–y* curves are presented in the following.

Matlock [16] proposed a *p–y* curve for a pile located in soft clay as a power relationship between the lateral resistance and the normalized lateral displacement:

$$p = 0.5 p_u \left(\frac{y}{y_c}\right)^{0.33} \tag{2}$$

y_c can be computed by:

$$y_c = 2.5 \varepsilon_{50} B \tag{3}$$

where ε_{50} is the axial strain corresponding to the 50% maximum principal stress difference in an undrained compression test and B is the diameter of the pile.

Jeanjean et al. [14] proposed a *p–y* curve function of the initial shear $G_{\max}$ and the undrained shear strength of the soil s_u:

$$p = p_u \tanh\left[\frac{G_{\max}}{100 s_u}\left(\frac{y}{B}\right)^{0.5}\right] \tag{4}$$

The *p–y* method is also adopted by the American Petroleum Institute (i.e., API) for a lateral loading pile design. According to API [25], the API *p–y* relationship is given as:

$$p = F p_u \tanh\left(\frac{kHy}{F p_u}\right) \tag{5}$$

where k is the initial modulus of subgrade reaction (kN/m^3), H is the depth (m), and F is a factor to account for cyclic or monotonic static loading conditions It can be evaluated by:

$F = 0.9$ for cyclic loading;

$F = \left(3.0 - 0.8\frac{H}{B}\right) \geq 0.9$ for monotonic loading.

In what follows, a finite element-based *p–y* curve deduction is presented, capable of accounting for the plastic behavior of the soil and of the soil–concrete interface. The results will be compared with the curves from the typical equations.

2.2. Drucker–Prager Model

The Drucker–Prager (DP) model [26] for the soil is implemented using the following MATLAB code, which performs the numerical simulation for calculating the *p–y* curves. The DP yield criterion is defined as [27]:

$$f = \sqrt{\frac{2}{3}}q - (A - Bp_1) = 0 \tag{6}$$

where p_1 is the mean normal stress, σ_1 is the major principal stress, σ_2 is the principal stress, and σ_3 is the minor principal stress. A and B are related to the cohesion c and friction angle φ in the Mohr–Coulomb (MC) criterion; q is the second invariant of the deviatoric stress [27]:

$$q = \sqrt{3J_2} = \frac{1}{\sqrt{2}}\left[(\sigma_1 - \sigma_2)^2 + (\sigma_2 - \sigma_3)^2 + (\sigma_1 - \sigma_3)^2\right] \tag{7}$$

As the Drucker–Prager model is represented in the stress plane by a cone with a circular cross section, it is possible to relate it to the Mohr–Coulomb model considering the section of the MC pyramid to be inscribed or circumscribed to the DP cone. If the circle of the DP cone passes through the tensile corners of the MC yield surface (the MC criterion is inscribed):

$$A = \frac{2\sqrt{6}c\cos\varphi}{3 + \sin\varphi}, B = \frac{2\sqrt{6}\sin\varphi}{3 + \sin\varphi} \tag{8}$$

the cohesion (c) and the friction angle (φ) can be obtained by direct shear tests [28–34]. If the circle of the DP cone passes through the compression corners of the MC yield surface (MC criterion is circumscribed):

$$A = \frac{2\sqrt{6}c\cos\varphi}{3 - \sin\varphi}, B = \frac{2\sqrt{6}\sin\varphi}{3 - \sin\varphi} \tag{9}$$

Furthermore, we can assume a plastic potential function of the form [27]:

$$g = \sqrt{\frac{2}{3}}q + bp_1 \tag{10}$$

where b is the constant plastic dilatancy parameter. Using a return-mapping algorithm in the stress invariant space, the discrete plastic multiplier is given as:

$$\Delta\lambda = \frac{\sqrt{2/3}q^{tr} - (A - Bp_1^{tr})}{2\mu + KBb} \tag{11}$$

where K and μ are the elastic bulk and shear moduli, respectively.

2.3. MATLAB Code

The MATLAB FE code was originally developed by Bonnet and Frangi [12]. Further improvements have been implemented in GeM (École Centrale de Nantes). The Drucker–Prager constitutive model has been implemented within the MATLAB FE code that considers the non-associate and associate flow rule. Although the code started from the relatively simple DP model, more complicated constitutive laws can be developed based on this primary work. The pre- and post-processing phases are performed with the freely available code Gmsh [35]. The link between the MATLAB FE code and the pre- and post-processing code for Gmsh, and several specific numerical tools, has been established as well.

3. 2-D *p–y* Curve Modeling

3.1. Geometry, Mesh, and Boundary Conditions

A 2-D circular geometry of the finite element model is presented in Figure 1. Due to the symmetry, only half of the domain is considered; the geometry consists of the soil, the pile, and the soil–pile interface. The pile is a rigid disc with diameter B = 1 m. The soil boundary at the bottom is made up with two parts in the horizontal direction, each part having a width W = 11.5B (11.5 m) from the pile surface point; therefore, the radius (the center of the circle is the pile center) of the largest half circle is H = 12B = 12 m, see Figure 1. The size of the circular domain is chosen to be large enough to avoid parasitic effects of the lateral boundaries. The outer boundary of the soil is fixed both in the x and y directions (Figure 1), while the bottom is fixed only in the vertical direction (i.e., y-axis in Figure 1). A horizontal displacement is applied to the pile section in the x direction in Figure 1.

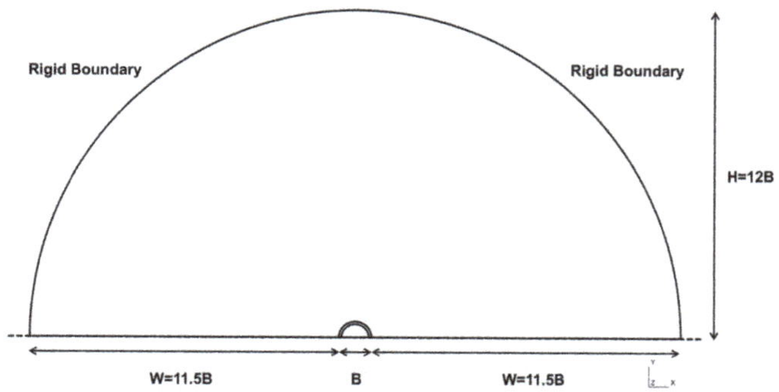

Figure 1. The circular geometry and boundary conditions.

The interface thickness in the geometry can be referred to as the typical experimental value, e.g., 5 ~ 14d_{50}. In previous modeling studies [36,37], it was set as 0.2 of the pile diameter B. In the following, a sensitivity analysis with respect to the interface thickness is provided, which ranges from 5 mm (which considers 20d_{50} the of Fontainebleau sand) to 5 cm. Constant-strain, three-node triangle elements (T3) are used in the mesh, while the mesh is refined near the interface (see Figure 2).

3.2. Initial Stress State

This starting numerical study works on a monophasic continuum, so it does not consider the presence of water, and this is the reason why effective stress is not introduced in the following work. Obviously as there is no water considered, the earth pressure can be expressed by total stresses rather than effective ones. The initial stress is generated with the K_0 condition according to the following relationship (Figure 3):

$$\begin{aligned} \sigma_v &= \sigma_{zz} = \gamma z \\ \sigma_n &= \sigma_{xx} = \sigma_{yy} = K_0 \sigma_{zz} = K_0 \gamma z \end{aligned} \tag{12}$$

in which γ is the unit weight of the soil, z is the depth from the pile section to the ground surface (Figure 3), and K_0 is the earth pressure coefficient. For the p–y modeling, a depth of 10 m and K_0= 0.5 are considered; the initial stress-state parameters are summarized in Table 1.

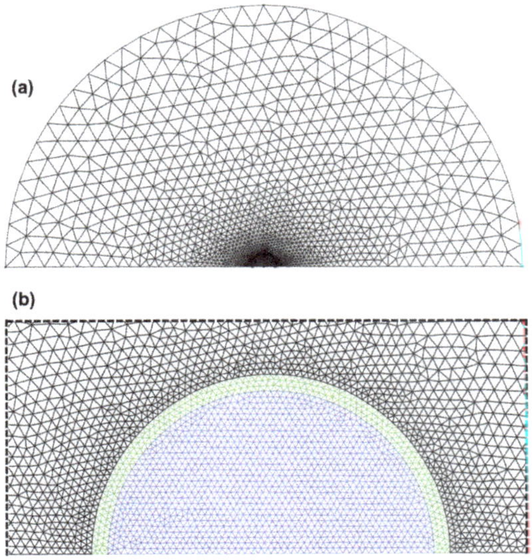

Figure 2. Example of the mesh with T3 elements: (**a**) the whole domain and (**b**) a zoom near the soil–pile interface zone.

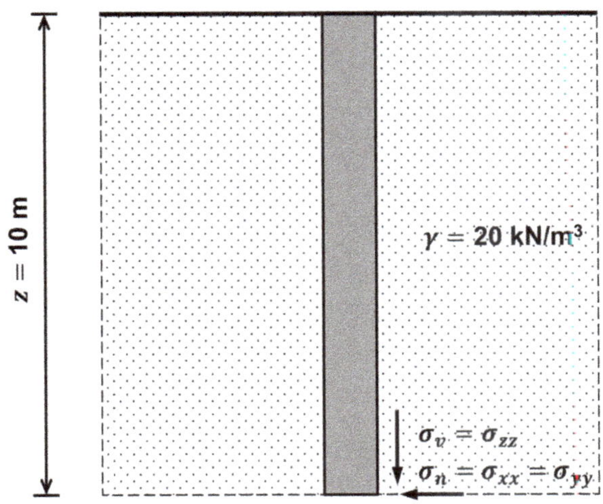

Figure 3. Initial stress state.

Table 1. Initial stress-state parameters.

γ (kN/m^3)	z (m)	K_0 (-)	σ_{zz} (kPa)	σ_{xx} (kPa)
20	10	0.5	200	100

3.3. Validation

To validate the calculation of the p–y curves, interface test results of soil consisting of 55% clay and concrete plate are selected as input parameters and the corresponding p–y curves are traced and compared with different design methods (API and Matlock).

A total of 11,390 T3 elements are used, and the interface width is equal to 5 mm. The properties of the interface, the soil, and the pile are given in Table 2. The concrete pile is considered a rigid material with a Young's modulus of 33 GPa (Table 2). The pile is loaded by stepwise loading (horizontal displacement) with increments of 0.002 m in the MATLAB code. The lateral capacity (p) of the pile is calculated by integrating the normal and tangential stresses along the circumference of the pile (or of the soil–pile interface).

Table 2. Parameters for validation that come from soil with 55% clay content.

-	Young's Modulus E (kPa)	Poisson's Ratio v (-)	Cohesion c (kPa)	Friction Angle φ (°)
Interface	6000	0.3	3.31	19.33
Soil	6000	0.3	10	21.81
Pile	33×10^6	0.2	-	-

The results are presented in terms of normalized p–y curves (Figure 4) using the shear strength of the soil and the pile diameter (see also [20,37]). More specifically, in this study, the soil is isotropic normally consolidated and the interface behavior is characterized by a direct shear test in drained condition, hence the critical drained interface shear strength (s_d) of the interface and the pile diameter B are used to normalize p so as to obtain the lateral bearing capacity factor $p/(s_d B)$, i.e., the N_p in Equation (1). The horizontal displacement is normalized by the pile diameter B.

The p–y curves calculated by the MATLAB code with DP model on the circular geometry are compared to the ones computed from the API and Matlock equations in Figure 4. The p–y curves from the DP model are lower than the API curve when the horizontal displacement is less than 0.35B (Figure 4). This indicates that at this horizontal displacement range ($y < 0.35B$) a smaller reaction force is mobilized by the DP model than the API model. The curve from the DP model is higher than that from the API when horizontal displacement is larger than 0.35B. From 0 to 0.4B, the MATLAB curve presents a similar shape to the Matlock curve, but with lower normalized p (Figure 4). The two curves become closer with increasing horizontal displacement. At $y = 0.4B$, the normalized p from the DP model in the MATLAB code stays between the API and Matlock values. The normalized p calculated from the MATLAB code is 8.58, which is 0.5 higher than the one of API (8.08) and 0.4 lower than the one from the Matlock empirical calculation (8.98), see Figure 4. As the horizontal displacement goes up to 0.45B, the result of the MATLAB code is equal to, then gradually becomes higher than that of Matlock. Finally, at 0.6B, the normalized lateral capacity of the MATLAB code is 10.57, which is 104.6% of that of Matlock (10.28).

In the MATLAB code, p is calculated by integrating the normal and tangential stresses along the circumference of the pile or the soil–pile interface. The two results are compared in Figure 5, which reveals that the p–y curves are the same before 0.4B and have a small difference (1.38 ~ 2.23%) at 0.4 ~ 0.6B.

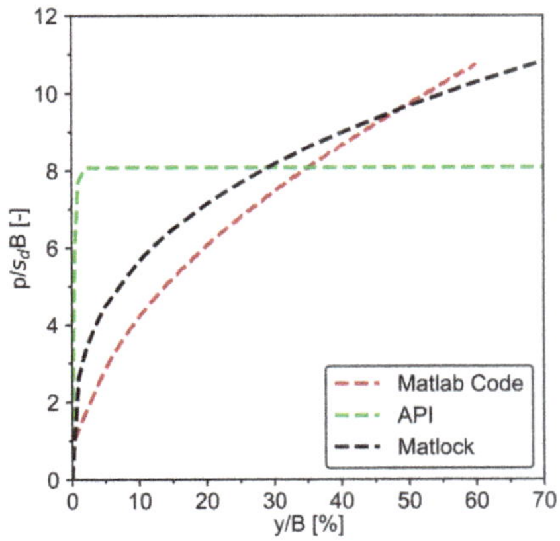

Figure 4. p–y curves from the MATLAB code and the design methods.

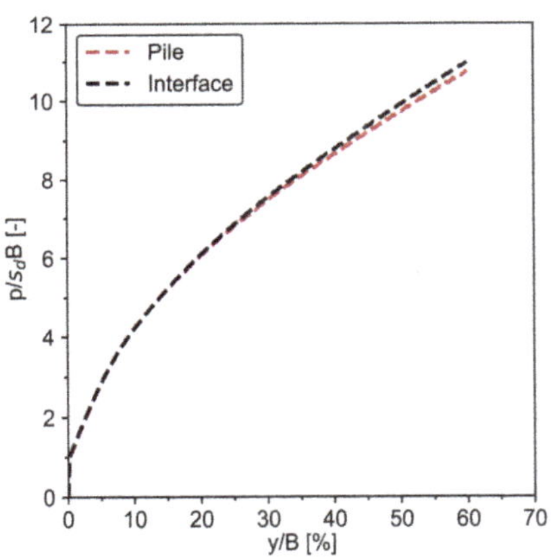

Figure 5. p–y curves integrating the stresses along the pile and the soil–pile interface

Even if the p–y curve obtained from the MATLAB code presents a shape which is not identical to those from the API and Matlock empirical equations, the normalized horizontal reaction force does not exceed the empirical values when $y < 0.35B$, then it is characterized with smaller difference to the Matlock curve up until $y/B = 0.6$. Furthermore, the validation procedure reveals that in engineering design, the DP model leads to lower p–y curves than the API and Matlock equations at $y < 0.35B$ (Figure 4). However, at horizontal displacement larger than $0.35B$, the MATLAB code generates lateral reaction force that is close to that of the Matlock formulation. The integrated results from the pile and soil–pile interface are the same. Both confirm that the p–y curves from the MATLAB code are reliable and consistent with the Matlock design method.

4. Sensitivity Analysis

4.1. Number of Elements

In order to check the spatial discretization, the influence of the number of elements is studied hereafter. The circular geometry with a 5 mm interface thickness is used for the FE number sensitivity analysis. More specifically 3600, 6300, and 11,000 T3 elements are considered. The input parameters of the simulations are listed in Table 2.

The normalized p–y curves are similar when the number of elements is 3600, 6300, and 11,000, see Figure 6. At 0.4B horizontal displacement, the normalized p values rank as 8.71, 8.69, and 8.58 for the three element numbers studied (3600, 6300, and 11,000). The p–y curves are fitted and presented in Table 3, indicating that no significant difference exists for the three element numbers. From a coarse mesh with 3600 elements to a dense mesh with 11,000 elements, the final normalized p (at 40% of y/B) decreases by a small percentage of 1.57%. Considering the necessary computational time, the element number for the p–y modeling can be chosen as 6000 ~ 10,000, to insure both effective calculation and enough elements on the interface zone.

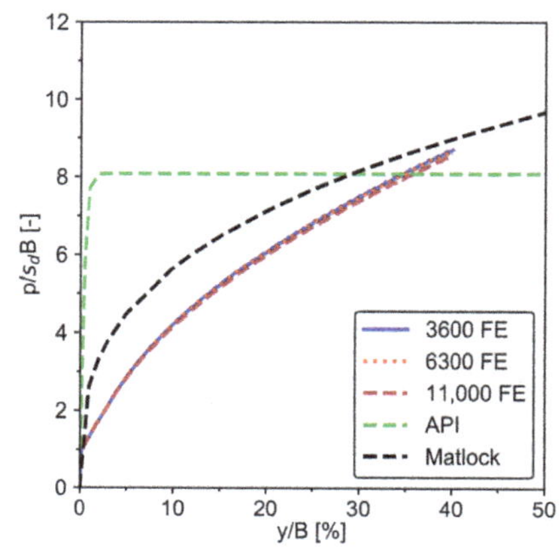

Figure 6. Influence of the number of elements.

Table 3. Fitting formulations of the p–y curves for different numbers of elements.

Element Number (-)	Fitting Formulation	R^2
3600	$\frac{p}{s_dB} = -0.0058\left(\frac{y}{B}\right)^2 + 0.4337\left(\frac{y}{B}\right)$	0.9524
6300	$\frac{p}{s_dB} = -0.0058\left(\frac{y}{B}\right)^2 + 0.4335\left(\frac{y}{B}\right)$	0.9520
11,000	$\frac{p}{s_dB} = -0.0058\left(\frac{y}{B}\right)^2 + 0.4280\left(\frac{y}{B}\right)$	0.9510

4.2. *Interface Thickness*

To check the influence of the interface thickness on the numerical results, the following different interface-thickness values are considered: 5 mm, 7 mm, 1 cm, 2.5 cm, 5 cm, and 10 cm. The input parameters of the simulations are listed in Table 2.

Figure 7 presents the effect of interface thickness on the soil reaction curves. All the p–y curves are independent of the interface thickness at the horizontal displacement considered 0 ~ 0.4B (Figure 7). From 5 mm to 10 cm interface width, the resulting normalized p–y curves have a close shape and almost the same value of the lateral bearing capacity factor at y = 0.4B: 8.58, 8.69, 8.55, 8.75, 8.81, and 8.67. The difference between the final normalized p of the six curves from the MATLAB code is 0.02 ~ 0.26. These values are higher than the one calculated by the API method (8.08) and lower than the one calculated by the Matlock equation (8.98), as shown in Figure 7.

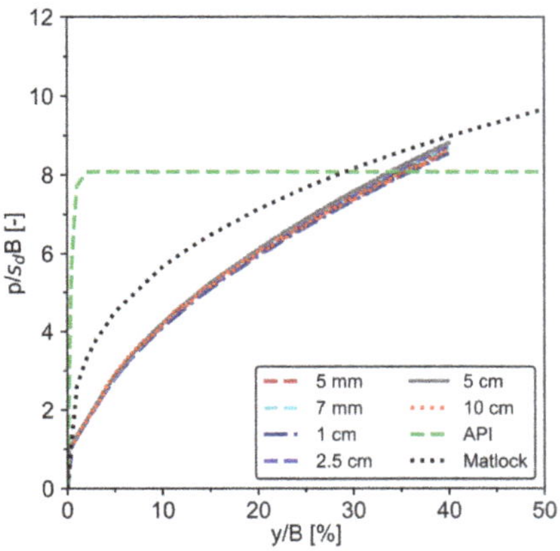

Figure 7. Influence of the interface thickness.

The obtained p–y curves are therefore independent of the interface thickness, from thickness ranges from 5 mm to 10 cm. This is also confirmed by the fitting formulations of the p–y curves as presented in Table 4.

Table 4. Fitting formulations of the p–y curves for different interface thickness.

Interface Thickness	Fitting Formulation	R^2
5 mm	$\frac{p}{s_d B} = -0.0058\left(\frac{y}{B}\right)^2 + 0.4280\left(\frac{y}{B}\right)$	0.9510
7 mm	$\frac{p}{s_d B} = -0.0058\left(\frac{y}{B}\right)^2 + 0.4338\left(\frac{y}{B}\right)$	0.9525
1 cm	$\frac{p}{s_d B} = -0.0056\left(\frac{y}{B}\right)^2 + 0.4227\left(\frac{y}{B}\right)$	0.9508
2.5 cm	$\frac{p}{s_d B} = -0.0059\left(\frac{y}{B}\right)^2 + 0.4393\left(\frac{y}{B}\right)$	0.9360
5 cm	$\frac{p}{s_d B} = -0.0057\left(\frac{y}{B}\right)^2 + 0.4335\left(\frac{y}{B}\right)$	0.9528
10 cm	$\frac{p}{s_d B} = -0.0058\left(\frac{y}{B}\right)^2 + 0.4316\left(\frac{y}{B}\right)$	0.9481

In the following, a circular mesh with 6000 T3 elements and an interface thickness of 5 mm is considered. This thickness also corresponds to the experimental laboratory results.

5. Effect of Clay Fraction on p–y Curves

The adhesion and friction angles from the interface direct shear results are used as input parameters for the p–y modeling to investigate how the clay fraction affects the curves. The influence of the clay content on the p–y behavior of pile with lateral loading is presented and discussed below.

The adhesion and friction angles from three clay fractions of 0%, 27.5%, and 55% are chosen as the input parameters for the p–y calculation (Tables 5 and 6). The parameters of the pile used are the same, as shown in Table 2. The horizontal loading is applied on the pile section considering 200 steps of 0.002 m each (i.e., in total 0.4B horizontal displacement). The residual drained shear strength, s_d, from the interface tests under 100 kPa normal stress and the pile diameter B are used to do the normalization.

Table 5. Parameters of the interface.

Clay Fraction (%)	Young's Modulus E (kPa)	Poisson's Ratio v (-)	Cohesion c (kPa)	Friction Angle δ (°)
0	8000	0.3	0	26.40
27.5	7500	0.3	2.22	22.91
55	6000	0.3	3.31	19.33

Table 6. Parameters of the soil.

Clay Fraction (%)	Young's Modulus E (kPa)	Poisson's Ratio v (-)	Cohesion c (kPa)	Friction Angle φ (°)
0	8000	0.3	0	34.11
27.5	7500	0.3	2.22	25
55	6000	0.3	10	21.81

The effect of clay content in the soil on the p–y curves is illustrated in Figure 8 and the reformulated functions are provided in Table 7. The curves show non-linearity shapes due to the non-linearity characteristics of the soil–pile interface. The overall shapes of the p–y curves in Figure 8 do not exhibit a strong asymptotic behavior in the horizontal displacement range considered (0 ~ 40 cm), which agrees with the results calculated using the Mohr–Coulomb model in [2] and [36]. When $y < 0.15B$, the same lateral capacity values are resulted at clay contents of 27.5% and 55%, and they are higher than the curve of the sand (Figure 8), but the differences are not significant. At the $0.15B$, the p is about 5.22 for the three clay contents presented. To conclude, clay content has nearly no effect on the p–y results when y is smaller than $0.15B$.

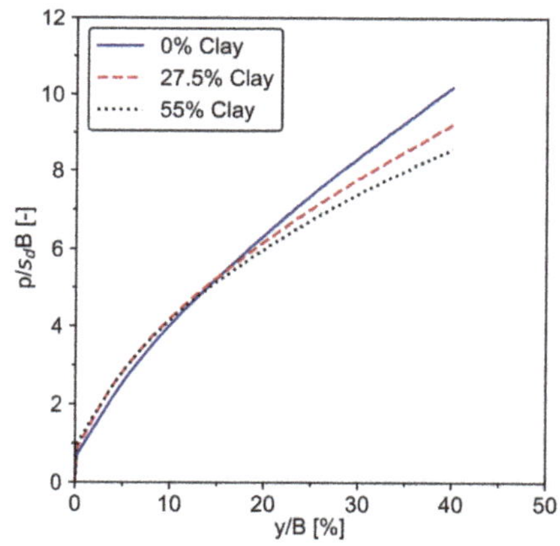

Figure 8. p–y curves as a function of clay content.

Table 7. Fitting formulations of the p–y curves at different clay contents.

Clay Content (%)	Fitting Formulation	R^2
0	$\frac{p}{s_d B} = -0.0042\left(\frac{y}{B}\right)^2 + 0.4109\left(\frac{y}{B}\right)$	0.9839
27.5	$\frac{p}{s_d B} = -0.0053\left(\frac{y}{B}\right)^2 + 0.4249\left(\frac{y}{B}\right)$	0.9668
55	$\frac{p}{s_d B} = -0.0058\left(\frac{y}{B}\right)^2 + 0.4280\left(\frac{y}{B}\right)$	0.9510

It follows from Figure 8 that the normalized p after $y > 0.15B$ is influenced by the clay content. The curve for 0% clay content (i.e., sand) becomes higher than the one for 27.5% clay content at each load step, see Figure 8. The curve of 27.5% clay content is also higher than the one for 55% clay content but with a smaller $p/(s_d B)$ gap (0.35) at each loading step compared to the one (0.46) for between 0% and 27.5% clay content. At a horizontal displacement of 40 cm (0.4B), the normalized p attains 10.18, 9.23, 8.58 for 0%, 27.5%, and 55% clay contents, respectively (Figure 8); this is because the adhesion, friction angles, and Young's modulus vary with the increasing clay content (Tables 5 and 6). The p–y curves are affected by the cohesion and friction angles, which are consistent with the literature [5,6,37].

6. Conclusions

This paper focuses on a numerical approach to finding the p–y curves for laterally loaded piles. The Drucker–Prager plastic model has been employed and the approach has been validated and compared to the API and Matlock design equations. A sensitivity analysis in terms of the number of elements and interface thickness has been presented. Finally, the influence of the clay content on the p–y behavior is presented and discussed. The main conclusions of this paper are summarized as follows:

1. The p–y results agree with the empirical results of Matlock. The validation reveals that the DP model leads to lower p–y curves with respect to those from Matlock and API when the horizontal displacement is less than 0.35B.
2. The number of elements has no important effect on the p–y curves. Considering the necessary computational time, the p–y modeling with the MATLAB code is effective when 6000 ~ 10,000 elements are adopted for the spatial discretization
3. The p–y curves are independent of interface thicknesses from 5 mm to 10 cm.
4. Clay content influences the p–y curve results. When $y < 0.15B$, the same lateral capacity values are resulted at clay contents of 27.5% and 55%, and they are higher than the curve of the sand. The normalized p–y curves show a decreasing trend with increasing clay content after $y > 0.15B$.

The primary results in this paper shed light on the relationship between p–y curves and interface parameters. However, the FE modeling of p–y curves in this study is performed by an implemented code using only the DP model, and the input parameters concern just three clay contents. Therefore, further numerical studies should involve more advanced constitutive laws, water, 3-D conditions, and the possibility to have gaps between the pile and the soil. A new function relating the lateral capacity and the horizontal displacement, as well as the clay content, should be proposed.

Author Contributions: Conceptualization, methodology, data curation, writing–original draft preparation, K.Y.; formal analysis, K.Y.; writing–review and editing, K.Y., L.L. and E.D.F. All authors have read and agreed to the published version of the manuscript.

Funding: This research received no external funding.

Institutional Review Board Statement: Not applicable.

Informed Consent Statement: Not applicable.

Data Availability Statement: The data presented in this study are available on request from the corresponding author.

Acknowledgments: Thanks for the editors' work and reviewers' comments.

Conflicts of Interest: The authors declare that they have no conflict of interest.

References

1. Zhang, Y.; Andersen, K.H.; Jeanjean, P. Cyclic py curves in clays for offshore structures. In Proceedings of the Offshore Technology Conference, Houston, TX, USA, 6–9 May 2019.
2. Wolf, T.K.; Rasmussen, K.L.; Hansen, M.; Ibsen, L.; Roesen, H. *Assessment of Py Curves from Numerical Methods for a Non-Slender Monopile in Cohesionless Soil*; Department of Civil Engineering, Aalborg University: Aalborg, Denmark, 2013.

3. Dash, S.; Rouholamin, M.; Lombardi, D.; Bhattacharya, S. A practical method for construction of py curves for liquefiable soils. *Soil Dyn. Earthq. Eng.* **2017**, *97*, 478–481. [CrossRef]
4. Liu, X.; Cai, G.; Liu, L.; Liu, S.; Duan, W.; Puppala, A.J. Improved py curve models for large diameter and super-long cast-in-place piles using piezocone penetration test data. *Comput. Geotech.* **2021**, *130*, 103911. [CrossRef]
5. Yang, Z.; Jeremić, B. Numerical analysis of pile behaviour under lateral loads in layered elastic-plastic soils. *Int. J. Numer. Anal. Methods Geomech.* **2002**, *26*, 1385–1406. [CrossRef]
6. Yang, Z.; Jeremić, B. Study of soil layering effects on lateral loading behavior of piles. *J. Geotech. Geoenvironmental Eng.* **2005**, *131*, 762–770. [CrossRef]
7. Wang, S.-F.; Tang, Y.; Wang, S.-Y. Influence of brittleness and confining stress on rock cuttability based on rock indentation tests. *J. Cent. South Univ.* **2021**, *28*, 2786–2800. [CrossRef]
8. Yu, Q.; Yin, K.; Ma, J.; Shimada, H. Vertical Shaft Support Improvement Studies by Strata Grouting at Aquifer Zone. *Adv. Civ. Eng.* **2018**, *2018*, 5365987. [CrossRef]
9. Wang, J.; Yang, S.; Wei, W.; Zhang, J.; Song, Z. Drawing mechanisms for top coal in longwall top coal caving (LTCC): A review of two decades of literature. *Int. J. Coal Sci. Technol.* **2021**, 1–26. [CrossRef]
10. Mikkelsen, A.T.; Nielsen, S.D.; Østergaard, M.U. Finite Element Modelling of p-y curves for Monopiles in Liquefied Soil. In Proceedings of the 29th International Ocean and Polar Engineering Conference, Honolulu, HI, USA, 16–21 June 2019.
11. Wang, S.-F.; Yu, T.; Li, X.-B.; Kun, D. Analyses and predictions of rock cuttabilities under different confining stresses and rock properties based on rock indentation tests by conical pick. *Trans. Nonferrous Met. Soc. China* **2021**, *31*, 31–1766. [CrossRef]
12. Bonnet, M.; Frangi, A. Analyse des structures mécaniques par la méthode des éléments finis. *Notes de Cours, Ecole Polytechnique*. 2005. Available online: http://nguyen.hong.hai.free.fr/EBOOKS/SCIENCE%20AND%20ENGINEERING/MATHEMATIQUE/PROGRAMMING/FINITE%20ELEMENTS%20METHODE/Analyse%20m%E9canique%20par%20FEM.Bonnet.pdf (accessed on 23 May 2021).
13. Bonnet, M.; Frangi, A.; Rey, C. *The Finite Element Method in Solid Mechanics*; McGraw Hill Education: New York, NY, USA, 2014; p. 365.
14. Jeanjean, P.; Zhang, Y.; Zakeri, A.; Andersen, K.; Gilbert, R.; Senanayake, A. A framework for monotonic py curves in clays. In Proceedings of the Offshore Site Investigation Geotechnics 8th International Conference Proceeding, London, UK, 12–14 September 2017; pp. 108–141.
15. Martin, C.; Randolph, M. Upper-bound analysis of lateral pile capacity in cohesive soil. *Géotechnique* **2006**, *56*, 141–145. [CrossRef]
16. Matlock, H. Correlations for design of laterally loaded piles in soft clay. In Proceedings of the Offshore Technology in Civil Engineering's Hall of Fame Papers from the Early Years, Houston, TX, USA, 1 April 1970; pp. 77–94.
17. McClelland, B.; Focht, J. Soil modulus for laterally loaded piles. *J. Soil Mech. Found. Div.* **1956**, *82*, 1081-1. [CrossRef]
18. Yang, M.; Ge, B.; Li, W.; Zhu, B. Dimension effect on py model used for design of laterally loaded piles. *Procedia Eng.* **2016**, *143*, 598–606. [CrossRef]
19. Nogami, T.; Otani, J.; Konagai, K.; Chen, H.-L. Nonlinear soil-pile interaction model for dynamic lateral motion. *J. Geotech. Eng.* **1992**, *118*, 89–106. [CrossRef]
20. Ahayan, S. *A constitutive Model for natural Clays: From Laboratory Testing to Modelling of Offshore Monopiles*; École centrale de Nantes, Université de Liège, Faculté des Sciences: Liège, Belgium, 2019.
21. Sjrensen, S.P.H.; Brjdbцk, K.T.; Mjller, M.; Augustesen, A.H. *Review of Laterally Loaded Mono-Piles Employed as the Foundation for Offshore Wind Turbines*; Department of Civil Engineering, Aalborg University: Aalborg, Denmark, 2012.
22. Ahayan, S.; Cerfontaine, B.; Collin, F.; Kotronis, P. Behaviour of laterally loaded pile. In Proceedings of the 9th European Conference on Numerical Methods in Geotechnical Engineering, NUMGE 2018, Porto, Portugal, 25–27 June 2018; pp. 1511–1518.
23. Suryasentana, S.K.; Lehane, B.M. Numerical derivation of CPT-based p-y curves for piles in sand. *Géotechnique* **2014**, *64*, 186–194. [CrossRef]
24. Zhang, Y.; Andersen, K.H. Scaling of lateral pile py response in clay from laboratory stress-strain curves. *Mar. Struct.* **2017**, *53*, 124–135. [CrossRef]
25. API. American petroleum institute recommended practice for planning, designing and constructing fixed offshore platforms-working stress design. In *API RP2A-WSD*; American Petroleum Institute: Washington, DC, USA, 2007.
26. Drucker, D.C.; Prager, W. Soil mechanics and plastic analysis or limit design. *Q. Appl. Math.* **1952**, *10*, 157–165. [CrossRef]
27. Borja, R.I. *Plasticity: Modeling & Computation*; Springer Science & Business Media: Berlin/Heidelberg, Germany, 2013.
28. Yin, K.; Liu, J.; Lin, J.; Vasilescu, A.-R.; Othmani, K.; Di Filippo, E. Interface Direct Shear Tests on JEZ-1 Mars Regolith Simulant. *Appl. Sci.* **2021**, *11*, 7052. [CrossRef]
29. Vasilescu, A.-R. Design and Execution of Energy Piles: Validation by In-Situ and Laboratory Experiments. Ph.D. Theis, École Centrale de Nantes, Nantes, France, 2019.
30. Dafalla, M.A. Effects of clay and moisture content on direct shear tests for clay-sand mixtures. *Adv. Mater. Sci. Eng.* **2013**, *2013*, 562726. [CrossRef]
31. Maghsoodi, S. Thermo-Mechanical Behavior of Soil-Structure Interface under Monotonic and Cyclic Loads in the Context of Energy Geostructures. Ph.D. Theis, Université de Lorraine, Nancy, France, 2020.
32. Maghsoodi, S.; Cuisinier, O.; Masrouri, F. Effect of Temperature on the Cyclic Behavior of Clay-Structure Interface. *J. Geotech. Geoenvironmental Eng.* **2020**, *146*, 04020103. [CrossRef]

33. Konkol, J.; Mikina, K. Some Aspects of Shear Behavior of Soft Soil–Concrete Interfaces and Its Consequences in Pile Shaft Friction Modeling. *Materials* **2021**, *14*, 2578. [CrossRef] [PubMed]
34. Yin, K.; Liu, J.; Vasilescu, A.-R.; Di Filippo, E.; Othmani, K. A Procedure to Prepare Sand–Clay Mixture Samples for Soil–Structure Interface Direct Shear Tests. *Appl. Sci.* **2021**, *11*, 5337. [CrossRef]
35. Geuzaine, C.; Remacle, J.F. Gmsh: A 3-D finite element mesh generator with built-in pre-and post-processing facilities. *Int. J. Numer. Methods Eng.* **2009**, *79*, 1309–1331. [CrossRef]
36. Østergaard, M.U.; Knudsen, B.S.; Ibsen, L.B. P-y curves for bucket foundations in sand using finite element modeling. In Proceedings of the 3rd Internationl symposium on Frontiers in offshore Geotechnics, Oslo, Norway, 10–12 June 2015; pp. 343–348.
37. Vethanayagam, V.; Ibsen, L.B. *Determination of py Curves for Bucket Foundations in Silt and Sand Using Finite Element Modelling*; Department of Civil Engineering, Aalborg University: Aalborg, Denmark, 2017.

mathematics

MDPI

Article

Analytical Stress Solution and Numerical Mechanical Behavior of Rock Mass Containing an Opening under Different Confining Stress Conditions

Lihai Tan [1,2], Ting Ren [2], Linming Dou [3], Xiaohan Yang [2,*], Gaofeng Wang [3] and Huaide Peng [1]

1 School of Resource Environment and Safety Engineering, University of South China, Hengyang 412001, China; lt716@uowmail.edu.au (L.T.); 2018000052@usc.edu.cn (H.P.)
2 School of Civil, Mining and Environmental Engineering, University of Wollongong, Wollongong, NSW 2522, Australia; tren@uowmail.edu.au
3 School of Mines, China University of Mining and Technology, Xuzhou 221116, China; lmdburst@163.com (L.D.); wanggaofengcumt@126.com (G.W.)
* Correspondence: xy987@uowmail.edu.au

Abstract: In this study, the triangle interpolation method for the calculation of mapping functions of plates containing an opening with arbitrary shapes is investigated with an improved method for point adjudgment during iterations. Afterwards, four kinds of openings with typical shapes are considered and the mapping functions for them are calculated, based on which the influence of calculation parameters such as iteration time and the number of terms on the accuracy of mapping function is discussed. Finally, the stress around an inverted U-shaped opening and around an arched opening under different far-field stress conditions is calculated and the effect of opening shape and lateral pressure coefficient on stress distribution and rock mechanical behaviors is further analyzed combined with the discrete element method (DEM) numerical simulation. The result shows that the stability and failure pattern of the rock mass is correlated with the stress around the opening, which is affected by the opening shape. The existence of opening also greatly reduces the enhancing influence of confining stress on rock specimen.

Keywords: complex variable method; conformal mapping; triangle interpolation; stress analytical solution; DEM numerical simulation

Citation: Tan, L.; Ren, T.; Dou, L.; Yang, X.; Wang, G.; Peng, H. Analytical Stress Solution and Numerical Mechanical Behavior of Rock Mass Containing an Opening under Different Confining Stress Conditions. *Mathematics* **2021**, *9*, 2462. https://doi.org/10.3390/math9192462

Academic Editor: Elena Benvenuti

Received: 1 September 2021
Accepted: 26 September 2021
Published: 2 October 2021

1. Introduction

There are an extensive range of underground openings such as roadways, tunnels, gas wells and so on in rock engineering. Stress distribution and failure characteristics around underground openings are important references in terms of the engineering design and stability assessment for such openings.

Currently, a great number of experimental studies have been conducted for the purpose of improving the understanding of the mechanical behavior of rock mass under different engineering conditions [1–5]. The mechanical behavior of underground engineering structures is usually studied via experiments on rock specimens containing one or more openings or joints under different stress conditions [6–8]. With the employment of the digital image correlation (DIC) technique, Zhou et al. [9] and Tan et al. [10] studied the mechanical behavior and crack propagation of rock specimens containing rectangular openings under static and dynamic loading, respectively. Wu et al. [11] processed rock specimens containing an opening with five presentive shapes and investigated the influence of an opening shape on the mechanical properties and fracture characteristics of rock specimens under uniaxial loading. These studies show that maniacal properties and fracturing behavior of rock specimens are tightly correlated with the loading condition and opening shape. The existence of openings significantly degrades the strength of specimens. The initial failure tends to appear at the top and bottom of the opening and dominated cracks always develop from opening corners. These influences of openings may be attributed to stress distribution. The mechanical behavior of rock mass containing defects also affects the confining stress. Wang et al. [12] reported that for rock specimens containing two flaws under compression tests, crack propagation and failure pattern of specimens are obviously affected by confining stress. The failure pattern of the specimen is transformed from vertical failure to horizontal failure with the increase in confining stress and is dominated by shear cracks under high confining stress conditions.

Experimental studies have significantly contributed to revealing the mechanical properties and failure behavior of rock mass containing defects. However, focusing on the phenomena analysis is less able to reveal the stress characteristics within the rock mass, which is essential for the prediction of potential risk [13]. Alternatively, with the development of computing capability, analytical and numerical methods have been widely employed for complex stress problems. Particularly, analytical studies promise high accuracy solutions and allow efficient parametric investigation to the analysis influence of engineering parameters on opening stability. With the employment of the complex variable method, stress solution for elastic plates containing an opening has been studied in a large amount of literature. Ukadgaonker and Awasare [14–17] conducted a series of studies on the analytical solutions for circular, elliptical, triangular, and rectangular openings in an infinite plane. Sharma [18] presented the general stress functions for determining the stress concentration around circular, elliptical and triangular openings with different opening orientation and far-field stress conditions. Wu et al. [19] calculated the stress concentration factor on the periphery of the horseshoe-shaped opening based on the analytical stress solution and analyzed the fracture response of specimens containing an opening with the combination of an analytical solution and experimental results. Zhao et al. [20] presented the analytical solution for rock stress around a square tunnel under different confining stress conditions. They found that with the increasing pressure coefficient, the boundary stress gradually converted from tensile stress to compressive stress for the two sidewalls while the opposite situation occurred for the roof and floor. According to the boundary conditions with the consideration of lining support force, Lv et al. [21] calculated the analytical solutions for a non-circular tunnel with closed support, which offers a perspective on the stress solution for supported openings at great depth. Recently, Setiawan and Zimmerman [22] revived a graphical approach proposed by Melentiev [23] and then proposed a new method for the calculation of in-plane stress around a hole with arbitrary shapes in isotropic or anisotropic materials.

In the above-mentioned analytical studies, the calculation of stress solution is based on conformal mapping, which allows an opening in a domain to be mapped into a unit circle in another domain via a mapping function. Therefore, the determination of the mapping function is preliminary and essential for the analytical solution based on the complex variable theory. The mapping function for a circular opening can be directly calculated according to its radius. For openings with simple shapes such as regular polygons, their mapping functions can also be easily calculated by given formulas [24,25]. However, for openings with complex shapes, mapping functions are usually characterized by a great number of terms and parameters, which makes it difficult to determine. Lv et al. [26] developed a general optimization method to calculate the mapping function parameters for plates containing an opening with arbitrary shapes, in which an objective function is proposed to calculate the optimal parameters by reducing the coordinate error of mapping points during iterations. Combined with Box's optimization method [27], this method is further improved by Tan et al. with a new objective function proposed for the optimization calculation [28,29]. Another method for the calculation of mapping functions is the triangle interpolation method. By repeating the mutual iteration of odd and even interpolation points, Zhu et al. [30] solved the mapping functions for a series of engineering openings with complex shapes based on the triangular interpolation theory. Compared with other methods for the calculation of mapping functions, this method is of high efficiency and allows more terms with coefficients in the form of complex numbers to obtain a high accurate solution for openings with complex shapes.

In this study, the procedure of calculating the mapping functions for openings with complex shapes using the triangular interpolation method was introduced and improved. The mapping functions of four openings with presentative shapes were calculated and the factors affecting their accuracy were discussed. In addition, the inverted U-shaped opening and arched opening were selected, and the stress solution around them with different lateral pressure coefficients were calculated. The influence of the opening shape and confining stress on the stress characteristics was studied. Furthermore, DEM numerical simulations were conducted with the failure patterns of openings at different confining stress levels presented. With the combination of analytical and numerical results, the correlation between stress distribution and failure patterns around the openings was discussed.

2. Determination of Mapping Function

2.1. Principles of Triangle Interpolation

Based on conformal mapping method, the plane containing an opening (z-plane) can be mapped to the plane containing a unit circle (ζ-plane), which is realized by the mapping function:

$$Z = \omega(\zeta) = \sum_{k=1}^{\infty} C_k \zeta^{2-k} \tag{1}$$

Take the leading m C_k of $\omega(\zeta)$, then it can be written as:

$$Z = \omega(\zeta) = \sum_{k=1}^{m} C_k \zeta^{2-k} \tag{2}$$

Usually, C_k are complex constants, which can be expressed as:

$$C_k = A_k + iB_k k = 1,\ 2,\ 3,\ \ldots,\ m \tag{3}$$

where both A_k and B_k are real constants.

For any point σ at the boundary of the unit circle in ζ-plane whose polar coordinate is $(1, \theta)$, it can be expressed as:

$$\sigma = \cos\theta + i\sin\theta \tag{4}$$

Similarly, for the mapping point t of σ at the boundary of the opening in z-plane, whose polar coordinate is (r, α), it can be expressed as:

$$t = r\cos\alpha + ir\sin\alpha \tag{5}$$

By substituting Equations (3)–(5) into Equation (2), we can find:

$$r\cos\alpha + ir\sin\alpha = \sum_{k=1}^{m} \left\{ \begin{array}{c} A_k\cos[(k-2)\theta] + B_k\sin[(k-2)\theta] \\ -iA_k\sin[(k-2)\theta] + iB_k\cos[(k-2)\theta] \end{array} \right\} \tag{6}$$

With the extraction of the real part and imaginary part, the following equation can be obtained:

$$\left\{ \begin{array}{l} r\cos\alpha = \sum_{k=1}^{m} \{A_k\cos[(k-2)\theta] + B_k\sin[(k-2)\theta]\} \\ r\sin\alpha = \sum_{k=1}^{m} \{-A_k\sin[(k-2)\theta] + B_k\cos[(k-2)\theta]\} \end{array} \right. \tag{7}$$

For m points at the unit circle in ζ-plane with coordinates of $(1, \theta_j)$, their mapping points in z-plane are (r_j, α_j). Based on the orthogonality of trigonometric functions, A_k and B_k can be expressed as:

$$\begin{cases} A_k = \frac{1}{m}\sum\limits_{j=1}^{m}\{r_j\cos\alpha_j\cos[(k-2)\theta_j] - r_j\sin\alpha_j\sin[(k-2)\theta_j]\} \\ B_k = \frac{1}{m}\sum\limits_{j=1}^{m}\{r_j\cos\alpha_j\sin[(k-2)\theta_j] + r_j\sin\alpha_j\cos[(k-2)\theta_j]\} \end{cases} \quad (8)$$

Then, $2n$ points are uniformly sampled at the unit circle in ζ-plane, which are divided into two groups $\sigma_{e,j}\ (1,\ \theta_{e,j})$ and $\sigma_{o,j}\ (1,\ \theta_{o,j})$:

$$\begin{cases} \theta_{e,j} = \frac{2\pi j}{n} \\ \theta_{o,j} = \frac{\pi(2j-1)}{n} \end{cases} \quad j = 1,\ 2,\ 3,\ \ldots,\ n \quad (9)$$

During the first iteration, n points $t_j\ (r_j, \alpha_j)$ ($j = 1, 2, 3, \ldots, n$) are randomly sampled at the boundary of the opening in z-plane. By substituting $\sigma_{e,j}$ and t_j into Equation (8), the initial solutions $A_k^{(0)}$ and $B_k^{(0)}$ can be obtained:

$$\begin{cases} A_k^{(0)} = \frac{1}{m}\sum\limits_{j=1}^{m}\{r_j\cos\alpha_j\cos[(k-2)\theta_{e,j}] - r_j\sin\alpha_j\sin[(k-2)\theta_{e,j}]\} \\ B_k^{(0)} = \frac{1}{m}\sum\limits_{j=1}^{m}\{r_j\cos\alpha_j\sin[(k-2)\theta_{e,j}] + r_j\sin\alpha_j\cos[(k-2)\theta_{e,j}]\} \end{cases} \quad (10)$$

Then the initial mapping function $\omega^{(0)}(\zeta)$ is determined with the substitution of $A_k^{(0)}$ and $B_k^{(0)}$ into Equation (2). Based on $\omega^{(0)}(\zeta)$, the mapping points $t_{o,j}$ in z-plane of $\sigma_{o,j}$ can be calculated by:

$$t_{o,j} = \omega^{(0)}(\sigma_{o,j})j = 1,\ 2,\ 3,\ \ldots,\ n \quad (11)$$

If the difference between t_j and $t_{o,j}$ is within tolerance, $A_k^{(0)}$ and $B_k^{(0)}$ are regarded as the optimal A_k and B_k, respectively. Otherwise, $t_{o,j}$ will be moved to the opening boundary and replaced by the original t_j. By substituting $\sigma_{o,j}$ and t_j into Equation (8), the solutions of A_k and B_k in the first iteration can be determined by:

$$\begin{cases} A_k^{(1)} = \frac{1}{m}\sum\limits_{j=1}^{m}\{r_j\cos\alpha_j\cos[(k-2)\theta_{o,j}] - r_j\sin\alpha_j\sin[(k-2)\theta_{o,j}]\} \\ B_k^{(1)} = \frac{1}{m}\sum\limits_{j=1}^{m}\{r_j\cos\alpha_j\sin[(k-2)\theta_{o,j}] + r_j\sin\alpha_j\cos[(k-2)\theta_{o,j}]\} \end{cases} \quad (12)$$

The mapping function $\omega^{(1)}(\zeta)$ in the first iteration can be determined by Equation (11) with the substitution of $A_k^{(1)}$ and $B_k^{(1)}$. Here the first iteration is completed. The accuracy of $\omega^{(1)}(\zeta)$ can be assessed by comparing t_j and the mapping points $t_{e,j}$ of $\sigma_{e,j}$ calculated by $\omega^{(1)}(\zeta)$; $\omega^{(1)}(\zeta)$ is employed as the optimal mapping function if its accuracy is satisfying. Otherwise, $t_{e,j}$ will be moved to the opening boundary and then become the new t_j. Then the next iteration calculation will be conducted. With the increase in iteration times, the solution accuracy will gradually increase and finally remain stable.

As for the movement of $t_{o,j}$ and $t_{e,j}$ into the opening boundary, the previous study [30] has realized this process in the Cartesian coordinate system. As shown in Figure 1, for a point z_0 in z-plane, its Cartesian coordinate is (x_0, y_0). The line through z_0 and the origin is described by function $g(z)$ and the opening boundary is described by function $f(z)$. Then, the corresponding point z_1 at the opening boundary for z_0 can be calculated by:

$$g(z_1) = f(z_1) \quad (13)$$

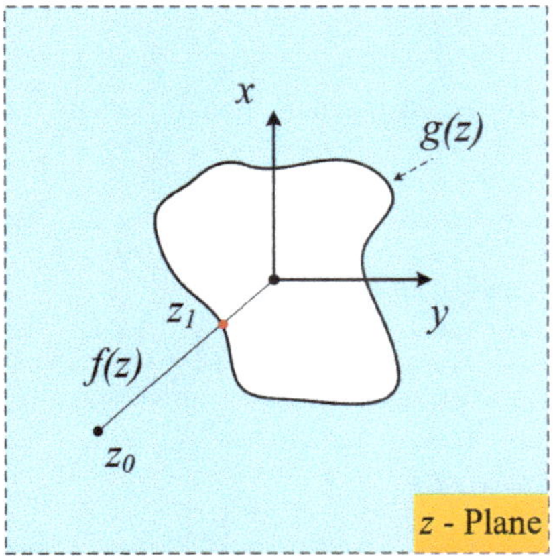

Figure 1. Schematic diagram of an opening in an infinite elastic plate.

Though it is feasible for openings with simple shapes, this method may be unavailable when the function describing the opening boundary is difficult to determine. To address this problem, reference points at opening boundary rather than the boundary function were used to moving an outside-opening point into the opening boundary. As shown in Figure 2, there are n_{ref} reference points uniformly distributed at the opening boundary in the polar coordinate system. For an outside-opening point t_0 (r_0, α_0), its corresponding point at the opening boundary is t_{new} (r_{new}, α_0). The reference points previous to and next to point t_0 are $p1(r_1,\alpha 1)$ and $p2(r_2,\alpha 2)$, respectively. Then r_{new} can be calculated by linear interpolation:

$$r_{new} = \frac{r_1 r_2(\cos\alpha_2 \sin\alpha_1 - \sin\alpha_1 \cos\alpha_2)}{(r_2 \sin\alpha_2 - r_1 \sin\alpha_1)\cos\alpha_0 - (r_2 \cos\alpha_2 - r_1 \cos\alpha_1)\sin\alpha_0} \tag{14}$$

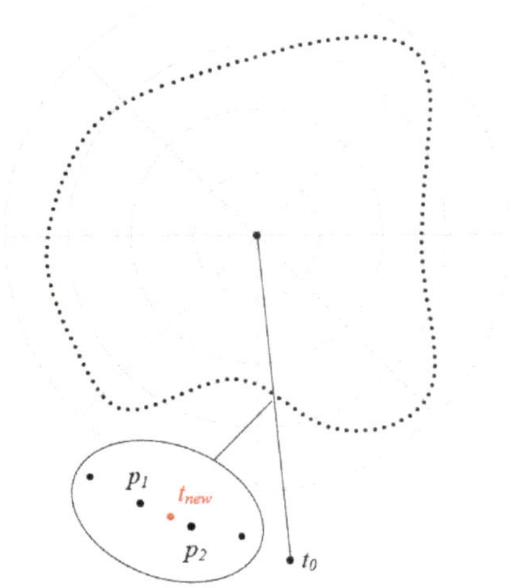

Figure 2. Schematic diagram of reference points at the opening boundary.

With Equation (14), the movement of $t_{o,j}$ and $t_{e,j}$ can be easily and accurately realized during each iteration. In this study, n_{ref} in all calculations for mapping functions is set as 1×10^4 for a high accurate solution.

2.2. The Determination of Mapping Functions

As shown in Figure 3, four openings with typical shapes were used to verify the availability of the triangle interpolation. Among them the trapezoidal opening and irregular inverted U-shaped opening are asymmetric openings while the inverted U-shaped opening and the arched opening are symmetric openings. For the trapezoidal opening (Figure 3a), *H1* = 2.50 m, *H2* = 3.70 m and *L1* = 4.00 m. For the irregular inverted U-shaped opening (Figure 3b), *H1* = 2.50 m, *H2* = 1.40 m, *H3* = 1.90 m, *L1* = 3.15 m and *L2* = 1.05 m. For the inverted U-shaped opening (Figure 3c), *H1* = 2.50 m, *H2* = 1.40 m, *H3* = 1.90 m and *L1* = 4.20 m. For the arched opening (Figure 3d), *R1* = 6.30 m, *R2* = 8.80 m, *R3* = 1.60 m, *R4* = 15.60 m, *L1* = 12.25 m, *L2* = 2.50 m, *L3* = 4.64 m, and *L4* = 0.96 m. The origin of all planes containing a single opening is the centroid of the opening.

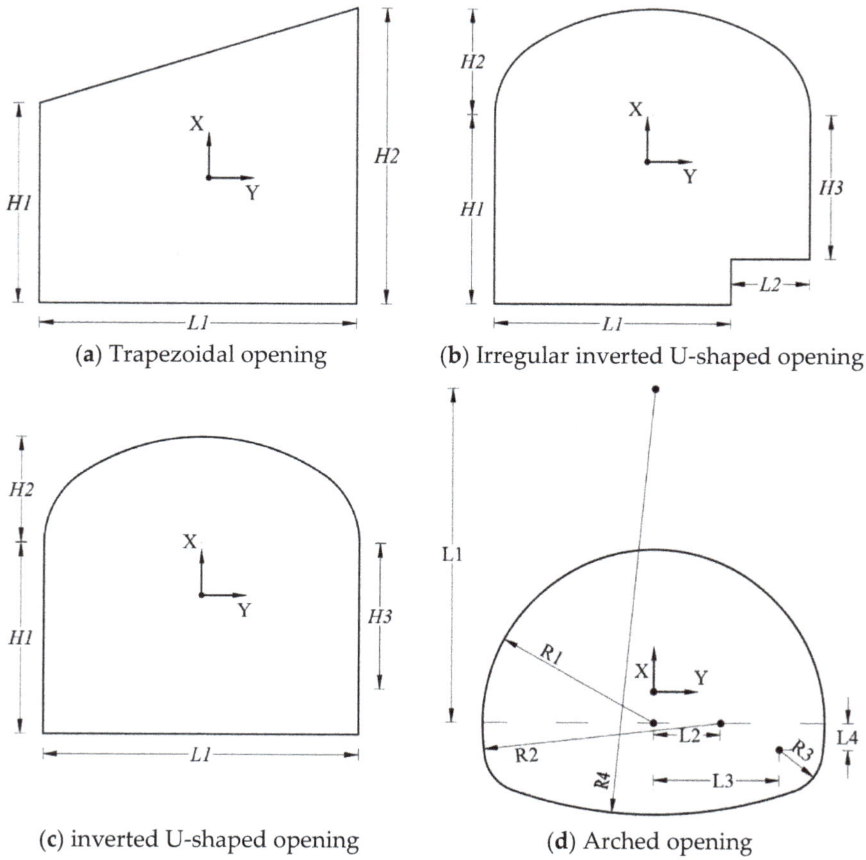

Figure 3. Four openings with typical shapes.

By uniformly sampling n points $\sigma_j\ (1,\ \theta_j)$ from $\sigma_1\ (1,\ 0)$, the mapping point of σ_j is $(r_\sigma,\ \alpha_\sigma)$ and its corresponding point at the opening boundary is $(r_\sigma^{(0)},\ \alpha_\sigma)$, the average absolute relative error (average ARE) of the n points is defined as:

$$f_{ebre} = \frac{1}{n}\sum_{j=1}^{n}\frac{\left|r_\sigma - r_\sigma^{(0)}\right|}{r_\sigma^{(0)}} \tag{15}$$

The change in the average ARE and the maximum absolute relative error (maximum ARE) of the mapping functions for each opening with the increase in iteration time are presented in Figure 4. In this study, n was set as 200 in all cases. It can be seen that both average ARE and maximum ARE decrease exponentially with the increase in iteration time. In all cases, average ARE decrease rapidly and then remains at a stable level close to zero with the increase in iteration time, indicating that the mapping functions for all openings are of high accuracy. However, the maximum ARE for all openings except the arched opening fails to decrease to zero but remains at a stable level greater than zero when the iteration time reaches a certain number. Especially for the irregular inverted U-shaped opening (Figure 4b), the maximum ARE is as high as 0.019. In contrast, the maximum ARE for the arched opening is almost closed to zero, showing little error of the mapping function. From Figure 4, the accuracy improvement of mapping functions becomes insignificant once iteration time reaches a certain value. In the following calculations, once error decrement, which is the difference between average ARE in the current iteration and in the prior iteration, is less than 1×10^{-6}, the calculation result is determined to be convergent, and the result of the last iteration is adopted.

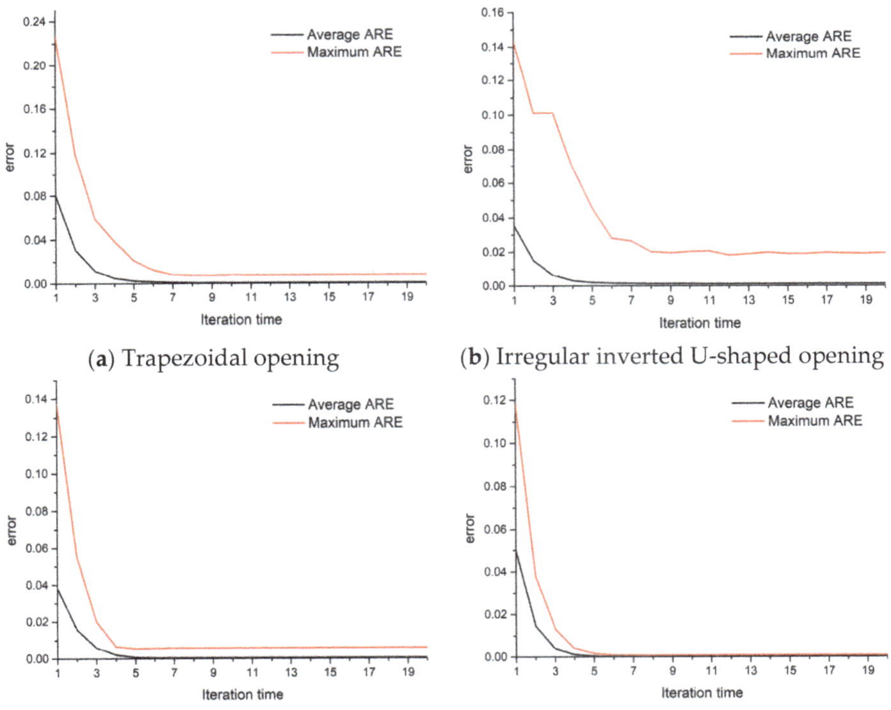

(**a**) Trapezoidal opening (**b**) Irregular inverted U-shaped opening (**c**) inverted U-shaped opening (**d**) Arched opening

Figure 4. The curves of average ARE and maximum ARE versus iteration time.

The increase in C_k number m contributes to the improvement of the accuracy of the mapping function. However, too many C_k terms may lead to a time-consuming calculation and high computational complexity. Therefore, the optimal C_k number is expected to be determined to achieve the balance between accuracy and efficiency. Accordingly, the relation curves between average ARE and m for the mapping functions of places with the four kinds of openings are presented in Figure 5. For the irregular inverted U-shaped opening, the accuracy of mapping function remains a stable level when m is greater than 60. For the others, mapping functions almost reach stable when m is greater than 20. Therefore, in this study, m was set as 60 for the irregular inverted U-shaped opening and 20 for the other openings. Figure 6 presents the iteration time for convergency with the increase of m for each opening. Overall, more iterations are required for the calculation convergency for the asymmetric openings (trapezoidal opening and irregular inverted U-shaped opening) than that for symmetric openings (inverted U-shaped opening and arched opening). This may be because C_k are complex numbers in the mapping functions for the former openings but are real number for the later openings, which makes the calculation for the former ones more complex and accordingly more iteration times for them are required.

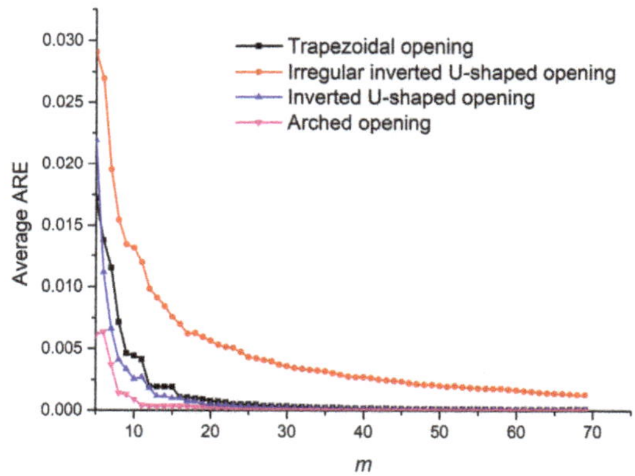

Figure 5. The relation curves between average ARE and m.

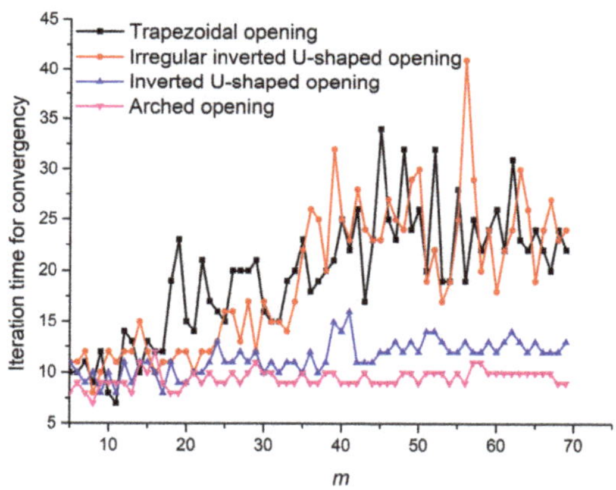

Figure 6. The relation curves between iteration time and m.

Figure 7 plots the comparisons between the mapping shape and original shape of all openings. We can find that obvious differences between the mapping shape and original shape exist at the lower right corners of the irregular inverted U-shaped opening though the whole mapping shape agree well with the original one. Comparing Figures 4 and 7 shows that for the arched opening without corners, the mapping function is perfectly accurate with little error. However, obvious error is more likely to happen at the corners of the other openings especially for those with complex shapes.

The parameters of mapping functions for some openings are listed in Table 1. As the mapping function of the plane containing the irregular inverted U-shaped opening has too many terms of C_k, its parameters are not presented. For the inverted U-shaped opening and the arched opening which are symmetrical about the x axis, C_k are real constants, namely $B_k = 0$.

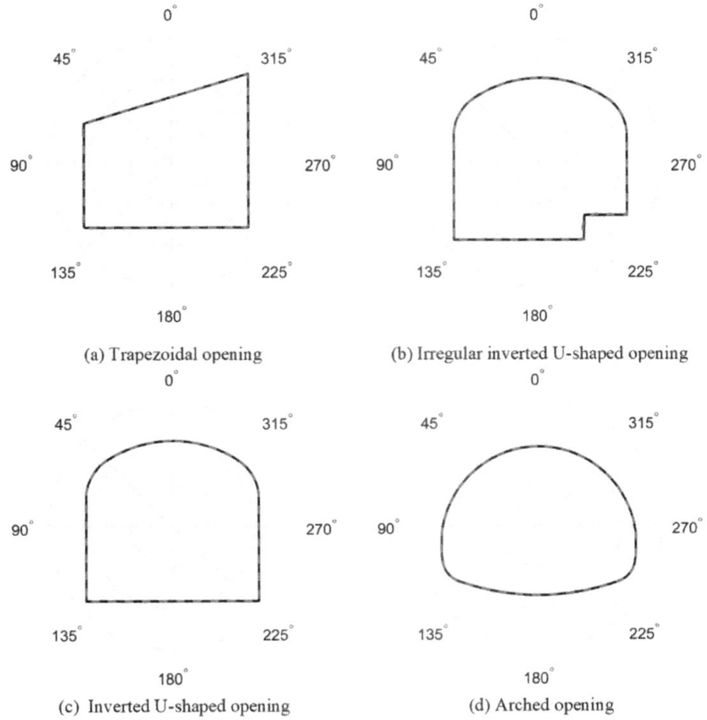

Figure 7. Comparisons between the mapping shape and original shape for different openings (black line indicates the original shape and red line indicates the mapping shape).

Table 1. Mapping function parameters of plates with different openings.

	Trapezoidal Opening		**Inverted U-Shaped Opening**	**Arched Opening**
k	A_k	B_k	C_k	C_k
1	2.108	−0.016	2.252	5.689
2	0.022	−0.027	−0.061	−0.071
3	−0.190	−0.233	−0.105	−0.760
4	0.054	−0.209	0.147	0.421
5	−0.295	−0.102	−0.237	−0.152
6	0.026	−0.035	0.073	−0.003
7	0.022	0.009	0.032	0.049
8	−0.016	0.026	−0.022	−0.033
9	0.017	0.014	0.013	0.004
10	−0.007	0.003	−0.008	0.009
11	−0.004	−0.005	−0.003	−0.007
12	0.010	−0.007	0.008	0.001
13	−0.001	−0.001	−0.006	0.000
14	0.001	0.001	−0.001	0.001
15	0.000	0.001	0.002	0.000
16	−0.005	0.001	−0.002	−0.002
17	0.000	−0.001	0.003	0.003
18	0.001	0.000	0.000	−0.001
19	0.000	0.001	−0.002	−0.001
20	0.002	0.000	0.001	0.002

3. Stress Solution of Plates Containing a Single Opening

3.1. Calculation Models

In practical engineering, the surrounding rock mass of the underground opening is usually in complex stress conditions. The far-field stress may have a significant influence on the stress distribution around the opening and thus may further affect the stability and failure pattern of the opening. To improve the understanding of stress distribution around the opening under different far-field stress conditions, the inverted U-shaped opening and arched opening are selected and the analytical stress solutions for them are calculated, based on which the effect of the opening shape and lateral pressure coefficient λ on stress distribution is investigated. To make the calculation more effective, principal stresses are suggested for the description the far-field stress condition, which can be realized by adjusting the orientation of the opening. In practical underground engineering, as the direction of vertical stress is usually close to or identical to that of a principal stress and the tunnel orientation is usually designed to be parallel to a principal stress direction to minimize the effect of antiplane stresses. The change in opening orientation is not considered in the following examples. As shown in Figure 8, the far-field vertical stress σ_x^∞ and the far-field horizontal stress σ_y^∞ are two principal stresses in the plate, which are set as p and λp, respectively. λ is the lateral pressure coefficient. In this study, seven levels of λ are considered, whose value is from 0 to 1.0 with an interval of 0.1. The detailed solution procedure for a plate containing an opening has been elaborated by many studies [28,31]. Therefore, it is not presented in this study.

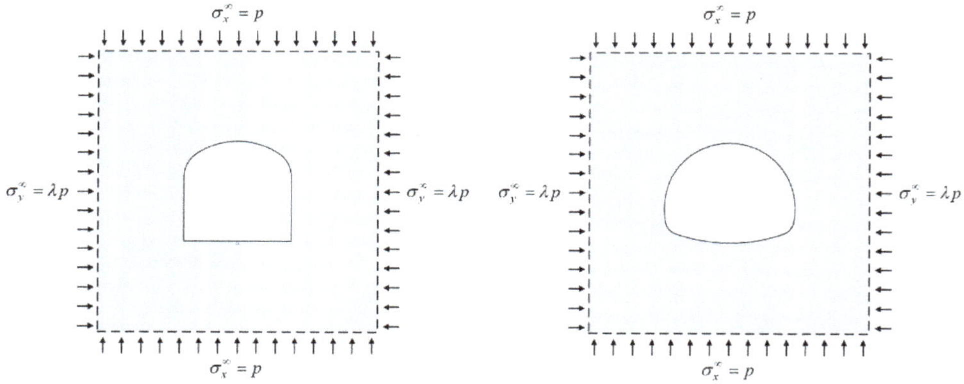

Figure 8. Schematic diagram of a plate containing an opening under far-field stress condition.

3.2. Stress Calculation for Plates

Figure 9 presents the analytical solutions of hoop stress σ_θ on the opening boundaries with some typical λ. The negative σ_θ is tensile stress and the positive σ_θ is compressive stress. Tensile stress appears in the roof and floor of the openings while compressive stress appears in other locations for both kind of openings under uniaxial stress (λ = 0), but the tensile stress gradually decreases and finally converts into compressive stress with the increase in confining stress. However, despite their similarity, significant difference is observed between the stress distribution around the inverted U-shaped opening and around the inverted U-shaped opening. The compressive stress concentration level at the corners of the inverted U-shaped opening is much higher than that at the corners of the arched opening. For example, the maximum hoop stress at the corners is 7.69p for the inverted U-shaped opening but is only 3.73p for the arched opening. In addition, the two kinds of openings also show different stress response to the change in λ. Figure 10 presents the change trend of the maximum hoop stress $\sigma_{\theta,max}$, the minimum hoop stress $\sigma_{\theta,min}$ and the difference between them ($\sigma_{\theta,max} - \sigma_{\theta,min}$) with the increase in λ. For both openings, $\sigma_{\theta,min}$ is close to -*p* when $\lambda = 0$ and linearly increase with the increase in λ (Figure 10b). When λ reaches a certain value, which is about 0.58 for the inverted U-shaped opening and 0.62 for the arched opening, $\sigma_{\theta,min}$ begin to be positive, which means there is no tensile stress appearing at the opening boundaries anymore. However, from Figure 10a we can see that $\sigma_{\theta,max}$ at the boundary of the inverted U-shaped opening increases linearly with λ but $\sigma_{\theta,max}$ at the boundary of the arched opening almost remains constant. In addition, Figure 10a shows that the gap between $\sigma_{\theta,max}$ and $\sigma_{\theta,min}$ is more and more greater for the inverted U-shaped opening but keeps narrowing for the arched opening.

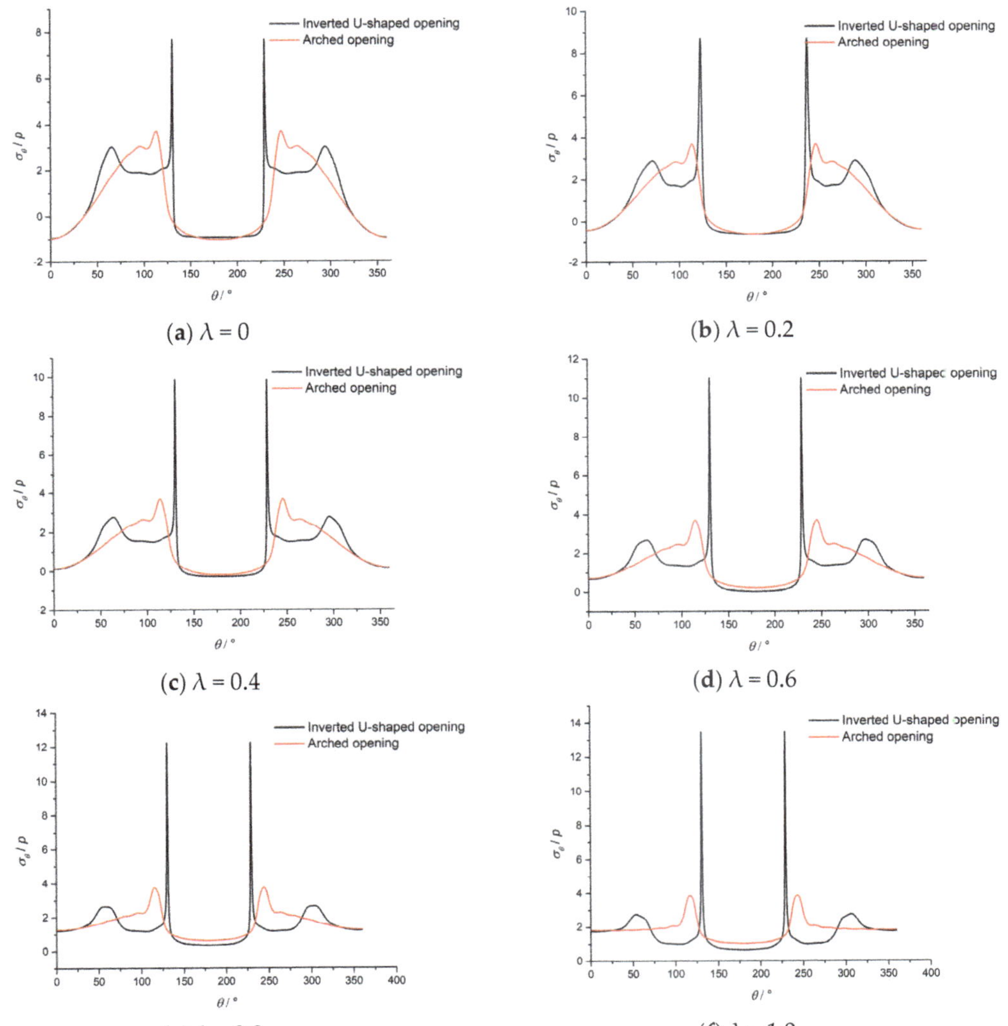

Figure 9. Analytical solution of hoop stress around openings with different λ.

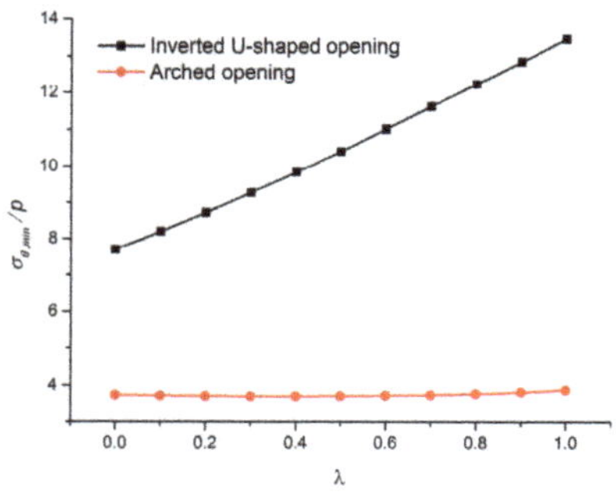

(**a**) $\sigma_{\theta,max}$

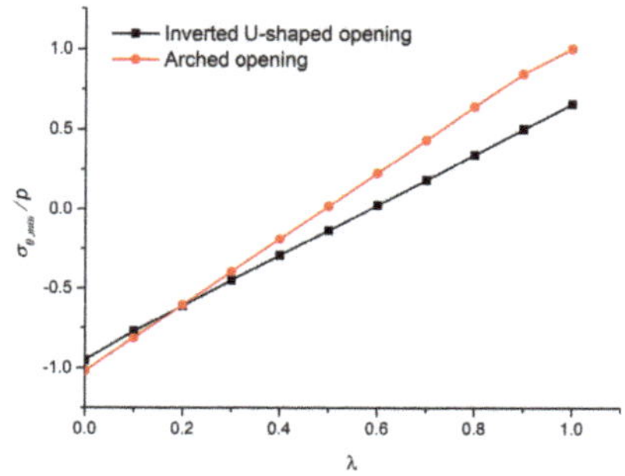

(**b**) $\sigma_{\theta,min}$

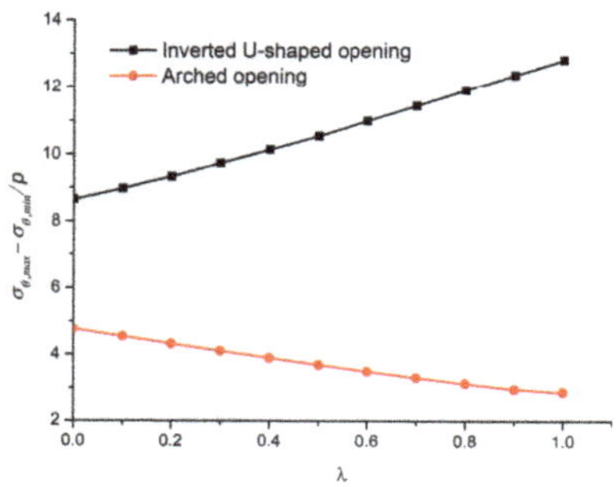

(**c**) $\sigma_{\theta,max} - \sigma_{\theta,min}$

Figure 10. Curves of $\sigma_{\theta,max}$, $\sigma_{\theta,min}$ and $(\sigma_{\theta,max} - \sigma_{\theta,min})$ versus λ.

4. DEM Numerical Simulation

4.1. Numerical Modelling

The mechanical properties and failure patterns of rock mass containing an opening under different stress conditions were further investigated by DEM numerical simulations via the commercial DEM code PFC2D. The balls within the models were bonded by contacts with the linear parallel bond model, which has been extensively used to study the mechanical behavior of rock mass by Itasca and many studies [32–35]. As shown in Figure 11, numerical specimens containing an inverted U-shaped opening and an arched opening were established. Axial stress σ_v was produced by axial displacement-control loading and confining stress σ_h was applied on both sides of the specimens. The meso-parameters of the numerical specimens were calculated by a series of numerical standard uniaxial compression tests based on a kind of sandstone with UCS of 49.5 MP and Young's Modulus of 7.2 GPa, respectively. To make the numerical results for them comparable, the parameter Ck for both of their mapping functions were scaled down by 1: C1 respectively to make sure the sizes of the two openings were at the same level. Then, the size of the scaled-down openings was used in the numerical modelling.

4.2. Mechanical Properties

Some representative curves of axial stress versus axial strain for numerical specimens are presented in Figure 12. The axial stress is computed by dividing sectional area of the specimen into the force on the loading plate. The axial strain is the ratio of the initial length of the specimen to the axial displacement of the loading plate. It should be noted that the computed stress and strain are used to demonstrate the effect of a hole on the mechanical behavior of surrounding rock mass rather than to quantify the stress and strain conditions within specimens as they are not uniform within the specimens containing holes.

For all intact specimens containing an opening, their stress-strain curves increase linearly during the early loading stage until close to their peak points. All the curves drop sharply during post-peak stage under uniaxial compressive loading (σ_h = 0 MPa), indicating strong brittle behavior. However, the plastic behavior appears under confining stress conditions, especially for the numerical specimens containing an inverted U-shaped opening where stress curves tend to be flat close to the peak point.

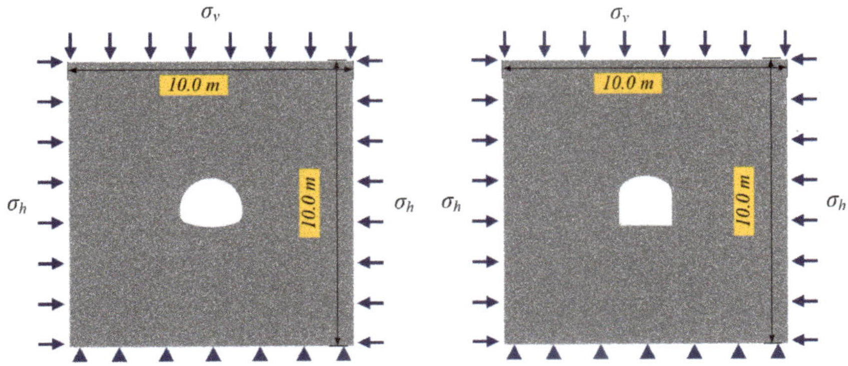

(**a**) Numerical specimen containing an inverted U-shaped opening (**b**) Numerical specimen containing an arched opening

Figure 11. Numerical modelling of rock specimens containing an opening.

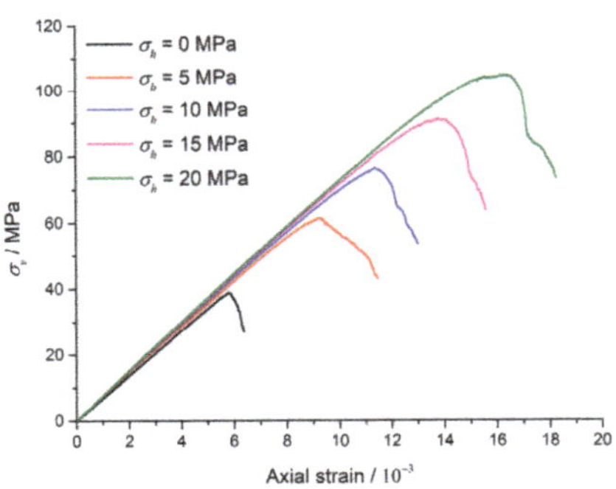

(**a**) Intact specimens

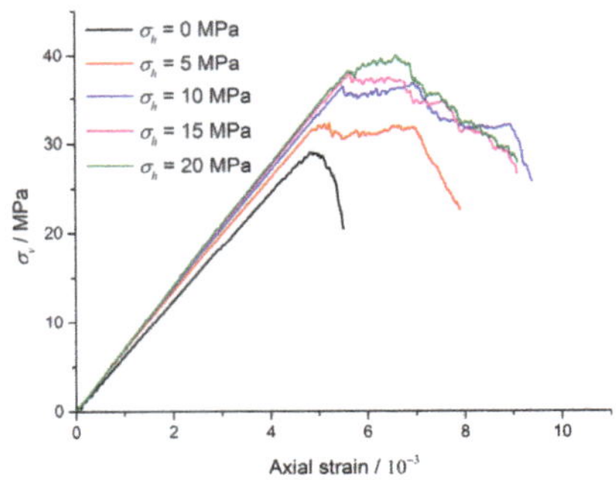

(**b**) Specimens containing an inverted U-shaped opening

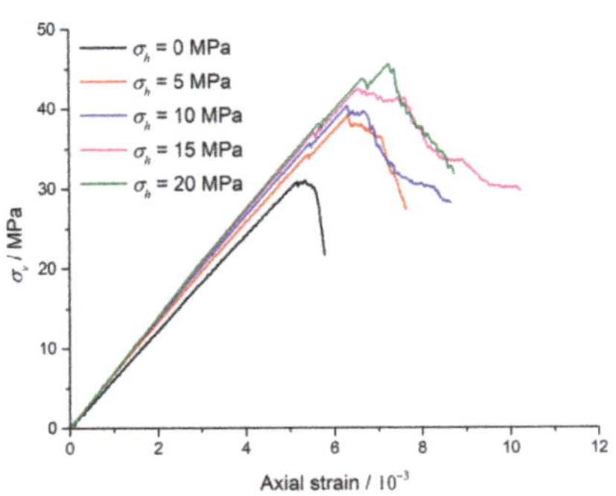

(**c**) Specimens containing an arched opening

Figure 12. Axial stress-strain curves for specimens with different openings.

The change trends of peak axial stress of numerical specimens with different confining stress are shown in Figure 13. For intact specimens, a near linear relation between peak axial stress and confining stress is observed. However, this relation is not linear for specimens containing an opening. Moreover, the peak axial stress of specimens containing an opening is much lower than that of intact specimens. Take the numerical specimen containing an arched opening for example, its peak axial stress under uniaxial compression condition is 30.94 MPa, which is 79.76% of that of the intact specimen. In comparison, when the confining stress reaches 20 MPa, peak axial stress of the intact specimen sharply increases to 104.74 MPa while that of the specimen containing an arched opening only slightly increases to 45.49 MPa, which is only 43.43% of the intact one, also indicates that the existence of the opening significantly suppresses the positive influence of confining stress on strength of specimens. In addition, the opening shape also has an influence on the specimen strength. The specimen containing an arched opening is able to bear higher axial stress than that containing an inverted U-shaped opening under the same confining stress condition.

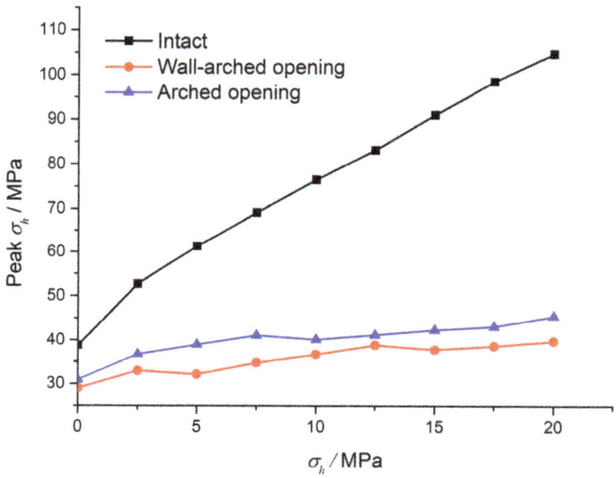

Figure 13. Peak axial stress of numerical specimens under different confining stress conditions.

4.3. Failure Patterns

Figures 14 and 15 present the failure patterns of specimens in some representative cases. The left part of each sub-figure plots the failure pattern of the specimen at the point of 95% peak axial stress in the pre-peak stage and the right part plots the failure pattern at the point of 70% peak axial stress in the post-peak stage, which is the end of the loading test. As revealed by Figure 12, specimens are mainly under elastic deformation in the pre-peak stage, therefore the failure that appears in this stage may tightly relate to the stress distribution determined by the analytical solution of the elastic plate containing the opening. For the specimen containing a single inverted U-shaped opening at the point of 95% peak axial stress in the pre-peak stage, failure concentrates the corners and the wall sides without confining stress. Meanwhile, a tensile crack parallel to the axial loading direction from the middle part of the roof, where the maximum tensile stress in the corresponding analytical stress solution, is observed. When there is confining stress, the characteristics of failure around the sidewalls and corners remains the same, but the tensile crack on the opening roof fails to appear anymore. It can be seen in all cases for the inverted U-shaped opening, failure forming a "V-shaped" wedge around the sidewalls, which are a compressive stress concentration area in corresponding cases. Similar failure patterns happened to the specimen containing a single arched opening in the pre-peak stage, where tensile cracks appear in the middle parts of the opening roof and floor under uniaxial compressive loading but do not appear under confining stress conditions.

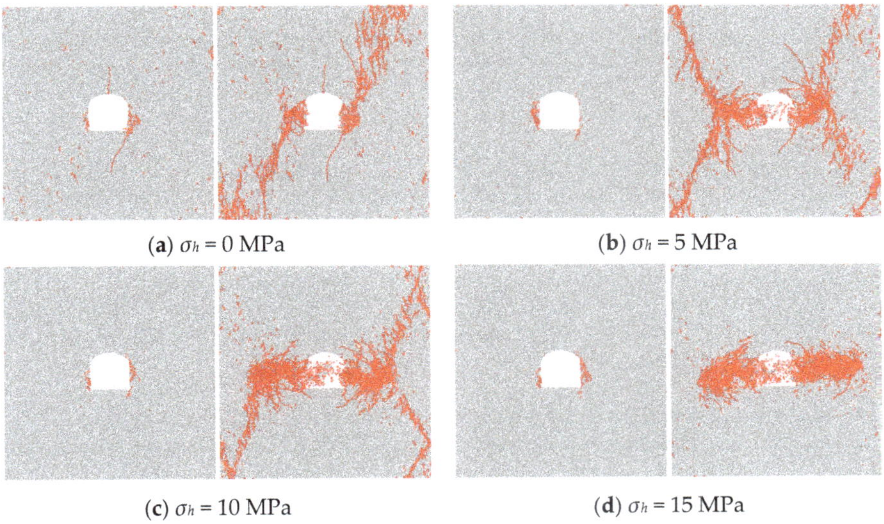

Figure 14. Failure patterns of numerical specimens containing an inverted U-shaped opening.

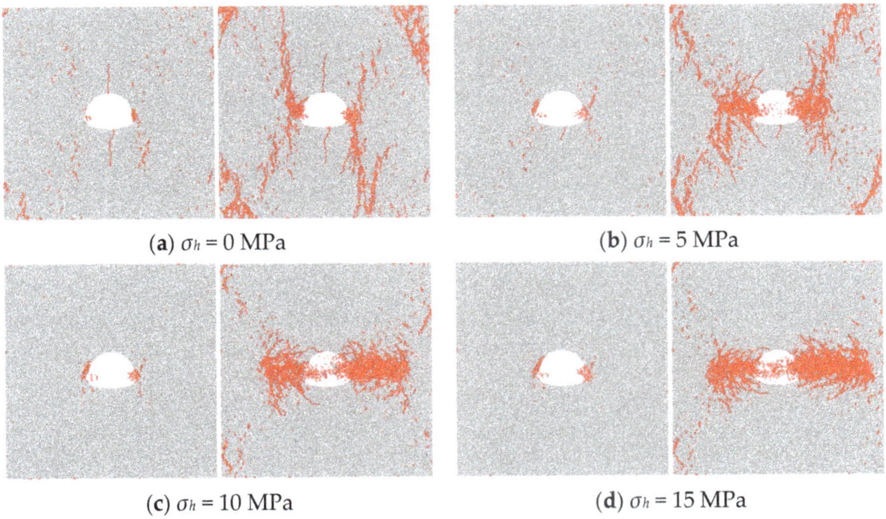

Figure 15. Failure patterns of numerical specimens containing an arched opening.

Combining the analytical solution and numerical results shows that the failure patterns around openings in the pre-peak stage are in accord with stress distribution characteristics in the corresponding analytical cases. With the loading going on, the stress distribution continuously changes with more and more cracks appearing. In the post-peak stage, the specimens are broken with intensive failure, where the patterns are not only led by initial cracks during elastic deformation but also affected by the interaction between the specimen boundaries and the openings. For the specimen containing an inverted U-shaped opening under uniaxial loading, two shear cracks appeared in the failure area around the opening sidewalls and extends to the upper right corner and the lower left corner respectively, which forms a diagonal failure area connected with the opening. When the confining stress reached 5 MPa, the specimen was fractured by four shear cracks connecting the opening and the specimen corners. Then, the shear cracks seem were suppressed with the confining stress increasing to 10 MPa. When the confining stress was 15 MPa, no shear cracks connecting the opening and the specimen boundary appeared. The specimen instability is dominated by intensive failure around both sidewall because of strong compressive stress concentration. Similar response of failure patterns to confining stress are also happened to the specimen containing an arched opening.

5. Discussion

5.1. Generality of the Improved Triangle Interpolation Method for the Determination of Mapping Function

The accuracy of the mapping function depends on the iteration time and the number of C_k. The optimal iteration time can be determined when the error decrement is less than a given value, which is 1×10^{-6} in this study. Generally, the optimal number of C_k depends on the complexity of the opening. The mapping function for an opening with a more complex shape requires more terms of C_k for a satisfying accuracy, but this study reveals that the accuracy will gradually converge on a constant value with the increase in C_k number, showing an exponential decrease trend. As too many terms of C_k may increase the difficulty of following the calculations for analytical stress solution, the minimum number of C_k satisfying the desired is suggested according to the practical engineering requirements.

The triangle interpolation method has been proved to be an efficient method to calculate the mapping functions for openings with arbitrary shapes. In the previous study, the adjustment of mapping points during iterations was conducted by means of the boundary curve function of the opening. However, the boundary curve function may be inaccurate and difficult to be determined for openings with complex shapes. Alternatively, by sampling enough reference points uniformly at the boundary, the adjustment of mapping points can be realized easily without an accuracy loss caused by the boundary curve function.

5.2. Influence of Opening Shape and Confining Stress on the Mechanical Behaviour of Specimens

Based on the analytical and numerical results, it can be inferred that for openings under low confining stress conditions, initial failure may always appear on the roof in the form of tensile cracks. However, under high confining stress, the tensile concentration will be reduced and even disappear. Fracturing, deformation and rock burst around the sidewalls and corners, which are high compressive stress concentration areas, are more likely to appear than a disaster led by tensile failure on the opening roof. Accordingly, the stability of sidewalls and corners deserve more attention than that of roof. Based on the stress variation laws revealed by the analytical solution, however, it is logical that the stress around the sidewalls will convert into tensile stress and compressive stress concentration will form in the roof and floor of the opening when the lateral pressure coefficient reaches a certain level. In such a case, vulnerable areas and failure patterns around the opening may be quite different and corresponding solutions and analyses should be further carried out.

Compared with the inverted U-shaped opening, the analytical stress solution shows that the compressive stress concentration around the arched opening is much lower under the same far-field stress condition. The stress distribution around the arched opening is more uniform and the difference between the maximum hoop stress and the minimum hoop stress reduces with the increase in lateral pressure coefficient, which is opposite to that for the inverted U-shaped opening. Numerical results also show that the strength of specimens contacting an arched opening is higher than that of specimens containing an inverted U-shaped opening under all test conditions. These comparisons seem to indicate that opening sections with corners are not suggested for long-term rock engineering underground.

6. Conclusions

In this study, the stress solution for plates containing an opening was studied based on conformal mapping. The triangle interpolation method for the determination of mapping functions was discussed and improved. The stress distribution and failure patterns of rock mass containing an opening under different confining stress conditions were further analyzed with the combination of analytical stress solutions and numerical simulations. The main conclusions of this paper include:

(1) Mapping function is essential for the stress solution of rock mass around underground openings. By keeping even and odd interpolation points iterating each other repeatedly, the mapping function for a given opening can be effectively determined by the triangle interpolation method. The key point of this method is to move the calculated mapping points into the opening boundary during each iteration. Compared with boundary curve function, the method of sampling reference points at the boundary combined with linear interpolation is suggested for this adjustment as it is easy to conduct and promise high accuracy.
(2) Stress distribution characteristics around the opening are significantly affected by the opening shape, which further affects the stability and failure pattern of the rock mass. The maximum hoop stress at the boundary of the inverted U-shaped opening is much higher than that at the boundary of the arched opening under the same far-field stress condition and shows a linear increasing trend with the increase in lateral pressure coefficient. However, the sensitivity of the maximum hoop stress of the later one to the lateral pressure coefficient is much less, remaining at a stable level despite of the varying lateral pressure coefficient.
(3) Combining the analytical stress solution and DEM numerical tests shows that the failure patterns of specimens in the pre-peak stage agree well with the analytical elastic stress solution. Under uniaxial stress conditions, initial failure is characterized by tensile cracks from the roof and floor of the openings, where there are tensile stress concentration areas, then failure from the sidewalls and corners caused by concentrated compressive stress is observed. Under biaxial stress conditions, the analytical stress solution reveals that tensile stress around the openings gradually decreases and finally coverts into compressive stress with the increase in the lateral pressure coefficient. Accordingly, the tensile failure is suppressed in the corresponding numerical cases.

Author Contributions: Conceptualization, methodology, software, writing—original draft preparation: L.T.; writing—review and editing, X.Y., G.W.; supervision, T.R., L.D.; funding acquisition, H.P., L.D. All authors have read and agreed to the published version of the manuscript.

Funding: This research is supported by financial grants from the National Natural Science Foundation of China (51934007, 51874292), Fundamental Research Funds for the Central Universities (2020QN44) and Research Foundation of Education Bureau of Hunan Province, China (19B479). The authors are very grateful to the financial contribution and convey their appreciation of the organization for supporting this basic research.

Data Availability Statement: Data sharing not applicable.

Conflicts of Interest: The authors declare no conflict of interest.

References

1. Zhou, Z.; Cai, X.; Li, X.; Cao, W.; Du, X. Dynamic Response and Energy Evolution of Sandstone Under Coupled Static–Dynamic Compression: Insights from Experimental Study into Deep Rock Engineering Applications. *Rock Mech. Rock Eng.* **2020**, *53*, 1305–1331. [CrossRef]
2. Cai, X.; Zhou, Z.; Zang, H.; Song, Z. Water saturation effects on dynamic behavior and microstructure damage of sandstone: Phenomena and mechanisms. *Eng. Geol.* **2020**, *276*, 105760. [CrossRef]
3. Wang, S.-F.; Tang, Y.; Li, X.-B.; Du, K. Analyses and predictions of rock cuttabilities under different confining stresses and rock properties based on rock indentation tests by conical pick. *Trans. Nonferr. Met. Soc. China* **2021**, *31*, 1766–1783. [CrossRef]
4. Wang, S.-F.; Tang, Y.; Wang, S.-Y. Influence of brittleness and confining stress on rock cuttability based on rock indentation tests. *J. Cent. South. Univ.* **2021**, 1–15. [CrossRef]
5. Song, Z.; Konietzky, H.; Cai, X. Modulus degradation of concrete exposed to compressive fatigue loading: Insights from lab testing. *Struct. Eng. Mech.* **2021**, *78*, 281–296.

6. Tan, L.; Ren, T.; Yang, X.; He, X. A numerical simulation study on mechanical behaviour of coal with bedding planes under coupled static and dynamic load. *Int. J. Min. Sci. Technol.* **2018**, *28*, 791–797. [CrossRef]
7. Cao, R.-H.; Yao, R.; Hu, T.; Wang, C.; Li, K.; Meng, J. Failure and mechanical behavior of transversely isotropic rock under compression-shear tests: Laboratory testing and numerical simulation. *Eng. Fract. Mech.* **2021**, *241*, 107389. [CrossRef]
8. Cai, X.; Zhou, Z.; Du, X. Water-induced variations in dynamic behavior and failure characteristics of sandstone subjected to simulated geo-stress. *Int. J. Rock Mech. Min. Sci.* **2020**, *130*, 104339. [CrossRef]
9. Zhou, Z.; Tan, L.; Cao, W.; Zhou, Z.; Cai, X. Fracture evolution and failure behaviour of marble specimens containing rectangular cavities under uniaxial loading. *Eng. Fract. Mech.* **2017**, *184*, 183–201. [CrossRef]
10. Tan, L.; Ren, T.; Dou, L.; Cai, X.; Yang, X.; Zhou, Q. Dynamic response and fracture evolution of marble specimens containing rectangular cavities subjected to dynamic loading. *Bull. Eng. Geol. Environ.* **2021**. [CrossRef]
11. Wu, H.; Zhao, G.; Liang, W. Mechanical properties and fracture characteristics of pre-holed rocks subjected to uniaxial loading: A comparative analysis of five hole shapes. *Theor. Appl. Fract. Mech.* **2020**, *105*, 102433. [CrossRef]
12. Wang, M.; Wan, W.; Zhao, Y. Experimental study on crack propagation and the coalescence of rock-like materials with two preexisting fissures under biaxial compression. *Bull. Eng. Geol. Environ.* **2020**, *79*, 3121–3144. [CrossRef]
13. Zhou, K.; Dou, L.; Gong, S.; Li, J.; Zhang, J.; Cao, J. Study of Rock Burst Risk Evolution in Front of Deep Longwall Panel Based on Passive Seismic Velocity Tomography. *Geofluids* **2020**, *2020*, 8888413. [CrossRef]
14. Ukadgaonker, V.; Awasare, P. A novel method of stress-analysis of an infinite-plate with rounded corners of a rectangular hole under uniform edge loading. *Indian J. Eng. Mater.* **1994**, *1*, 17–25.
15. Ukadgaonker, V.; Awasare, P. A novel method of stress analysis of an infinite plate with elliptical hole with uniform loading at infinity. *J. Inst. Eng. (India)* **1993**, *73*, 309–311.
16. Ukadgaonker, V.; Awasare, P. A novel method of stress-analysis of infinite-plate with circular hole with uniform loading at infinity. *Indian J. Eng. Mater. Sci.* **1994**, *1*, 17–25.
17. Ukadgaonker, V.; Rao, D. Stress distribution around triangular holes in anisotropic plates. *Compos. Struct.* **1999**, *45*, 171–183. [CrossRef]
18. Sharma, D.S. Stress distribution around circular/elliptical/triangular holes in infinite composite plate. *Eng. Lett.* **2012**, *20*, 1.
19. Wu, H.; Ma, D. Fracture response and mechanisms of brittle rock with different numbers of openings under uniaxial loading. *Geomech. Eng.* **2021**, *25*, 481–493.
20. Zhao, G.; Yang, S. Analytical solutions for rock stress around square tunnels using complex variable theory. *Int. J. Rock Mech. Min. Sci.* **2015**, *80*, 302–307. [CrossRef]
21. Lu, A.-z.; Zhang, N.; Kuang, L. Analytic solutions of stress and displacement for a non-circular tunnel at great depth including support delay. *Int. J. Rock Mech. Min. Sci.* **2014**, *70*, 69–81. [CrossRef]
22. Setiawan, N.B.; Zimmerman, R.W. A unified methodology for computing the stresses around an arbitrarily-shaped hole in isotropic or anisotropic materials. *Int. J. Solids Struct.* **2020**, *199*, 131–143. [CrossRef]
23. Melentiev, P. *Several New Methods and Devices for Approximate Computations*; Nordhoff: Groningen, The Netherlands, 1953.
24. Sharma, D.S. Stress distribution around polygonal holes. *Int. J. Mech. Sci.* **2012**, *65*, 115–124. [CrossRef]
25. Balakrishnan, A.; Joines, W.T.; Wilson, T.G. Air-gap reluctance and inductance calculations for magnetic circuits using a Schwarz-Christoffel transformation. *IEEE Trans. Power Electron.* **1997**, *12*, 654–663. [CrossRef]
26. Lv, A.Z.; Wang, Q.W. New Method of Determination for the Mapping Function of Tunnel with Arbitrary Boundary Using Optomozation Techniques. *Chin. J. Rock Mech. Eng.* **1995**, *014*, 269–274.
27. Holland, J.H. *Adaptation in Natural and Artificial Systems: An Introductory Analysis with Applications to Biology, Control, and Artificial Intelligence*; The University of Michigan Press: Ann Arbor, MI, USA, 1975.
28. Tan, L.; Ren, T.; Dou, L.; Yang, X.; Qiao, M.; Peng, H. Analytical stress solution and mechanical properties for rock mass containing a hole with complex shape. *Theor. Appl. Fract. Mech.* **2021**, *114*, 103002. [CrossRef]
29. Tan, L.; Ren, T.; Dou, L.; Yang, X.; Cai, X.; Qiao, M. Analytical stress solution for rock mass containing two holes based on an improved Schwarz alternating method. *Theor. Appl. Fract. Mech.* **2021**, *116*, 103092. [CrossRef]
30. Zhu, J.H.; Yang, J.H.; Shi, G.P.; Wang, J.; Cai, J.P. Calculating method on the conformal mapping from the exterior of unit circle to the exterior of cavern with the arbitrary excavation cross section. *Rock Soil Mech.* **2014**, *35*, 175–183.
31. Wu, H.; Kulatilake, P.H.S.W.; Zhao, G.; Liang, W.; Wang, E. A comprehensive study of fracture evolution of brittle rock containing an inverted U-shaped cavity under uniaxial compression. *Comput. Geotech.* **2019**, *116*, 103219. [CrossRef]
32. Itasca, C. PFC (particle flow code in 2 and 3 dimensions), version 5.0 [User's manual]. *Numer. Anal. Methods Geomech.* **2014**, *32*, 189–213.
33. Wu, N.; Liang, Z.; Zhou, J.; Zhang, L. Energy evolution characteristics of coal specimens with preformed holes under uniaxial compression. *Geomech. Eng.* **2020**, *20*, 55–66.
34. Vahab, S.; Tabaroei, A. Numerical simulation of the influence of interaction between Qanat and tunnel on the ground settlement. *Geomech. Eng.* **2020**, *23*, 455–466. [CrossRef]
35. Fan, H.; Sun, S.; Wei, J.; Le, H.; Zhu, F.; Wang, W.; Liu, Y.; Wang, J. Shear Behavior of Soil–Rock Mixture Composed of Diorite-Porphyrite Considering Weathering Sequence. *Int. J. Geomech.* **2021**, *21*, 04021122. [CrossRef]

mathematics

Article

Closed-Form Solutions for Locating Heat-Concentrated Sources Using Temperature Difference

Daoyuan Sun [1], Yifan Wu [2], Longjun Dong [1] and Qiaomu Luo [1,*]

[1] School of Resources and Safety Engineering, Central South University, Changsha 410083, China
[2] College of Foreign Languages and Cultures, Xiamen University, Xiamen 361005, China
* Correspondence: lqm6012@csu.edu.cn

Abstract: The closed-form solution, one of the effective and sufficient optimization methods, is usually less computationally burdensome than iterative and nonlinear minimization in optimization problems of heat source localization. This work presents two-dimensional, closed-form solutions for locating heat-concentrated sources using temperature differences for known and unknown temperature gradient systems. The nonlinear location equations for heat-concentrated source location are simplified to linear equations, and they are solved directly to obtain the analytical solution. To validate the accuracy of the proposed analytical solutions, three numerical examples of heat source localization were conducted. Results show that the proposed analytical solutions have a higher accuracy than iterative results by Levenberg–Marquardt. The locating accuracy for the three sources using AS-KTG improved by 94.82%, 90.40%, and 92.77%, while the locating accuracy for the three sources using AS-UTG improved by 68.94%, 16.72%, and 46.86%, respectively. It is concluded that the proposed method can locate the heat sources using temperatures and coordinates of sensors without the need for a heat transfer coefficient, a heat transfer rate, and thermal conductivity. These proposed analytical solutions can provide a new approach to locating heat sources for more complicated conditions using temperature differences, such as the localization of geothermal sources and nuclear waste leak points.

Keywords: heat-concentrated source; optimization of heat source location; temperature gradient; closed-form solution; temperature difference

MSC: 65H10; 65J15; 65H04

Citation: Sun, D.; Wu, Y.; Dong, L.; Luo, Q. Closed-Form Solutions for Locating Heat-Concentrated Sources Using Temperature Difference. *Mathematics* **2022**, *10*, 2843. https://doi.org/10.3390/math10162843

Academic Editors: Andrey Amosov and Manuel Pastor

Received: 23 May 2022
Accepted: 15 July 2022
Published: 10 August 2022

1. Introduction

The demand for energy increases rapidly with the development of human society. Geothermal resources, as a kind of clean energy, have attracted the attention of countries all over the world. Exploring the distribution of geothermal resources for geothermal development is an important topic [1]. Electrical resistivity tomography [2,3] and magnetotelluric methods [4,5] are useful methods to investigate the geothermal reservoir and to predict the storage of geothermal sources. Traveltime tomography [6,7], as an effective means to understand underground structures, can also provide valuable information for geothermal resource development. Moreover, the deep rockmass is under high stress conditions [8,9], which will affect the safety and economy of geothermal resource exploitation. Identification of the principal stress directions [10,11] and localization of the abnormal regions in underground structures [12,13] is crucial to obtain the distribution of geothermal resources. If the location of the heat-concentrated source can be obtained in advance, it will provide a great convenience for geothermal reservoir exploration.

The solution to the problem of locating a heat source has a wide range of applications. The stable temperature of a certain material is closely related to its properties [14]. A high temperature will seriously affect the comfort and the work efficiency of workers, and further influence their health condition [15]. Nonlinear fitting methods [16,17], such as the Levenberg–Marquardt method (LM), the Monte Carlo technique [18,19], Nelder Mead's simplex method [20], the Gauss–Newton method, the gradient descent method, and dynamic programming, were widely used to determine solutions to the inverse problem. Drag and Styczeń [21] transformed the temperature distribution control problem into a nonlinear optimization problem and designed an interior-point algorithm to solve dynamic optimization tasks. Similar works have been discussed with regard to the localization of acoustic emission sources. The traditional localization methods usually applied the time difference of arrival (TDOA) to solve the results of the sources [22–24]. Analytical solutions and iterative solutions are innovatively combined for solving the microseismic source coordinates in a complex mining environment [25]. The influence of premeasured wave velocity errors and abnormal arrivals are eliminated, and the localization accuracy is greatly improved. Moreover, the A* path search algorithm is improved to obtain the real wave propagation paths, which are further applied to conduct source localization in complex structures [26]. Additionally, the proposed A* based localization method is applicated to the localization of autonomous driving vehicles in underground intelligent mines [27]. This indicates that the localization method has great potential for applications.

Accuracy, efficiency, and economy are mainly considered factors in practical engineering [28]. Compared with iterative solutions, analytical solutions have the advantage of requiring fewer operations due to their high computational efficiency, and they could achieve good accuracy [29]. Recently, nonlinear problems have been simplified into linear equations, and they have achieved good results both in accuracy and in efficiency [22,30,31]. In this paper, nonlinear localization equations are simplified into linear equations, and two-dimensional analytical solutions are developed for locating heat-concentrated sources. This work is expected to provide beneficial information for the localization of geothermal sources and nuclear waste leak points.

2. Iterative Method

Fourier's law regulates the rate of heat transfer due to conduction [32,33]. Thermal conductivity is a material-related property which indicates the ease of heat transfer through a certain material. The general form of Fourier's law can be written as Equation (1) [34,35] because the heat flux is a vector quantity.

$$\dot{q} = -k\nabla T = -k\left(\vec{i}\frac{\partial T}{\partial x} + \vec{j}\frac{\partial T}{\partial y} + \vec{k}\frac{\partial T}{\partial z}\right) \tag{1}$$

where ∇, indicating the temperature gradient, is the three-dimensional del operator and T is the scalar temperature field.

Under steady-state and isotropic heat transfer systems, there are three important assumptions: (1) The thermal conductivity system consists of an isotropic homogeneous material; (2) The thermal conductivity, density, and specific heat of the material are constants; and (3) A heat source exists inside the thermal conductivity system. Therefore, the simplified form of heat conduction is Equation (2) [34].

$$\frac{\partial}{\partial x}\left(k\frac{\partial T}{\partial x}\right) + \frac{\partial}{\partial y}\left(k\frac{\partial T}{\partial y}\right) + \frac{\partial}{\partial z}\left(k\frac{\partial T}{\partial z}\right) + \frac{dq}{dt} = 0 \tag{2}$$

Locating a heat-concentrated source, which is an inverted solution problem, is to calculate the coordinated x_0, y_0, z_0 of P, according to a point that meets the following objective function as Equation (3).

$$\min J(x_i, y_i, z_i) = \frac{1}{N}\sum_{1}^{N}(T_i(x_i, y_i, z_i) - T_c(x_c, y_c, z_c)) \tag{3}$$

where n is the number of measurement points, $T_i(x_i, y_i, z_i)$ is the coordinate of measurement points, and $T_c(x_c, y_c, z_c)$ is the calculated coordinated, according to the inverted location of the heat source.

The objective function is changing with the change of locations of heat sources. Thus, the locations of heat sources can be calculated through boundary conditions, the temperatures at measurement points, and interactive algorithms. However, the problems involved in the existing method with iterative algorithms are: (1) The location accuracy is influenced by initial values; and (2) The existence and the multiplicity of solutions are induced by iterative calculations. To remove the above two defects and to improve locating accuracy, in this work, two closed-form solutions were developed to locate the heat-concentrated sources.

3. Closed-Form Solution for Known Temperature Gradient

In a steady-state and isotropic heat transfer system of two-dimensional circular and quadrate plates with adiabatic boundary and displacement components of 0, as shown in Figure 1, the method to locate internal heat sources with a constant temperature is discussed in this paper. A heat-concentrated source is located at $S(x_0, y_0, T)$ and the four receivers are located at $R_1(x_1, y_1)$, $R_2(x_2, y_2)$, $R_3(x_3, y_3)$, and $R_4(x_4, y_4)$.

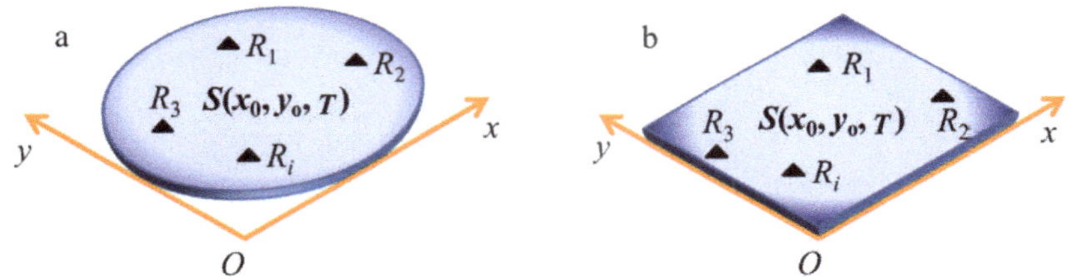

Figure 1. Heat sources in circular and quadrate plates with the adiabatic boundary and displacement components of 0: (**a**) the circular plate; and (**b**) the quadrate plate.

The temperature of the nearest receiver is the largest among the four receivers. After identifying the nearest receiver and calculating the temperature difference between the nearest receiver and other receivers, the four receivers are numbered. For record purposes, the nearest receiver from the source is noted as receiver 1 (R_1). According to Fourier's Law and the definition of the temperature gradient, the equations governing the position of heat-concentrated sources and receivers are:

$$(x_1 - x_0)^2 + (y_1 - y_0)^2 = \left(1/\nabla^2\right)T_0^2 \tag{4}$$

$$(x_2 - x_0)^2 + (y_2 - y_0)^2 = \left(1/\nabla^2\right)(T_0 + T_{12})^2 \tag{5}$$

$$(x_3 - x_0)^2 + (y_3 - y_0)^2 = \left(1/\nabla^2\right)(T_0 + T_{13})^2 \tag{6}$$

$$(x_4 - x_0)^2 + (y_4 - y_0)^2 = \left(1/\nabla^2\right)(T_0 + T_{14})^2 \tag{7}$$

where T_0 is the temperature difference between the heat-concentrated source and the nearest receiver. T_{12}, T_{13}, and T_{14} are the temperature differences between receiver 1 and receivers 2, 3, and 4, respectively; ∇ is a constant temperature gradient.

The receivers are located in the center of the circles determined by Equations (4)–(7). These circles pass through the certain source. It is important to note that any two circles of Equations (4)–(7) intersect, and the source is located on the intersecting line. The equation of intersecting lines of Equations (4) and (5) is given in Equation (8) where l_i is a constant.

$$2(x_2 - x_1)x_0 + 2(y_2 - y_1)y_0 + 2\left(1/\nabla^2\right)T_{12}T_0 = l_1 \tag{8}$$

where $l_1 = x_2^2 - x_1^2 + y_2^2 - y_1^2 - (1/\nabla^2)T_{12}^2$.

Similarly, the equations for the intersecting lines for circles (4) and (6) and for circles (4) and (7) are obtained by taking the difference of Equations (4) and (6), as well as (4) and (7), respectively. These equations are given in Equations (9), and (10), respectively.

$$2(x_3 - x_1)x_0 + 2(y_3 - y_1)y_0 + 2\left(1/\nabla^2\right)T_{13}T_0 = l_2 \tag{9}$$

$$2(x_4 - x_1)x_0 + 2(y_4 - y_1)y_0 + 2\left(1/\nabla^2\right)T_{14}T_0 = l_3 \tag{10}$$

where $l_2 = x_3^2 - x_1^2 + y_3^2 - y_1^2 - (1/\nabla^2)T_{13}^2$, $l_3 = x_4^2 - x_1^2 + y_4^2 - y_1^2 - (1/\nabla^2)T_{14}^2$.

Equations (8)–(10) constitute a set of linear equations, and they can be rewritten as:

$$l_4x_0 + l_5y_0 + l_6T_0 = l_1 \tag{11}$$

$$l_7x_0 + l_8y_0 + l_9T_0 = l_2 \tag{12}$$

$$l_{10}x_0 + l_{11}y_0 + l_{12}T_0 = l_3 \tag{13}$$

where $l_4 = 2(x_2 - x_1)$, $l_5 = 2(y_2 - y_1)$, $l_6 = 2T_{12}(1/\nabla^2)$, $l_7 = 2(x_3 - x_1)$, $l_8 = 2(y_3 - y_1)$, $l_9 = 2T_{13}(1/\nabla^2)$, $l_{10} = 2(x_4 - x_1)$, $l_{11} = 2(y_4 - y_1)$, $l_{12} = 2T_{14}(1/\nabla^2)$.

By solving the linear equations, the x_0, y_0 can be obtained and simplified as:

$$T_0 = \frac{l_{11}l_2l_4 - l_{10}l_2l_5 - l_1l_{11}l_7 + l_1l_{10}l_8 - l_3l_4l_8 + l_3l_5l_7}{l_9l_{11}l_4 - l_9l_{10}l_5 - l_{12}l_4l_8 + l_{12}l_5l_7 - l_{11}l_6l_7 + l_{10}l_6l_8} \tag{14}$$

$$x_0 = \frac{l_1l_9l_{11} + l_{12}l_2l_5 - l_9l_3l_5 - l_{11}l_2l_6 - l_1l_{12}l_8 + l_3l_6l_8}{l_9l_{11}l_4 - l_9l_{10}l_5 - l_{12}l_4l_8 + l_{12}l_5l_7 - l_{11}l_6l_7 + l_{10}l_6l_8} \tag{15}$$

$$y_0 = -\frac{(l_1l_9l_{10} + l_{12}l_2l_4 - l_9l_3l_4 - l_{10}l_2l_6 - l_1l_{12}l_7 + l_3l_6l_7)}{l_9l_{11}l_4 - l_9l_{10}l_5 - l_{12}l_4l_8 + l_{12}l_5l_7 - l_{11}l_6l_7 + l_{10}l_6l_8} \tag{16}$$

The linear system defined by Equations (11)–(13) can be written as:

$$AS = B \tag{17}$$

where $A = \begin{bmatrix} l_4 & l_5 & l_6 \\ l_7 & l_8 & l_9 \\ l_{10} & l_{11} & l_{12} \end{bmatrix}$, $S = \begin{bmatrix} x_0 \\ y_0 \\ t_0 \end{bmatrix}$ and $B = \begin{bmatrix} l_1 \\ l_2 \\ l_3 \end{bmatrix}$.

Now, the x_0, y_0 and T_0 can be solved. The method is named AS-KTG (Analytical Solution-Known temperature gradients).

4. Closed-Form Solution for Unknown Temperature Gradient

A heat-concentrated source is also located at $S(x_0, y_0, T)$, and the five receivers are located at $R_A(x_{w1}, y_{w1})$, $R_B(x_{w2}, y_{w2})$, $R_C(x_{w3}, y_{w3})$, $R_D(x_{w4}, y_{w4})$, and $R_E(x_{w5}, y_{w5})$. The temperature of the nearest receiver is the largest among the five receivers. The processes of identifying the receivers are similar, as introduced in AS-KTG. The nearest receiver from the source is numbered as sensor R_A. T_{w12}, T_{w13}, T_{w14}, and T_{w15} are the temperature differences between receiver 1 and receivers 2, 3, 4, and 5, respectively. The equations governing the position of the heat source and receivers are:

$$(x_{w1} - x_0)^2 + (y_{w1} - y_0)^2 = \left(1/\nabla_w^2\right)T_{w0}^2 \tag{18}$$

$$(x_{w2} - x_0)^2 + (y_{w2} - y_0)^2 = \left(1/\nabla_w^2\right)(T_{w0} + T_{w12})^2 \tag{19}$$

$$(x_{w3} - x_0)^2 + (y_{w3} - y_0)^2 = \left(1/\nabla_w^2\right)(T_{w0} + T_{w13})^2 \tag{20}$$

$$(x_{w4} - x_0)^2 + (y_{w4} - y_0)^2 = \left(1/\nabla_w^2\right)(T_{w0} + T_{w14})^2 \tag{21}$$

$$(x_{w5} - x_0)^2 + (y_{w5} - y_0)^2 = \left(1/\nabla_w^2\right)(T_{w0} + T_{w15})^2 \tag{22}$$

where T_{w0} is the temperature difference between the heat-concentrated source to the nearest temperature sensor; ∇ is the unknown temperature gradient.

The intersecting line for the Equations (18) and (19) can be obtained by taking the difference. This equation of the intersecting line is given in Equation (23). In the following equations, l_i is a constant.

$$2(x_{w2} - x_{w1})x_0 + 2(y_{w2} - y_{w1})y_0 + 2T_{w12}\left(1/\nabla_w^2\right)T_0 + \left(1/\nabla_w^2\right)T_{w12}^2 = l_{w1} \tag{23}$$

where $l_{w1} = x_{w2}^2 - x_{w1}^2 + y_{w2}^2 - y_{w1}^2$.

Similarly, the equations for the intersecting lines for circles (18) to (20), (18) to (21), and (18) to (22) are obtained by taking the difference of Equation (18) and Equations (20)–(22), respectively. These equations are given in Equations (24)–(26), respectively.

$$2(x_{w3} - x_{w1})x_0 + 2(y_{w3} - y_{w1})y_0 + 2T_{w13}\left(1/\nabla_w^2\right)T_{w0} + \left(1/\nabla_w^2\right)T_{w13}^2 = l_{w2} \tag{24}$$

$$2(x_{w4} - x_{w1})x_0 + 2(y_{w4} - y_{w1})y_0 + 2T_{w14}\left(1/\nabla_w^2\right)T_{w0} + \left(1/\nabla_w^2\right)T_{w14}^2 = l_{w3} \tag{25}$$

$$2(x_{w5} - x_{w1})x_0 + 2(y_{w5} - y_{w1})y_0 + 2T_{w15}\left(1/\nabla_w^2\right)T_{w0} + \left(1/\nabla_w^2\right)T_{w15}^2 = l_{w4} \tag{26}$$

where $l_{w2} = x_{w3}^2 - x_{w1}^2 + y_{w3}^2 - y_{w1}^2$, $l_{w3} = x_{w4}^2 - x_{w1}^2 + y_{w4}^2 - y_{w1}^2$, $l_{w4} = x_{w5}^2 - x_{w1}^2 + y_{w5}^2 - y_{w1}^2$.

Equations (23)–(26) constitute a set of linear equations, which can be rewritten as follows:

$$l_{w5}x_0 + l_{w6}y_0 + l_{w7}D + l_{w8}V = l_{w1} \tag{27}$$

$$l_{w9}x_0 + l_{w10}y_0 + l_{w11}D + l_{w12}V = l_{w2} \tag{28}$$

$$l_{w13}x_0 + l_{w14}y_0 + l_{w15}D + l_{w16}V = l_{w3} \tag{29}$$

$$l_{w17}x_0 + l_{w18}y_0 + l_{w19}D + l_{w20}V = l_{w4} \tag{30}$$

where $D = 2T_{w0}(1/\nabla_w^2)$, $V = 1/\nabla_w^2$, $l_{w5} = 2(x_{w2} - x_{w1})$, $l_{w6} = 2(y_{w2} - y_{w1})$, $l_{w7} = 2T_{w12}$, $l_{w8} = T_{w12}^2$, $l_{w9} = 2(x_{w3} - x_{w1})$, $l_{w10} = 2(y_{w3} - y_{w1})$, $l_{w11} = 2T_{w13}$, $l_{w12} = T_{w13}^2$, $l_{w13} = 2(x_{w4} - x_{w1})$, $l_{w14} = 2(y_{w4} - y_{w1})$, $l_{w15} = 2T_{w14}$, $l_{w16} = T_{w14}^2$, $l_{w17} = 2(x_{w5} - x_{w1})$, $l_{w18} = 2(y_{w5} - y_{w1})$, $l_{w19} = 2T_{w15}$, and $l_{w20} = T_{w15}^2$.

Therefore, the coordinates of x_0 and y_0, which can be obtained by solving the function of Equations (27)–(30), are as follows:

$$\begin{aligned}
x_0 = {} & [l_{w1}(l_{w10}l_{w16}l_{w19} - l_{w11}l_{w16}l_{w18} - l_{w12}l_{w14}l_{w19} + l_{w12}l_{w15}l_{w18} - l_{w10}l_{w15}l_{w20} + l_{w11}l_{w14}l_{w20}) \\
& + l_{w6}(l_{w15}l_{w2}l_{w20} - l_{w16}l_{w19}l_{w2} + l_{w12}l_{w19}l_{w3} - l_{w11}l_{w20}l_{w3} + l_{w11}l_{w16}l_{w4} - l_{w12}l_{w15}l_{w4}) \\
& + l_{w7}(l_{w16}l_{w18}l_{w2} - l_{w14}l_{w2}l_{w20} - l_{w12}l_{w18}l_{w3} - l_{w10}l_{w20}l_{w3} - l_{w10}l_{w16}l_{w4} + l_{w12}l_{w14}l_{w4}) \\
& + l_{w8}(l_{w14}l_{w19}l_{w2} - l_{w15}l_{w18}l_{w2} - l_{w10}l_{w19}l_{w3} + l_{w11}l_{w18}l_{w3} + l_{w10}l_{w15}l_{w4} - l_{w11}l_{w14}l_{w4})] / [l_{w5}(l_{w10}l_{w16}l_{w19} \\
& - l_{w11}l_{w16}l_{w18} - l_{w12}l_{w14}l_{w19} + l_{w12}l_{w15}l_{w18} - l_{w10}l_{w15}l_{w20} + l_{w11}l_{w14}l_{w20}) \\
& + l_{w6}(l_{w11}l_{w16}l_{w17} + l_{w12}l_{w13}l_{w19} - l_{w12}l_{w15}l_{w17} - l_{w11}l_{w13}l_{w20}) \\
& + l_{w7}(l_{w10}l_{w13}l_{w20} - l_{w10}l_{w16}l_{w17} - l_{w12}l_{w13}l_{w18} + l_{w12}l_{w14}l_{w17}) \\
& + l_{w8}(l_{w10}l_{w15}l_{w17} - l_{w10}l_{w13}l_{w19} + l_{w11}l_{w13}l_{w18} - l_{w11}l_{w14}l_{w17}) \\
& + l_{w9}(l_{w15}l_{w6}l_{w20} - l_{w19}l_{w16}l_{w6} + l_{w7}l_{w16}l_{w18} - l_{w20}l_{w14}l_{w17} + l_{w19}l_{w14}l_{w8} - l_{w15}l_{w18}l_{w8})]
\end{aligned} \tag{31}$$

$$
\begin{aligned}
y_0 = {} & [l_{w1}(l_{w11}l_{w16}l_{w17} - l_{w12}l_{w13}l_{w19} - l_{w12}l_{w15}l_{w17} + l_{w11}l_{w13}l_{w20} - l_{w16}l_{w19}l_{w9} + l_{w15}l_{w19}l_{w20}) \\
& + l_{w5}(l_{w16}l_{w2}l_{w19} - l_{w5}l_{w20}l_{w2} + l_{w12}l_{w19}l_{w3} - l_{w11}l_{w20}l_{w3} + l_{w11}l_{w16}l_{w4} - l_{w12}l_{w15}l_{w4}) \\
& + l_{w7}(l_{w16}l_{w17}l_{w2} - l_{w13}l_{w2}l_{w20} - l_{w12}l_{w17}l_{w3} - l_{w12}l_{w13}l_{w4}) \\
& + l_{w9}(l_{w20}l_{w3}l_{w7} - l_{w16}l_{w4}l_{w7} - l_{w19}l_{w8}l_{w3} - l_{w15}l_{w14}l_{w8}) \\
& + l_{w8}(l_{w13}l_{w19}l_{w2} - l_{w15}l_{w17}l_{w2} - l_{w11}l_{w17}l_{w3} + l_{w11}l_{w13}l_{w4})] / [l_{w5}(l_{w10}l_{w16}l_{w19} \\
& - l_{w11}l_{w16}l_{w18} - l_{w12}l_{w14}l_{w19} + l_{w12}l_{w15}l_{w18} - l_{w10}l_{w15}l_{w20} + l_{w11}l_{w14}l_{w20}) \\
& + l_{w6}(l_{w11}l_{w16}l_{w17} + l_{w12}l_{w13}l_{w19} - l_{w12}l_{w15}l_{w17} - l_{w11}l_{w13}l_{w20}) \\
& + l_{w7}(l_{w10}l_{w13}l_{w20} - l_{w10}l_{w16}l_{w17} - l_{w12}l_{w13}l_{w18} + l_{w12}l_{w14}l_{w17}) \\
& + l_{w8}(l_{w10}l_{w15}l_{w17} - l_{w10}l_{w13}l_{w19} + l_{w11}l_{w13}l_{w18} - l_{w11}l_{w14}l_{w17}) \\
& + l_{w9}(l_{w15}l_{w6}l_{w20} - l_{w19}l_{w16}l_{w6} + l_{w7}l_{w16}l_{w18} - l_{w20}l_{w14}l_{w17} + l_{w19}l_{w14}l_{w8} - l_{w15}l_{w18}l_{w8})]
\end{aligned} \quad (32)
$$

The method is defined as AS-UTG (Analytical Solution-Unknown temperature gradient). The linear system defined by Equations (27)–(30) can be written as:

$$A_w S = B_w \quad (33)$$

where $A_w = \begin{bmatrix} l_{w5} & l_{w6} & l_{w7} & l_{w8} \\ l_{w9} & l_{w10} & l_{w11} & l_{w12} \\ l_{w13} & l_{w14} & l_{w15} & l_{w16} \\ l_{w17} & l_{w18} & l_{w19} & l_{w20} \end{bmatrix}$, $S = \begin{bmatrix} x_0 \\ y_0 \\ D \\ V \end{bmatrix}$, and $B_w = \begin{bmatrix} l_{w1} \\ l_{w2} \\ l_{w3} \\ l_{w4} \end{bmatrix}$.

5. Validated Examples and Discussion

In practical engineering or laboratory experiment, thermocouple, thermal resistance, infrared thermal imaging camera, and other temperature measuring devices can be employed to measure the temperature data. In this work, numerical examples are conducted to validate the accuracy of the proposed methods. Three heat source location systems are used to validate the localization accuracy of the proposed analytical solutions. Each system has one heat source. The Cartesian coordinates of five temperature sensors are $R_1(110, 180)$, $R_2(20, 30)$, $R_3(20, 160)$, $R_4(180, 20)$, and $R_5(170, 130)$ in the three systems. The unit is cm. The three heat sources are $S_1(80, 60)$, $S_2(120, 90)$, and $S_3(110, 130)$, and the corresponding temperature gradients are 10, 18, and 21 °C/m. The temperature data were numerically calculated according to the distance between the three heat sources and the receivers R1 to R5 and the corresponding temperature gradients. The temperatures of heat sources S_1, S_2, and S_3, as well as their recording temperatures of five temperature sensors, are listed in Table 1. The iterative results were also calculated using Levenberg–Marquardt to compare the locating accuracy between the proposed analytical solutions and the iterative method. The Levenberg–Marquardt algorithm [36,37], which combines the advantages of the Gauss–Newton algorithm and the gradient descent method by modifying the parameters during computation, can provide a minimum numerical solution for a number of nonlinear equations and improve the shortcomings of the Gauss–Newton algorithm and the gradient descent method.

Table 1. Heat source locations and their recorded temperatures.

Source				Recorded Temperatures (°C)				
No.	T_0 (°C)	X (cm)	Y (cm)	R_1	R_2	R_3	R_4	R_5
S_1	160.00	80	60	147.63	153.29	148.34	149.23	148.60
S_2	180.00	120	90	163.70	159.01	158.03	163.40	168.47
S_3	210.00	110	130	199.50	181.75	190.08	182.62	197.40

For known temperature gradient, the coordinates of the first four receivers, the recorded corresponding temperatures, and the temperature gradients of the three heat sources are known. The coordinates of three heat sources and corresponding temperatures are unknown. The problems are to locate the heat sources and to calculate the temperatures at source points. The known data were utilized to validate the proposed AS-KTG. The coordinates of receivers and their recorded temperatures as well as the temperature gradient were used to calculate $l_i(i = 1, 2, \ldots, 12)$. Then, the calculated $l_i(i = 1, 2, \ldots, 12)$ were substituted into Equations (14)–(16) or Equation (17) to resolve the heat source location. The calculation of the location results could be performed using Matlab, Python, and even Excel since it involves only simple calculation processes. The calculated results are shown in Figure 2. It can be seen from Figure 2 that the calculated results of AS-KTG are consistent with the authentic results. The locating errors of the three sources using AS-KTG are 0.18 cm, 0.46 cm, and 0.52 cm, while the locating errors using Levenberg–Marquardt are 3.54 cm, 4.82 cm, and 7.17 cm, respectively. Compared to the localization results of Levenberg–Marquardt, the locating accuracy for the three sources using AS-KTG is improved by 94.82%, 90.40%, and 92.77%, respectively.

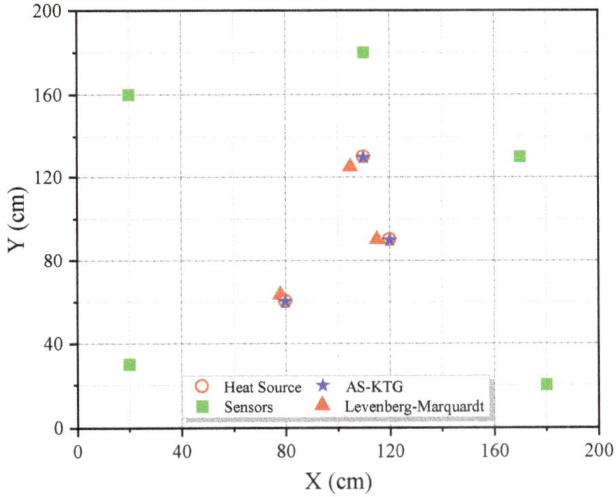

Figure 2. Locations of receivers and calculated results of AS-KTG and inverse results by Levenberg–Marquardt.

For an unknown temperature gradient, the coordinates of five receivers, and the recorded corresponding temperatures are known. The coordinates of three heat sources, and corresponding temperatures, as well as temperature gradients of three heat sources, are unknown. The problems are to locate the heat sources and to calculate the temperatures at the source point. The known data were used to validate the proposed AS-UTG. The coordinates of temperature sensors (TS) and their recorded temperatures were used to calculate $l_i(i = 1, 2, \ldots, 20)$. Then, the calculated $l_i(i = 1, 2, \ldots, 20)$ were substituted into Equations (31) and (32) or Equation (33) to resolve the heat source locations. The calculated results are shown in Figure 3. It can be seen from Figure 3 that the locating errors of AS-UTG are obviously smaller than the errors of Levenberg–Marquardt, which means that the analytical solutions have a higher accuracy than iterative results by Levenberg–Marquardt. The locating errors of the three sources using AS-UTG are 1.38 cm, 3.21 cm, and 3.89 cm, while the locating errors using Levenberg–Marquardt are 4.44 cm, 3.85 cm, and 7.32 cm, respectively. Compared to the localization results of Levenberg–Marquardt, the locating accuracy for the three sources using AS-UTG is improved by 68.94%, 16.72%, and 46.86%, respectively.

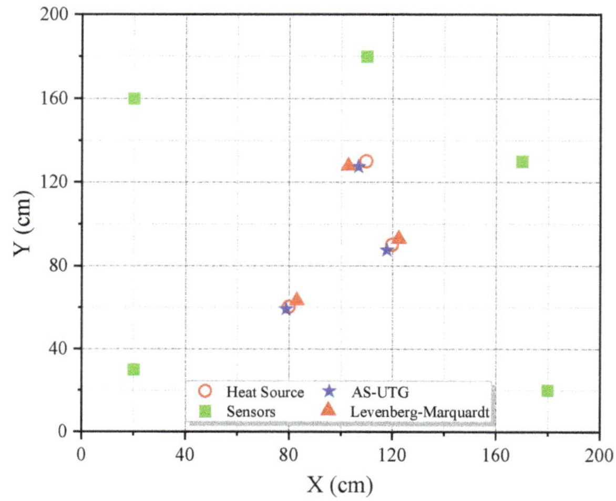

Figure 3. Locations of receivers and calculated results of AS-UTG and inverse results by Levenberg–Marquardt.

By comparing the locating results of AS-KTG and ASUTG, it can be noticed that, for the three sources, the locating accuracy of AS-KTG is 86.73%, 85.57%, and 86.67%, which is better than the locating accuracy of Levenberg–Marquardt, respectively. It should be noted that the high accuracy of the numerical examples for two-dimensional hat source localization is obtained under steady-state and isotropic systems. The locating accuracy in practical situations with complex conditions may be reduced.

6. Conclusions

The nonlinear location equations for heat sources were simplified to linear equations in a steady-state and isotropic heat transfer system of circular or quadrate plates, with the adiabatic boundary, as well as internal heat sources with a constant temperature. Based on simplified linear equations, two-dimensional analytical solutions were obtained for heat source locations using temperature differences. The problems of the existence and the multiplicity induced by calculations of the iterative method were solved successfully.

The proposed method can locate the heat sources using only temperatures and coordinates of sensors, without the need for coefficients of heat transfer, heat quantity, and thermal conductivity. Numerical examples are conducted to study the algorithms' performance and compare them with the existing techniques. Results show that, compared to the localization results of Levenberg–Marquardt, the locating accuracy for the three sources using AS-KTG improved by 94.82%, 90.40%, and 92.77%, while the locating accuracy for the three sources using AS-UTG improved by 68.94%, 16.72%, and 46.86%, respectively.

The proposed method can not only avoid the problems of initial values—the existence and the multiplicity of solutions induced by iterative calculations in existing methods—but also locate the source in unknown temperature gradient systems. It should be noted that though this work presents two analytical solutions for two particular cases of heat source localization, it can provide a new approach to locating heat sources for more complicated conditions using temperature differences, such as the localization of geothermal sources and nuclear waste leak points.

Author Contributions: Conceptualization, D.S. and L.D.; methodology, L.D.; validation, D.S., Y.W., L.D. and Q.L.; formal analysis, D.S.; investigation, Y.W. and Q.L.; resources, L.D.; writing—original draft preparation, D.S., L.D. and Q.L.; writing—review and editing, D.S., Y.W. and Q.L.; supervision, L.D.; project administration, L.D.; funding acquisition, L.D. All authors have read and agreed to the published version of the manuscript.

Funding: This research was funded in part by the International (Regional) Cooperation and Exchange Program of National Natural Science Foundation of China, grant number 52161135301 and in part by the Special Fund for Basic Scientific Research Operations in Universities, grant number 2282020cxqd055.

Institutional Review Board Statement: Not applicable.

Informed Consent Statement: Not applicable.

Data Availability Statement: Not applicable.

Acknowledgments: The authors would like to thank the editor and reviewers in advance for carefully dealing with this manuscript.

Conflicts of Interest: The authors declare no conflict of interest.

Nomenclature

A	area, m^2
k	thermal conductivity, W/m·K
N	number of measurement points
Q	heat transfer rate, W
Q_x	heat transfer rate of x coordinate, W
q''_x	heat flux of x coordinate, W/m^2
$\dot{q}$	heat transfer rate per unit volume of the medium, W/m^3
t_{12}	temperature difference between the receivers 1 and 2 in known temperature gradient systems, °C
t_{13}	temperature difference between the receivers 1 and 3 in known temperature gradient systems, °C
t_{14}	temperature difference between the receivers 1 and 4 in known temperature gradient systems, °C
t_{w0}	temperature difference between the heat concentrated source to the nearest receiver in unknown temperature gradient systems, °C
t_0	temperature difference between the heat concentrated source to the nearest receiver in known temperature gradient systems, °C
t_{w12}	temperature difference between the receivers 1 and 2 in unknown temperature gradient systems, °C
t_{w13}	temperature difference between the receivers 1 and 3 in unknown temperature gradient systems, °C
t_{w14}	temperature difference between the receivers 1 and 4 in unknown temperature gradient systems, °C
t_{w15}	temperature difference between the receivers 1 and 5 in unknown temperature gradient systems, °C
x_i, y_i	coordinate of measurement points in known temperature gradient systems, m
x_{wi}, y_{wi}	coordinate of measurement points in unknown temperature gradient systems, m
x_0, y_0	coordinate of heat sources in known temperature gradient systems, m
ΔT	difference in temperature over which heat is transferred, K
Δx	distance, m
∇	temperature gradient in known temperature gradient systems, °C/m
∇_w	temperature gradient in unknown temperature gradient systems, °C/m

References

1. Afandi, A.; Lusi, N.; Subono; Febriani, S.D.A. Prediction of the distribution of geothermal sources based on the geothermal temperature gradient in the Blawan Bondowoso. *Case Stud. Therm. Eng.* **2021**, *25*, 100931. [CrossRef]
2. Hermans, T.; Irving, J. Facies discrimination with electrical resistivity tomography using a probabilistic methodology: Effect of sensitivity and regularisation. *Near Surf. Geophys.* **2017**, *15*, 13–25. [CrossRef]
3. Hermans, T.; Paepen, M. Combined Inversion of Land and Marine Electrical Resistivity Tomography for Submarine Groundwater Discharge and Saltwater Intrusion Characterization. *Geophys. Res. Lett.* **2020**, *47*, e2019GL085877. [CrossRef]
4. Cheng, Y.; Pang, Z.; Kong, Y.; Chen, X.; Wang, G. Imaging the heat source of the Kangding high-temperature geothermal system on the Xianshuihe fault by magnetotelluric survey. *Geothermics* **2022**, *102*, 102386. [CrossRef]

5. Mekkawi, M.M.; Abd-El-Nabi, S.H.; Farag, K.S.; Abd Elhamid, M.Y. Geothermal resources prospecting using magnetotelluric and magnetic methods at Al Ain AlSukhuna-Al Galala Albahariya area, Gulf of Suez, Egypt. *J. Afr. Earth Sci.* **2022**, *190*, 104522. [CrossRef]
6. Dong, L.; Pei, Z.; Xie, X.; Zhang, Y.; Yan, X. Early Identification of Abnormal Regions in Rock-Mass Using Traveltime Tomography. *Engineering* **2022**. [CrossRef]
7. Dong, L.; Tong, X.; Ma, J. Quantitative Investigation of Tomographic Effects in Abnormal Regions of Complex Structures. *Engineering* **2021**, *7*, 1011–1022. [CrossRef]
8. Wang, S.; Li, X.; Yao, J.; Gong, F.; Li, X.; Du, K.; Tao, M.; Huang, L.; Du, S. Experimental investigation of rock breakage by a conical pick and its application to non-explosive mechanized mining in deep hard rock. *Int. J. Rock Mech. Min. Sci.* **2019**, *122*, 104063. [CrossRef]
9. Wang, S.; Sun, L.; Li, X.; Zhou, J.; Du, K.; Wang, S.; Khandelwal, M. Experimental investigation and theoretical analysis of indentations on cuboid hard rock using a conical pick under uniaxial lateral stress. *Geomech. Geophys. Geoenerg. Georesour.* **2022**, *8*, 34. [CrossRef]
10. Dong, L.; Chen, Y.; Sun, D.; Zhang, Y. Implications for rock instability precursors and principal stress direction from rock acoustic experiments. *Int. J. Min. Sci. Technol.* **2021**, *31*, 789–798. [CrossRef]
11. Dong, L.; Chen, Y.; Sun, D.; Zhang, Y.; Deng, S. Implications for identification of principal stress directions from acoustic emission characteristics of granite under biaxial compression experiments. *J. Rock Mech. Geotech. Eng.* **2022**. [CrossRef]
12. Dong, L.; Tao, Q.; Hu, Q.; Deng, S.; Chen, Y.; Luo, Q.; Zhang, X. Acoustic emission source location method and experimental verification for structures containing unknown empty areas. *Int. J. Min. Sci. Technol.* **2022**, *32*, 487–497. [CrossRef]
13. Dong, L.; Tong, X.; Hu, Q.; Tao, Q. Empty region identification method and experimental verification for the two-dimensional complex structure. *Int. J. Rock Mech. Min. Sci.* **2021**, *147*, 104885. [CrossRef]
14. Ghasemi, M.H.; Hoseinzadeh, S.; Memon, S. A dual-phase-lag (DPL) transient non-Fourier heat transfer analysis of functional graded cylindrical material under axial heat flux. *Int. Commun. Heat Mass* **2022**, *131*, 105858. [CrossRef]
15. Dong, L.; Zhou, Y.; Deng, S.; Wang, M.; Sun, D. Evaluation methods of man-machine-environment system for clean and safe production in phosphorus mines: A case study. *J. Cent. South Univ.* **2021**, *28*, 3856–3870. [CrossRef]
16. Jarny, Y.; Ozisik, M.N.; Bardon, J.P. A general optimization method using adjoint equation for solving multidimensional inverse heat conduction. *Int. J. Heat Mass Tran.* **1991**, *34*, 2911–2919. [CrossRef]
17. Alifanov, O.M. *Inverse Heat Transfer Problems*; Springer Science & Business Media: Cham, Switzerland, 2012.
18. Yan, F.; Farouk, B. Computations of low pressure fluid flow and heat transfer in ducts using the direct simulation Monte Carlo method. *J. Heat Transf.* **2002**, *124*, 609–616. [CrossRef]
19. Modest, M.F. Backward Monte Carlo simulations in radiative heat transfer. *J. Heat Transf.* **2003**, *125*, 57–62. [CrossRef]
20. Nelder, J.A.; Mead, R. A simplex method for function minimization. *Comput. J.* **1965**, *8*, 308–313. [CrossRef]
21. Drąg, P.; Styczeń, K. A general optimization-based approach for thermal processes modeling. In Proceedings of the 2017 Federated Conference on Computer Science and Information Systems, Prague, Czech Republic, 3–6 September 2017; Volume 11, pp. 1347–1352.
22. Dong, L.; Sun, D.; Li, X.; Du, K. Theoretical and Experimental Studies of Localization Methodology for AE and Microseismic Sources Without Pre-Measured Wave Velocity in Mines. *IEEE Access* **2017**, *5*, 16818–16828. [CrossRef]
23. Dong, L.; Li, X.; Zhou, Z.; Chen, G.; Ma, J. Three-dimensional analytical solution of acoustic emission source location for cuboid monitoring network without pre-measured wave velocity. *Trans. Nonferrous Met. Soc. China* **2015**, *25*, 293–302. [CrossRef]
24. Oğuz-Ekim, P. TDOA based localization and its application to the initialization of LiDAR based autonomous robots. *Robot. Auton. Syst.* **2020**, *131*, 103590. [CrossRef]
25. Dong, L.; Zou, W.; Li, X.; Shu, W.; Wang, Z. Collaborative localization method using analytical and iterative solutions for microseismic/acoustic emission sources in the rockmass structure for underground mining. *Eng. Fract. Mech.* **2019**, *210*, 95–112. [CrossRef]
26. Dong, L.; Hu, Q.; Tong, X.; Liu, Y. Velocity-Free MS/AE Source Location Method for Three-Dimensional Hole-Containing Structures. *Engineering* **2020**, *6*, 827–834. [CrossRef]
27. Dong, L.; Sun, D.; Han, G.; Li, X.; Hu, Q.; Shu, L. Velocity-Free Localization of Autonomous Driverless Vehicles in Underground Intelligent Mines. *IEEE Trans. Veh. Technol.* **2020**, *69*, 9292–9303. [CrossRef]
28. Brandstein, M.S.; Adcock, J.E.; Silverman, H.F. A closed-form location estimator for use with room environment microphone arrays. *IEEE Trans. Speech Audio Process.* **1997**, *5*, 45–50. [CrossRef]
29. Mellen, G.; Pachter, M.; Raquet, J. Closed-form solution for determining emitter location using time difference of arrival measurements. *IEEE Trans. Aerosp. Electron. Syst.* **2003**, *39*, 1056–1058. [CrossRef]
30. Li, X.; Dong, L. An efficient closed-form solution for acoustic emission source location in three-dimensional structures. *AIP Adv.* **2014**, *4*, 027110. [CrossRef]
31. Dong, L.; Shu, W.; Li, X.; Han, G.; Zou, W. Three Dimensional Comprehensive Analytical Solutions for Locating Sources of Sensor Networks in Unknown Velocity Mining System. *IEEE Access* **2017**, *5*, 11337–11351. [CrossRef]
32. Slifka, A.J.; Filla, B.J.; Phelps, J.M.; Bancke, G.; Berndt, C.C. Thermal Conductivity of a Zirconia Thermal Barrier Coating. *J. Therm. Spray Technol.* **1998**, *7*, 43–46. [CrossRef]
33. Tsai, M.Y.; Chen, C.H.; Kang, C.S. Thermal measurements and analyses of low-cost high-power LED packages and their modules. *Microelectron. Reliab.* **2012**, *52*, 845–854. [CrossRef]

34. Columbia University, Laser Machining Processes, Level 2 Chapter 3: Energy Transfer and Modeling, S3.4 Modeling of Laser Machining Process (1). Available online: http://aml.engineering.columbia.edu/ntm/level2/ch03/html/l2c03s04.html (accessed on 20 May 2022).
35. Both, S.; Czél, B.; Fülöp, T.; Gróf, G.; Gyenis, Á.; Kovács, R.; Ván, P.; Verhás, J. Deviation from the Fourier law in room-temperature heat pulse experiments. *J. Non-Equilib. Thermodyn.* **2016**, *41*, 41–48. [CrossRef]
36. Levenberg, K. A Method for the Solution of Certain Non-Linear Problems in Least Squares. *Q. Appl. Math.* **1944**, *2*, 164–168. [CrossRef]
37. Marquardt, D.W. An Algorithm for Least-Squares Estimation of Nonlinear Parameters. *J. Soc. Ind. Appl. Math.* **1963**, *11*, 431–441. [CrossRef]

MDPI
St. Alban-Anlage 66
4052 Basel
Switzerland
Tel. +41 61 683 77 34
Fax +41 61 302 89 18
www.mdpi.com

Mathematics Editorial Office
E-mail: mathematics@mdpi.com
www.mdpi.com/journal/mathematics

MDPI

www.ingramcontent.com/pod-product-compliance
Lightning Source LLC
Chambersburg PA
CBHW061542220926
51850CB00005B/230
9783036557618